# DIFFERENTIAL EQUATIONS LABORATORY WORKBOOK

A Collection of Experiments, Explorations and Modeling Projects for the Computer

# DIFFERENTIAL EQUATIONS LABORATORY WORKBOOK

A Collection of Experiments, Explorations and Modeling Projects for the Computer

**Robert L. Borrelli**
*Harvey Mudd College*

**Courtney S. Coleman**
*Harvey Mudd College*

**William E. Boyce**
*Rensselaer Polytechnic Institute*

**John Wiley & Sons, Inc.**
New York • Chichester • Brisbane • Toronto • Singapore

ISBN 0-471-55142-2

Printed in the United States of America

10 9 8 7 6 5 4 3 2 1

# Contents

# List of Experiments

Experiments listed in **boldface** are team experiments.

# Preface

Differential equations arise in connection with mathematical models in diverse settings in mathematics, engineering, the sciences (physics, chemistry, biology, geology, astronomy, etc.)—and even in the social sciences. Since solutions of **o**rdinary **d**ifferential **e**quations (**ODEs**) define curves, much can be learned about the behavior of solutions by graphing them. Such graphs, generated with easy-to-use interactive numerical solvers, provide compelling visual evidence of theoretical deductions and an understanding of the qualitative properties of solutions. The main focus of this workbook is on computer experiments that support and amplify the topics usually found in introductory ODE texts. The workbook is intended as a supplement, not as a textbook.

### *Prerequisites*

The material in the workbook presumes a knowledge of single and multivariable calculus and some linear algebra. It is expected that the student is concurrently enrolled in a course involving ODEs, and that the student has available a software package that can numerically solve systems of differential equations and present the solutions in graphical form on suitable hardware platforms. Many excellent ODE solvers do not require knowledge of computer programming, and little or none is presumed. The experiments do not usually require special features of any particular text or platform.

### *Mathematical Models and Computers*

The essential ingredient in the application of mathematical techniques to the real world is the construction of a mathematical model for the system of interest. Modeling practitioners believe that a "good" model actually "describes" the system they wish to study, and since mathematical models are amenable to mathematical analysis, great progress can be made in understanding the system. Before the computer age, modelers were forced to keep their models simple enough to allow solution by the analytical techniques of the day (many quite ingenious and sophisticated). Modern computers and software have liberated modelers from this constraint. Models need no longer be analytically tractable, hence no longer require unrealistic or artificial assumptions. Computers can numerically solve (with remarkable accuracy and speed) complicated systems of ODEs and, together with striking graphical displays, allow modelers an insightful look at complex systems.

### *Overview*

The approach in the workbook closely parallels what goes on in science and engineering laboratories. Each **computer problem set** is a combination of pencil-and-paper and computer work selected from the tasks given in an experiment. The work may be straightforward and explicit, but the approach is often open-ended and exploratory. The computer experiments provide students with "hands-on" experience in the behavior of solutions of ODEs and require the student to communicate the results via a worksheet (included), which should be filled out and handed in with attached computer graphics hard copy. The worksheets are arranged for convenient use. Some experiments are designed for student teams and require more extensive reports. Appendix A contains some features to aid students in writing team reports. Appendix C is a compendium of plots of solutions of ODEs (called the **Atlas**) that are referenced throughout this workbook.

Although not the main focus, modeling projects appear in experiments throughout the workbook. Models are drawn from rate processes, electrical circuits, and mechanics. Appendix B gives a brief overview of these modeling environments.

### *Structure of Experiments*

Every **experiment** in the workbook has the same format. First a **cover sheet** outlines the problem area and gives a quick introduction and a few hints. The cover sheet spells out the mathematical topics and modeling areas touched upon. Then one or more experiments follow. Some space is provided on each **laboratory worksheet** for written answers, if appropriate. Additional sheets for longer answers, scratch work, and hard copy of any graphics should be attached to the worksheet before it is handed in. A **computer problem set** consists of a selection of the tasks given on the worksheet of an experiment. The cover sheets and the Atlas have examples that serve as a kind of calibration in preparation for the main goals of the experiment. The cover sheet for each experiment follows a fixed format:

- **Purpose.** Brief statement of goal of the experiment.
- **Keywords.** Main concepts touched upon in the experiment.
- **See also** (optional). References to the Atlas and to other related experiments. As a matter of policy no standard textbooks are referenced.
- **Background.** Brief discussion of the ideas involved in the experiment.
- **Observations** (optional). Presentation of useful techniques or hints for the experiment.
- **Examples** (optional). Illustrations of the concepts underlying the experiments.

To aid the experimenter in recognizing the kind of effort expected, one or more **icons** may appear in the margin by each task:

When the thrust of the task mainly involves the computer

For tasks involving significant mathematical pencil-and-paper work

* For tasks at a more advanced level

At least once in each chapter the following "team experiment" icon will appear next to an experiment name on the worksheet, but not on the cover sheet:

For an experiment that requires the effort of a team of students

### *The Figures in This Workbook*

The headers of the workbook figures give information about the ODE being solved. In some cases only the ODEs are listed. In others, more information is given. For example, the header in Figure 0.1 lists three coupled ODEs (the Lorenz system), a user-defined function $f$ employed in the first ODE, two sets of initial data, the initial and final solve times, and the number of solution points used to create the graphs. Different line styles mark the graphs of the respective $x$-components of the two solutions corresponding to the two data sets. A Lorenz system is also treated in Figure 0.2, but with variables $x_1, x_2, x_3$, and a parameter $a_1$. Different line styles mark the graph of the respective $x_1$-components of the two solutions corresponding to the two listed values of $a_1$.

### *Hardware/Software Issues*

There is an abundance of available solver packages and hardware platforms that are adequate for most of these experiments. For effective use of this workbook, students should have convenient access to:

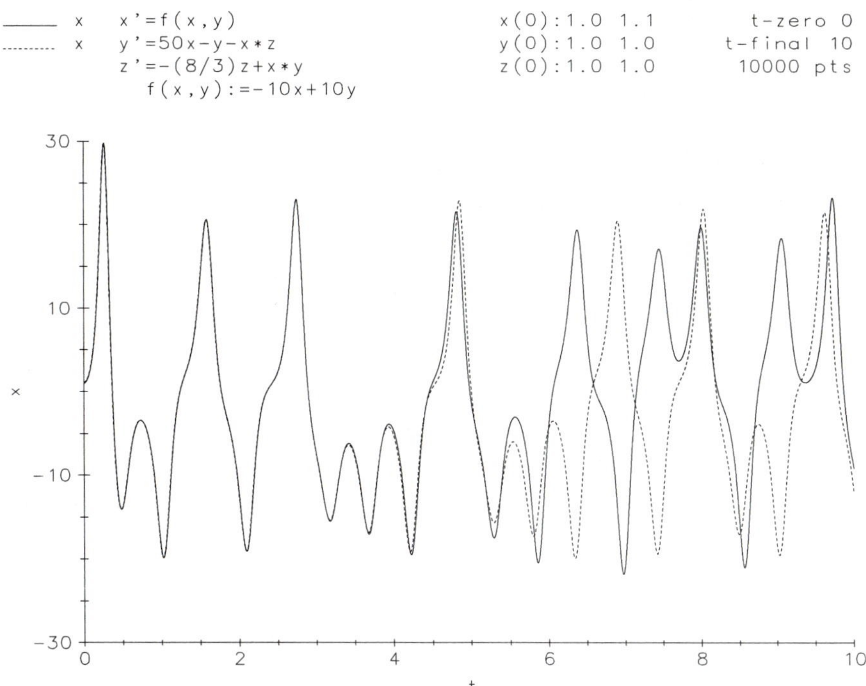


*Figure 0.1* *Sample graph showing overlay of x-components for two sets of initial data.*

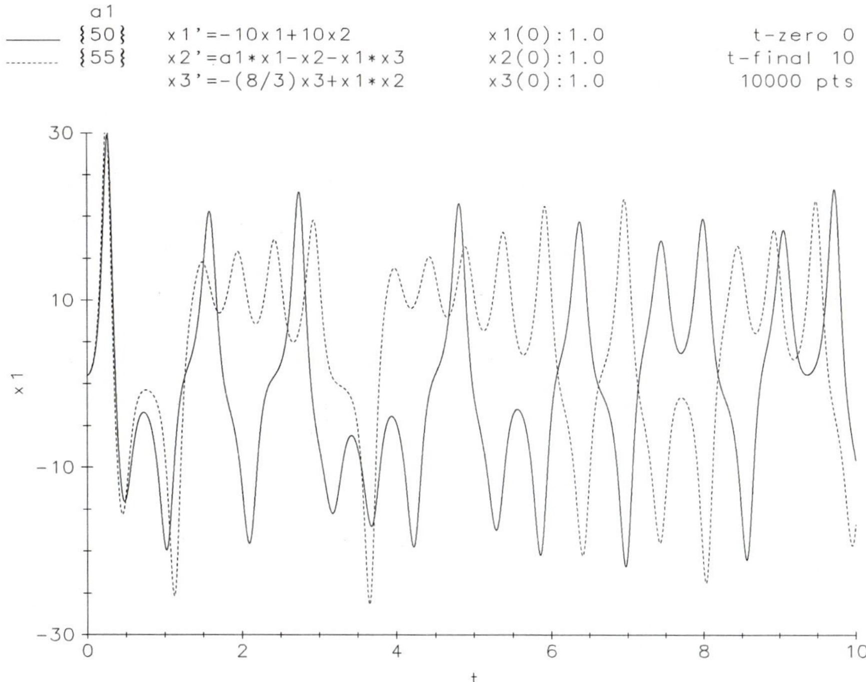


*Figure 0.2* *Sample graph showing sensitivity of the $x_1$-component to parameter changes.*

- A robust and reliable ODE solver with a fairly wide selection of predefined functions. It would be nice (but not crucial) if the solver accepted user-defined functions and handled systems with at least three state variables.
- A hardware platform with two- and three-dimensional graphics capabilities for displaying the output of the solver.
- A printer for producing hard copy of graphics displays.

A few tasks may require a higher level of computer hardware or software performance—these are marked with an asterisk. Students are encouraged to use their own platform to reproduce as many of the numerous workbook plots as time allows. The required data is given on every plot (or in the adjoining text).

### *Hardware/Software Used for This Workbook.*

All ODEs in this workbook were solved using DEQSOLVE, an ODE solver from the MATHLIB[1] software package. DEQSOLVE is based on LSODA, a descendant of C. W. Gear's DIFFSUB,[2] that is part of the package ODEPACK developed by Alan Hindmarsh at Lawrence Livermore National Laboratories with contributions by Linda Petzold and Jeffrey Painter. DEQSOLVE was written in PASCAL by Kevin Carosso,[3] Ned Freed, and Dan Newman[3] of Harvey Mudd College, and runs under the VAX/VMS[4] operating system. DEQSOLVE switches automatically, as necessary, between methods appropriate for stiff and non stiff problems. Adams-Moulton formulas are used for non stiff problems, and backward differentiation formulas for stiff problems. By default, the solver uses double-precision numerics and a relative local error of one part in $10^6$ for large values of the state variables and otherwise an absolute error of $10^{-6}$. Numerical solutions are converted to graphic displays by TEMPLATE.[5]


### *Acknowledgments*

Very special thanks goes to Tony Leneis who was the principal author of the software package ODETOOLKIT, a command script and user interface over DEQSOLVE. This workbook was designed in LaTeX by Don Hosek, and mainly set by Dave Richards, whose LaTeX talents and meticulous attention to detail deserve our warmest appreciation. Our thanks go to the the students in our courses who have tested various versions of these experiments. We thank Professors Cave, Di Franco, Farmer, Karukstis, Leader, Molinder, Van Hecke, West, and many other colleagues for their suggestions. We are especially grateful to Professor Shampine for a number of corrections and improvements. Our appreciation also goes to Professor Moody, whose calculus laboratory workbook influenced the format and design of this workbook.

The authors are especially indebted to Mr. J.E. Jonsson for bringing Harvey Mudd College and the Rensselaer Polytechnic Institute together for cooperative projects, and for his generous support of these ventures. This workbook is an outgrowth of such a cooperative effort. And last but not least, the authors are grateful for help from their home institutions.


R.L. Borrelli
C.S. Coleman
W.E. Boyce

---

1. Registered trademark of Innosoft International Inc., Claremont, CA.
2. Gear, C. W., *Numerical Initial Value Problems in Ordinary Differential Equations*. Prentice-Hall, Englewood Cliffs, NJ, 1971.
3. Now with Innosoft International Inc., Claremont, CA.
4. Registered trademark of Digital Equipment Corporation, Maynard, MA.
5. Registered trademark of TGS, San Diego, CA.

# Notes for the Instructor

The integration of computers with the standard symbolic and theoretical approach of the classroom requires some thought and organization beforehand, especially where the support facilities are concerned. There are important choices to be made in advance: What hardware/software platforms should be used? Which experiments are appropriate for the platforms? How can modeling and computer-based experiments be presented in a first course in ODEs in a way that is challenging, but appropriate to the current stage of student development? This workbook was designed to facilitate these choices and to provide a coherent, organized structure for a "differential equations laboratory."

## *Style and Scope*

The experiments generally follow the order of presentation of topics in a traditional first course in ODEs, but no references are made to standard textbooks. There is more than enough material for a one-credit-hour laboratory course.

All the experiments require the use of computers at certain stages. Some experiments are straightforward and some may require substantial effort and ingenuity. Some experiments involve the construction of a mathematical model but many do not.

The experiments fall into two categories. Most can be done by individual students, but some experiments are extensive in nature and are more appropriate for a student team.

This workbook is organized and written to permit students to benefit from its use without extensive guidance from the instructor. A lab assistant can handle questions on the use of software and hardware, but some follow-up in class is recommended.

Ironically, the introduction of computers into a course often leads to heightened interest in the theory rather than less. As some experiments in this workbook show, visual displays of solutions of ODEs can be misleading or difficult to interpret unless the basic theory is well in hand. Interestingly enough, the same claim can be made about modeling areas that generate the ODEs. For these reasons, this workbook includes relevant highlights of the theory and major modeling environments (albeit in condensed form).

## *Organization of This Workbook*

The first chapter explores the power and the limitations of the user's platform and introduces some basic ideas and techniques. After this introductory chapter, Chapters 2–6 roughly parallel the logical development of topics in a typical ODE course:

- **Chapter 2** treats numerical solutions of ODEs, reinforces the graphical underpinnings of solution curves and orbits, and examines properties of solutions of simple ODEs.
- **Chapter 3** goes into more detail on the properties of solutions of first order ODEs.
- **Chapter 4** treats second order linear ODEs.
- **Chapter 5** delves into the behavior of solutions of planar systems (i.e., first order systems with two state variables), both linear and nonlinear.
- **Chapter 6** covers the properties of solutions of higher dimensional systems.

Three appendices complement the experiments in many important ways. For example, the experiments reference the appendices for background information and illustrations.

- **Appendix A:** *How to Write Team Reports.* Experiments specifically designed for student teams (indicated by a "team" icon) are more difficult to complete because they are open-ended in nature. This appendix contains some hints on how to work in a team mode, and suggests an outline for the team report.
- **Appendix B:** *Modeling.* The sections on modeling are useful starting points for experiments that involve the construction of a mathematical model.
- **Appendix C:** *Atlas.* The Atlas consists of a series of plates giving solution curves and orbits of a large variety of ODEs and systems of ODEs. The plates, which constitute a handy visual reference, are often mentioned in the experiments. They are also useful in "tuning up" the user's hardware/software platform. Plates are accompanied by brief descriptions of the underlying system. Experiments throughout the workbook elaborate on Atlas entries—for example, by asking the student to describe the effect on solution plots when varying a parameter that appears in the ODE.

Each chapter is introduced by a brief summary defining terms and concepts. Every chapter contains experiments that guide the student in visualizing important concepts or applications involving ODEs. The Notes at the end of some chapters elaborate at greater length on the mathematical ideas in a chapter, and should be helpful in doing some of the experiments.

## *Modes of Use*

A great deal of flexibility has been designed into the computer experiments. In particular, each experiment consists of several tasks, generally arranged in ascending order of difficulty. As a rule, there are more tasks and questions in a single experiment than are intended to be assigned (a feature shared by traditional problem sets in textbooks). Thus, the level of difficulty of an assigned computer problem set may be adjusted by appropriate selection of questions and tasks. Advanced-level tasks are marked with an asterisk.

Most computer problem sets can be done by students individually, but some are more appropriate for a team (with a common laboratory report—see Appendix A). If access to platforms is restricted, computer problem sets intended for individuals can be conducted by a team but with individual write-ups.

Instructors may lecture beforehand on the material relevant to assigned computer problem sets, if it is deemed appropriate. This workbook was designed to allow students to use it on their own with occasional advice from laboratory assistants. The "keywords" listed on the cover sheet of an experiment are a guide to the mathematical prerequisites of that experiment. A summary of relevant concepts is included in the cover sheet. Instructors can use the cover sheets (as well as appendices, chapter introductions, and chapter notes) as lecture resources.

The workbook can be used in a variety of modes because of the special features incorporated into its design. It is not expected that all of the features of the workbook will be used in any one mode:

- **Homework set replacement**. A computer problem set can replace a homework set from time to time in any course involving ODEs. The table on the facing page lists possible computer problem set assignments for a one-semester first course in ODEs. As in any lab course, it is not particularly important to synchronize computer work with the material covered in the lectures (indeed, the lack of sufficient hardware may dictate staggering the due dates of computer problem sets). A brief follow-up in class is recommended.
- **Lab text.** The workbook may serve as a "text" for a separate laboratory course parallel to a traditional course involving ODEs. Science and engineering courses have long

# Suggestions for Computer Problem Sets for an Introductory ODE Course

The syllabi below assume a three-credit-hour, semester-long course with an assigned textbook. One mode for using this workbook has the following features:

- **A computer problem set (CPS)** is a selection of tasks from an experiment. Normally, one CPS per week is assigned, and is collected, graded, and returned just as a regular homework set.
- **Time required** to complete a CPS averages two hours (team experiments take longer).
- **Traditional homework sets** are still assigned, taking into account the time needed to complete a CPS.
- **Tutorial on local hardware/software platforms** is available to the students.
- **Laboratory assistants** hold office hours in the computer laboratory to assist students with hardware/software problems.
- **Tutors** (upperclassmen or graduate assistants) provide help on a CPS just as on homework sets.

In Syllabus I, only one of the two listed computer problem sets is assigned each week. Note that different sections need not be assigned the same CPS. Syllabus II is appropriate for an honors section.

| Week | Syllabus I | Syllabus II |
|---|---|---|
| 1. | No Experiments | No Experiments |
| 2. | **Exp 1.1:** 1, or **1.1:** 2 | **Exp 2.5:** 1a, b, 2, 3a, b, 4 |
| 3. | **Exp 2.1:** 1, 2, 3a, 5, or **2.2:** 1A | **Exp 3.3:** 1a, b, 2, 3a, b, c |
| 4. | **Exp 3.3:** 1, 2, 3a, or **3.5:** 1, 2, 3 | **Exp 4.1:** 2, 3Aa, b, c, 4A |
| 5. | **Exp 1.11**, **3.8** or **4.9** (team) | **Exp 1.11**, **3.8** or **4.9** (team) |
| 6. | Exam Week | Exam Week |
| 7. | **Exp 4.2:** 1, 2A, 3, or **4.3:** 1, 2A, 3 | **Exp 4.11:** 3A, C, 4 |
| 8. | **Exp 4.5:** 1, 2, 3, or **4.6:** 1, 4 | **Exp 5.1:** 1a, b, c, 2Aa, b, c |
| 9. | No Experiments | **Exp 5.3:** 1a, b, c, d, 2 |
| 10. | **Exp 4.7:** 1, 2, 3, or **4.8:** 1, 2, 4 | **Exp 5.6:** 1, 2 |
| 11. | Exam Week | Exam Week |
| 12. | No Experiments | **Exp 5.10**, or **6.12** (team) |
| 13. | **Exp 5.1:** 1, 2A, or **5.2:** 1, 2, 3a, 4a | **Exp 6.5:** 1, 3b |
| 14. | **Exp 6.1:** 1, 2, or **6.2:** 1a, b, 2 | **Exp 6.6:** 1, 2, 3, 4, 5 |

operated in this mode. The variety and level of experiments involving models makes this workbook appropriate for use in conjunction with a wide range of courses.

- **Project handbook.** This workbook can be used as a "handbook for projects" in any course that touches upon ODEs (for example, Calculus, Advanced Engineering Mathematics, Theoretical Mechanics, Physical Chemistry, or Population Dynamics). The terminology used in building the model ODEs is consistent with current usage by professionals in those fields.
- **Validation set.** The plots and graphs in this workbook can serve as a "validation set" for the user's own platform or for user-written ODE solvers. All of the nearly 200 plots of solutions of ODEs in the workbook are described and labeled, and may be useful for benchmarking or "tuning" local platforms.
- **Reference/review.** The many annotated graphs, the careful (but brief) overviews of theory, the clear derivation of all model ODEs, and a detailed index all combine with the user's own platform to produce an effective reference and review tool (a kind of "futuristic Schaum's Outline").

### *Selection of Hardware/Software Platforms*

The choice of a suitable hardware/software platform depends on the local environment, local resources, and the computing expertise of the users. There are currently many hardware platforms that support software suitable for carrying out the experiments in this workbook. There are vast differences among platforms, and the first task of the instructor is to find a combination that is easy enough for nonprogrammers to use yet robust enough to carry out accurately the steps called for in the experiments.

Usually there is little choice in which hardware platforms are available at a given site. Many institutions already have a "PC-Lab" or "Work Station Lab." If a hardware choice is possible, technical support may become an issue. Software selection, however, can be a many-splendored thing. Characteristics of software to look for include the following:

- Interactive, easy to use (no knowledge of programming languages necessary)
- Quality of documentation and support
- Accurate and transparent operation
- A large collection of predefined functions
- Excellent graphics, 2D and 3D
- Support for hard copy facilities

The rapid growth of hardware/software platforms accessible to the general user for solving and plotting the solutions of ODEs shows no sign of letting up. For this reason, it makes no sense to attempt to describe any of these solvers here. Reviews of mathematical software appear regularly in many technical journals.

Software packages have their own peculiarities depending on the actual ODE solver algorithms employed, the hardware platforms, and so on. Some are menu-driven and interact with the user through elaborate displays, whereas others interact in a linear command/response stream, and some combine aspects of these two extremes. Some packages act like oracles (i.e., giving no hint that a mistake has occurred), whereas others print warnings and/or supply information on what went wrong. However, some peculiarities common to all solver/platform combinations affect the accuracy of the output. Users should be aware of these peculiarities and recognize them when they occur; solvers cannot be run mindlessly and results always accepted without question. Listed below are some problems that occur often enough to merit comment.

- **Clipping.** A graphics screen can display only a finite portion of space. Some solvers will "shut down" when orbits leave the region represented by the screen. Thus orbits that would have returned to the screen after exiting will not be picked up. Zoom features may also be affected, depending on how clipping is implemented.

- **Default settings.** The number of points plotted on an orbit segment or a solution curve may have been selected by default. Thus if the represented screen size is small and the rate functions are large, too few solution points will appear on the screen, with the result that the orbit or solution curve looks like a broken line graph. Rate functions should be carefully examined before setting the screen size, the solution time interval, or the number of plotted solution points.
- **Screen resolution.** Only a finite number of points (called **pixels**) on any screen can be turned "on." Hence orbits of a differential system may appear to run together when they actually do not. A zoom feature may resolve orbits in a given region. Also, direction field lines that appear to be distorted on the screen, may look normal when printed on a high-resolution printer.
- **Overflow/underflow.** Numerical results that are too large will overflow their storage registers. Numerical results whose magnitudes are too small produce underflow errors.

A brief tutorial for students on the use of the local computing facilities with some examples is generally beneficial and is strongly recommended. Students may not have access to platforms that will produce all the graphs called for in an experiment. They should be encouraged to be creative and find a way to do the best job possible with the graphing (and hard copy) tools at hand.

### *Other Design Features*

This workbook incorporates a number of additional design features that are worth keeping in mind:

- No extraordinary pains are taken to treat first order ODEs before higher order equations and systems of ODEs or linear ODEs before nonlinear ones—those distinctions do not always matter to numerical solvers anyway.
- Sensitivity of solutions of initial value problems as the data change is examined throughout, not treated in just one location.
- It is important to know that computers have limitations. Some experiments were designed to highlight this fact.
- Redundancies often occur in the workbook, permitting students to corroborate their results (or obtain hints on what to look for or do).
- Each chapter contains a few experiments that involve the construction and analysis of mathematical models for applications treated in most ODE textbooks. The modeling areas covered are mechanics, electrical circuits, and rate processes. Appendix B gives a brief introduction to each of these areas and a discussion of the modeling process itself.
- In the modeling experiments, the actual design of an environment depends not only on the modeler's breadth of knowledge, but also on time constraints and the computer (and other) resources available. This workbook tries, insofar as is possible, to reflect awareness of this fact.
- Scaling and nondimensionalization are important techniques in the practical world of computation. Appendix B.5 gives a brief introduction to these techniques, and some experiments give examples, but there is no substitute for experience.
- A detailed index references key terms, definitions, and concepts.

## *Graphical Dilemma*

Extra dimensions of state space not only permit a richness of orbital geometry, but also lead to a pictorial dilemma. How can orbits in a multidimensional state space be portrayed on a two-dimensional graphics screen? Depending upon the available hardware and software, three-dimensional perspectives, planar plots of each pair of state variables, and plots of the individual state variables against time (component graphs) may be used to "see" an orbit or a solution curve through its projections. Figures 6.1.1, 6.1.3, 6.2.1, 6.3.1, and Atlas Plates `Autocatalator A`, `B`, and `Lorenz A-C` display some of the graphical possibilities.

## *Effect of Workbook on Course*

How will using this workbook change the way a course is taught? Some changes are bound to occur, even though users may not be aware of them.

- The graphical visualization approach used in the workbook gives an intuitive sense of how solutions of ODEs behave.
- A new tool and approach will inevitably change the questions posed in the course. The questions that arise naturally in the new environment will somehow become the "right" questions, and some old familiar questions will no longer assume the importance they once had.
- The problems and projects posed in the experiments are "hands-on" and by nature focused and goal-oriented. Hence students will likely feel the pride of achievement on completing an experiment.
- Computer problem sets may replace some textbook problem sets. Eventually, the laboratory portion of the course may evolve into a separate, required one-unit course (as has been the case in the sciences for many years).
- Some topics not usually covered in lectures can be (perhaps more appropriately) investigated through computer experimentation.
- In a strange turn of events, interest in the basic theory of ODEs and details of modeling environments will be heightened rather than diminished after the introduction of the computer into the course. This comes about naturally because visual displays of solutions of ODEs can sometimes be misleading or difficult to interpret without the extra insight gained from these sources. (Examples are given in Experiment 1.11 and elsewhere in the workbook.)

## *A Concluding Word*

The formalized nature of the experiments in this workbook provides a template for instructors to add further experiments of their own choosing. The support features built into the workbook make this process convenient and effective. The authors welcome comments on the nature and content of these experiments (comments may be sent by E-mail to ODEWORKBOOK@SIF.CLAREMONT.EDU). The reporting of typographical (or other) errors and suggestions for future editions are always appreciated. Because the workbook is 100% electronic, corrections and modifications are easy to incorporate. New experiments for inclusion (with attribution) in future editions are particularly welcome.

CHAPTER 1

# Introduction:<br>Learning About Your Hardware/Software

At the heart of the experimental approach to the study of ordinary differential equations (abbreviated **ODEs**) is the hardware/software platform, a tool that vastly expands our ability to visualize solutions. Therefore, the first step must be familiarization with the features and limitations of the platform of choice. Experiments in this chapter address this task. Limitations are inherent in any hardware/software platform, and the experiments in this chapter illustrate the most important ones. The following introductory examples and telegraphic review of standard terminology may be used for reference.

## *Introduction*

Suppose that all that is known about a function is a formula for its rate of change (i.e., for its derivative) and the coordinates of one point on its graph. Can the function be determined from that information alone? For example, suppose that the unknown function of time is denoted by $x(t)$, and that the following information is given:

$$x'(t) = 4t - 3$$
$$x(0) = 0$$

where $x'(t)$ denotes $dx(t)/dt$, and the initial condition $x(0) = 0$ requires that at $t = 0$ the value of $x$ be zero (i.e., $(0, 0)$ is a point on the graph of $x = x(t)$). Antidifferentiating $x'(t) = 4t - 3$, we have that

$$x(t) = 2t^2 - 3t + C$$

where $C$ is a constant of integration. The initial condition is used to find $C$:

$$x(0) = 0 = 0 + 0 + C$$

hence, $C = 0$, and the unknown solution $x(t)$ has been determined. Graphs of solutions in $tx$-space are called solution curves. Different initial conditions yield different values of $C$, hence different solution curves. Figure 1.1 displays solution curves for a variety of initial conditions.

The rate of change in the above example is a function of time $t$ only and is not a function of the unknown state $x$. However, examples abound in contemporary engineering and science in which the rate of change of the state $x(t)$ of a system at time $t$ depends not only upon time but also upon the current value of the state. For example, it may be known that

$$x'(t) = 1 - t^2 + \sin(t\,x(t))$$
$$x(0) = 3$$

The right-hand side of the above differential equation is a complicated rate function of $t$ and $x$, and there is no explicit formula for the solution $x(t)$. Nevertheless, a computer solver can approximate the solution and draw its graph. Figure 1.2 shows the graph of the computed solution curve; the initial point $(0, 3)$ is boxed.

Even though modern solvers enable users to graph solutions of very complicated initial value problems, they should be used with care. This chapter explores the use of computer solvers to numerically approximate solutions of initial value problems and discusses some

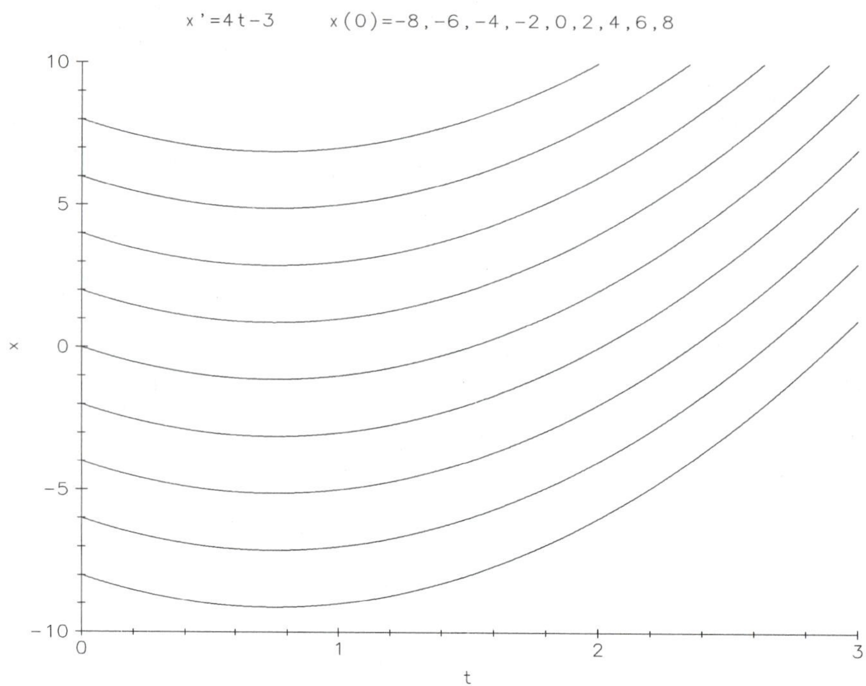


*Figure 1.1 Solution curves for an initial value problem using various initial conditions.*

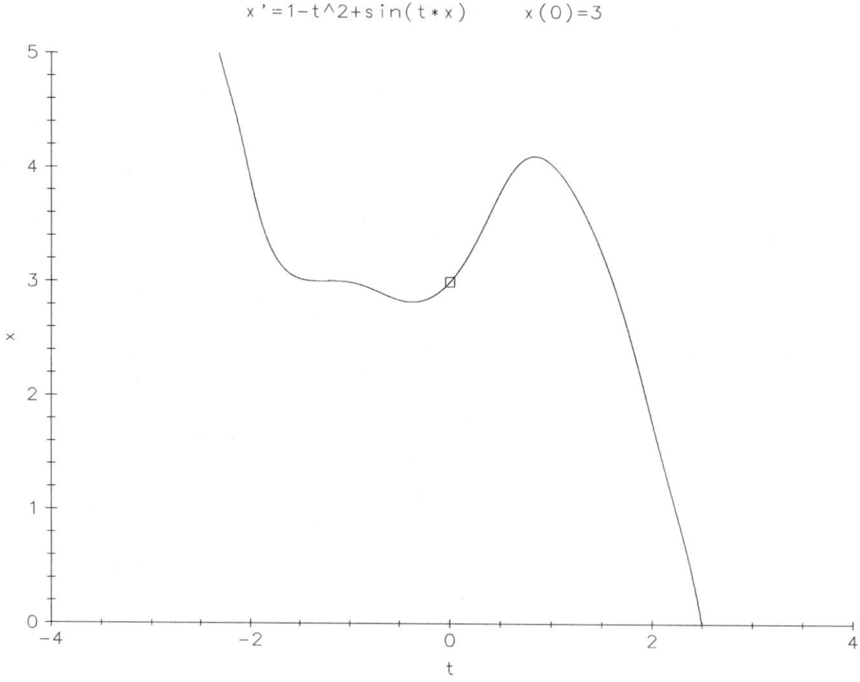


*Figure 1.2 The solution curve for an initial value problem.*

limitations of such solvers. First however, some notation suggested by the two preceding examples needs to be introduced and made a part of the operating language for the experiments of the chapter.

### *First Order ODEs*

Let the function $f(t, x)$ be defined in some region $R$ of the $tx$-plane. Suppose that a differentiable function $x(t)$, defined on a $t$-interval $I$, is such that the curve $(t, x(t))$ lies in $R$ for $t$ in $I$, and is such that $x'(t) = f(t, x(t))$ for all $t$ in $I$. Then $x(t)$ is a **solution of the first order ODE**

$$x' = f(t, x) \tag{1}$$

Another notation for $x'$ is $dx/dt$. The graph of the solution $x(t)$ in the $tx$-plane is called a **solution curve** for the ODE (1). The function $f(t, x)$ is called the **rate function** for the ODE (1) because it defines the rate of change of any solution $x(t)$ as it arrives at a point $(t, x)$ on its solution curve. If $(t_0, x_0)$ is a point in the region $R$ then the equation pair

$$\begin{aligned} x' &= f(t, x) \\ x(t_0) &= x_0 \end{aligned} \tag{2}$$

is called an **initial value problem** (abbreviated **IVP**) for the ODE (1). A function $x(t)$ defined on an interval $I$ is a **solution of the IVP** (2) if $x(t)$ is a solution of the ODE (1) for all $t$ in $I$, if $t_0$ is in $I$, and if $x(t_0)$ has the value $x_0$. For a solution $x(t)$ of the IVP (2), the value of $x(t)$ is said to be the **state** of the "system" described by the ODE (1) at the time $t$. The requirement $x(t_0) = x_0$ in the IVP (2) is called an **initial condition**, $t_0$ the **initial time**, and $x_0$ the **initial state** or **initial data**. Figure 1.1 displays several solution curves of an ODE. An example of an IVP and its solution curve may be seen in Figure 1.2.

### *First Order Systems*

The ODE (1) involves only one **state variable** called $x$, which is differentiated only once. One way to increase the number of state variables in an IVP, say to two—call them $x_1$ and $x_2$—is as follows: take two rate functions denoted by $f_1$ and $f_2$, both defined on some common region $R$ in $tx_1x_2$-space, and set up the **first order system**

$$\begin{aligned} x_1' &= f_1(t, x_1, x_2) \\ x_2' &= f_2(t, x_1, x_2) \end{aligned} \tag{3}$$

An IVP based on the system (3) is constructed by choosing any point $(t_0, x_1^0, x_2^0)$ in $R$ and then writing

$$\begin{aligned} x_1' &= f_1(t, x_1, x_2) \\ x_2' &= f_2(t, x_1, x_2) \\ x_1(t_0) &= x_1^0, \quad x_2(t_0) = x_2^0 \end{aligned} \tag{4}$$

The pair of functions, $x_1(t), x_2(t)$, defined over a common $t$-interval $I$, is a **solution of the system** (3) if for all $t$ in $I$, the equations (3) are satisfied with the functions $x_1(t), x_2(t)$ substituted for the state variables $x_1, x_2$. If $x_1(t), x_2(t)$, is a solution pair for the system (3), then the graph of the points $(t, x_1(t), x_2(t))$, $t$ in $I$, in $tx_1x_2$-space is called a **solution curve for the system** (3). The planar graphs of the points $(t, x_1(t))$ and $(t, x_2(t))$ are called the **component graphs** of the solution. Another kind of graph for systems with more than one state variable is very important. For a solution pair $x_1(t), x_2(t)$, the graph $G : (x_1(t), x_2(t))$, $t$ in $I$, in $x_1x_2$-space (called the **state space** of the system), is an **orbit** of the system (3). Examples of solution curves, component graphs, and orbits of first order systems appear in Figures 5.1 and 5.2 and in the Atlas (Atlas Plates `Planar Systems B-D`).

A system of ODEs whose rate functions depend explicitly *only* on the state variables is said to be an **autonomous system**. For example, $x' = x(x - 1)$ is autonomous, but

$x' = x^2 + t$ is not. An **equilibrium point** of an autonomous system is a point in the state space where all the rate functions are zero. Thus, the ODE $x' = x(x-1)$ has two equilibrium points, $x^* = 0$ and $x^* = 1$.

Generalizing the definition of first order systems and their associated IVPs to $n$ state variables, $x_1, x_2, \ldots, x_n$, is straightforward. When $n = 2$, however, the state variables are often called $x$ and $y$, instead of $x_1$ and $x_2$, in order to avoid subscripts. Similarly, if $n = 3$, then $x$, $y$, and $z$ are often used.

## *Higher Order ODEs*

Let $F$ be a function of three variables. Then a **second order ODE** has the form

$$x'' = F(t, x, x') \tag{5}$$

A function $x(t)$, $t$ in $I$, is a **solution of the second order ODE** (5) if $x''(t) = F(t, x(t), x'(t))$, for all $t$ in $I$. If the point $(t_0, x_0, x_0')$ is in the domain of $F$, then an **IVP for the second order ODE** is

$$\begin{aligned} &x'' = F(t, x, x') \\ &x(t_0) = x_0, \quad x'(t_0) = x_0' \end{aligned}$$

ODEs of $n^{\text{th}}$ order, $n = 3, 4, \ldots$, and their associated IVPs are defined similarly. Examples of higher order ODEs and their solution curves are given in the Atlas (for example, Atlas Plate `Duffing A`). It is not difficult to show that if $x(t)$ is a solution of the second order ODE (5), then setting $x_1 = x$, $x_2 = x' = x_1'$, and noting that $x_2' = x''$, we see that

$$\begin{aligned} x_1' &= x_2 \\ x_2' &= F(t, x_1, x_2) \end{aligned} \tag{6}$$

Conversely, if $x_1(t)$, $x_2(t)$ is a solution pair for the first order system (6), then $x(t) = x_1(t)$ is a solution of the second order ODE (5). Thus, there is an equivalence between first order systems of the form (6) and higher order ODEs. Hence, numerical ODE solvers, which address only first order systems, may be used to solve higher order ODEs as well, provided we first convert them to their equivalent first order systems.

## *Numerical Solutions of IVPs*

Most ODE numerical solver software packages are designed to solve IVPs for first order systems. Such solvers can also be used to solve IVPs for higher order ODEs (or even *systems* of higher order ODEs) by converting them to their equivalent form as IVPs for first order systems. Most often this conversion must be done by hand (it is straightforward), but occasionally a software package has an interface that will accept second order ODEs directly—a convenience owing to the importance of second order ODEs in modeling physical systems. It is worth pointing out, however, that a number of programs have been written for the special second order ODE $x'' = F(t, x)$ because of its importance in mechanics.

# Plotting Orbits and Solution Curves

***Purpose*** To interpret geometrically the relation between the solution curves of the first order ODE, $x' = f(t, x)$, and the rate function $f(t, x)$. The relation between the rate functions and orbits of a planar autonomous system will also be explored.

***Keywords*** Slope of a line, direction field, first order ODE, normal form for ODEs, planar autonomous system, orbits

***See also*** Chapter 1 Notes; Atlas Plates `First Order A-F`, `Planar System A-I`

***Background*** Let the rate function $f(t, x)$ in the first order ODE

$$x' = f(t, x) \tag{1}$$

be defined on the rectangle $R$ in the $tx$-plane characterized by the inequalities $a \le t \le b$, $c \le x \le d$. A **solution curve** in $R$ for ODE (1) is a function $x = x(t)$ whose graph is in $R$ and for which $x'(t) \equiv f(t, x(t))$, for all $t$ in some interval $I$ contained in $a \le t \le b$. From calculus we know that if $(t_0, x_0)$ in $R$ is a point on the solution curve $x = x(t)$ for equation (1), then $x'(t_0)$ (i.e., $f(t_0, x_0)$) is the slope of the tangent line to the curve $x = x(t)$ at the point $(t_0, x_0)$. In other words, the line

$$x - x_0 = f(t_0, x_0)(t - t_0) \tag{2}$$

is tangent to the solution curve $x = x(t)$ at $(t_0, x_0)$, as indicated in Figure 1.1.1. This observation gives a way to "construct" solution curves for (1). First, define a uniformly spaced grid of points in the rectangle $R$ using equally-spaced points in the intervals $a \le t \le b$, $c \le x \le d$. Then at each grid point $(t_0, x_0)$ draw a short segment of the line (2) whose center is at the grid point $(t_0, x_0)$. Of course, the lengths of these line segments should be adjusted so that no two of them overlap. This collection of line segments in $R$ is called a **direction field** for ODE (1). If the grid is fine enough, then the resulting direction field may "suggest" curves in $R$ with the property that the tangent line to the curve at each point lies along the direction field line at that point. From the above observations, these curves are close to being solution curves for ODE (1). The finer the grid for the direction field, the

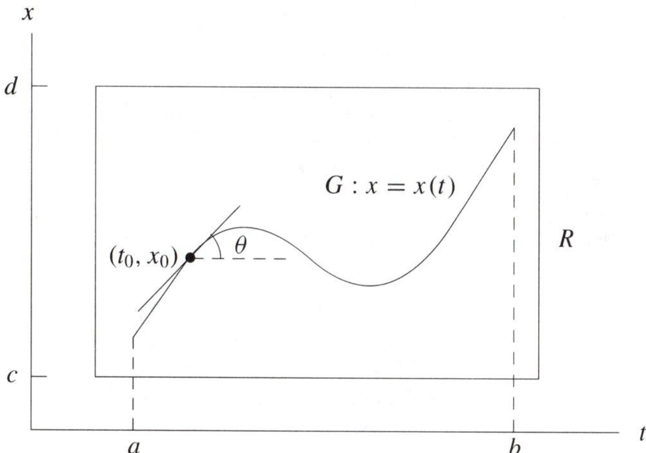


*Figure 1.1.1 Solution curve for $x' = f(t, x)$. Note that $f(t_0, x_0) = \tan\theta$.*

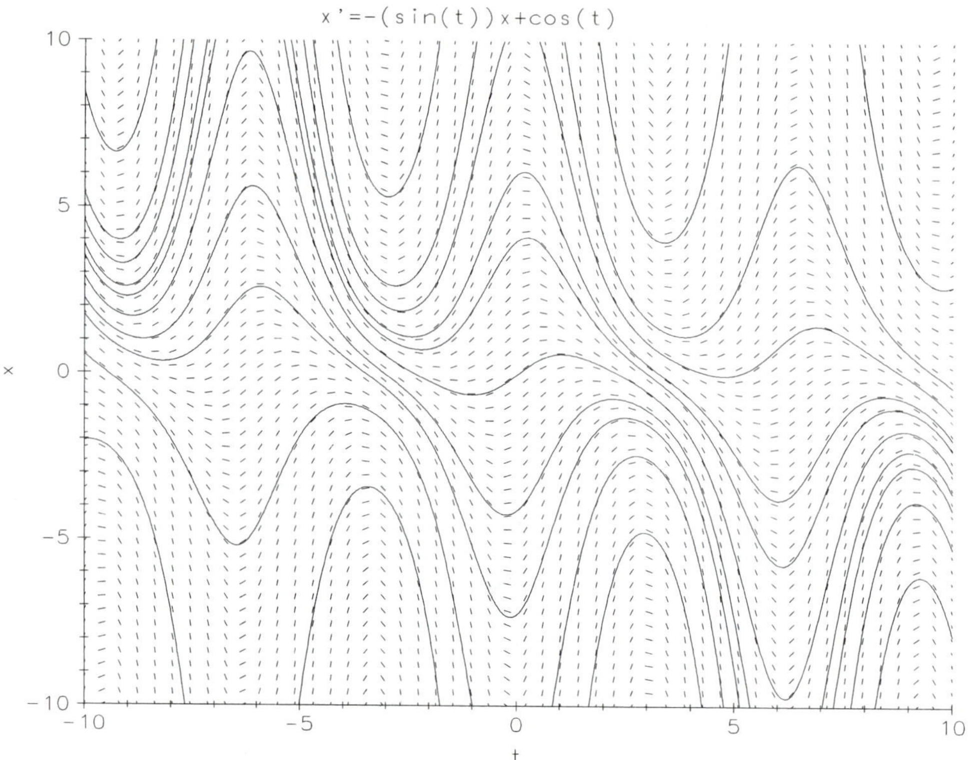


*Figure 1.1.2 Solution curves and a direction field.*

better the drawn curves can be made to "fit" the field in the above sense. Many ODE solvers are capable of automatically generating direction fields for first order ODEs.

*Example 1* **(Isoclines)** Before plotting solution curves of the single, first order linear ODE:

$$x' = -(\sin t)x + \cos t \tag{3}$$

it is helpful to create a direction field plot in the selected rectangle $R$ in the $tx$-plane. Using the direction field, a rough idea of the shape of solution curves can be ascertained. Figure 1.1.2 shows some solution curves of ODE (3) plotted by a solver. Notice how the solution curves "fit" the direction field.

The rate function for ODE (3) is $f(t, x) = -(\sin t)x + \cos t$. Observe that the curve $f(t, x) = 0$ in the $tx$-plane is given by $x = \cot t$ and has the property that when a solution curve crosses this curve it does so with a horizontal tangent line (because at that point $x' = 0$). These curves are called **0-isoclines** because of this property. Isoclines for other slope values are defined similarly. Observe that the concavity of solution curves can also be ascertained directly from an ODE without knowing explicit solution formulas. Indeed, if $x(t)$ is a solution of ODE (3), then differentiating the ODE we obtain $x''(t) = (\sin^2 t - \cos t)x - (1 + \cos t)\sin t$, where we have used ODE (3) to replace $x'$. Thus, in regions of the $tx$-plane where $x'' > 0$, the solution curves must be concave upward and where $x'' < 0$, concave downward. Arcs of the curve $(\sin^2 t - \cos t)x - (1+\cos t)\sin t = 0$ separate these two regions. The arcs are called **inflection curves** for this reason. Figure 1.1.3 shows both the 0-isoclines and the inflection curves for ODE (3).

*Example 2* **(Direction fields for planar autonomous systems)** This example illustrates that the direction field of a planar autonomous system is intimately connected to the orbits of the system

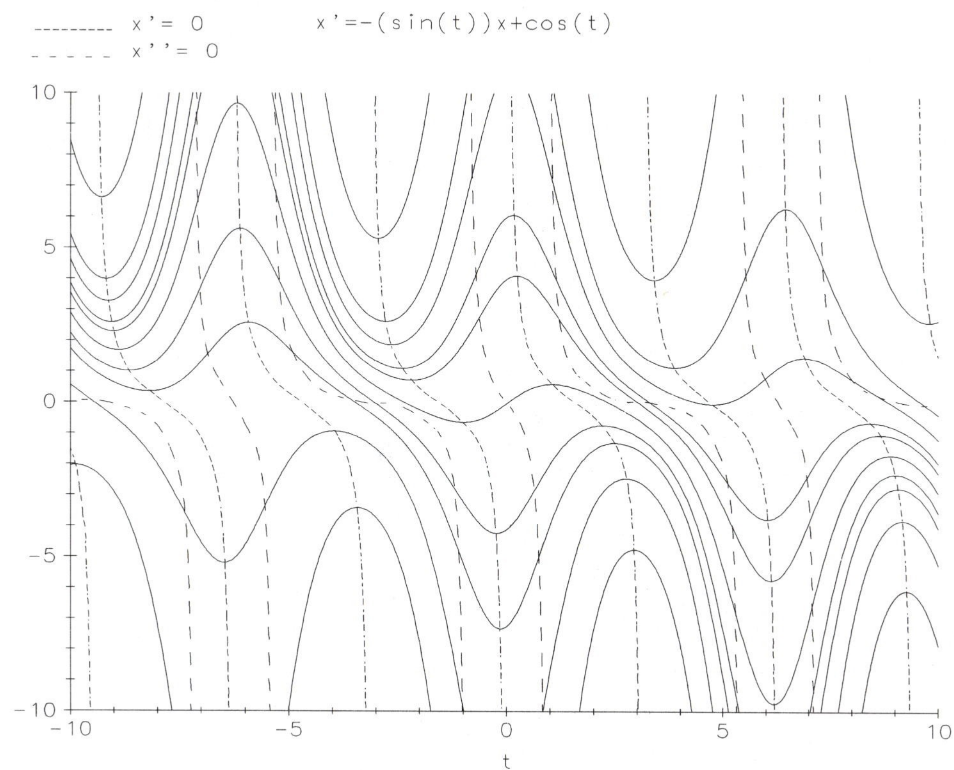


*Figure 1.1.3 0-isoclines and inflection curves.*

and plays an important role in plotting these orbits. Consider the following ODE based on the **Rayleigh Equation** (named after the English physicist):

$$x'' - x' + \tfrac{1}{3}(x')^3 + x = 0 \tag{4}$$

To write (4) as an equivalent planar system, set $y = x'$ and obtain

$$\begin{aligned} x' &= y \\ y' &= y - \tfrac{1}{3}y^3 - x \end{aligned} \tag{5}$$

The rate functions in (5) evaluated at any point $(x_0, y_0)$ define the tangential direction of the (unique) orbit of the planar system (5) through that point. To see this, consider the parametrically defined curve traced out by the position vector $\mathbf{r}(t) = x(t)\hat{\mathbf{i}} + y(t)\hat{\mathbf{j}}$, where $\hat{\mathbf{i}}$ and $\hat{\mathbf{j}}$ are unit vectors on the $x$ and $y$-axes, respectively. Recall from calculus that the vector $\mathbf{r}'(t) = x'(t)\hat{\mathbf{i}} + y'(t)\hat{\mathbf{j}}$ has the property that $\mathbf{r}'(t_0)$ is tangent to the curve traced out by $\mathbf{r}(t)$ at the point $\mathbf{r}(t_0)$. Thus a direction field can be plotted in $xy$-space by using the system rate functions, and this direction field is everywhere tangent to the orbits of the system. Caution: Do not confuse these direction fields with those defined for first order scalar ODEs as in Example 1. The latter are drawn on the $tx$-plane. See Figure 1.1.4 for a direction field plot of the planar system and orbits defined by the ODE (5). Many ODE solvers automatically produce direction fields for autonomous planar systems. In plotting orbits of such a system, it is helpful to identify any equilibrium points that appear on the screen. The system (5) has only one equilibrium point, the origin.

***Observation*** **(Time's arrow)** It is often helpful to draw arrowheads on the orbits of the planar system $x' = f(x, y)$, $y' = g(x, y)$, to indicate the direction of increasing time. To do this, it is first

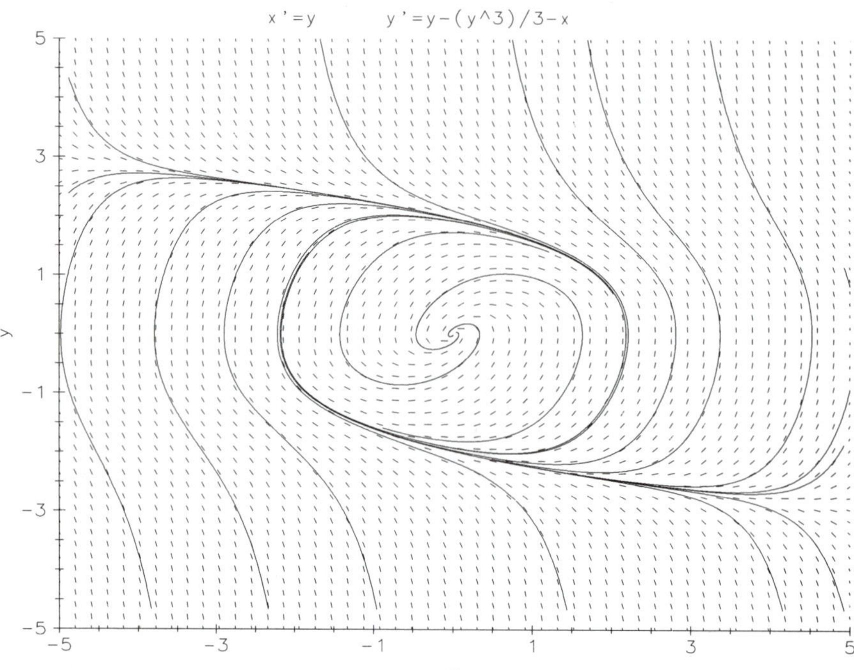


*Figure 1.1.4 Direction field and orbits for ODE in (5).*

useful to draw on the direction field element through $(x_0, y_0)$ an arrowhead that points in the direction of the vector $f(x_0, y_0)\hat{\mathbf{i}} + g(x_0, y_0)\hat{\mathbf{j}}$. Note that the arrowheads on the orbits are consistent with the arrowheads on the direction field elements.

***Observation*** **(Normal form for ODEs)** In converting an $n^{\text{th}}$ order ODE to a first order system as often required by solvers, the first step is to write the ODE in normal form. An $n^{\text{th}}$ order ODE is in **normal form** if it has been (algebraically) solved for the highest order derivative. For example, $(x')^2 - \sin(tx) = 0$ is not in normal form, nor is the second order ODE, $x' = \exp(t^2 + x'')$. On the other hand, $x' = (\sin(tx))^{1/2}$ and $x' = -(\sin(tx))^{1/2}$ are both in normal form. Note that the function $x(t)$ solves the ODE, $(x')^2 - \sin(tx) = 0$, if it solves either of the normalized ODEs, $x' = \pm(\sin(tx))^{1/2}$ (with either sign). A system of ODEs is in normal form if it has already been solved (algebraically) for the highest derivatives of the state variables. For example, the system

$$\begin{aligned} x'' &= 3yx' - \sin t + (y')^2 \\ y'' &= \exp(x' + y') + txy \end{aligned}$$

is in normal form. The conversion of ODEs (or systems of ODEs) in normal form into a first order system in normal form is a straightforward process. The system above of two second order equations becomes the system of four first order equations

$$\begin{aligned} x' &= u \\ y' &= v \\ u' &= 3yu - \sin t + v^2 \\ v' &= \exp(u + v) + txy \end{aligned}$$

First order systems have the same number of state variables as ODEs (four in this case).

## *1.1 Direction Fields and Solution Curves*

***Abstract*** The close relationship between solution curves of a first order ODE and the direction field it generates in the $tx$-plane is examined (see Figures 1.1.1 and 1.1.2).

**1.** A direction field plot for the ODE $x' = x - t^2$ appears in Figure 1.1.5. Do the tasks indicated below. Write directly on the figure where appropriate.

**(a)** Using a colored pencil, draw a curve on the figure with the property that if a solution curve $x(t)$ intersects this curve at $(t_0, x_0)$, then $x'(t_0) = 0$. Explain the significance of this curve.

**(b)** In a different color, draw a curve with the property that if a solution curve $x(t)$ of the ODE intersects this curve at $(t_0, x_0)$, then $x''(t_0) = 0$. Explain the significance of this curve. (Hint: Calculate $x''$ by differentiating the ODE.)

**(c)** Sketch solution curves on Figure 1.1.5, taking note of the curves drawn in (a) and (b). Verify your sketches of the solution curves by using a solver. Explain any differences.

**2.** Make a hard copy of a direction field for the ODE $x' = tx + 1$. Repeat Problem 1, using the scales on the $t$ and $x$-axes that appear in Figure 1.1.5.

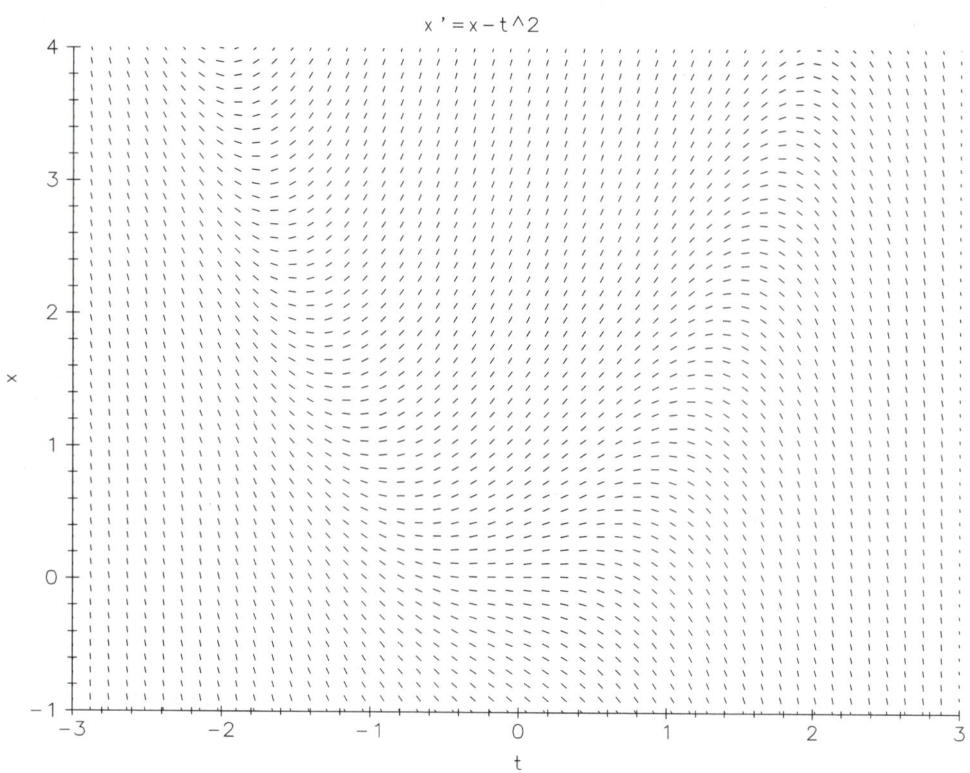


*Figure 1.1.5 Direction field for Problem 1.*

hardware / software

name / date

Answer questions in the space provided,
or on attached sheets or carefully labeled graphs.

course / section

## *1.2 ODEs in Non Normal Form*

***Abstract*** The behavior of solution curves of first order ODEs with discontinuous rate functions is examined. Examples appear in Atlas Plates `First Order` E, F.

**1.** Textbooks consider the **normal form** $x' = f(t, x)$ of a first order ODE for convenience in stating theoretical properties of IVPs. Numerical solvers require the users to convert ODEs to normal form before entering data. A direction field plot for the ODE $tx' - x = t^3$ appears in Figure 1.2.1. Note that direction field lines over $t = 0$ have been avoided, since the rate function $f(t, x) = (1/t)x + t^2$ is undefined on the $x$-axis.

**(a)** Sketch some solution curves using the direction field plot. Do any curves cross the $x$-axis?

**(b)** Use a solver/grapher package to verify your sketch in (a).

* **(c)** Find a formula for all solutions. (Hint: Rewrite the ODE as $(x/t)' = t$). Does any solution satisfy the initial condition $x(0) = x_0$ for $x_0 \neq 0$? Is there any solution that is defined on a $t$-interval centered at $t = 0$?

* **2.** Plot a direction field for the ODE $tx' + 2x = 3t$ (which can be rewritten as $(t^2x)' = 3t^2$). Repeat Problem 1.

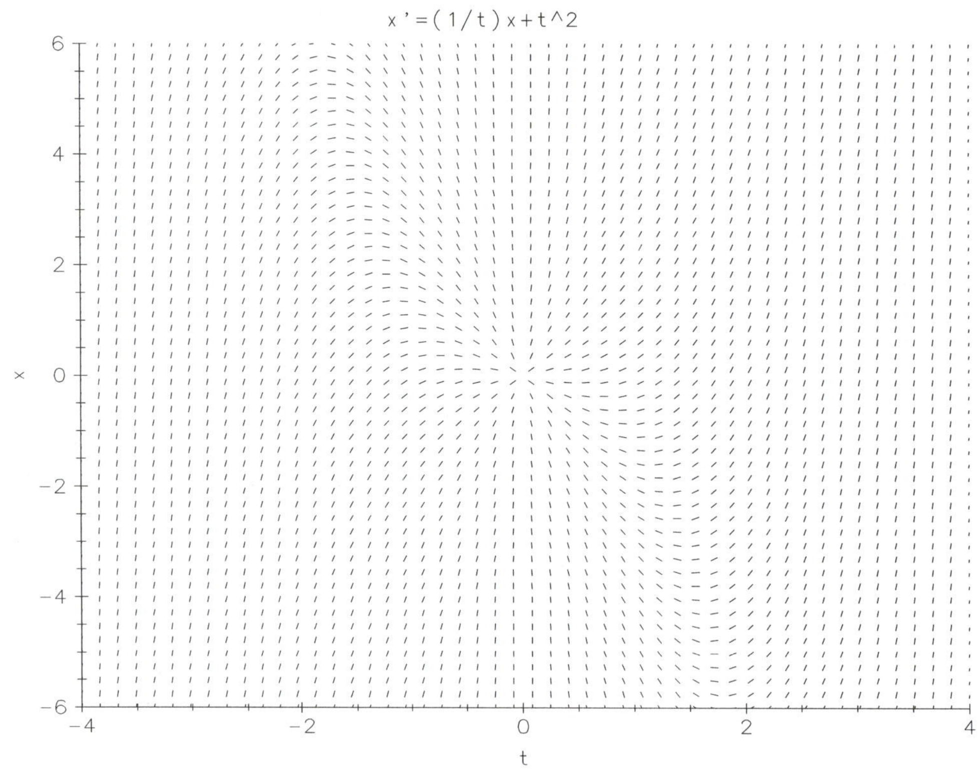


*Figure 1.2.1 Direction field for Problem 1.*

| hardware / software | name / date |
|---|---|
| Answer questions in the space provided, or on attached sheets or carefully labeled graphs. | course / section |

## *1.3 Direction Fields and Orbits*

***Abstract*** The rate functions for planar autonomous systems generate a direction field in the system state space whose close relationship with the orbits of the system will be explored. (See Example 2 for a description of this connection.) Warning: The notion of a direction field of a first order ODE as described in Example 1 should not be confused with the notion of a direction field used in this experiment.

**1.** Figure 1.3.1 shows a direction field plot for the planar autonomous system $x' = y^2 - 1$, $y' = x^2 - 1$. Perform the tasks below. Write directly on the figure.

**(a)** Draw arrowheads on a representative sample of the direction field lines. Interpret these arrowheads. (See Example 2 and the observation that follows it.) Sketch some orbits on Figure 1.3.1. Use arrowheads on the orbits to show increasing time.

**(b)** Use your solver to plot some representative orbits of the given planar system. Draw arrowheads on each orbit to indicate increasing time. Describe any unusual features.

**(c)** Compare the plot obtained in (b) with the Atlas Plate `Planar System D`. Explain any differences.

**2.** Make hard copy of a direction field for the system $x' = -4y$, $y' = x$. Repeat (a) and (b) of Problem 1, using the same $x$, $y$-scales as in Figure 1.3.1.

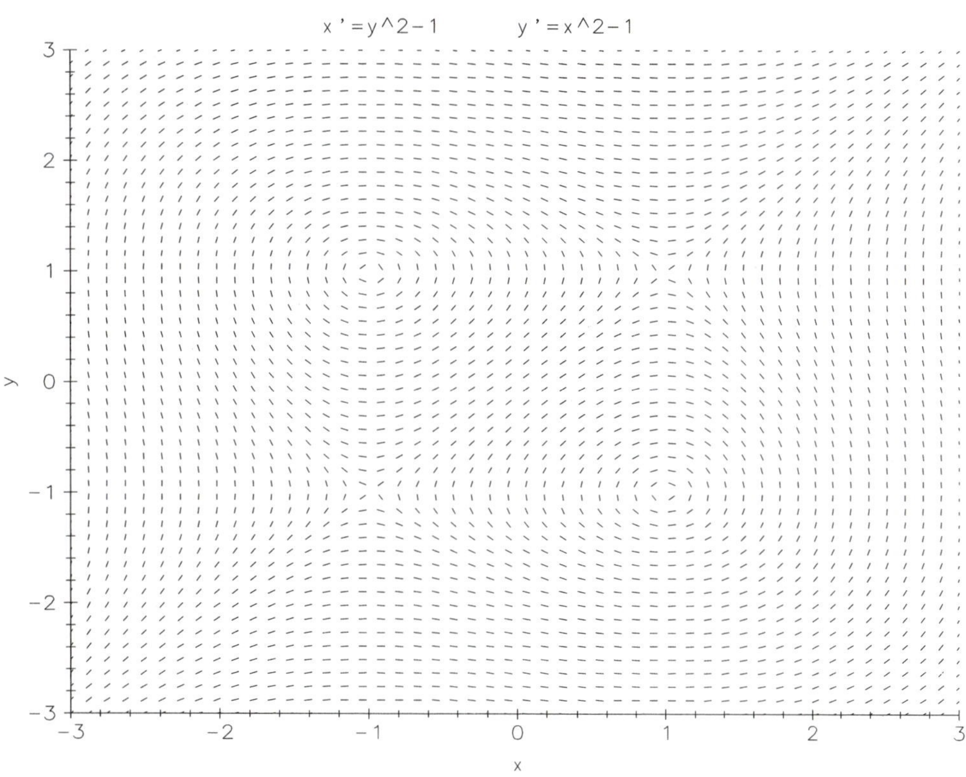


*Figure 1.3.1 Direction field for Problem 1.*

hardware / software name / date

Answer questions in the space provided, or on attached sheets or carefully labeled graphs.

course / section

## *1.4 Solvers and IVPs*

***Abstract*** Solvers are used to examine both the relation between direction field and the solution curves for a first order ODE and the connection between the orbits of an autonomous planar system of ODEs and the direction field defined by the system rate functions.

**1.** Over the time interval $0 \le t \le 7.5$ solve the IVP

$$x' = 3x\sin(x) - t$$
$$x(0) = 0.5$$

Plot this solution in the $tx$-plane. (Hint: See Atlas Plate `First Order A`.)

**2.** Graph some orbits for the Rayleigh system (5) in the $xy$-plane. Draw arrowheads on the orbits to indicate the direction of increasing time. Use the scaling $-5 \le x \le 5$, $-5 \le y \le 5$ for your plots. Describe in words what the orbits are doing as $t \to +\infty$ and as $t \to -\infty$ (write directly on your graph). Plot some component graphs ($x(t)$ versus $t$).

**3.** Over the time interval $0 \le t \le 10$, solve the IVP

$$x'' + x'|x'| + x = 0$$
$$x(0) = -\tfrac{1}{2}, \quad x'(0) = \tfrac{3}{4}$$

Plot this solution in the $xx'$-plane as well as in the $tx$-plane. Draw arrowheads on the orbits to indicate the direction of increasing time. Estimate $x(5)$ from the graph.

* **4.** Write the following IVP

$$(1+t^2)x'' + 3ty(x')^2 = t\cos t$$
$$y'' + (x^2+y^2)y' + t^2x' = 0$$
$$x(3) = 0, \quad x'(3) = 1, \quad y(3) = 1, \quad y'(3) = 0$$

as an IVP involving four normal first order ODEs. Graph the solution $(x(t), y(t))$ in the $xy$-plane over a time interval centered at $t_0 = 3$. Draw arrowheads on these curves to indicate the direction of increasing time.

# Generating Atlas Plots

***Purpose*** To familiarize the experimenter with the features and limitations of laboratory hardware/software platforms by reproducing a variety of plots in the Atlas (Appendix C).

***Keywords*** **O**rdinary **D**ifferential **E**quations (**ODEs**), **I**nitial **V**alue **P**roblems (**IVPs**), hardware/software platforms for solving IVPs

***See also*** Appendix C (the Atlas); Chapter 1 Introduction

***Background*** The first order of business for an experimenter is to become comfortable with the laboratory "equipment" and tools as well as a few "tricks of the trade." The main tool in the ODE laboratory is a software package, which resides on a hardware platform with some sort of printing device attached. The software may have already been installed on the hardware platform, but if your laboratory is not a dedicated facility, then the laboratory assistant will aid users in the installation procedure. Instructions on the use of hard copy devices will be provided.

From the experimenter's viewpoint, the platform can be viewed as a "black box" with various "switches," "buttons," and "levers" that the user needs to identify and learn to operate. The questions below will aid the experimenter in this process.

- **Functions recognized**. What predefined functions are available on the solver? Are user-defined functions possible? How long are user-defined functions retained?
- **Operating characteristics**. Does use of the solver require the knowledge of a set of commands? A programming language? An operating system? How easy is the solver to use? Is it interactive? Is documentation required? If so, are user guides available? Does your hardware platform support a mouse (or other analog input device)?
- **ODE input**. How does the solver like to have ODEs specified? How many first order ODEs can it handle at one time?
- **Accuracy**. How accurate is your solver? Can the user set the number of solution points and limits on errors? Are there any bounds on these settings? Are there any default settings that affect accuracy?
- **Robustness**. How hard is it to crash the solver? Does the solver ever hang up when rules of operation are followed? How are error messages (if any) handled? How easy is it to recover from an erroneous command?
- **Graphics**. What graphics hard copy devices are available to you? What is the resolution on your graphics screen? On your hard copy device? Does your solver support 3D as well as 2D graphics? Is there a color graphics screen? How many colors are there? Can graphs be overlaid on the screen for comparison? What information is printed automatically on all graphs? Is there automatic scaling?
- **Special features**. Does your software platform have a reference library of plots? Are there "walk-through" examples activated by default? Can your software interface with other programs?

***Observation*** In trying to reconstruct the atlas plates of Appendix C, the user will have to discover answers to many of these questions.

_____ / _____
hardware software

_____ / _____
name date

Answer questions in the space provided,
or on attached sheets or carefully labeled graphs.

_____ / _____
course section

## *1.5 Generating Atlas Plots*

***Abstract*** Graphical displays lie at the heart of the experiments in this workbook. These displays on a graphics terminal, PC, or work station and the corresponding hard copy are powerful tools for visualizing properties of solutions of ODEs. However, the creation of a good graphical display is as much an art as anything else, and the experimenter needs practice to master these skills on available platforms. The atlas plates are examples of what can be done. These plates are arranged alphabetically by title.

**1.** Describe your platform. What features does the solver have? What are the solver's limitations? What printers do you have available for producing graphs? Use the questions on the cover sheet as a guide.

**2.** Reproduce each of the atlas plates listed below on your own platform. Explain any differences. In (b), (c), and (d) draw arrowheads on the orbits to indicate the direction of increasing time. Change the number of solution points on solution curves and orbits, and note any changes in your plots. Repeat this process on another platform (if available) and compare to earlier results. Make hard copies of your plots, and write comments directly on the graphs.

**(a)** `First Order A`. Fill in some solution curves in the blank areas.

**(b)** `Planar System D`. Use $t_0 = 0$ when generating the orbits. Fill in some orbits in the central blank area. Are there any solutions that repeat themselves in time (so-called **periodic solutions**)? If so, identify a few and calculate their **periods** (the minimum time required for orbits to repeat themselves). Find all equilibrium points of the system. Is the system autonomous? Use $t_0 = 1$ to generate orbits on the same screen. How do these orbits differ from those in `Planar System D`? Try the initial condition $x_0 = -2$, $y_0 = 1$, at $t_0 = 1$.

**(c)** `Pendulum A`. Are there any periodic solutions? Any equilibrium points? Is the system autonomous?

**(d)** `Pendulum (Upended) A`. Generate the orbit originating at $x_0 = -1.4$, $y_0 = 1$, at the initial time $t_0 = 2$. How does this orbit compare with the orbit in `Pendulum (Upended) A`?

* **(e)** `Lorenz A`. If your platform allows, choose several eye positions from which to view this 3D plot.

# First Order Rate Laws

***Purpose*** To examine changing populations in diverse settings where rate laws are first order.

***Keywords*** First order growth process, radioactive decay, radioactive dating, first order rate laws, doubling time, half-life, Balance Law

***See also*** Appendix B.1, B.3, B.5 (Example 1); Atlas Plate `Compartment Model C`

***Background*** The size of a population of a given species at time $t$ is a nonnegative quantity measured as a density or a concentration of that species at the given time or "the number of individuals in the population at time $t$." Denoting the population size by $x(t)$, the **Balance Law** in Appendix B.1 implies that $x'(t) =$ rate in (at time $t$)$-$ rate out (at time $t$). The "rate in" term is the sum of the **birth rate** and the **immigration rate** (rate at which individuals enter the population from the "outside"), while the "rate out" term is the sum of the **death rate** and the **emigration rate** (rate at which individuals leave the population, but not by death). Accumulated evidence shows that the birth and death rates at a given time are often proportional to the population size at that time. When this is so, there are positive constants $b$ and $d$, called **rate constants** such that $b \cdot x(t)$ and $d \cdot x(t)$ are the respective birth and death rates at time $t$. The constants $b$ and $d$ measure the individual's contribution to birth and death rates; they have the units of (time)$^{-1}$. If $f(t)$ denotes the **migration rate** (immigration rate $-$ emigration rate), then $x(t)$ solves the ODE

$$x' = (b - d)x + f(t) \tag{1}$$

The rate function in ODE (1) is a **first order rate law**, that is, an ODE of the form $x' = \alpha x + \beta$, where $\alpha$ and $\beta$ are constants or functions of $t$ only. Some applications of first order rate laws in diverse settings follow.

- **Exponential growth**. Suppose that a population is closed off from outside influence (i.e., no immigration or emigration) and that the only change in the population is due to births and deaths, which follow a first order rate law. Then the size of the population $x(t)$, with initial size $x_0$, satisfies the initial value problem (IVP)

$$x' = (b - d)x, \qquad x(0) = x_0 \tag{2}$$

  The solution of the IVP (2) is $x(t) = x_0 e^{(b-d)t}$. Thus if the birth rate exceeds the death rate then the population "explodes" exponentially in time, whereas if the death rate is larger than birth rate then the population "dies out" exponentially in time.

- **Substance accumulation**. Suppose we wish to model the accumulation of a substance in a container in which the only change is due to migration. Specifically, suppose the container is filled with a liquid of known initial volume and the substance $S$ is dissolved in the liquid. Suppose that a solution of $c(t)$ pounds of $S$ per gallon pours into the container at the constant rate of $g$ gal/min, and that liquid flows out the bottom at the same rate. Assuming that the solution in the container is thoroughly mixed at all times, and letting $x(t)$ denote the pounds of $S$ in the container at time $t$, we see that the "rate in" of $S$ is given by $c(t)g$. Since the volume of liquid in the container is a constant at all times, say $V_0$ gallons, it follows that the "rate out" is $(x(t)/V_0)g = (g/V_0)x(t)$. (Note that the "rate out" term appears to follow a first order rate law.) Thus $x(t)$ satisfies the ODE

$$x' = c(t)g - (g/V_0)x \tag{3}$$

  See Experiments 3.14–3.15 for specific examples. For an example involving several containers in which all migrations follow a first order rate law, see Experiment 6.5.

- **Radioactive decay**. If $x(t)$ tracks the population size of a radioactive substance, then there are only deaths and no births or migration. Decay in a radioactive substance follows a first order rate law. Hence, if the initial size of the population is $x_0$, then $x(t)$ satisfies the IVP

$$x' = -kx, \qquad x(0) = x_0 \tag{4}$$

where $k$ traditionally denotes the rate constant. It is easy to see that $x(t) = x_0 e^{-kt}$ solves IVP (4), and the radioactive substance decays exponentially in $t$. It is an interesting fact that the time required for the population to decrease by half its current size is constant throughout the whole decay process. To see this, observe that $x(t_0 + T)/x(t_0) = 1/2$ implies that $e^{-kT} = 1/2$, or $T = \ln 2/k$. This time $T$ is called the **half-life** of the radioactive substance and is often denoted by $t_{1/2}$. Thus the decay process described by IVP (4) has the half-life $t_{1/2} = \ln 2/k$.

***Example 1*** **(Determining the rate constant)** The rate constant $k$ in the radioactive decay process (4) can be determined if the initial amount $x_0$ and the remaining amount of the substance, $x(t_1)$, at some later time $t_1$, are known. Thus, if 500 grams of the substance is present initially and after 10 years only 300 grams remains, then $300 = 500e^{-10k}$, and so $-10k = \ln(3/5)$, or approximately, $k = 0.0511$. Of course, with this determination of $k$, then $t$ in the solution formula $x(t) = x_0 e^{-kt}$ must be measured in years, and $x(t)$ has the units of $x_0$. Since $-k = \ln(3/5)^{1/10}$, it follows that $x(t) = x_0 e^{-kt} = x_0 (3/5)^{t/10}$.

***Example 2*** **(Radiocarbon dating)** Living cells absorb carbon directly or indirectly from carbon dioxide ($CO_2$) in the air. Some of the carbon atoms in this $CO_2$ are a radioactive form of carbon, $^{14}C$, rather than the common isotope $^{12}C$. The $^{14}C$ nuclei decay to nitrogen atoms following a first order rate law. Thus all living things, or things that were once alive, contain some radioactive carbon nuclei, $^{14}C$. In any living organism the ratio of $^{14}C$ to $^{12}C$ in the cells is the same as that in the air. If the ratio in the air is constant in time and location, then so is the ratio in living tissue. When the organism dies, ingestion of $CO_2$ ceases and only the radioactive decay continues. A careful measurement of the $^{14}C$ decay rate in a fragment of dead tissue can be used to determine the number of years since its death.[1] The theory is straightforward. Let $x(t)$ denote the amount of $^{14}C$ per gram of carbon at time $t$ (measured in years). If $t = 0$ is the present time, and the living tissue died at time $T$ (thus $T < 0$), then $x(t)$ decays according to a first order rate law for $t \geq T$. If $x(T) = x_T$, then $x(t) = x_T e^{-k(t-T)}$ solves the IVP $x' = -kx$, $x(T) = x_T$, for all $t \geq T$. If $x_0 = x(0)$ then $x_0/x_T = e^{kT}$, or, using $t_{1/2} = \ln 2/k$

$$T = \frac{1}{k}\ln\left(\frac{x_0}{x_T}\right) = \frac{t_{1/2}}{\ln 2}\ln\left(\frac{x_0}{x_T}\right) \tag{5}$$

The ratio $x_0/x_T$ can be determined by using a Geiger counter to measure the number of disintegrations per minute per gram of carbon for both living tissue and the fragment of dead tissue. Indeed, since $x(t)$ satisfies a first order rate law for $t \geq T$, it follows that $-kx_0 = x'(0)$ and $-kx_T = x'(T)$, hence $x_0/x_T = x'(0)/x'(T)$. Since the number of disintegrations per minute recorded by a Geiger counter is proportional to the rate $x'(t)$, the value $x_0/x_T$ is the ratio of the Geiger counter readings.

***Example 3*** **(Dating a cave painting)** A Geiger counter is used to measure the current decay rate of $^{14}C$ in charcoal fragments found in the cave of Lascaux, France, where there are prehistoric wall paintings. The counter recorded about 1.69 disintegrations per minute per gram of carbon, while for living tissue the number of disintegrations was measured in 1950 to be 13.5 per minute per gram of carbon. How long ago was the charcoal formed (and, presumably, the

1. Willard Libby demonstrated this in the early 1950s. He received a Nobel Prize for his work.

paintings painted)? Using the half-life of $^{14}C$ as $5568 \pm 30$ years and the Geiger counter readings, we see that $x_0/x_T = 1.69/13.5$, hence from (5)

$$T = \frac{5568 \pm 30}{\ln 2} \ln\left(\frac{1.69}{13.5}\right) \approx -16,692 \pm 90 \text{ years}$$

***Example 4*** **(Potassium-argon dating)** Potassium-argon-calcium decay is a simple branching process of radioactive disintegration, a process used to date relatively young volcanic geological strata (in geologic time). Potassium-40 ($^{40}K$) is a radioactive form of potassium that decays to argon-40 ($^{40}Ar$) and to calcium-40 ($^{40}Ca$) by the branching cascade illustrated in Figure 1.6.1. Potassium decays to calcium by emitting a $\beta$ particle (i.e., an electron). Some of the potassium atoms, however, decay to argon by capturing an extranuclear electron and emitting a gamma ray. Both decay processes follow first order rate laws, but with differing rate constants. In contrast to carbon-14 dating, which is based on the disappearance of radioactive "parent" atoms, the potassium-argon method of dating geological strata depends on measuring the accumulation of "daughter" argon atoms. The possibility of using potassium-argon decay as a dating process depends on the fact that the argon gas is trapped in the lattices of a cool rock, but escapes if the rock is molten. The rate equations for this decay process may be written in terms of the masses $K(t)$, $A(t)$, and $C(t)$ of potassium, argon, and calcium in a sample of rock from the stratum:

$$\begin{aligned} K' &= -(k_1 + k_2)K \\ A' &= k_1 K \\ C' &= k_2 K \end{aligned} \tag{6}$$

and are valid for all $t$ after the volcanic stratum was deposited. Let $t = 0$ be the current time and $T$ the time in the past (thus $T < 0$) that the stratum was deposited in a molten state. Putting $k = k_1 + k_2$, $K(T) = K_T$, $A(T) = A_T$ and $C(T) = C_T$, we solve (6) with the initial data to find

$$K(t) = K_T e^{-k(t-T)}, \quad A(t) = \frac{k_1 K_T}{k}(1 - e^{-k(t-T)}) + A_T,$$
$$C(t) = \frac{k_2 K_T}{k}(1 - e^{-k(t-T)}) + C_T$$

for $t \geq T$. From an earlier remark we see that $A_T = 0$ and so

$$\frac{A(0)}{K(0)} = \frac{k_1}{k}(e^{-kT} - 1) \tag{7}$$

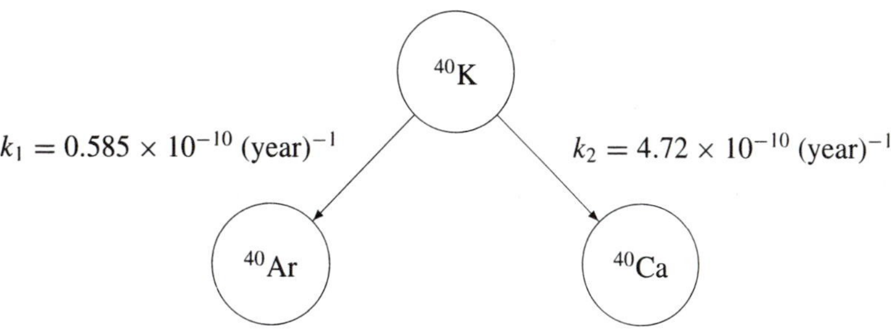


*Figure 1.6.1 Potassium-argon and potassium-calcium decay.*

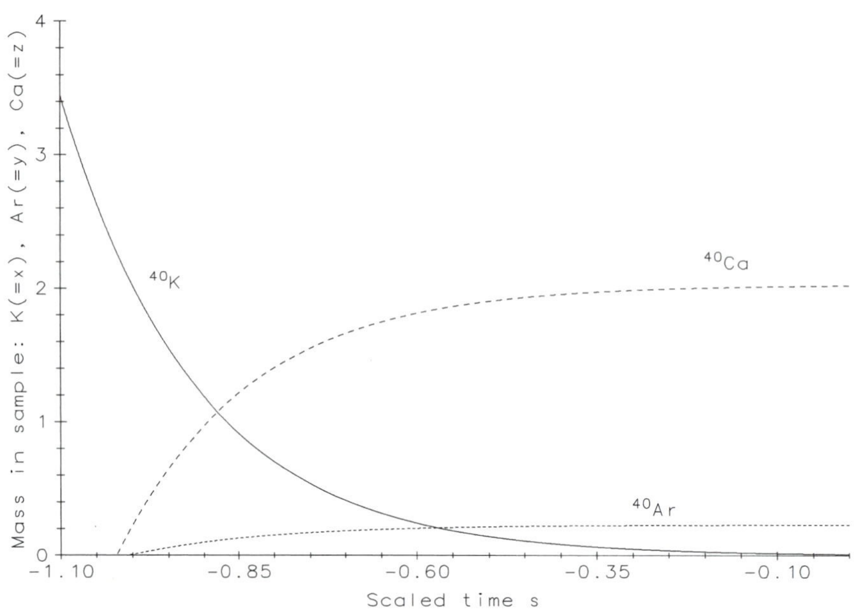


*Figure 1.6.2 Potassium-argon-calcium branching cascade.*

Solving (7) for $T$ we see that

$$T = -\frac{1}{k}\ln\left(\frac{k}{k_1}\frac{A(0)}{K(0)} + 1\right) \tag{8}$$

Thus, the current value $A(0)/K(0)$ determines the age of the stratum from (8).

In 1959, Mary and Louis Leakey uncovered in the Olduvai Gorge, Tanzania, a fossil hominid skull and primitive stone tools of obviously great age, older by far than any hominid remains found up to that time. Carbon-14 dating methods are inappropriate for a specimen of that age and nature. It is fortunate that the strata both above and below the find were largely volcanic materials which had formed and then cooled quickly (in geological time), thereby starting the potassium-argon clock. Using $A(0)/K(0)$ measurements made at the University of California for the Olduvai samples and formula (8), an approximate age of 1.75 million years was determined.

***Example 5*** **(Rescaling time)** The coefficients $k_1$ and $k_2$ in system (6) are so small that it is hard to compute the solutions. Rescaling time by setting $s = t/10^{10}$ transforms (6) into

$$\begin{aligned} dK/ds &= -5.305K \\ dA/ds &= 0.585K \\ dC/ds &= 4.72K \end{aligned} \tag{9}$$

System (9) is solved backward from $s = 0$ to $s = -1.1$ with the initial data listed in Figure 1.6.2. One of the curves in that figure shows that the argon was trapped in the sample at $s \approx -1$ (i.e., at $T \approx 10$ billion years ago). Before that time the sample must have been in a molten state with potassium decaying, calcium accumulating, and argon escaping.

_______________/_______________ _______________/_______________
hardware software name date

Answer questions in the space provided, or on attached sheets or carefully labeled graphs.

_______________/_______________
course section

## *1.6 Population Growth and Decay*

*Abstract* The Balance Law is used to model changing populations in various settings.

**1.** Radium-226 is a radioactive substance with a half-life of 1620 years.

**(a)** Write an ODE describing the decay of a sample of radium-226, and graph a solution. From your graph of sample mass versus time, find the time required for the sample to decrease to 25% of its original size.

**(b)** Compute this time from a solution formula for the sample size and compare with your result in (a).

**2.** A mold whose growth follows a first order rate law grows from an initial mass of 1 gram to a mass of 3 grams in 1 day.

**(a)** Write an ODE describing the rate of change of the mass of the mold. Graph the mass versus time and find the mass after one week.

**(b)** Compute this mass from a solution formula and compare with your result in (a).

**3.** Suppose that a population has a birth rate $b$ and a death rate $d$ so that the growth of the population is $dN/dt = (b - d)N$. After 15 years of steady growth, the population members simply stop reproducing; that is, $b$ becomes zero. If initially, $b = 0.06$ births/year and $d = 0.04$ deaths/year, use a graphical procedure to determine how long after reproduction stops will it take for the population

**(a)** to return to its original level

**(b)** to reach 50% of its original level

**(c)** to decrease to 30% of its population at the time $b$ became zero

**4.** Nutrients flow into a cell at a constant rate of $r$ molecules per unit time and leave it at a rate proportional to the concentration, with constant of proportionality $k$. Let $N$ be the concentration at time $t$. Assume that the cell volume $V$ is fixed.

**(a)** Write the mathematical expression for the rate of change of the concentration of the nutrients in the process.

**(b)** Sketch graphs of $N(t)$ for $t \geq 0$ for various values of $N(0) \geq 0$ including $N(0) = r/k$. Will the concentration of nutrients reach an equilibrium?

**5.** A 1000 gallon tank contains 100 gallons of water with 50 pounds of salt dissolved in it (this liquid is called **brine**). Suppose pure water runs into the tank at 3 gal/min and brine runs out at 2 gal/min (assume that the concentration of salt in the brine is kept uniform by stirring).

* **(a)** Derive an ODE for the amount of salt in the tank for any $t \geq 0$ before the tank overflows.

* **(b)** Produce a graph for the amount of salt in the brine for $0 \leq t \leq 2$ hours. How much salt is left in the tank at the end of one hour?

* **(c)** Find a solution formula for the amount of salt in the brine for $t \geq 0$ before the tank overflows.

**6.** The first order rate laws described by the ODEs below have nonconstant rate coefficients which, however, tend to zero as $t \to \infty$. Choose one IVP below and solve the problems that follow:

☐ **A.** $x' = \pm x/(1+t) \qquad x(0) = x_0$

☐ **B.** $x' = \pm tx/(1+t^2) \qquad x(0) = x_0$

**(a)** Using the "plus" sign in the ODE, graph solution curves for the IVP for various values of $x_0$, both in the forward direction and the backward direction. Describe the behavior of the solution curves. Why is this behavior to be expected?

**(b)** Repeat (a) for the "negative" sign in the ODE.

**(c)** Using the "plus" sign in the ODE, find a solution formula for the IVP.

**7.** Consider the IVP $x' = 2|x|$, $x(0) = x_0$.

**(a)** Construct a direction field for the IVP.

**(b)** Graph some solution curves for the IVP using various values of $x_0$.

**(c)** Find a solution formula for the above IVP and compare it to the solution formula for the IVP $x' = 2x$, $x(0) = x_0$.

___________/___________ ___________/___________

hardware software name date

Answer questions in the space provided, or on attached sheets or carefully labeled graphs.

___________/___________

course section

## *1.7 Radioactive Decay: Carbon-14 Dating*

***Abstract*** Graphical methods are explored for radiocarbon dating problems. A nongraphical approach to finding the time of death $T$ of the sample is given by formula (5).

***Special instructions*** Using the notation and concepts developed in the cover sheet, the tasks below describe several methods for the graphical solution of the following radiocarbon dating problem.

*In 1977, the rate of* $^{14}$C *radioactivity of a piece of charcoal found at Stonehenge was* 8.2 *disintegrations per minute per gram. Given that the rate of* $^{14}$C *radioactivity of a living tree is* 13.5 *disintegrations per minute per gram and assuming that a tree was burned to produce the charcoal during the construction of Stonehenge, estimate the construction date.*

**1. Method I (Scaling, solving a backward IVP)**

**(a)** Let $x(t)$ denote the amount of $^{14}$C per gram of carbon in the sample at time $t$. Put current time at $t = 0$, and denote by $T$ that time in the past when the sample ceased to be living tissue. Denote by $x_0$ and $x_T$ the respective amounts of $^{14}$C per gram of carbon in the living tissue and the sample and let $k = (\ln 2)/t_{1/2}$. Show that $x(t)$ solves the backward IVP

$$x' = -kx, \qquad x(0) = x_0, \qquad x(T) = x_T$$

**(b)** Show that the scaling $y = x/x_0$ in the backward IVP in (a) nondimensionalizes the state variable (refer to Example 1 of Appendix B.5) and yields the backward IVP

$$y' = -ky, \qquad y(0) = 1$$

whose graphical solution can be produced by a solver. Show that $T$ satisfies the equation $y(T) = 13.5/8.2$, where $y(t)$ solves this IVP. Determine $T$ graphically.

**(c)** Calculate $T$ via the formula (5) and compare it with the result derived graphically above. Try to improve your graphical determination of $T$ by scaling the independent variable in (b) by the substitution $s = t/t_{1/2}$ where $t_{1/2}$ is the half-life of $^{14}$C.

**2. Method II (Redefining the rate function)**

**(a)** Show that the state variable $x(t)$ defined in Problem 1(a) can be extended into $t \le T$ as a solution of the backward IVP

$$x' = \begin{cases} 0, & \text{for } t \le T \\ -kx, & \text{for } T \le t \le 0 \end{cases} \qquad x(0) = x_0$$

**(b)** Using the scale transformation $y = x/x_0$, show that the IVP in (a) is equivalent to the backward IVP

$$y' = \begin{cases} 0, & \text{for } y > \frac{13.5}{8.2} \\ -ky, & \text{for } y \le \frac{13.5}{8.2} \end{cases} \qquad y(0) = 1$$

* **(c)** Use the conditionally-defined function

$$f(y) = \text{if } y \le 13.5/8.2 \text{ then } -(\ln 2)y/5568 \text{ else } 0$$

to rewrite the IVP in (b) as the backward IVP $y' = f(y)$, $y(0) = 1$. Use the graphical solution of this IVP to determine $T$. Compare with determination of $T$ via (5).

_______________/_______________ hardware software

_______________/_______________ name date

Answer questions in the space provided, or on attached sheets or carefully labeled graphs.

_______________/_______________ course section

## *1.8 Potassium-Argon Dating*

***Abstract*** Using a backward IVP for a planar system, a graphical method is explored for solving the potassium-argon dating problem. A nongraphical approach to finding the time that a volcanic stratum was deposited is provided by formula (8).

***Special instructions*** Using the concepts and notation developed in the cover sheet, the tasks below outline a graphical solution technique for the following potassium-argon dating problem.

*The potassium-argon method has recently been used to date the occasional reversals of the earth's magnetic field over geologic time. The fraction of argon to potassium in a sample of lava containing reversely magnetized minerals is 0.000146. About how long ago did a reversal of the earth's magnetic field take place?*

**1.** Duplicate Atlas Plate `Compartment Model C`. Highlight the argon curve and estimate the current amount of argon in the sample. Note that in this model time "begins" when the sample was last in a molten state.

**2.** Let $x(t)$ denote the amount of $^{40}$K and $y(t)$ the amount of $^{40}$Ar per gram of the sample at time $t$. Put the current time at $t = 0$, and denote by $T$ the most recent time in the past when the sample was in a molten state.

**(a)** Denoting by $x_0$, $y_0$ the values of $x(t)$, $y(t)$ at $t = 0$, and using the values of $k_1$, $k_2$ listed in Figure 1.6.1, verify that $x(t)$, $y(t)$ solve the backward IVP

$$x' = -(5.305 \times 10^{-10})x, \qquad x(0) = x_0$$
$$y' = (0.585 \times 10^{-10})x, \qquad y(0) = y_0$$

and that $T$ solves the equation $y(T) = 0$.

**(b)** Show that the scaling $x^* = x/x_0$, $y^* = (y/x_0) \times 10^5$ converts the IVP in (a) to the backward IVP

$$dx^*/dt = -(5.305 \times 10^{-10})x^*, \qquad x^*(0) = 1$$
$$dy^*/dt = (0.585 \times 10^{-5})x^*, \qquad y^*(0) = 14.6$$

whose graphical solution can be produced by a solver. Show that if $x^*(t)$, $y^*(t)$ solves this IVP for $t \leq 0$, then $T$ satisfies the equation $y^*(T) = 0$.

**(c)** Scale the independent variable in the IVP in (b) by the substitution $t = s \times 10^4$. Show that this IVP then becomes

$$dx^*/ds = -(5.305 \times 10^{-6})x^*, \qquad x^*(0) = 1$$
$$dy^*/ds = (0.585 \times 10^{-1})x^*, \qquad y^*(0) = 14.6$$

Find $T$ from a graphical solution of this IVP.

**(d)** Calculate $T$ from formula (8) and compare it with the result derived graphically in (c).

# Falling Bodies

***Purpose*** To examine the motion of a body moving along the earth's local vertical.

***Keywords*** Position and velocity of a body, Newton's Laws of Motion, mass, gravitational forces, damping forces, air resistance

***See also*** Appendix B.2, Appendix B.5

***Background*** Choosing an origin, $O$, on the earth's surface, Newton's Second Law can be used to chart the evolution of the position vector $\mathbf{R}(t)$ of a moving body if we know the forces acting on that body. The only forces considered here are the forces due to gravity and to air resistance. Orient a coordinate frame, as in Figure 1.9.1, with one axis (say the $x$-axis) coinciding with the local vertical at the origin $O$, and assume that this frame is inertial (so that Newton's Laws of Motion apply). Now the force of gravity always acts downward along the local vertical and the force due to air resistance acts along the direction of the velocity, $\mathbf{v} = d\mathbf{R}/dt$, so as to oppose the motion (i.e., in the direction of $-\mathbf{v}$). It follows directly from Newton's Second Law that if the initial position vector $\mathbf{R}(0)$ and initial velocity vector $\mathbf{v}(0)$ of the body are parallel to the local vertical, the motion takes place completely on the local vertical (until the body crashes into the ground). Since the initial position and velocity are always assumed here to have this property, we may as well write $\mathbf{R}(t) = x(t)\hat{\mathbf{i}}$, and so $\mathbf{v}(t) = \mathbf{R}'(t) = x'(t)\hat{\mathbf{i}}$. Using the notation $v(t) \equiv x'(t)$, Newton's Second Law is then expressed by the ODE

$$mv'(t) = F \tag{1}$$

where $m$ is the **mass** of the body, and $F$ is the resultant external force acting on the body at the time $t$. Various settings for the model ODE appear below.

- **Motion in a vacuum, near the earth's surface**. The force of gravity on a body of mass $m$ near the earth's surface has the essentially constant magnitude of $mg$ **newtons**, where $g = 9.8$ m/s$^2$ and $m$ is measured in kilograms. This gravitational force is directed downward toward the center of the earth. If the mass is released at the height $x_0$ with initial velocity $v_0$ in the vertical direction, we see using (1) that $v(t)$ satisfies the IVP

$$mv' = -mg, \qquad v(0) = v_0 \tag{2}$$

The position of the body on the $x$-axis, $x(t)$, solves the IVP

$$x'(t) = v(t), \qquad x(0) = x_0 \tag{3}$$

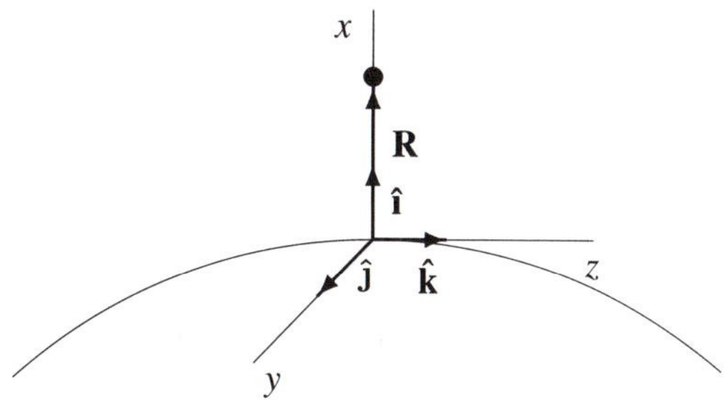


*Figure 1.9.1 Geometry for falling bodies.*

where $v(t)$ is the solution to IVP (2). The solution of IVP (2) is given by an easy antidifferentiation to be $v(t) = -gt + v_0$, and substituting this for $v(t)$ on the right-hand side of IVP (3) yields

$$x(t) = -\tfrac{1}{2}gt^2 + v_0 t + x_0$$

Notice that when $v > 0$, the body moves upward and when $v < 0$, downward.

- **Linear drag, near the earth's surface.** Experiments show that for a body of low density and extended rough exterior moving near the earth's surface (e.g., a feather or a snowflake), the resistive force on the body is proportional to the body's velocity but acts opposite to the direction of motion. Thus if a body of mass $m$ is released at height $x_0$ with initial velocity $v_0$ in the vertical direction (up or down), then using (1) we see that $v(t)$ satisfies the IVP

$$mv' = -mg - kv, \qquad v(0) = v_0 \tag{4}$$

where $k > 0$ is the **drag** coefficient. The position $x(t)$ of the body on the $x$-axis solves the IVP

$$x' = v, \qquad x(0) = x_0 \tag{5}$$

where $v(t)$ is the solution of IVP (4). An explicit solution formula for $v(t)$ can be derived from (4) by separating the variables but is omitted. It will be seen in the experiment that follows that a body described by IVP (4) approaches a finite limiting velocity as it falls.

- **Quadratic drag, near the earth's surface.** When a dense body (e.g., a raindrop, baseball, or bullet) moves near the earth's surface, the resistive force of the air is proportional to the square of the speed and acts opposite to the direction of motion. This is known as **Newtonian damping**. Thus, if the body is released at height $x_0$ with an initial velocity $v_0$ in the vertical direction (up or down), then using (1) we see that $v(t)$ satisfies the IVP

$$mv' = -mg \pm kv^2, \qquad v(0) = v_0 \tag{6}$$

where $k > 0$ is the drag coefficient; the upper sign $(+)$ is chosen if the body is falling, and the lower sign $(-)$ if the body is rising. See Example 1.

- **No drag, not restricted to being near the earth's surface.** Motion of a body on the local vertical, not limited to being near the earth's surface and not subjected to air resistance, can be modeled by ODE (1), where the force $F$ is given by **Newton's Law of Universal Gravitation**. Thus if a body of mass $m$ is released at height $x_0$ with an initial velocity $v_0$ in the vertical direction, then the position $x(t)$ of the body satisfies the IVP

$$mx'' = -mMG/(x+R)^2, \qquad x(0) = x_0, \qquad x'(0) = v_0 \tag{7}$$

where $M$ is the earth's mass, $G$ is the universal gravitational constant, and $R$ is the radius of the earth (in consistent units, of course). IVP (7) is an example of a second order ODE that cannot be reduced to a single first order ODE by the substitution $v = x'$, as was done in the cases considered earlier. In this case the substitution $v = x'$ gives rise to an IVP for a first order system:

$$\begin{aligned} v' &= -MG/(x+R)^2, & v(0) &= v_0 \\ x' &= v, & x(0) &= x_0 \end{aligned} \tag{8}$$

The ODEs are a normalized first order system whose associated IVP can be treated directly by numerical solvers. Note also that the ODEs in (8) are a planar autonomous system. The interesting problem posed by (8) is this: Is there an initial velocity $v_0$ that allows the body to "escape" in the sense that it will never return to the earth? We shall see that such **escape velocities** do indeed exist for any celestial body. See Example 2 and Experiment 1.10.

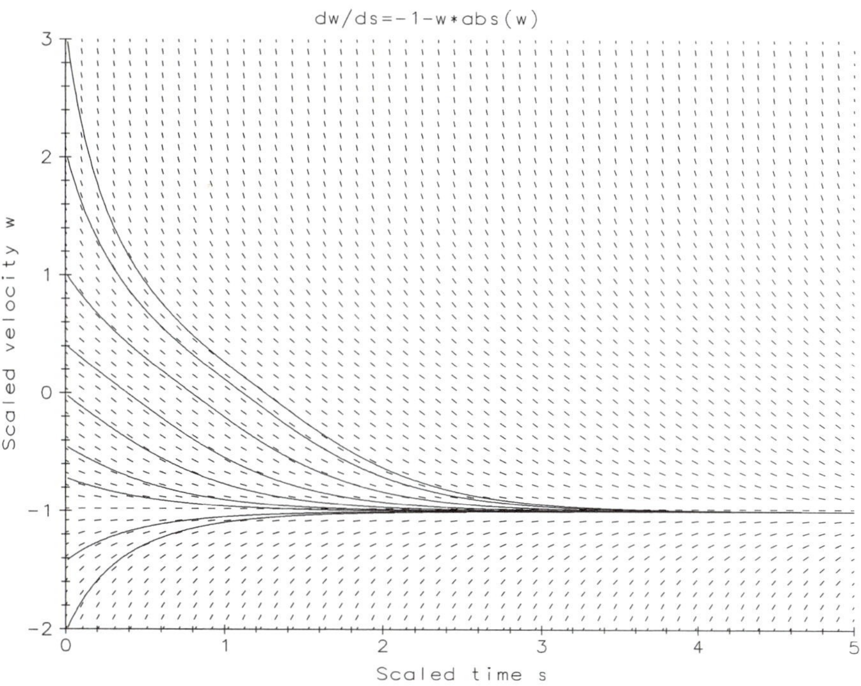


*Figure 1.9.2 Scaled velocity of raindrop moving along the local vertical.*

***Example 1*** **(Falling raindrop)** The IVP (6) describes the velocity $v(t)$ of a raindrop that moves near the earth's surface along the local vertical, with initial velocity $v_0$, acted upon by air resistance and gravitational forces. To take into account upward *and* downward motion, IVP (6) can be written with a single ODE as follows:

$$v' = -g - \frac{k}{m} v|v|, \qquad v(0) = v_0 \tag{9}$$

Scaling the variables $t$ and $v$ as the new variables $s$ and $w$ with

$$t = \left(\frac{m}{kg}\right)^{1/2} s, \qquad v = \left(\frac{mg}{k}\right)^{1/2} w \tag{10}$$

converts the ODE in (9) into the equivalent ODE

$$\frac{dw}{ds} = -1 - w|w| \tag{11}$$

whose solution curves appear in Figure 1.9.2. Notice that all the solution curves of the ODE in (11) appear to stream together toward the equilibrium solution $w \equiv -1$. Thus using (10) we can infer that no matter what the initial velocity is, it is always the case that the velocity of the raindrop $v(t) \to -(mg/k)^{1/2}$, as $t \to \infty$. In other words, raindrops eventually fall with the terminal velocity $(mg/k)^{1/2}$, unless they hit the ground first.

***Example 2*** **(Escape velocity)** There is a way to convert IVP (8) into a first order ODE by using $v$ as the dependent variable and $x$ as the independent variable. Indeed, suppose that $v(t)$, $x(t)$ solves the IVP given in (8) and that $x'(t) \neq 0$ over the time interval $0 \leq t \leq T$. Then the equation $x = x(t)$ can be solved for $t$ in terms of $x$, and the resulting function $t = t(x)$ substituted for $t$ in the equation $v = v(t)$. Thus, $v$ is a function of $x$ given by $v = v(t(x))$

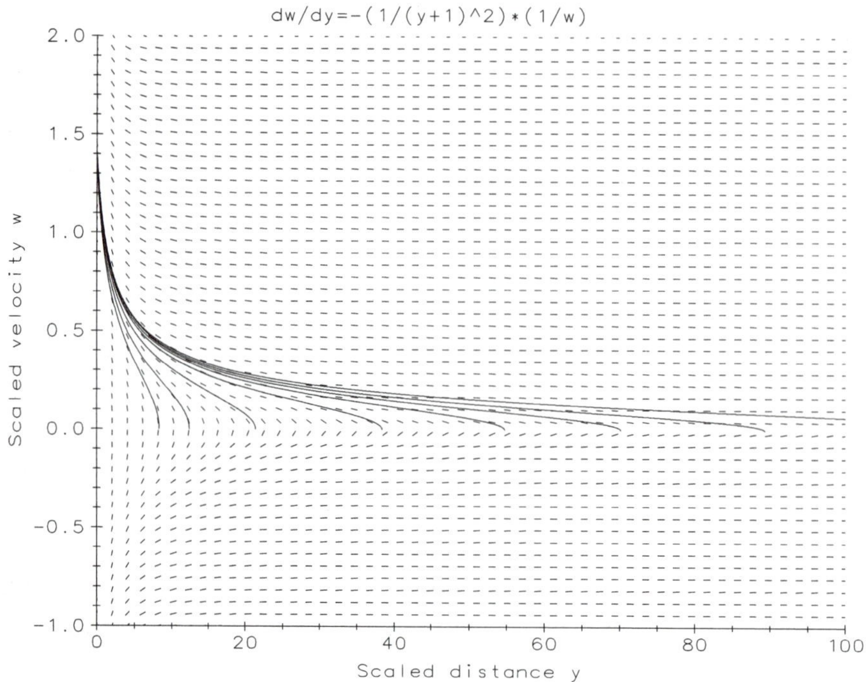


*Figure 1.9.3 Simulation to find escape velocity.*

and the Chain Rule yields $dv/dx = (dv/dt)(dt/dx)$. Recalling $dt/dx = (dx/dt)^{-1}$, we see from (8) that the composite function $v(x)$ satisfies the first order ODE

$$\frac{dv}{dx} = -\frac{MG}{(x+R)^2}\frac{1}{v} \tag{12}$$

To solve (12) numerically we first do some scaling. First put $x = Ry$ so that (12) becomes

$$\frac{dv}{dy} = -\left(\frac{MG}{R}\right)\frac{1}{(y+1)^2}\frac{1}{v} \tag{13}$$

Next put $v = (MG/R)^{1/2}w$, so that (13) becomes

$$\frac{dw}{dy} = -\frac{1}{(y+1)^2}\frac{1}{w} \tag{14}$$

See Figure 1.9.3 for solution curves of ODE (14). Although the figure does not prove it conclusively, it would appear that solution curves originating at $y_0 = 0$ and any $w_0$ larger than about 1.5 do not ever cross the $w = 0$ line. We accept this as a "fact" for the moment. Recalling how $v$ and $x$ are defined from $w$ and $y$, we then see that the solution of the IVP

$$\frac{dv}{dx} = -\frac{MG}{(x+R)^2}\frac{1}{v}, \qquad v(0) = v_0 \tag{15}$$

is such that $v(x) > 0$ for all $x > 0$, provided $v_0$ exceeds $(1.5)(MG/R)^{1/2}$. In other words, the body of mass $m$ can escape the earth's gravitational field if it is hurled upward with an initial velocity $v_0 \geq (1.5)(MG/R)^{1/2}$. In any case, note that the "escape velocity" (if it exists) does not depend on the mass $m$. Experiment 1.10 gives the precise result.

_______________/_______________
hardware software

_______________/_______________
name date

Answer questions in the space provided, or on attached sheets or carefully labeled graphs.

_______________/_______________
course section

## *1.9 Falling Bodies Near the Earth's Surface*

***Abstract*** The motion of bodies on the local vertical near the earth's surface is explored.

**1.** At time $t = 0$ a parachutist of mass 80 kg opens his parachute at the height of 3 km when his downward velocity is 30 m/s. The force of air resistance is given by $20v(t)$ newtons (recall that a newton has the units kg·m/s$^2$), where $v(t)$ is the velocity of the parachutist at time $t$.

**(a)** Write out and justify an IVP for $v(t)$, and another IVP for $x(t)$, the height of the parachutist above the ground at time $t$.

**(b)** Create a graph of $v$ versus $t$, $0 \leq t \leq 1$ min. What value does $v$(20 seconds) have?

**(c)** When will the parachutist hit the ground? (Hint: The IVPs (2) and (3) need not be solved in serial order to find $x(t)$. Together (2) and (3) are an IVP for a planar autonomous system in the state variables $x$ and $v$ and can be solved directly with a solver package.)

**(d)** What will the parachutist's velocity be when he hits the ground?

**2.** Suppose that a projectile of mass 1 kg is shot vertically upward from the earth's surface with initial velocity $v_0 = 100$ m/s. Assuming no air resistance answer the following questions.

**(a)** Write out an IVP that describes the evolution of the velocity $v(t)$ and position $x(t)$ at time $t$ for this projectile.

**(b)** Find the time required for the projectile to reach the top of its path.

**(c)** Find the time when the projectile hits the ground.

**(d)** Find the velocity of the projectile when it hits the ground.

**3.** Answer the questions of Problem 2 if the projectile has mass $m = 10$ grams and encounters air resistance of $(0.001)v(t)$ newtons at time $t$. (Hint: The hint of Problem 1c also holds for IVPs (4) and (5).)

* **4.** Answer the questions of Problem 2 if the projectile has mass $m = 100$ grams and encounters air resistance of $(0.001)\,|v(t)|\,v(t)$ newtons at time $t$.

_______________/_______________
hardware software

_______________/_______________
name date

Answer questions in the space provided,
or on attached sheets or carefully labeled graphs.

_______________/_______________
course section

## *1.10 Escape Velocities*

***Abstract*** To show that there is an escape velocity for vertical motion under gravitational forces.

**1. (Existence of escape velocity)** Computer simulation was used unsuccessfully in Example 2 in an attempt to demonstrate the existence of an escape velocity for a body moving on the local vertical. The velocity $v_{esc}$ is said to be the **escape velocity** if $v_{esc} > 0$ is the smallest value of $v_0$ for which IVP (8) with initial data $v(0) = v_0$, $x(0) = 0$ has a solution for which $v(t) > 0$ for all $t > 0$.

**(a)** Reproduce Figure 1.9.3. Why does Example 2 fail to be very convincing that an escape velocity, in fact, exists?

**(b)** Show that (14) can be written as $(d/dy)\{(w^2/2) - (y+1)^{-1}\} = 0$ for $y$ such that $w(y) > 0$.

**(c)** Use (b) to show that the solution $w(y)$ of ODE (14) with the initial data $w(0) = w_0$ satisfies the identity $\frac{1}{2}(w(y))^2 = 1/(y+1) + (w_0^2/2) - 1$, for all $y$ for which $w(y) > 0$.

**(d)** Use (c) to show that $w_0 = \sqrt{2}$ is the smallest positive value of $w_0$ for which the IVP $dw/dy = -(y+1)^{-2}w^{-1}$, $w(0) = w_0$ has a solution for which $w(y) > 0$ for all $y > 0$. Use this fact to find the escape velocity of the earth[1]. Does it depend on $m$? Compare with the result in Example 2.

**2. (Escape velocity for any celestial body)** Proceed as follows to find the escape velocity for any celestial body, given the earth's escape velocity.

**(a)** Let $R_0$, $M_0$ be the radius and the mass of the earth. Show that for any values $R, M > 0$ if $r = R/R_0$, $n = M/M_0$, then the substitution $x = ry$ converts the planar system of IVP (8) into

$$v' = -\frac{nM_0G}{(y+R_0)^2}\frac{1}{r^2}, \qquad y' = \frac{v}{r}$$

**(b)** Show that the variables change $v = (n/r)^{1/2}w$, and $t = r(r/n)^{1/2}s$ converts the planar system in (a) into $dw/ds = -M_0G(y+R_0)^{-2}$, $dy/ds = w$. Show that this system defines the escape velocity problem for the earth (which was solved in Problem 1). Use this fact to find the escape velocity for any celestial body.

1. Refer to *Handbook of Chemistry and Physics* (CRC Press) for planetary data.

# Aliasing and Other Phenomena

***Purpose*** To examine some peculiarities of graphical representations of numerical solutions of ODEs on a graphics display device.

***Keywords*** Aliasing, aspect ratio, sinusoid, influence of solver parameters

***Background*** Some inherent problems arise in trying to display numerical solutions of ODEs graphically. These problems range from slightly bothersome to highly irritating. Of course there are ways to work around them, but the first step is to be aware that they exist. However a numerical ODE solver computes a numerical approximation to the solution of an IVP, the output often is plotted for a prescribed number of equally-spaced $t$-values in a given $t$-interval (as is the case here). A graphics device connects these approximate values with straight line segments to create the display. Therefore, the closer the solution points, the smoother the solution curve appears.

***Example 1*** **(Aspect ratio)** Let $x(t)$ be any solution of the linear second order ODE $x'' + x = 0$. Multiplying the ODE through by $2x'(t)$ we obtain the identity $((x')^2 + x^2)' = 0$, for all $t$. Thus, $(x'(t))^2 + (x(t))^2 =$ constant for all $t$, hence describes a circle in the $xx'$-plane. But if the experimenter numerically solves the IVP $x'' + x = 0$, $x(0) = 1$, $x'(0) = 1$, and plots the solution $x(t)$ against $x'(t)$ in the $xx'$-plane for $0 \leq t \leq 10$, chances are that the output viewed on a graphics screen will be an ellipse, not a circle. See Figure 1.11.1. This is because the units on the axes are not the same size as measured on the screen. The ratio of these unit sizes, when the same range is displayed on each axis, is called the **aspect ratio** and is usually inherent in the shape of the display device. If this geometrical distortion is not too severe, the eye quickly becomes accustomed to it.

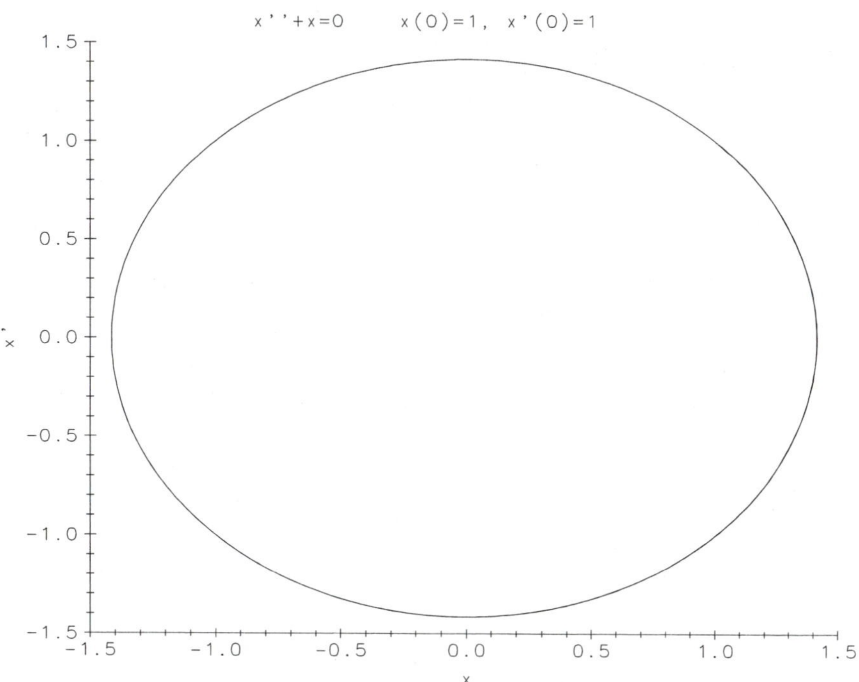


*Figure 1.11.1 Illustration of aspect ratio.*

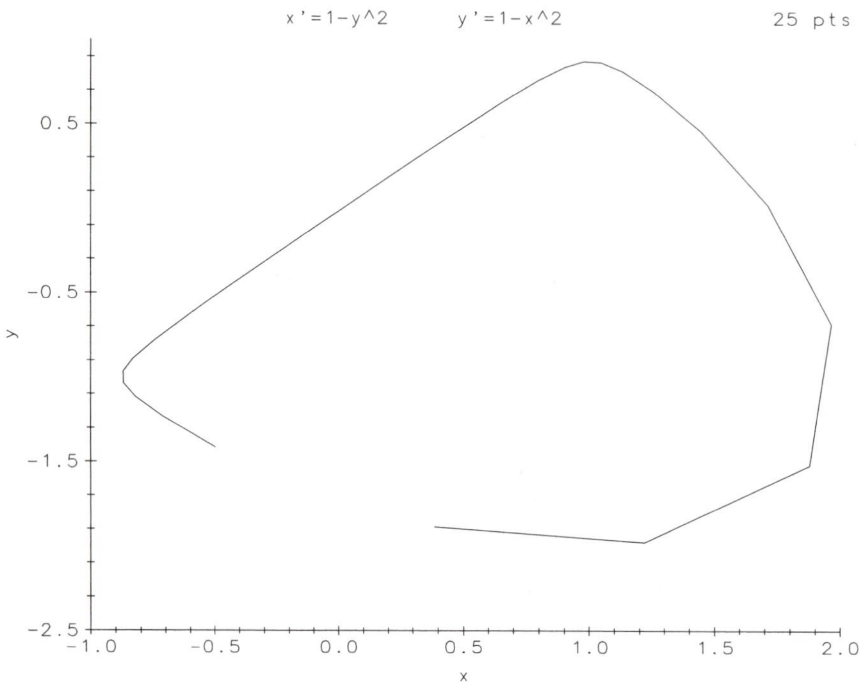


*Figure 1.11.2 Effect of using too few solution points.*

***Example 2*** **(Large rates)** The rate functions in Atlas Plate `Planar Systems D` are autonomous (i.e., do not explicitly contain the independent variable $t$) and depend quadratically on the state variables $x$ and $y$. This means that points that trace out orbits move faster and faster (quadratically in this case) with increasing distance from the origin. This fact has two ramifications. First, if too few solution points are used, the orbit will have some "flat" portions because high rates affect the proximity of solution points on the orbit. See Figure 1.11.2 (cf. Atlas Plate `Planar Systems D`). Second, high rates may cause orbits to leave a prescribed viewing screen long before the calculation is complete. Some solvers will automatically shut down when orbits leave the screen, possibly missing the reentry of an orbit into the viewing screen once it has exited. This **clipping** phenomenon is particularly bothersome when searching for periodic orbits.

***Example 3*** **(Aliasing** and **beats)** All solutions of the ODE $x'' + 4x = 0$ can be expressed in the form $x = A\sin(2t + a)$, where $A$ and $a$ are arbitrary constants. Thus for any choice of initial data $x(0) = x_0, x'(0) = x'_0$ an ODE solver should return a sinusoid of period $\pi$. In practice, however, the graphics output depends strongly on the number of solution points per unit time (the "sampling rate") used by the solver. It is not hard to see that even though the accuracy of the numerical solution at the solution points is very high, at least two solution points per period are necessary to reproduce accurately the correct oscillatory character of the solution. (This observation is closely related to what communication engineers call the **Sampling Theorem**) When the number of solution points per period drops below two, the solver output may look oscillatory, but the frequency will be different from the correct one (see Figure 1.11.5). Engineers use the term **aliasing** in reference to this phenomenon to indicate that the correct behavior of the output is masquerading as a periodic signal with a different frequency. By way of illustration, solve the IVP:

$$\begin{aligned} &x'' + 4x = 0 \\ &x(0) = 0, \quad x'(0) = 1 \end{aligned} \tag{1}$$

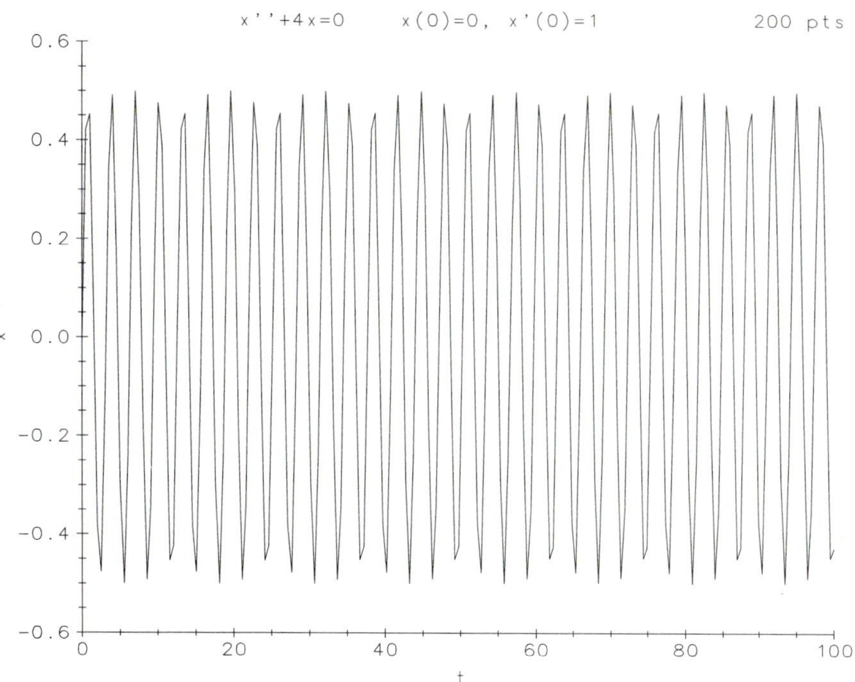


*Figure 1.11.3 Solving IVP (1): using 200 solution points.*

over the time interval $0 \leq t \leq 100$ using 200, 100 and then 50 solution points. Close inspection of Figures 1.11.3–1.11.5 reveals that these graphs exhibit a kind of distortion best described as "modulated" sinusoids. This distortion bears a resemblance to what happens when signals of two frequencies are combined—a phenomenon that scientists call **beats**. The character of the modulation changes with the number of solution points, but tends to disappear as the number of solution points increases. In fact, if the reader solves IVP (1) with 1000 solution points over $0 \leq t \leq 100$, no modulation is observed, and there is some confidence that this plot is a fairly accurate representation of the sinusoid solution of the given IVP. Moreover if this accurate plot is produced at the same scale as Figures 1.11.3–1.11.5 and held up to the light against each of those plots, it can be observed that the values at the solution points in Figures 1.11.3–1.11.5 agree very closely with the "true" values. Thus, the "aliasing" and "modulation" phenomena observed above are *not* the result of a bad solver.

To "explain" the modulation phenomenon observed in Figure 1.11.3, we might proceed as follows: If the IVP: $x'' + 4x = 0$, $x(0) = 0$, $x'(0) = 1$, is solved with 1000 solution points over $0 \leq t \leq 100$, we obtain a good representation of the expected sinusoidal solution with period $\pi$ seconds. Now if this solution is sampled by only 200 solution points over $0 \leq t \leq 100$, then the time between two sample points is $\frac{1}{2}$ second. Thus each period of the signal is "sampled" by six points, enough to avoid aliasing, but apparently not enough to avoid "modulation." Say that a sample point lands on the crest of the sinusoidal signal. Representing one period on a circle of circumference $= \pi$ with the top point of the circle as the point where the signal is a maximum, we see that successive sample points merely travel around the circle (and values of the signal at these sample points can be read off this diagram). If the top point of the circle is bracketed with a centered arc of length $\frac{1}{2}$, then clearly two successive sample points will not fit in this interval. Starting with the first sample point at the top of the circle (where the signal is a maximum), we track the sequence of sample points that fall into the bracketed interval because these points define the tops of

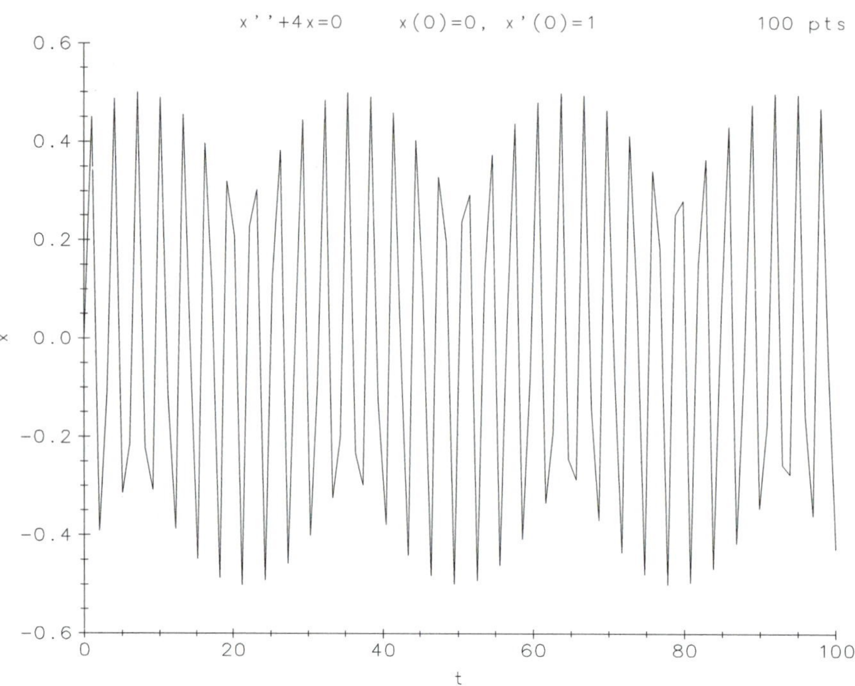


*Figure 1.11.4 Solving IVP (1): using 100 solution points.*

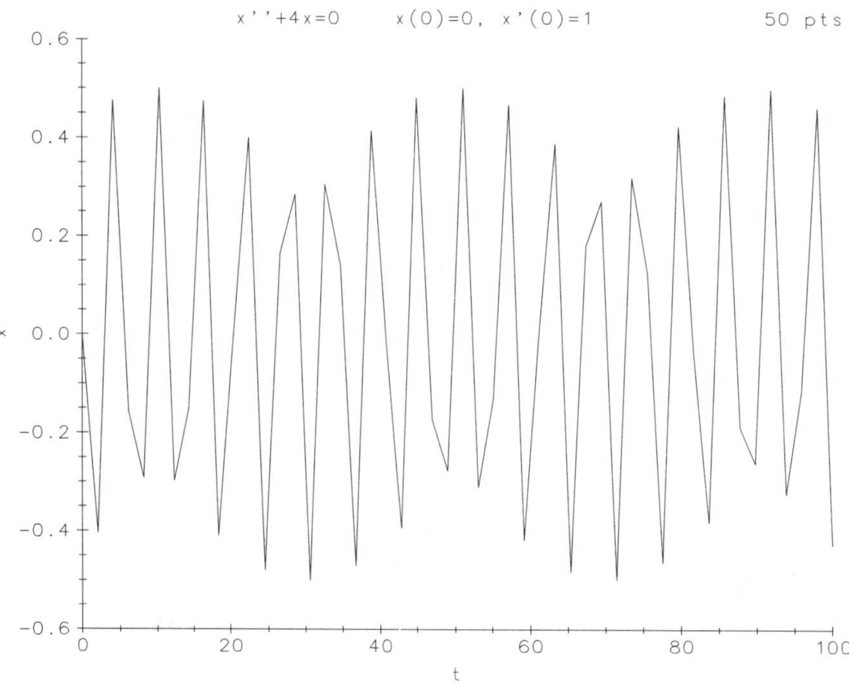


*Figure 1.11.5 Solving IVP (1): using 50 solution points.*

the modulated signal. Figure 1.11.6 identifies these sample points.

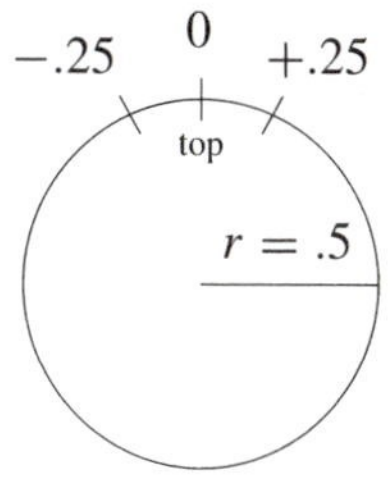

$-.25$ 0 $+.25$
top
$r = .5$

| Sample Point # | Location on Circle |
|---|---|
| 0 | 0 |
| 6 | −0.1416 |
| 13 | 0.2168 |
| 19 | 0.0752 |
| 25 | −0.0663 |

*Figure 1.11.6 Location of sample points in a period.*

The 25th sample point is close to the top point on the circle, and thus a "beat" of this modulated signal has a period of $25 \times \frac{1}{2} \approx 12$ seconds. Reading the period of the modulated signal from Figure 1.11.3, we obtain the approximate value of 13 seconds, hence our calculation was fairly accurate. To compute the amplitude of the modulation, first observe that the amplitude of the "true" signal is 0.5 (read it off the graph of IVP (1) with 1000 solution points which the reader was asked to produce earlier), and the amplitude of the modulation is $\frac{1}{2}[\frac{1}{2}\sin 2(\frac{\pi}{4}) - \frac{1}{2}\sin 2(\frac{\pi}{4} + \frac{1}{4})] \approx 0.03$. Reading this amplitude from Figure 1.11.3 yields approximately 0.025, not far from the predicted value.

***Example 4*** **(Solver error bounds)** Some ODE solvers let the user change the internal solver error bounds. Generally, setting these error bounds to smaller values gives improved accuracy, but at the cost of processor speed. A slight change in error bounds may sometimes cause a dramatic change in a computed solution. The IVP in Figures 1.11.7–1.11.8 models the angular position $x(t)$ of a swinging pendulum subject to damping, gravity, and a periodic driving force. At $t = 0$ the pendulum is vertically downward ($x = 0$) and has initial angular velocity $x' = 2$ rad/s. Experiments 5.7–5.9 give more details of this model. If $|x'(0)|$ is large enough, one expects the pendulum to go over the top and eventually settle into oscillations about some downward position $x = 2N\pi$. Figures 1.11.7–1.11.8 show a solution to this problem with two different solver error bounds. Which picture is correct?

***Example 5*** **(Plotted solution points)** Some solvers let the user choose the number of solution points plotted on the screen. Figures 1.11.9–1.11.10 show the startling effect that may result from a small change in that number. In Figures 1.11.9 and 1.11.10 the only differences between the two sets of component graphs is a single solution point, but the effect is dramatic. The system in these plots models an oscillating chemical reaction. Experiment 6.12 includes discussion of the model equations in these figures. Is the difference in the figures some artifact of the platform? Is it is due to some extreme sensitivity of the IVP? More importantly, if you had to interpret these graphs, which would you believe and why? Clearly one cannot accept computer output blindly. Some mathematical analysis *by hand* is unavoidable. It also helps to know something about the modeling environment that gives rise to the ODEs.

***Observation*** A possible explanation for the difference between Figures 1.11.9 and 1.11.10 is that the solver that produced these figures generates plotted points by integration over adaptively determined steps. These steps may vary considerably in size and can vastly exceed the distance between solution points, thus requiring interpolation when plotting. The extra plotted point in Figure 1.11.10 influences the step size just enough that the solver "notices" the oscillation. Well written adaptive solvers allow the user to set the maximum step size. Step sizes cannot be permitted to be too large if the solver is to recognize phenomena that occur on a time scale that is small compared to the interval of integration.

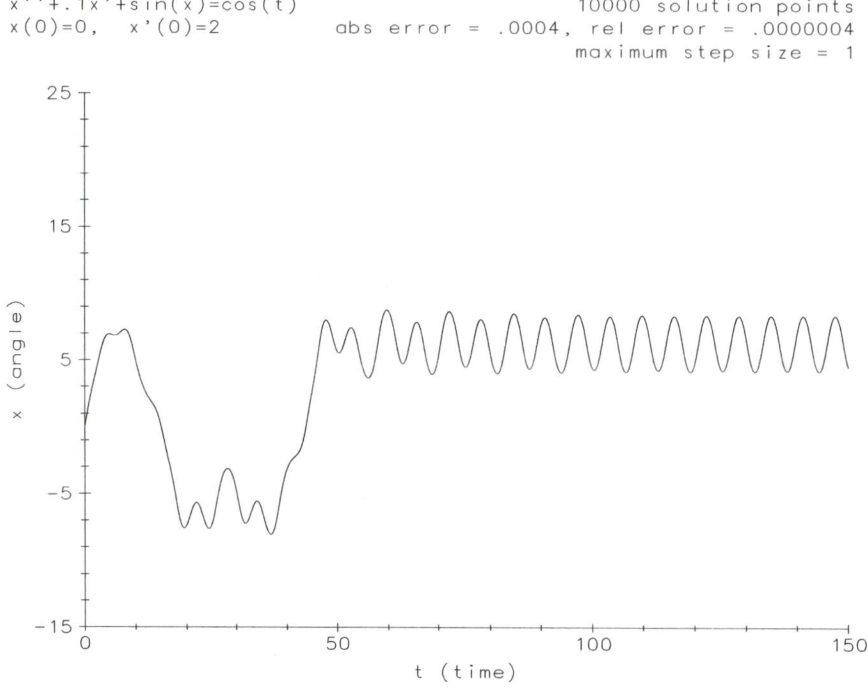


*Figure 1.11.7 Pendulum appears to settle down to an oscillation about $x = 2\pi$.*

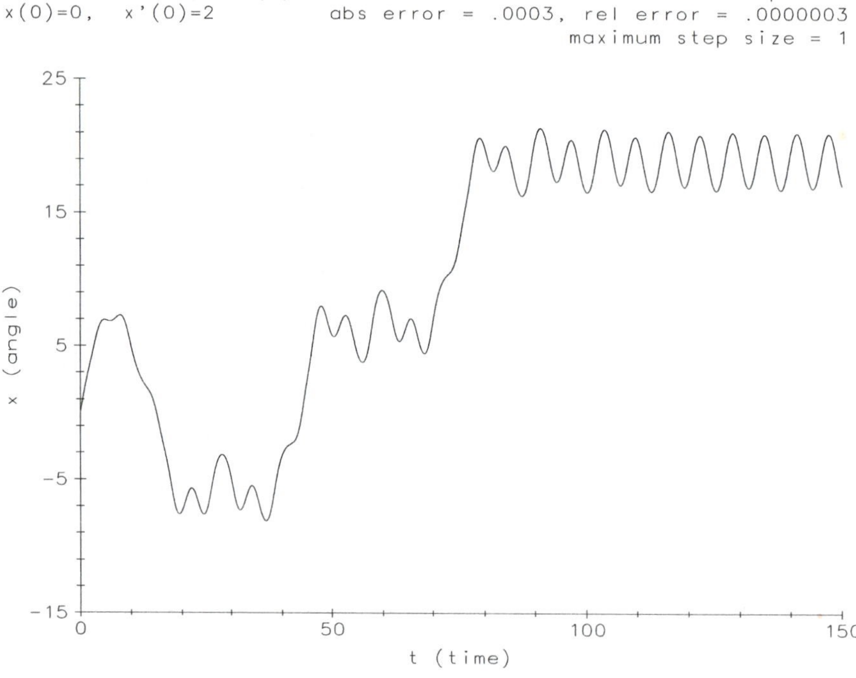


*Figure 1.11.8 Or does it?*

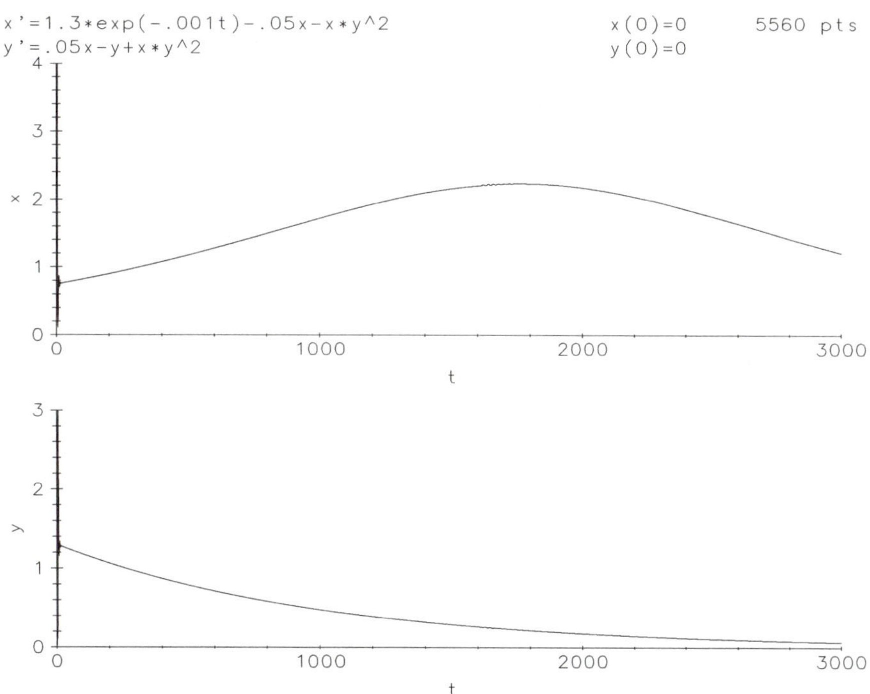


*Figure 1.11.9 Component plots of an autocatalator with 5560 solution points.*

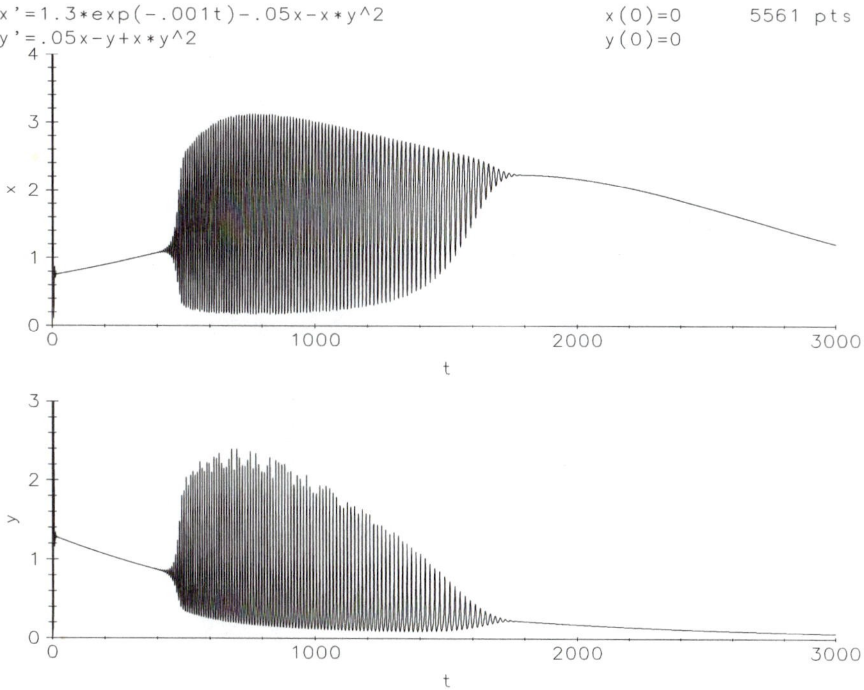


*Figure 1.11.10 The same system as Figure 1.11.9 but with 5561 solution points.*

________________ / ________________
hardware software

________________ / __________
name date

Answer questions in the space provided,
or on attached sheets or carefully labeled graphs.

________________ / __________
course section

## 1.11 *Aliasing and Other Phenomena*

***Abstract*** This experiment illustrates some of the peculiarities that can occur when numerical solutions of ODEs are displayed graphically. With some understanding of the origin of these peculiarities, the experimenter can learn to work around them.

***Special instructions*** The subtle distortions of reality caused by aliasing, modulation, and sensitivity to the number of plotted points imply that this experiment may be best approached from several different angles. This suggests that it would best be done by a team. Appendix A provides ideas for writing team reports.

---

Study aliasing, modulation, sensitivity to small changes in data, and other output distorting phenomena. You may want to touch on the following points.

- What are the aspect ratios in Figure 1.11.1, Figures 1.1.1–1.1.2, and the Atlas Plates `Planar System D` and `Pendulum B`, and your own platform? Can you change the aspect ratio of your platform?
- Solve numerically the IVP $x'' + x = 0$, $x(0) = 1$, $x'(0) = 1$, $0 \leq t \leq 10$, using 10, 20, 30, and 500 solution points. Graph orbits in the $xx'$-plane. Overlay the graphs, hold them up to the light, and explain what you see.
- Reproduce all or part of Atlas Plate `First Order G`. (Depending on your platform, it could be a time-consuming process). Then plot orbits with $y_0 = -2.8$, $7 \leq x_0 \leq 14$, moving $x_0$ from left to right. Is this the way that orbits really behave? How can you "fix things up" to improve the accuracy of the orbits?
- Reproduce Atlas Plates `Pendulum (Upended) B` and `C` and explain what might cause the marked orbital difference even though the only difference is a small change in the initial data.
- Measure the amplitude and period of modulation of the graph in Figure 1.11.4. Explain why these values are to be expected. What happens to the modulation as the number of solution points increases? Explain. Compare graphs obtained using 500, 1000, and 5000 solution points. Repeat with 50 and finally 10 solution points.
- Graph the orbit in the $xy$-plane generated by the IVP $x' = 5y$, $y' = -4y - 3x + t \sin t$, $x(0) = y(0) = 0$ using 500 solution points over the interval $0 \leq t \leq 1000$. Does this orbit really look like this? What happens to this graph as the number of solution points increases? Repeat with $tx$-plots. Explain what you see.
- Read Examples 4 and 5 and try to explain what you see in Figures 1.11.7–1.11.10. Reproduce the graphs on your system. What happens? Experiment with other values of the parameters and with other numbers of plotted solution points. (Suggestion: Changing the solver parameters may introduce just enough error to "push" the pendulum over the top and cause it to oscillate around $x = 2N\pi$ for a different value of $N$.)

# CHAPTER 1 NOTES

Ordinary differential equations (**ODEs**) have been studied for more than 300 years; hence a great deal is known about them. The purpose of the experiments in this workbook is to portray visually many of the properties of ODEs through graphical representations of numerical approximations of their solutions. Because ODEs have been amazingly successful in the construction of mathematical models for the real world, there are a large number of ODE applications from which to choose.

## *Basic Properties of IVPs*

Mathematical models of real-world systems often involve initial value problems (**IVPs**) for ODEs, and the data come from the system under scrutiny. Thus, the more we know about the behavior of the solutions of IVPs, the better able we are to infer properties of the systems associated with them. Some basic properties of IVPs concern questions of:

**Existence.** What conditions on the data guarantee that the IVP has at least one solution?

**Uniqueness.** What conditions on the data imply that the IVP has no more than one solution?

**Solvability.** Assuming that the IVP has a unique solution, how can we find it?

**Sensitivity.** Assuming that the IVP has a unique solution for data within a given class, how does the solution change as the data change within that class?

Under mild conditions on the rate function $f(t, x)$, it can be shown that the IVP

$$\begin{aligned} x' &= f(t, x) \\ x(t_0) &= x_0 \end{aligned}$$

has a unique solution $x(t)$ on some interval $I$ containing $t_0$. This result is known as the **Existence and Uniqueness Theorem for IVPs**.

The basic properties and questions listed above apply equally well to IVPs based on first order systems for any number of state variables. The terminology introduced for first order ODEs carries over to first order systems in the obvious way.

## *Linear ODEs*

A rate function $f(t, x)$ in one state variable is **linear** if it has the form $a(t)x + b(t)$, where the coefficients $a$, $b$ are functions of $t$ alone. If $f(t, x)$ is a linear rate function the ODE $x' = f(t, x)$ is called a **linear ODE**. For some examples see Atlas Plates `First Order D-F`. Similarly, a rate function $f(t, x_1, x_2)$ in two state variables is linear if it has the form $a(t)x_1 + b(t)x_2 + c(t)$, where the coefficients $a, b, c$ are functions of $t$ alone. If the rate functions $f_1$, $f_2$ are linear in the two state variables, then the system of ODEs, $x_1' = f_1(t, x_1, x_2)$, $x_2' = f_2(t, x_1, x_2)$, is called a **linear system**. For example, see Atlas Plate `Planar System A`. Linear systems have many special properties, which will be examined in later chapters.

## *Approach Used in This Workbook*

First courses in differential equations treat the basic properties of IVPs mostly through the construction of a solution formula from a "general" formula for all solutions of the associated ODE. The simple fact is, however, that except for a handful of special (but important) rate functions, there are no known procedures for finding solution formulas for IVPs. Thus, other approaches to examining the properties of IVPs have been devised, approaches that use numerical approximations and graphical techniques. This workbook mostly adopts the latter approach and leaves the solution formula approach to the classroom.

We assume that the student has access to a hardware/software platform that merely requires the user to enter the rate functions of the first order system, the initial data, the dimensions of the viewing window, and the solution time interval, and perhaps to set some solver parameters. We also assume that the output of the solver is displayed on some hardware device. Many platforms commercially available today perform these tasks simply and efficiently. Some are interactive with the user and allow a high level of flexibility and utility. The point is that almost no expertise is required of users to operate their platforms. This fact makes possible an approach to the study of ODES radically different from the traditional approach of the classroom. In particular,

- There is no need to start with simple first order ODEs and progressively build up to more complex ones, since data for one IVP are no harder to input than data for another.
- Linear and nonlinear ODEs can be treated together from the start. The only exception is Chapter 4 where, because of their importance, linear second order ODEs are treated. First order systems whose rate functions are linear have many special properties, hence are taken up early.
- ODEs can be chosen for examination in whatever way we like because finding exact solutions of IVPs is not required. It is not always an advantage to have a symbolic expression for the solution of an IVP. Sometimes a solution is defined only implicitly from an equation and is next-to-impossible to visualize. Even if the solution of an IVP is given explicitly by formula, it still may be very difficult to visualize.
- Well-chosen graphs of solutions of ODEs can be the beginning point of a lecture.

### *A Simple Solution Technique*

Sometimes a simple trick is successful in finding a formula that determines all solutions (albeit implicitly) of a first order ODE. Suppose that a function $F(t, x)$ can be found such that when $x$ is replaced by $x(t)$ the ODE can be written in the form $(d/dt)F(t, x(t)) = 0$ with the help of the **Chain Rule**. Then the **Mean Value Theorem** implies that for any solution $x(t)$ of the ODE there is a constant $C$ such that $F(t, x(t)) \equiv C$, for all $t$ where $x(t)$ is defined. The converse is also true: If $C$ is any constant and $x(t)$ is any differentiable function for which $F(t, x(t)) \equiv C$, then $x(t)$ is a solution of the ODE.

***Example 1*** The ODE, $tx' + x = 2t$ (displayed in Atlas Plate `First Order F`), can be written as $(tx)' = 2t = (t^2)'$, and ultimately in the form $(tx - t^2)' = 0$. Thus the function $F(t, x) = tx - t^2$ is such that for any constant $C$, any differentiable solution $x(t)$ of the equation $tx - t^2 = C$ is a solution of the ODE. Solving explicitly for $x$, we see that for any constant $C$, $x = t + C/t$ is a solution of the ODE on any interval not containing the origin. If $C = 0$ then $x = t$ is a solution of the ODE for all $t$ including $t = 0$.

CHAPTER 2

# Solution Curves and Numerical Methods

IVPs for first order ODEs were briefly treated in Chapter 1. In this chapter, we explore some qualitative properties of the behavior of solutions of IVPs before looking into procedures for calculating numerical approximations to these solutions.

## *Introduction*

Let $f(t, x)$ be defined in some region $R$ of the $tx$-plane and suppose that $(t_0, x_0)$ is a point in $R$. Consider the IVP

$$\begin{aligned} x' &= f(t, x) \\ x(t_0) &= x_0 \end{aligned} \tag{1}$$

The fundamental questions of **existence**, **uniqueness**, and **sensitivity** of solutions of IVP (1) will be addressed. The question of **solvability** of IVP (1) is examined via numerical approximation techniques. Details are found in the experiments and the Chapter 2 Notes.

## *Existence and Uniqueness*

Simple conditions on the function $f(t, x)$ produce satisfactory answers to all four questions. These conditions are merely that $f$ and $\partial f/\partial x$ are continuous functions of $t$ and $x$ in the region $R$. If these conditions hold, then every point in $R$ has a solution curve of the ODE passing through it, and moreover no two solution curves of the ODE can intersect in $R$. So we have a fairly good idea of what the screen will look like when solutions of IVP (1) are generated on a solver platform for various initial points $(t_0, x_0)$. See, for example, Atlas Plates `First Order A-D`. There is one subtle matter, however, that needs to be addressed. The notion of a solution $x(t)$ of IVP (1) puts no size requirement on the interval $I$ where $x(t)$ is defined other than that $t_0$ belongs to $I$. So, an important question is this: Can a solution curve "die" *inside* $R$? That is, can a solution curve have a "final" or "initial" point *inside* $R$? Solvers "answer" this question by using larger and larger solution times for IVP (1) and observing the solution curves on the screen. If $f$ and $\partial f/\partial x$ are continuous in the region defined by the screen, then invariably what is observed is that the solution curve eventually leaves the screen both in *forward* time ($t > t_0$) and in *backward* time ($t < t_0$). To state this property more precisely, we need the concept of an extension of a solution. See Experiment 2.4 and the Chapter 2 Notes for a discussion of this important idea.

## *Sensitivity*

A question that every modeler must ultimately ask about IVP (1) is this: Is it possible to find bounds on the determination of the data $f(t, x)$ and $x_0$ which will guarantee that the corresponding solution will be within prescribed error bounds over a given $t$-interval? If this question can be answered in the affirmative, one consequence is that any "small" change in the data of an initial value problem produces only a "small" change in the solution. If this is so, then models involving IVPs are not invalidated simply because empirically-determined elements do not have their "true" values.

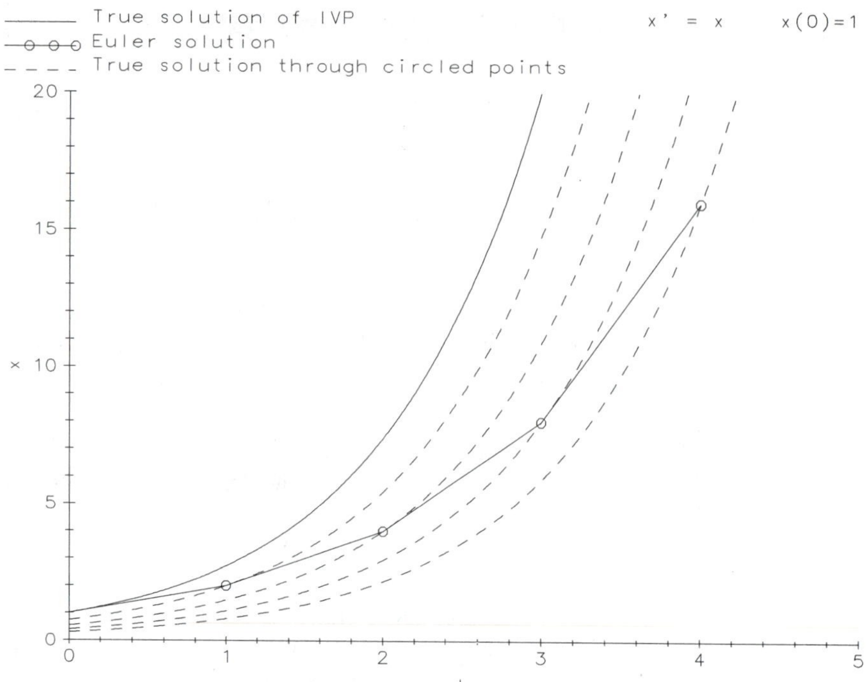


*Figure 2.1* *Geometry of an Euler solution for the IVP* $x' = x$, $x(0) = 1$: $h = 1$.

## *Solvability, Numerical Solutions*

The direction field approach used in Chapter 1 to provide an intuitive notion of a solution of a first order ODE also suggests techniques for finding approximate numerical solutions for IVP (1). Below, we describe one such method called **Euler's method** , but leave the details to the Chapter 2 Notes and the experiments. Say we wish to approximate $x(T)$, where $T$ is a $t$-value $> t_0$. First, partition the interval $t_0 \leq t \leq T$ with $N$ equal **steps** $t_n = t_0 + nh$, $n = 1, 2, \ldots, N$, where the **step size** is $h = (T - t_0)/N$. We know that $(t_0, x_0)$ is on the solution curve. To find an approximation to $x(t_1)$ just follow the (known) tangent line to the solution curve through $(t_0, x_0)$ out to $t_1$. Since the slope of the tangent line to the solution curve at $(t_0, x_0)$ is $f(t_0, x_0)$, we see that $x_1 = x_0 + hf(t_0, x_0)$ is a reasonable approximation to $x(t_1)$ if $h$ is small. Using $(t_1, x_1)$ as a base point, and pretending that $(t_1, x_1)$ is on the desired solution curve, we may construct an approximation $x_2$ to $x(t_2)$ in the same way: $x_2 = x_1 + hf(t_1, x_1)$. From the description above, we have

$$x_{n+1} = x_n + hf(t_n, x_n), \quad n = 0, 1, \ldots, N - 1 \tag{2}$$

Of course, since $(t_1, x_1)$ is not on the desired solution curve, the calculated value $x_2$ acquires an error from this source as well. If the function $f(t, x)$ is smooth enough, the total error in calculating $x(t_2)$ can be made small if $h$ is small. This simple calculation can be repeated as often as desired to produce a broken line approximation to the solution curve through a given point of the $tx$-plane. See Figure 2.1.

There are many methods to find approximate numerical solutions of IVPs. Euler's method has the virtue of being simple to visualize and implement, but it is not the most practical choice for IVP solvers. The fourth order Runge-Kutta method (defined in the Chapter Notes) provides a combination of accuracy and efficiency which makes it a better choice than Euler's method. Commercial solvers (like the one used to produce the plots in this workbook) generally use a variety of advanced methods in an adaptive environment.

# Properties of Orbits and Solution Curves

***Purpose*** To examine the behavior of orbits and solution curves of ODEs in a variety of settings and explore the capacity of your platform to portray these curves.

***Keywords*** Normal form for ODEs, solution curves, orbit, initial conditions, initial value problems (IVPs), Existence and Uniqueness Principles for IVPs, extension of solutions, autonomous ODEs, equilibrium solutions of an autonomous ODE, sensitivity of solutions

***See also*** Chapters 1–2 Introductions; the solution curves and orbits portrayed in the Atlas; Experiments 1.1–1.4

***Background*** Listed below are some properties of solutions, solution curves, and orbits of ODEs in normal form.

- **Nonintersection of solution curves**. The Uniqueness Principle states that under appropriate smoothness conditions on the rate function, no two solution curves of a first order ODE in normal form can intersect. The same is true for solution curves of systems of ODEs in normal form. A glance at Atlas Plates `First Order A-D` shows that solution curves sometimes appear to flow together very strongly, but in spite of this, we know that no two can intersect, since each of the rate functions is smooth in the region defined by the screen. This is not the case for Atlas Plate `First Order E`. Why not?
- **Extension of solutions**. Using the geometric conceptualization of a solution curve for the ODE $x' = f(t, x)$, discussed in Experiment 1.1, we can "see" that IVPs have the following property: Let $f(t, x)$ satisfy the usual smoothness conditions, and let $x(t)$ solve the IVP $x' = f(t, x)$, $x(t_0) = x_0$, on the interval $t_0 \le t \le t_1$, and suppose that $y(t)$ solves the IVP $y' = f(t, y)$, $y(t_1) = x(t_1)$, for $t_1 \le t \le t_2$. Then the solution $x(t)$ can be extended to $t_0 \le t \le t_2$, and $x(t) = y(t)$, for $t_1 \le t \le t_2$. The second IVP for $y(t)$ can even be solved *backward* in time from the initial point$(t_1, x(t_1))$ and the same result is true (if the inequalities on the $t$-interval are adjusted appropriately). In particular, this means if $x(t)$ solves the IVP $x' = f(t, x)$, $x(t_0) = x_0$, forward to $(t_1, x_1)$ and $y(t)$ solves the IVP $x' = f(t, x)$, $x(t_1) = x_1$, backward to $t_0$, $y(t_0) = x_0$. In other words, if a dynamical system is run forward for $T$ units of time and backward for the same amount of time, then the state of the system returns to its original value (but see Problem 1 in Experiment 2.1).
- **Solution curves for autonomous ODEs**. If the rate function $f(t, x)$ does not depend explicitly on $t$, it is usually written $f(x)$ and the ODE is said to be **autonomous**. Similarly, if the rate functions of a first order system of ODEs in normal form depend only on the state variables, then the system is autonomous. If $x(t)$ solves the autonomous ODE $x' = f(x)$, over the interval $t_0 \le t \le t_1$, then the function $x(t + T)$ solves the *same* ODE over the interval $t_0 - T \le t \le t_1 - T$ for *any* constant $T$. Clearly, the solution curve for $x(t + T)$ is just the solution curve $x(t)$ shifted $T$ units backward along the $t$-axis if $T > 0$ and forward along the $t$-axis if $T < 0$. Thus the solution of an autonomous ODE does not depend on the starting time in any essential way. An interesting consequence of this observation is that by shifting one solution curve along the $t$-axis, *all* the solution curves in a region of $tx$-space can be generated. See Figure 2.4.1 for an example.
- **Sign analysis**. Consider the autonomous ODE $x' = f(x)$, where $f$ and $df/dx$ are continuous over an interval $I$. Let $x_2 > x_1$ be two consecutive zeros of $f$ in $I$. Then $f(x)$ must be of one sign in $x_1 < x < x_2$. Say, for purposes of discussion, that $f(x) > 0$ for $x_1 < x < x_2$. Note that $x \equiv x_1$ and $x \equiv x_2$ are solution curves for ODEs which are defined on the entire $t$-axis. These solutions are called **equilibrium solutions** of the ODE because the system is at rest in those states. Now the solution for an ODE that

originates at any point $(t_0, x_0)$, $x_1 < x_0 < x_2$, must be increasing as $t$ increases as long as the solution curve remains in the strip $x_1 < x < x_2$. But to leave this strip, the solution curve would have to cross the equilibrium solution curve $x \equiv x_2$, and the Uniqueness Principle implies that this cannot happen. The Extension Principle easily shows that our solution curve is defined for all $t \geq t_0$. Similarly, we see that for backward time our solution curve falls for decreasing $t$, but it never crosses the equilibrium solution $x \equiv x_1$; hence the Extension Principle shows again that the solution is defined for all $t \leq t_0$. Thus it follows that the limits

$$\lim_{t\to\infty} x(t) = b \leq x_2 \text{ and } \lim_{t\to-\infty} x(t) = a \geq x_1$$

both exist. In fact, $a = x_1$ and $b = x_2$, as we now show. By shifting this solution curve in $t$ we can completely fill up the strip $a < x < b$ with the solution curves of the ODE (this follows from the paragraphs above). Now if $b < x_2$, then as noted above, the solution curve defined by the IVP $x' = f(x)$, $x(t_0) = b$, can be extended to the entire $t$-axis. This solution must cross the line $x = b$ because $f(b) > 0$, hence must intersect a solution curve in the strip $a < x < b$, in contradiction to the Uniqueness Principle. Thus, it must be that $b = x_2$. Similarly, it can be shown that $a = x_1$.

- **Orbits of autonomous systems**. For ease of presentation a planar system will be considered, but the results apply equally well to any autonomous system of ODEs in normal form. As above, we note that if $(x(t), y(t))$, $t_0 \leq t \leq t_1$, is a solution of an autonomous system, then so is $(x(t+T), y(t+T))$, but over the shifted interval $t_0 - T \leq t \leq t_1 - T$. Now the orbit of a system of ODEs is a curve in the $xy$-plane, the state space of the system. Hence it is clear that the orbit itself does not change when the $t$-interval is shifted—only the $t$-values attached to the points on the orbit are shifted. Hence orbits of autonomous systems do not care when the clock is started. Next, we observe that two orbits of an autonomous system can never intersect. To show this, say that the autonomous planar system $x' = f(x, y)$, $y' = g(x, y)$ has two orbits $(x_1(t), y_1(t))$ and $(x_2(t), y_2(t))$ that intersect at a point $(x_0, y_0)$; hence, $x_1(t_1) = x_2(t_2) = x_0$ and $y_1(t_1) = y_2(t_2) = y_0$, for two times $t_1$ and $t_2$ (we cannot assume that $t_1 = t_2$). Shifting the second orbit in time by $T = t_2 - t_1$, we find that $(x_1(t), y_1(t))$ and $(x_2(t+T), y_2(t+T))$ are both solutions of the same IVP

$$\begin{aligned} x' &= f(x, y), \qquad & x(t_1) &= x_0 \\ y' &= g(x, y), \qquad & y(t_1) &= y_0 \end{aligned}$$

  Hence from the Uniqueness Principle the two orbits must be identical. The Atlas Plates `Planar System A-C` and `Pendulum A-C` illustrate this property. Atlas Plates `Planar System I` and `Pendulum (Upended) A-C` show that orbits of nonautonomous systems may intersect without being identical.

***Observation*** **(Rescaling time on orbits)** An orbit of the system

$$\frac{dx}{dt} = f(x, y), \qquad \frac{dy}{dt} = g(x, y)$$

is also an orbit of the system

$$\frac{dx}{dt} = r(x, y)f(x, y), \qquad \frac{dy}{dt} = r(x, y)g(x, y)$$

for any positive function $r(x, y)$. An example of this is provided by the Atlas Plates `Limit Set A, B`. Note that the component graphs are very different in the two cases, but nevertheless, they trace out exactly the same orbit. As shown in the Chapter 5 Notes the effect of the factor $r(x, y)$ is to change the parameterization on the orbit—i.e., to "rescale time."

***Observation*** **("Velocity" of a point tracing out an orbit)** Consider a planar system $x' = f(t, x, y)$, $y' = g(t, x, y)$, and think of the orbit $(x(t), y(t))$ as being traced out in the $t$-variable, thought of as time. Then using the unit vectors $\hat{\mathbf{i}}$ and $\hat{\mathbf{j}}$ pointing in the positive $x$ and $y$-directions, respectively, we have the position vector $\mathbf{R} = x(t)\hat{\mathbf{i}} + y(t)\hat{\mathbf{j}}$ in $xy$-space. We see that $\mathbf{R}' = x'(t)\hat{\mathbf{i}} + y'(t)\hat{\mathbf{j}}$ is the velocity vector of the point tracing out the orbit. The $x$ and $y$-components of the velocity vector are clearly $f(t, x, y)$ and $g(t, x, y)$ at the point $(x, y)$ and time $t$. This explains why graphs of numerical approximations of orbits sometimes have flat portions. The velocities are sometimes so high that the points in state space corresponding to consecutive solution points $t_n$, $t_{n+1}$, are far away from one another. A word of warning though: Numerical algorithms have their own peculiar sense of "time," which may have more to do with function evaluations and control of errors than anything else. Thus, the plotting the experimenter sees taking place on the screen is often only loosely related to the plotting velocity of orbits as characterized above. From the observation above, one device for slowing down orbit tracing for autonomous systems is to multiply both rate functions of the system by the same (appropriately chosen) nonnegative function $r(x, y)$.

***Example 1*** **(Sensitivity to parameter change)** Figure 2.1.1 displays the angular position $x(t)$ of a swinging pendulum for three slightly different values of the damping coefficient $a$. Observe that the solutions of the corresponding ODEs track each other very closely for a brief time, but eventually the differences in the values of $a$ lead to big differences in the values of $x(t)$. See Example 4 in Experiment 1.11 for an illustration of the sensitivity of the swinging pendulum ODE to changes in solver parameters.

***Example 2*** **(Suggestion for sensitivity studies)** In examining the sensitivity of the solution of a first order ODE to changes in a parameter that appears in the rate function, the natural approach is to overlay plots of $x$ versus $t$ as the parameter changes. Not all solvers support features that do this directly. But you may be able to "trick" your solver into doing it. In Example 3 of the Chapter 2 Notes we consider the sensitivity of the IVP

$$\begin{aligned} x' &= -kx \\ x(0) &= x_0 \end{aligned} \tag{1}$$

to changes in the parameters $k$ and $x_0$. For ease of presentation we consider only changes in $k$ here. By adjoining the ODE, $dk/dt = 0$ to the ODE in (1) we form the following IVP for a planar autonomous system

$$\begin{aligned} x' &= -kx \qquad & x(0) &= x_0 \\ k' &= 0 \qquad & k(0) &= k_0 \end{aligned} \tag{2}$$

Overlaying the orbits of (2) for $x_0 = 1$, and $k_0 = 0.5, 0.7, 0.9, 1.1, 1.3$, and $1.5$, and then plotting the component graphs $x$ versus $t$ we obtain Figure 2.1.2.

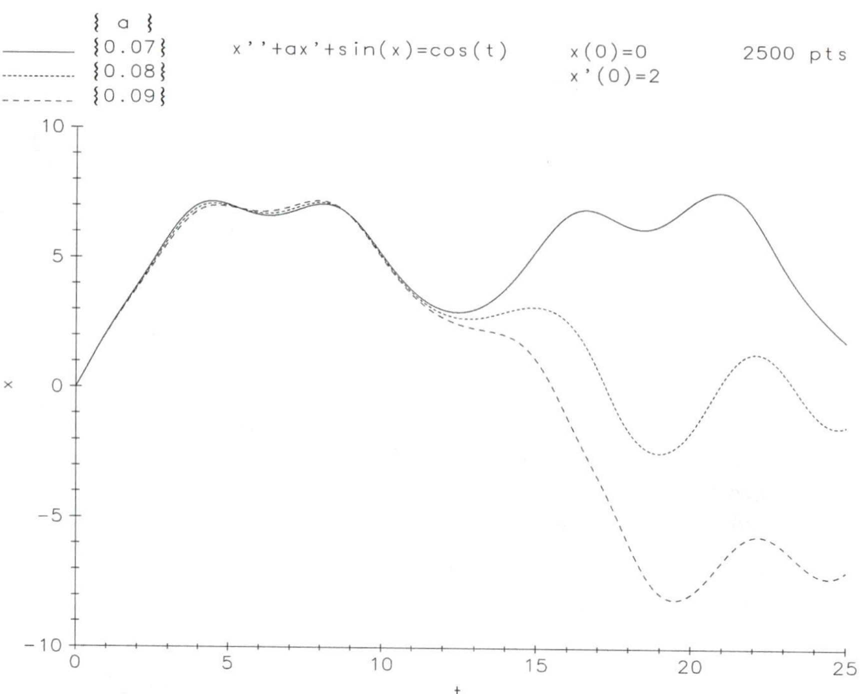


*Figure 2.1.1 Sensitivity study for changes in a damping coefficient.*

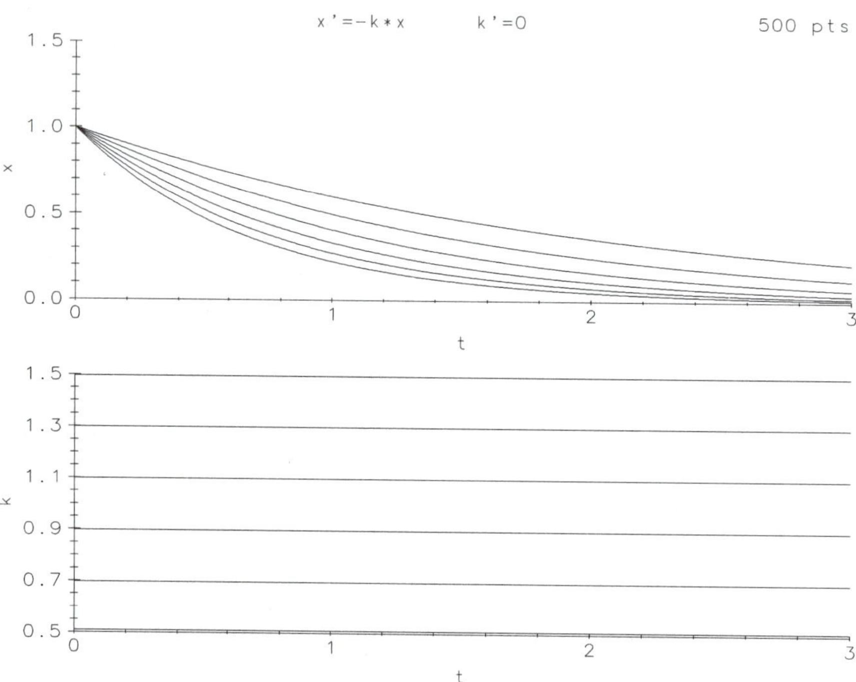


*Figure 2.1.2 Sensitivity study for IVP (1).*

_______________/_______________
hardware software

_______________/_______________
name date

Answer questions in the space provided,
or on attached sheets or carefully labeled graphs.

_______________/_______________
course section

## 2.1 *Fundamentals*

***Abstract*** Properties of solutions of first order ODEs and planar systems will be examined along with inherent problems in graphical representations of numerical approximations.

**1.** Consider the ODE $x' = 3x\sin(x) - t$.

**(a)** Reproduce Atlas Plate `First Order A` on your own platform.

**(b)** Is this ODE autonomous? Explain why the solution curves in the Atlas Plate appear to flow together, but actually do not.

**(c)** Solve the IVP $x' = 3x\sin(x) - t$, $x(0) = 0.4$, over the interval $0 \le t \le 7$. Save the numerical approximation $x^*$ for $x(7)$ and then solve and plot the solution of the *backward* IVP $x' = 3x\sin(x) - t$, $x(7) = x^*$, over $0 \le t \le 7$. Compare the graphs of the forward and backward IVPs and explain any differences. Write directly on your graph.

**2.** Write the IVP $x'' + 25x = \sin(5t)$, $x(0) = 1$, $x'(0) = 1$, as an IVP for a first order system and solve it over $0 \le t \le 20$. Plot $x$ versus $x'$ over $0 \le t \le 20$ in $xx'$-space. Does this orbit intersect itself? Explain. Change the initial conditions to $x(2) = 1$ and $x'(2) = 1$ and repeat. Does the same orbit result? Explain. Write directly on your graphs.

**3.** Find the maximally extended solutions of the IVPs $x' = 2t^{1/2}/(3\cos(x))$, $x(1) = .5$ and $x' = (1 - x^2)^{1/2}$, $x(0) = 0$. Discuss the validity of your results.

**4.** Duplicate Figures 2.1.1 and 2.1.2 (or equivalent forms appropriate for your solver). Plot curves corresponding to enough other values of the parameters that you can make conjectures about the general nature of the curves as the parameter increases. Explain the reasons behind your conjectures. Write on the graphs.

**5.** Reproduce the Atlas Plates indicated below and explain why the orbit plots do not violate the nonintersection property at the indicated points. Write on the graphs.

**(a)** The Atlas Plate `Planar System B` at $(-1, 1)$.

**(b)** The Atlas Plate `Planar System E` at $(0, \sqrt{3\pi/2})$, and $(0, -\sqrt{3\pi/2})$.

* **6.** Reproduce the Atlas Plates `Limit Set A, B`. Explain why the orbits appear to be identical, while the component plots of $x$, $y$ versus $t$ are not. Describe the "velocity" of the "motion" in plotting the orbits in both Atlas Plates. Write directly on the graphs.

_______________ / _______________
hardware software

_______________ / _______________
name date

Answer questions in the space provided,
or on attached sheets or carefully labeled graphs.

_______________ / _______________
course section

## 2.2 *Sign Analysis*

***Abstract*** The relationship between equilibrium solutions and solution curves of an autonomous first order ODE is examined.

**1.** Chose one of the ODEs listed below and answer the questions that follow:

- ☐ **A.** $x' = (x-1)^2(x+1)$
- ☐ **B.** $x' = (1-x)(x+1)$
- ☐ **C.** $x' = x^3 - 3x^2 + 4x - 2$
- ☐ **D.** $x' = \sin x$

**(a)** What are the equilibrium solutions of the ODE?

**(b)** Describe in qualitative terms the way the solution curves of the ODE behave. Are any solution curves of the ODE defined over the entire $t$-axis? Explain. Describe the character of solution curves that originate near each of the equilibrium solutions.

**(c)** Plot enough solution curves of the ODE to see whether this plot is consistent with your descriptions in (b). Explain. Write directly on your graph.

**2.** Choose one function $f(x)$ below. Do the problems that follow using the ODE $x' = f(x)$.

- ☐ **A.** $f(x) = x\sin(1/x)$, if $x \neq 0$, $f(0) = 0$
- ☐ **B.** $f(x) = (1-x^2)^{1/2}$.

* **(a)** Find all the equilibrium solutions of the ODE.

* **(b)** Describe the solution curves of $x' = f(x)$. Are all the solution curves of this ODE defined over the entire $t$-axis? Are any solution curves unbounded? Do any solution curves reach equilibrium values in finite time? Explain.

* **(c)** Plot some representative solution curves of $x' = f(x)$. Is your plot consistent with your description in (b)? Write directly on your graph.

* **(d)** In case A, does the IVP $x' = f(x)$, $x(1) = 0$, have a unique solution? Is the Uniqueness Principle satisfied? If not, does the IVP have more than one solution? Explain. Answer these questions in case B for the IVP $x' = (1-x^2)^{1/2}$, $x(0) = 1$.

# Equilibrium Solutions and Sensitivity

***Purpose*** To show how the equilibrium points of a first order nonlinear ODE bifurcate as a parameter in the rate function is changed.

***Keywords*** Equilibrium points, pitchfork bifurcation, attractor, repeller, sensitivity

***See also*** Experiment 2.2

***Background*** The location and nature of the equilibrium points of an autonomous equation, $x' = f(x, a)$, may be quite sensitive to changes in the parameter $a$. For example, the equilibrium points $x_1 = 0, x_2 = \sqrt{a}, x_3 = -\sqrt{a}$, of the ODE $x' = x(a - x^2)$ change in a drastic way as $a$ increases through zero. If $a < 0$, then $x_2$ and $x_3$ are pure imaginary and are not visible in real state space. As $a$ increases through 0, $x_2$ and $x_3$ suddenly emerge from $x_1$ in a **pitchfork bifurcation** and move out on the arms of the parabola $x^2 = a$ in $ax$-space. The graph of the real equilibrium points is the pitchfork in Figure 2.3.1. Using sign analysis, we see that $x_1$ is a **global attractor** if $a \leq 0$ (i.e., all orbits of the ODE tend to $x_1$ as $t \to \infty$), but $x_1$ is transformed into a **local repeller** if $a > 0$ (all nearby orbits move away from $x_1$ as $t$ increases from $-\infty$). On the other hand, if $a > 0$, $x_2$ is a **local attractor** (all nearby orbits tend to $x_2$ as $t \to \infty$); $x_3$ is also a local attractor.

***Observation*** The $x$-state space in the pitchfork example is one-dimensional, which is why the orbits in Figure 2.3.1 lie on vertical lines. The pitchfork bifurcation is one of many examples of a sudden change in orbital structure of an ODE as a parameter changes.

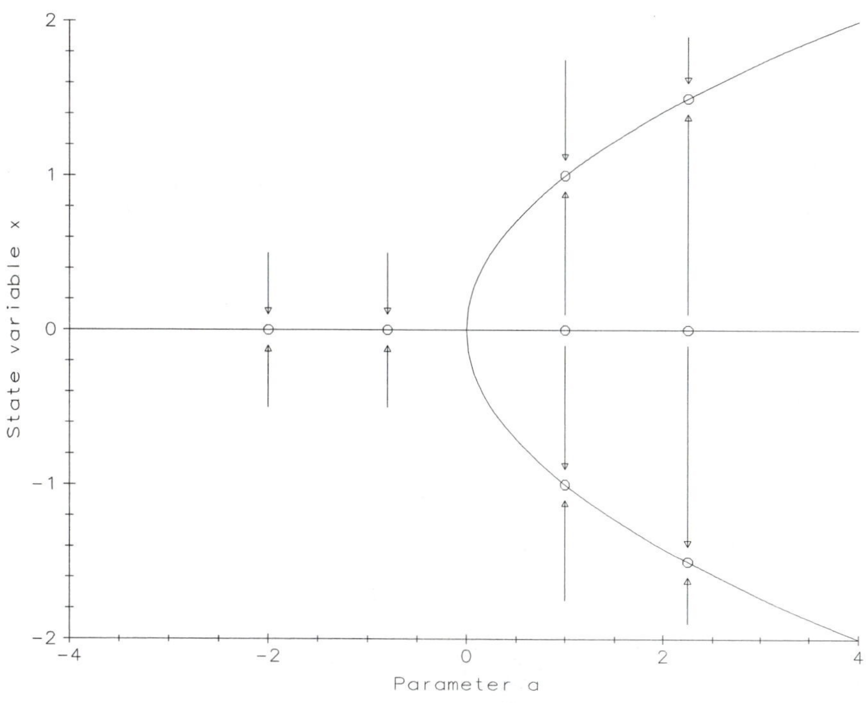


*Figure 2.3.1 The pitchfork bifurcation for $x' = x(a - x^2)$.*

_____________/_____________ hardware software

_____________/_____________ name date

Answer questions in the space provided, or on attached sheets or carefully labeled graphs.

_____________/_____________ course section

## 2.3 *Pitchfork Bifurcation*

***Abstract*** The pitchfork and other bifurcations are studied for first order nonlinear ODEs whose rate functions involve a parameter.

***Special instructions*** Use a plotting routine to draw the bifurcation curves. If such a routine is not available, rename the bifurcation parameter $a$ as a second state variable $y$ and set $y' = 0$. After solving $x' = f(x, y)$, $y' = 0$ for an appropriate set of initial conditions and a time range, hand sketch the curve $f(x, y) = 0$ over the orbits in the $xy$-state plane.

**1.** Reproduce Figure 2.3.1 and explain what you see.

**2.** Select one of the ODEs below and perform the following tasks.

☐ **A.** $x' = x(a - x)$ (**transcritical bifurcation**)

☐ **B.** $x' = a - x^2$ (**saddle-node bifurcation**)

☐ **C.** $x' = x(a - x^2)(4a - x^2)$ (**five-pronged pitchfork**)

☐ **D.** $x' = x(x^2 - 4a^2 + 4a^6)$

☐ **E.** $x' = x(\sin^2 a - x^2)$

**(a)** In the region below, sketch the curve of the equilibrium points in the $ax$-plane. Using sign analysis, draw the (vertical) orbits with an arrowhead giving the direction of time.

**(b)** Now use a computer to draw the curve of equilibrium points and some of the vertical orbits. Compare with the sketches in (a).

# Solutions That Escape to Infinity

***Purpose*** To investigate solutions of ODEs that escape to infinity in finite time.

***Keywords*** First order ODEs, solution curves for ODEs, finite escape time

***See also*** Experiment 2.2

***Background*** A solution $x(t)$ of the first order ODE $x' = f(t, x)$ sometimes exhibits the following peculiar property: There is a point $T$ such that $x(t)$ is defined on an interval of the form $a \leq t < T$ and $x(t) \to +\infty$ (or $-\infty$) as $t$ approaches $T$ from the left. Or perhaps $x(t)$ is defined on an interval of the form $T < t \leq b$ and $x(t) \to +\infty$ (or $-\infty$) as $t$ approaches $T$ from the right. When this happens, the solution $x(t)$ is said to **escape to infinity in finite time** and $T$ is called the **escape time.** As we shall see, the rate function $f(t, x)$ may be extremely well-behaved and give no indication that some of the solutions escape to infinity in finite time.

***Example 1*** **(Escape to infinity)** A look at solution curves in any bounded region of the $tx$-plane is insufficient to determine the existence of solutions that escape to infinity. Indeed the solution curve plot of the ODE $x' = x^2$ (Figure 2.4.1) reveals nothing, and yet *every* solution of this equation escapes to infinity (except for the solution, $x \equiv 0$). A simple trick shows this. Making the change of dependent variable $y = 1/x$, we see that $y(t)$ satisfies the ODE $y' = -1$. Thus $x(t)$ is the solution of the IVP $x' = x^2$, $x(0) = x_0$, if $x(t) = 1/y(t)$, where $y(t)$ solves the IVP $y' = -1$, $y(0) = 1/x_0$. Now since every solution, $y = -t + C$, of $y' = -1$ reaches $y = 0$ at some finite value $T$, it follows that every nonconstant solution of $x' = x^2$ escapes to infinity.

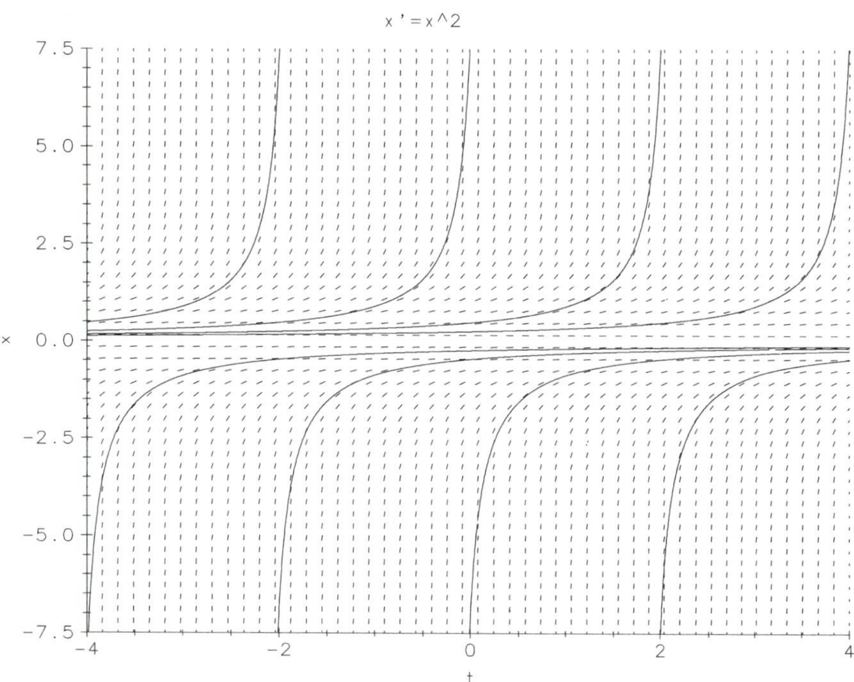


*Figure 2.4.1 Solution curves of $x' = x^2$.*

***Observation*** **(Can a solver "see" infinity?)** Many commercially available solvers "know" enough not to integrate through a time when a solution escapes to infinity. On the other hand, some solvers may well integrate right through a time at which the solution has escaped to infinity (giving no indication that a mistake has occurred!).

***Example 2*** **(Finite escape times for systems)** Solutions of systems of ODEs also exhibit the finite escape time property. For example, consider the system,

$$\begin{aligned} x' &= y \\ y' &= -x \end{aligned} \tag{1}$$

which has the general solution $x = A\sin(t+a)$, $y = A\cos(t+a)$, where $A$ and $a$ are arbitrary constants whose values are determined by the initial conditions. Solutions are bounded for all time and do not escape to infinity. Take the initial conditions $x(0) = 3\sin(\pi/3)$, $y(0) = 3\cos(\pi/3)$. Then the unique solution of system (1) that satisfies these initial conditions is:

$$\begin{aligned} x &= 3\sin(t+\pi/3) \\ y &= 3\cos(t+\pi/3) \end{aligned} \tag{2}$$

Note that this orbit is a circle of radius 3, which crosses the $y$-axis at $t_1 = 2\pi/3$. Now if we replace the dependent variable, $x$, in the system (1) by $1/z$ we obtain an equivalent system

$$\begin{aligned} z' &= -yz^2 \\ y' &= -1/z \end{aligned} \tag{3}$$

whose orbital solution corresponding to (2) is

$$\begin{aligned} z &= \frac{1}{(3\sin(t+\pi/3))} \\ y &= 3\cos(t+\pi/3) \end{aligned} \tag{4}$$

Observe that the orbit (4) escapes to infinity in finite time, $t_1 = 2\pi/3$. Note that all orbits of system (3) escape to infinity (in the $zy$-plane) in finite time because all nonconstant orbits of the system (1) cross the $y$-axis (in the $xy$-plane). Thus whether an orbit escapes to infinity in finite time (or at all) depends on the choice of state variables. This is hardly surprising because "infinity" itself depends on this choice.

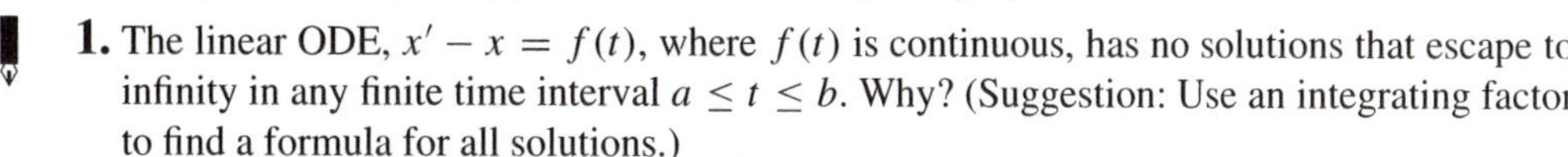

## 2.4 *Solutions That Escape to Infinity*

***Abstract*** Solutions of $x' = f(t, x)$ may escape to infinity in finite time (but not if $f$ is linear in $x$). Escape times may be approximated from computer graphs.

**1.** The linear ODE, $x' - x = f(t)$, where $f(t)$ is continuous, has no solutions that escape to infinity in any finite time interval $a \leq t \leq b$. Why? (Suggestion: Use an integrating factor to find a formula for all solutions.)

**2.** Consider the IVP $x' = x^3$, $x(0) = x_0 \neq 0$.

**(a)** What is the escape time of the solution to this IVP when $x_0 = 1$? Are there other solutions of $x' = x^3$ that escape to infinity in finite time? Describe and plot them. (Hint: First solve the IVP explicitly by changing to a new variable $y$, where $y = 1/x^2$. Show that $y' = -2$.)

* **(b)** Suppose that $f(t, x) > g(t, x) > 0$ for all $x \geq 0$, $t \geq 0$. Show that if $x = u(t)$ solves $x' = f(t, x)$ and $x = v(t)$ solves $x' = g(t, x)$, and if $u(t_0) = v(t_0)$ for some $t_0 > 0$, then $u(t) > v(t)$ for all $t > t_0$ where both are defined. (Hint: What happens if $u(t_1) = v(t_1)$, for some $t_1 > t_0$?) Use this result and the answer to (a) in searching for finite escape times for solutions of the ODE $x' = x^3 + t^2$.

**3.** Select one of the IVPs below and do (a) and (b) for that IVP.

☐ **A.** $x' = 1 - tx^2, \quad x(0) = x_0$

☐ **B.** $x' = x^2 + t, \quad x(0) = x_0$

**(a)** For what values of $x_0$ does your IVP have a solution that escapes to infinity in finite time? Use the computer to estimate these values and write your answers on the graph.

**(b)** Use the computer to find approximate escape times (if any) for your IVP for several values of $x_0$. Write your answers on the graph.

**4.** Consider the IVP $x' = \sin(x^2 + t) - 2t$, $x(0) = x_0$.

**(a)** Solve the IVP using various values of $x_0$ and $0- \leq t \leq 10$.

**(b)** Find solution formulas for the two IVPs $x' = \pm 1 - 2t$, $x(0) = x_0$, and plot their graphs for $0 \leq t \leq 10$ and the same values of $x_0$ as in (a). For each $x_0$, why does the graph found in (a) lie between the corresponding graphs found here?

**(c)** Explain why no solution of $x' = \sin(x^2 + t) - 2t$ escapes to infinity in finite time.

* **5.** Suppose that $f(t, x)$ is continuous for all $t$ and $x$. Suppose that for every $t_0$ and $x_0$ the IVP $x' = f(t, x), x(t_0) = x_0$, has exactly one solution (maximally extended in time). Explain why if the IVP has one solution with finite escape time, it has infinitely many such solutions. (Hint: Every solution can be extended forward or backward from $t_0$ until $|t|$ or $|x(t)|$ (or both) tend to $\infty$. No two maximally extended solutions touch.) Create your own IVP such that all solutions with $|x_0| > 2$ have finite escape times while solutions with $|x_0| \leq 2$ are defined for all time. Show this graphically. Explain your system.

# Picard Process for Solving IVPs

***Purpose*** To examine the convergence of the Picard iterates in the process of solving an IVP for a first order ODE.

***Keywords*** Existence Principle, convergence of a sequence of functions

***Background*** Referring to Figure 2.N.1 in the Chapter 2 Notes, we consider the forward IVP

$$\begin{aligned} x' &= f(t, x) \\ x(t_0) &= x_0 \end{aligned} \tag{1}$$

where $f$ and $\partial f/\partial x$ are continuous on the rectangle $S$. The Picard process iteratively constructs a sequence of functions—called **Picard iterates**—which converge to the unique solution of IVP (1) on some interval $t_0 \le t \le t_0 + c$. If $x(t)$ solves IVP (1) on this interval, then integration of the ODE in (1) shows that $x(t)$ solves the integral equation

$$x(t) = x_0 + \int_{t_0}^{t} f(s, x(s))\, ds, \quad t_0 \le t \le t_0 + c \tag{2}$$

Conversely, if a continuous function $x(t)$ satisfies the integral equation (2), it can be shown that $x(t)$ also solves the forward IVP (1).

The Picard iterates $x_0(t), x_1(t), x_2(t), \ldots$, are generated from (2) as follows. Set $x_0(t) \equiv x_0$, where $x_0$ is the initial data. Now select a positive constant $a$ (for example, the size of the $t$-range of $S$ in the IVP (1)) such that the sequence of functions

$$x_n(t) = x_0 + \int_{t_0}^{t} f(s, x_{n-1}(s))\, ds, \quad n = 1, 2, \ldots \tag{3}$$

is defined over the interval $I$: $t_0 \le t \le t_0 + a$. The Picard sequence $\{x_n(t)\}$ can be shown to converge uniformly on some interval $t_0 \le t \le t_0 + c$, where $0 < c < a$. In practice, the constant $a$ is not hard to find, but it should not be assumed that convergence of the Picard sequence takes place on $I$.

***Example 1*** **(Convergence of Picard iterates)** The solution of the IVP $x' = -x$, $x(0) = 3$, is $x(t) = 3e^{-t}$. The first few Picard iterates are

$$\begin{aligned} x_0(t) &= 3 \\ x_1(t) &= 3 - \int_0^t x_0(s)\, ds = 3 - \int_0^t 3\, ds = 3 - 3t \\ x_2(t) &= 3 - \int_0^t x_1(s)\, ds = 3 - \int_0^t (3 - 3s)\, ds = 3 - 3t + 3t^2/2 \\ x_3(t) &= 3 - \int_0^t x_2(s)\, ds = 3 - \int_0^t (3 - 3s + 3s^2/2)\, ds = 3 - 3t + 3t^2/2 - t^3/2 \end{aligned}$$

Observe that for this particular IVP the process generates a Taylor expansion of the solution $3e^{-t} = 3(1 - t + t^2/2 - t^3/6 + \cdots)$. See Figure 2.5.1 for graphs of the first ten iterates over the interval $0 \le t \le 5$. The convergence of the iterates to the solution appears to be quite rapid over any fixed interval.

In practice, however, iterates cannot be produced by a symbolic integration. Instead, an interval $I$: $t_0 \le t \le t_0 + a$, is selected and the iterates are generated from (2) by numerical integration and graphed on a common set of axes. Figures 2.5.2–2.5.4 illustrate this process; note that the iterates in the figures are denoted by $x^{(k)}$.

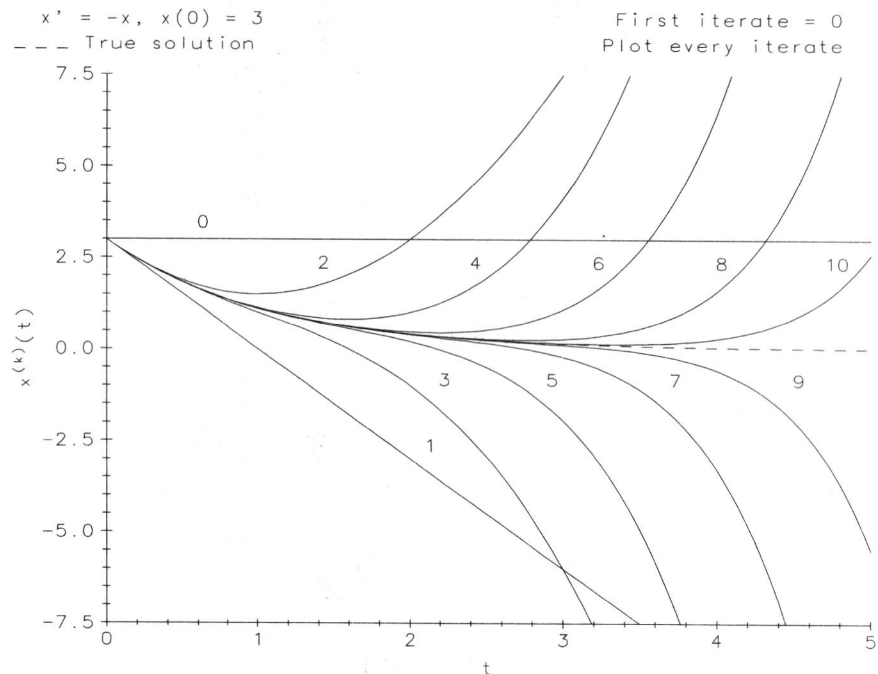


*Figure 2.5.1 Picard iterates bracket the true solution.*

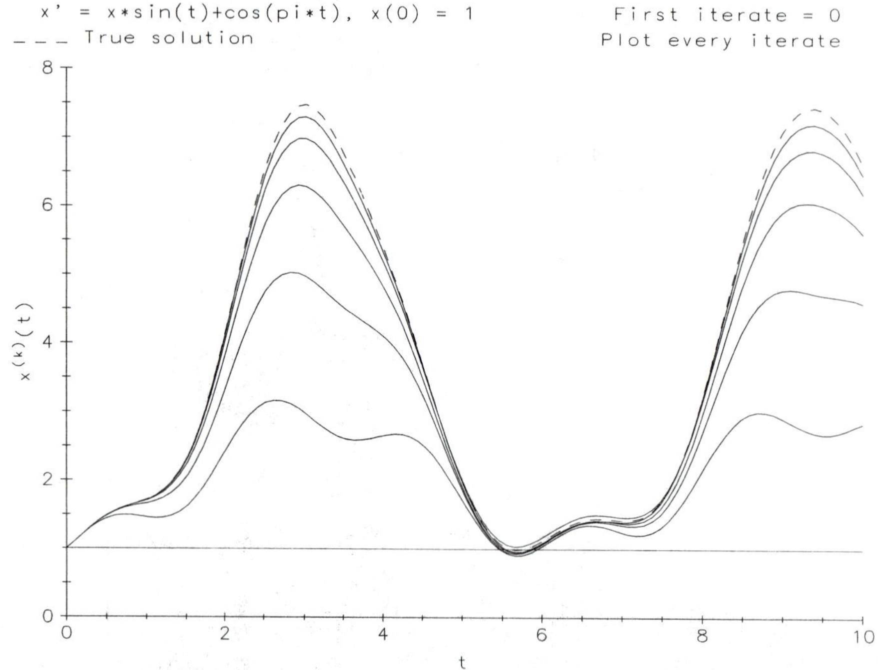


*Figure 2.5.2 Iterates rapidly converge to the true solution.*

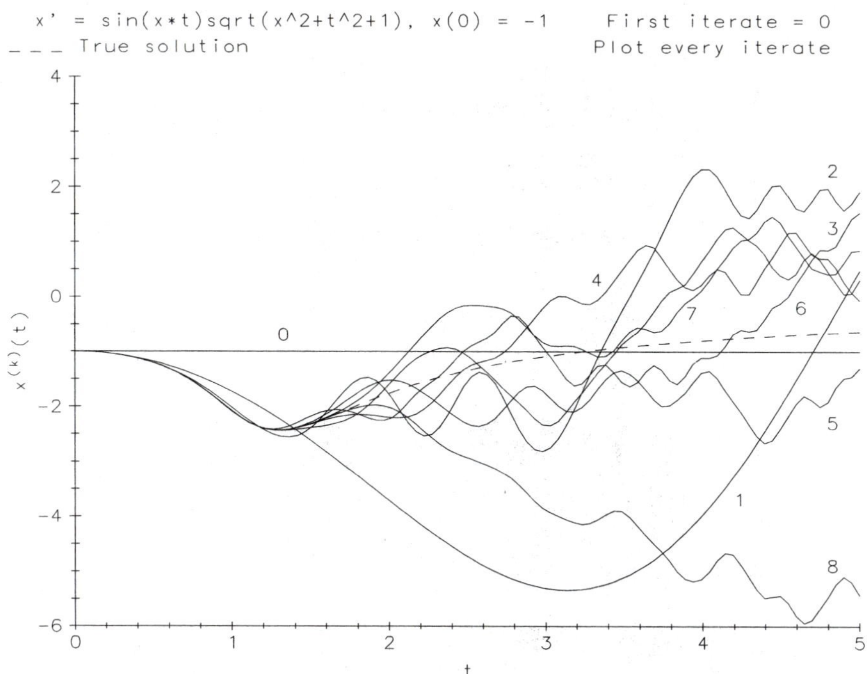


*Figure 2.5.3 Picard iterates wander around the true solution.*

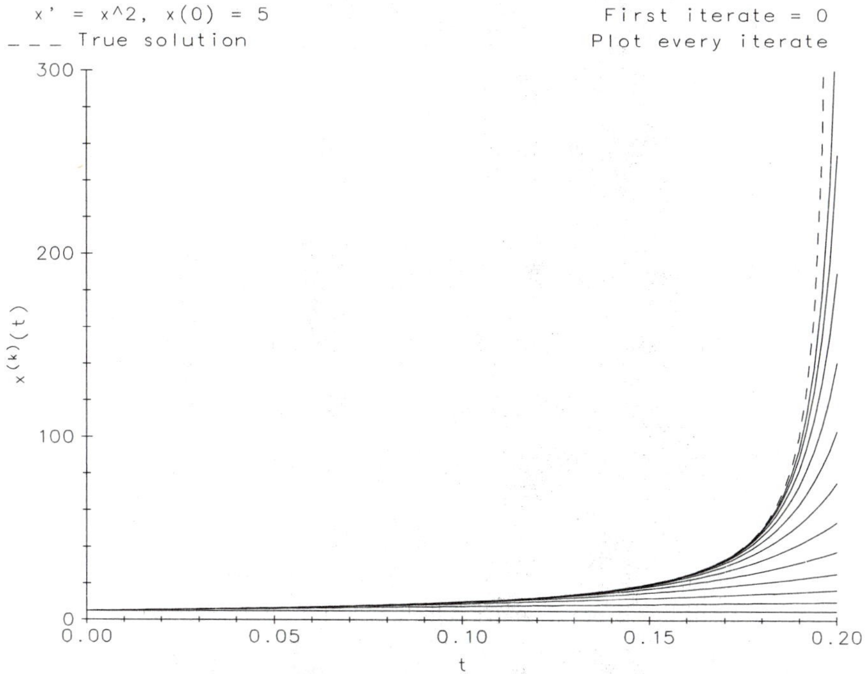


*Figure 2.5.4 True solution escapes to infinity at $t = 0.20$.*

__________/__________ hardware software

__________/__________ name date

Answer questions in the space provided, or on attached sheets or carefully labeled graphs.

__________/__________ course section

## 2.5 *Picard Process for Solving IVPs*

***Abstract*** Picard iterates for IVPs are constructed and graphed. The iterates are compared to the true solutions.

***Special instructions*** A numerical integration routine is required to carry out these problems.

**1.** Consider the IVP $x' = -x$, $x(0) = 3$.

**(a)** Reproduce Figure 2.5.1 on your platform. Can you discern any pattern in the way the iterates converge to the actual solution? Explain.

* **(b)** In the Existence Principle for the forward IVP that goes with Figure 2.N.1 it is stated that the solution of the IVP is defined at least on $t_0 \leq t \leq t_0 + c$, where $c = \min\{a, b/M\}$, and $M$ is a constant such that $|f(t, x)| \leq M$ for all $(t, x)$ in $S$. Calculate $c$ for the IVP $x' = -x$, $x(0) = 3$. Looking at the iterates in Figure 2.5.1, do you think this estimate for the interval where the iterate converges is generous? Explain.

**2.** Reproduce Figures 2.5.2–2.5.3 on your platform. Describe the pattern of convergence of the iterates, if any. Write directly on the graphs.

**3.** Consider the IVP $x' = x^2$, $x(0) = 5$.

**(a)** Reproduce Figure 2.5.4 on your platform. Describe the pattern of convergence of the iterates. Estimate the interval of convergence $0 \leq t \leq c$ for the iterates.

**(b)** Find a closed-form solution of the IVP.

**4.** Find the first seven Picard iterates for the IVP $x' = 1 - t\sin(x)$, $x(0) = 1$, over the interval $0 \leq t \leq 30$. Discuss convergence of the iterates. Write directly on your graph.

# Euler's Method for Solving IVPs

***Purpose*** This experiment continues the discussion of Euler's method for finding approximate numerical solutions of IVPs and explores some properties of such approximate solutions.

***Keywords*** First order ODEs, first order systems of ODEs, difference equations, numerical solution of an IVP

***Background*** Using Euler's method to solve the IVP

$$\begin{aligned} x' &= f(t, x) \\ x(t_0) &= x_0 \end{aligned} \tag{1}$$

produces a sequence of approximations $x_1, x_2, \ldots, x_N$ to the true solution values at the points $t_j = t_0 + jh$, $j = 1, 2, \ldots, N$, where $h$ is the step size. Starting with the given $x_0$, the values for $x_{j+1}$ are calculated by the algorithm

$$x_{j+1} = x_j + hf(x_j, t_j), \quad j = 0, 1, 2, \ldots \tag{2}$$

Connecting the points $(t_0, x_0), (t_1, x_1), \ldots, (t_N, x_N)$ in the $tx$-plane by straight line segments gives an approximation, called here an **Euler solution,** to the solution curve of IVP (1). Interpreting the algorithm (2) geometrically, we see that the approximating Euler solution actually steps from one solution curve to another, each segment being tangent to the solution curve through the left-hand endpoint of the segment. See Figure 2.1.

How good an approximation to the true solution $x(t)$ of IVP (1) are the values $x_1, x_2, \ldots, x_N$ generated by the Euler method? Suppose that the rate function $f(t, x)$ is continuously differentiable over a rectangle $R$ in the $tx$-plane and that the solution curve $x(t)$ for IVP (1) lies in $R$ over the interval $t_0 \le t \le T$. Then it can be shown that there exists a positive constant $M$, depending only on $R$ and $f$, such that

$$\left|x(t_0 + jh) - x_j\right| \le Mh, \quad j = 1, 2, \ldots, N \tag{3}$$

where $h = (T - t_0)/N$ for any positive integer $N$. Thus, it would appear that the values generated by Euler's method all become better approximations to the solution of IVP (1) at the grid points as the step size $h$ becomes smaller. In fact, if $M$ were known we could choose a step size $h$ such that the values generated by Euler's method would be within any given error bound from the exact values of the solution of IVP (1). The estimate (3) assumes that *exact arithmetic* is used (i.e., that decimal strings are not chopped or rounded).

***Observation*** **(Order of a method, discretization error)** Euler's method is said to be of **first order** because the error bound (3) contains the step size $h$ to the first power. The fourth order Runge-Kutta method has order four (see the Chapter 2 Notes). Most computer solvers use methods of at least fourth order. As mentioned above, the error estimate (3) assumes that the exact values are used for $x(t_0 + jh)$ and $x_j$. The exact value $\left|x(t_0 + jh) - x_j\right|$ is called the **discretization error**. Of course exact calculations are impossible to do in practice and so cumulative round-off errors must be considered. In an example below we show, in fact, that Euler approximations to IVP (1) do not always improve as $h$ is chosen to be smaller and smaller.

***Example 1*** **(An Euler solution)** For the IVP $x' = x \sin 3t$, $x(0) = 1$, over the $t$-interval, $0 \le t \le 4$, choose step size $h = 0.2$ and $N = 20$. Euler's algorithm in this case becomes

$$x_0 = 1, \quad x_{j+1} = x_j + 0.2x_j \sin(3t_j), \quad t_j = 0.2j$$

In Figure 2.6.1 the Euler solution is the solid broken line; the true solution is the dashed curve.

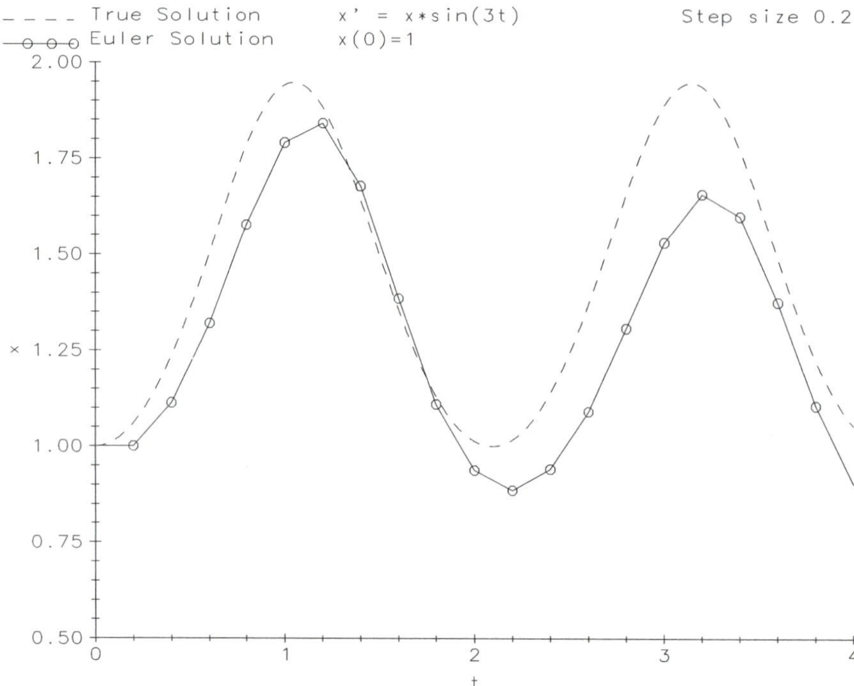


*Figure 2.6.1 Comparison of the true solution to the Euler solution.*

***Example 2*** **(Round-off errors)** Suppose rounding is used in Euler's method to estimate $x(3)$, where $x(t)$ is the solution of the initial value problem $x' = x$, $x(0) = 1$. Using smaller and smaller step sizes (hence, more and more steps), the total error at first decreases. However, as seen in Figure 2.6.2, if single precision arithmetic is used (1 part in $10^8$), the error stops decreasing when the number of steps exceeds approximately $10^4$. The total number of operations and accompanying round-off errors overwhelms the drop in discretization error.

***Example 3*** **(Long-term degradation of Euler solutions)** If the IVP $x' = 1 - t\sin x$, $x(0) = 1$, is solved over the interval $0 \le t \le 30$ using Euler's method with step size $h = 0.15$ and graphed in the $tx$-plane, a strange phenomenon is observed. See Figure 2.6.3. Comparing the Euler solution to the "actual" solution (represented by the dashed curve in the figure), we see that initially the agreement is not good, but for a while it improves dramatically with increasing time. Then at about $t = 24$ the accuracy of the Euler solution seems to degenerate badly. Note that the estimate (3) does not apply to the *long-term* accuracy of Euler's method. Choosing smaller step sizes only seems to delay the onset of degeneration of the Euler solution.

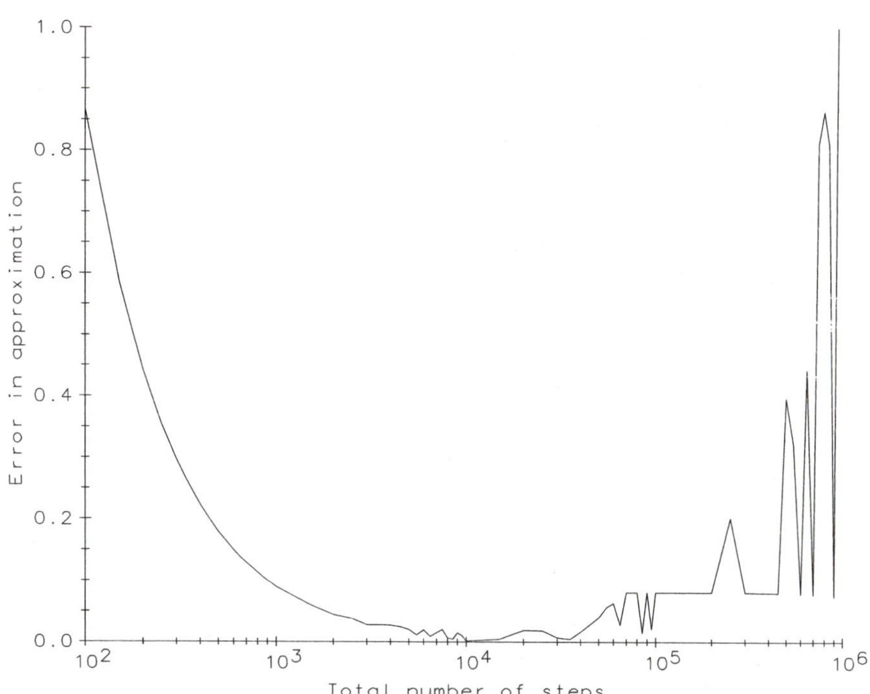


*Figure 2.6.2 Decline and rise in total error (see Example 2).*

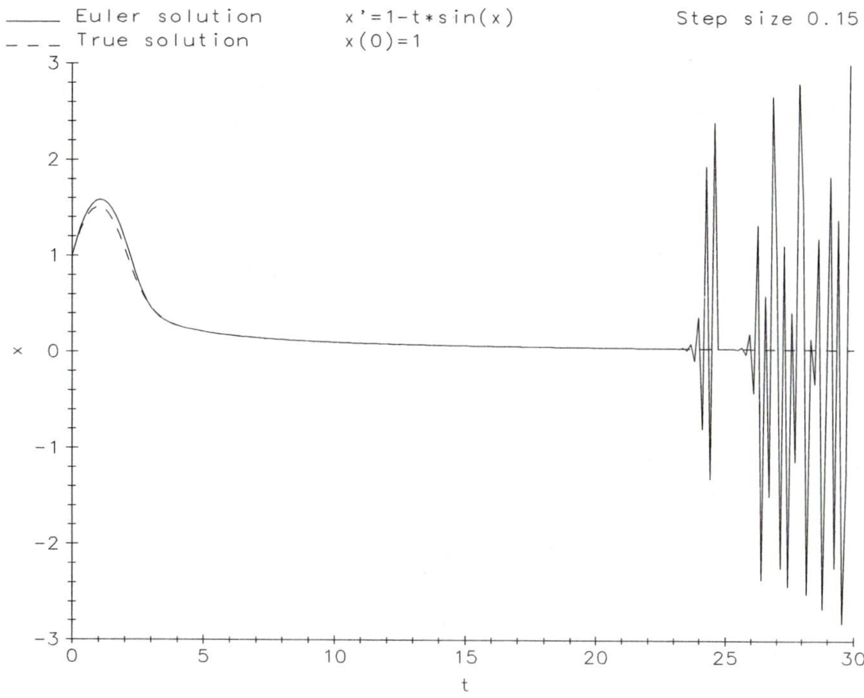


*Figure 2.6.3 Euler solution accuracy in the long term (see Example 3).*

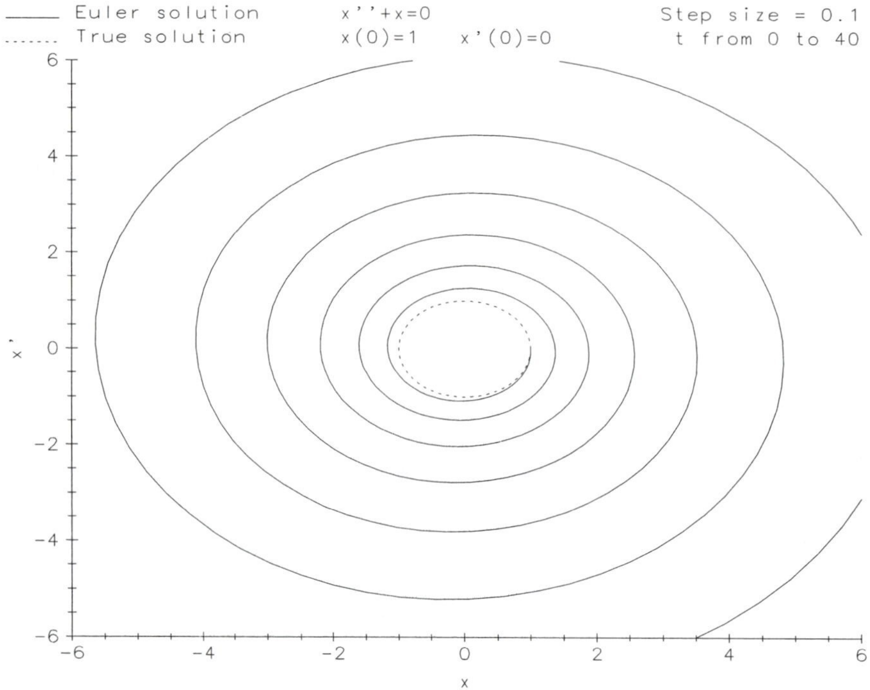


*Figure 2.6.4 Euler solution accuracy for a planar system.*

***Observation*** **(Euler's method for planar systems)** Euler's method applies equally well to first order systems of ODEs. As an example, for the IVP based on a planar system

$$\begin{aligned} x' &= f(t, x, y), \qquad & x(t_0) &= x_0 \\ y' &= g(t, x, y), \qquad & y(t_0) &= y_0 \end{aligned}$$

Euler's method takes the follow form:

$$\begin{aligned} x_{j+1} &= x_j + hf(t_j, x_j, y_j), \qquad & j &= 0, 1, 2, \ldots \\ y_{j+1} &= y_j + hg(t_j, x_j, y_j), \qquad & j &= 0, 1, 2, \ldots \end{aligned}$$

where $h$ is the step size, and $t_j = t_0 + jh$, $j = 0, 1, 2, \ldots$.

***Example 4*** **(Behavior of Euler solution of a system)** Before using Euler's method to solve the IVP $x'' + x = 0$, $x(0) = 1$, $x'(0) = 0$, it is necessary to replace it by the equivalent IVP

$$\begin{aligned} x' &= y \\ y' &= -x \\ x_0 &= 1, \qquad y_0 = 0 \end{aligned}$$

for a planar system. Solving this IVP over the interval $0 \leq t \leq 40$ using Euler's method with step size $h = 0.1$ and graphing the orbit in the $xx'$-plane, another odd phenomenon is observed. Figure 2.6.4 shows how the Euler solution moves away from the true orbit (represented by the dashed unit circle—see Example 1 in Experiment 1.11) in ever widening spirals. Decreasing the step size makes the spirals tighter but does not eliminate them.

______/______ hardware / software

______/______ name / date

Answer questions in the space provided, or on attached sheets or carefully labeled graphs.

______/______ course / section

## 2.6 *Euler's Method and Explicit Solutions*

***Abstract*** Euler solutions of an IVP are compared with the "true" solution of the IVP.

**1.** Consider the IVP $x' = \sin(t^2 + x^2)$, $x(0) = 0$. Carry out the procedure below to find a reasonably accurate approximate solution of the IVP over $0 \le t \le 4$ by using Euler's method.

**(a)** Find and graph the Euler solutions of the IVP using the step sizes $h_1 = 0.2$, $h_2 = 0.1$, $h_3 = 0.05$. Compare and describe the graphs of these Euler solutions. Write directly on the graphs.

**(b)** Use your solver package to graph the solution of the IVP over $0 \le t \le 4$. Compare this solution to the last computed Euler solution in (a). Account for unusual differences (if any).

**2.** Repeat Problem 1 for the IVP $x' = x^2 - t$, $x(2) = 1.5$, over the interval $2 \le t \le 6$.

**3.** Find and graph Euler solutions of the IVP $x' = 1 - xt$, $x(0) = 0$, over the interval $0 \le t \le 20$ using the step sizes $h_1 = 0.2$, $h_2 = 0.1$, $h_3 = 0.05$. Use your solver to find the true solution of the IVP. Describe and compare the graphs of the Euler solutions.

**4.** Select one of the IVPs below and complete (a)–(c).

| | | | |
|---|---|---|---|
| ☐ **B.** | $x' = x/4$, | $x(0) = 1$, | $0 \le t \le 2$ |
| ☐ **C.** | $x' = 3 - 2t + 2x$, | $x(0) = 1$, | $0 \le t \le 1$ |
| ☐ **D.** | $x' = x \sin 3t$, | $x(0) = 1$, | $0 \le t \le 4$ |

**(a)** Find the exact (explicit) solution to the indicated IVP.

**(b)** Solve the IVP over the given interval using Euler's method for 8 steps, then 16 steps, and finally 24 steps. Graph these approximate solutions and compare them with the graph of the exact solution. Calculate the error in the Euler solutions from the exact value of the solution at the right-hand endpoint of the interval.

**(c)** Find the error in the Euler solutions of the IVP at the endpoint of the interval using $N =$ 500 steps, 1000 steps, 2000 steps, 3000 steps. Does the error always diminish as the number of steps increases?

hardware / software — name / date

Answer questions in the space provided, or on attached sheets or carefully labeled graphs.

course / section

## 2.7 *Limitations of Euler's Method*

***Abstract*** The accuracy of Euler's method is examined in a case where the actual solution of an IVP either escapes to infinity in finite time or exhibits other singular features.

**1.** Consider the IVP $x' = -x^2, \quad x(0) = -1/2$.

**(a)** Find the Euler solution of this IVP over the interval $0 \leq t \leq 2.5$, using the step size $h = 0.15$.

**(b)** Find the exact solution of this IVP and compare its graph to the graph of the Euler solution in (a). Explain the differences. Write directly on your graph.

**(c)** Does the phenomenon observed in (b) go away as the step size is reduced? Explain.

**2.** Consider the IVP, $x' = x^2 - t$, $x(-2) = -3$ and perform the tasks below. See also Atlas Plate `First Order C`. (For an extensive discussion of the numerical solution for this ODE see Hubbard, J.H. and West, B.H. *Differential Equations: A Dynamical Systems Approach, Part I: Ordinary Differential Equations.* Springer-Verlag, New York, 1991.)

**(a)** Graph the solution of the IVP with your solver on the interval $-3 \leq t \leq 35$. Then solve the IVP using Euler's method with various values for the step size $h$. Does the Euler solution "approach" the real solution as $h \to 0$? Justify your claim.

**(b)** The **improved Euler method** is given by

$$x_{j+1} = x_j + \frac{h}{2}\{f(t_j, x_j) + f(t_j + h, x_j + hf(t_j, x_j))\}, \quad j = 0, 1, 2, \ldots$$

Give a simple description (in prose) of the geometry of this method. Graph the improved Euler approximation for the IVP using the same values of the step size $h$ as in (a). Compare results with those from (a) and discuss differences.

**(c)** The **modified Euler method** is given by

$$x_{j+1} = x_j + hf(t_j + \frac{h}{2}, x_j + \frac{h}{2}f(t_j, x_j)), \quad j = 0, 1, 2, \ldots$$

Repeat part (b) using the modified Euler method in place of the improved Euler method.

* **3.** Use Euler's method to solve the system in Figures 1.11.9 and 1.11.10. Discuss differences, if any.

# Euler Solutions to the Logistic Equation

***Purpose*** These experiments illustrate some limitations of Euler's method in predicting the nature of solutions of the logistic equation, especially as $t \to \infty$. Along the way, the experiments also show that there cannot be any universal notion of "smallness" of step size in one-step numerical approximations of IVPs.

***Keywords*** First order ODEs, separable ODEs, difference equations, population growth models, Euler method for solution of an IVP

***See also*** Experiments 2.2, 2.6–2.7; Appendix B.1

***Background*** The **logistic equation** $x' = rx(1-x)$, where $r$ is a positive constant, governs the evolution of the size of a single population, $x(t)$, taking into account certain natural factors that affect population growth. Euler's method applied to the initial value problem

$$\begin{aligned} x' &= rx(1-x) \\ x(0) &= x_0 \end{aligned} \tag{1}$$

generates the sequence of approximations $x_1, x_2, \ldots,$ given by

$$x_{n+1} = x_n + hrx_n(1-x_n), \qquad n = 0, 1, 2, \ldots \tag{2}$$

where $h$ is the step size and $x_0$ is the given initial value of the state variable $x$. In this experiment the behavior of all Euler solutions of IVP (1) for initial values $x_0$ near 1 and for all step sizes $h > 0$ will be examined. These Euler solutions exhibit some surprising properties as $h$ changes.

To analyze the Euler solutions it is helpful to put $a = rh$ and rewrite the right-hand side of (2) by completing the square in the quadratic expression for $x_n$, obtaining

$$x_{n+1} = -a\left(x_n - \frac{1+a}{2a}\right)^2 + \frac{(1+a)^2}{4a} \tag{3}$$

The graph of (3) is the parabola that appears in Figure 2.8.1. Note that $x_n^1$ and $x_n^2$ in the figure are the values of $x_n$ that make $x_{n+1} = 1$.

Observe that if $0 < a \le 3$ then $(1+a)^2/4a \le (1+a)/a$; hence from Figure 2.8.1 it follows that the value $x_n$ in the interval $[0, (1+a)/a]$ returns the value $x_{n+1}$ in the same interval. Thus, if $0 \le x_0 \le (1+a)/a$, then Euler's method (3) produces values $x_1, x_2, \ldots,$ all of which lie in the interval $[0, (1+a)/a]$ as well.

There is a simple geometric way of "seeing" what the Euler solution will look like from a given starting point $x_0$ in $[0, (1+a)/a]$ when $0 < a < 3$. First, superimpose the line $x_{n+1} = x_n$ over the graph of the parabola in Figure 2.8.1. Note that the point $(1, 1)$ on the parabola occurs to the *left* of the high point if $0 < a < 1$ and to the *right* of the high point if $a > 1$. Now, depending on the size of $a$, we describe a graphical procedure for constructing the Euler sequence $x_1, x_2, \ldots,$ from the initial choice $x_0$ (refer to Figure 2.8.2, 2.8.3, or 2.8.4): follow the vertical dashed line upward from $x_0$ to the parabola to find $x_1$, and then horizontally over to the line $x_{n+1} = x_n$ and then vertically to the parabola again to find $x_2$, and so on. Note, however, that this graphical process for generating the Euler solution changes significantly when $a > 1$. The constructions and graphs in Figures 2.8.2, 2.8.3, and 2.8.4 are typical for the three cases $0 < a < 1$, $1 < a < 2$, and $2 < a < 3$, respectively. A great deal about the behavior of the Euler sequence $x_1, x_2, \ldots$ can be inferred from these graphs.

***Observation*** **(Reliability)** Approximate numerical solutions of IVPs are not always reliable for predicting the behavior of solutions as $t \to \infty$. Such techniques are designed to provide increasingly

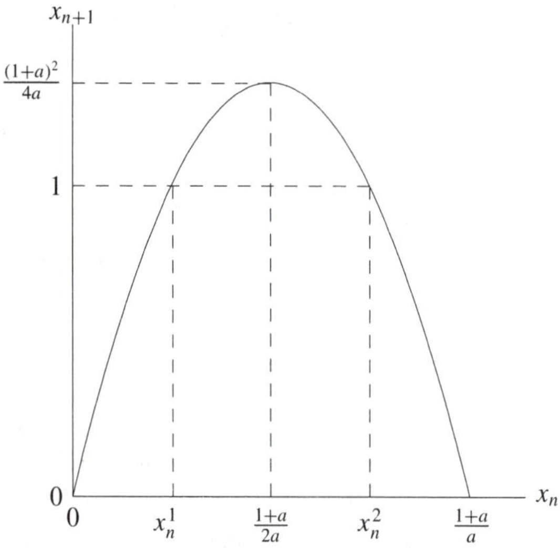


*Figure 2.8.1 Graph of iteration scheme (3): $a = rh$.*

better numerical solutions with decreasing step size but only over fixed, bounded $t$-intervals. This experiment demonstrates this fact convincingly.

***Observation*** **(Equilibrium solutions)** Note that $x \equiv 1$ and $x \equiv 0$ are two equilibrium solutions of the autonomous ODE $x' = rx(1-x)$. Thus the behavior of solution curves is an easy consequence of sign analysis. See Experiment 2.2.

***Observation*** **(Explicit solution)** To find an explicit solution formula for IVP (1), ignore the equilibrium solution $x \equiv 0$ and change the dependent variable to $z = 1/x$, to obtain the linear equation in $z$:

$$z' = -x'/x^2 = (-1/x^2)rx(1-x) = -rz + r \tag{4}$$

Solving (4), we find that $z = 1 + Ce^{-rt}$ where $C$ is an arbitrary constant and so $x = (1+Ce^{-rt})^{-1}$ gives all solutions of the logistic equation (except for the equilibrium solution $x \equiv 0$). Imposing the initial condition in IVP (1) yields that $C = 1/x_0 - 1$, and so

$$x = \frac{x_0}{x_0 + (1-x_0)e^{-rt}} \tag{5}$$

(Note that even the equilibrium solution $x \equiv 0$ is included in formula (5).) No use of this solution formula will be made in this experiment, other than to observe that for any $x_0 > 0$ the corresponding solution $x(t)$ in (5) tends monotonically to 1 as $t \to \infty$.

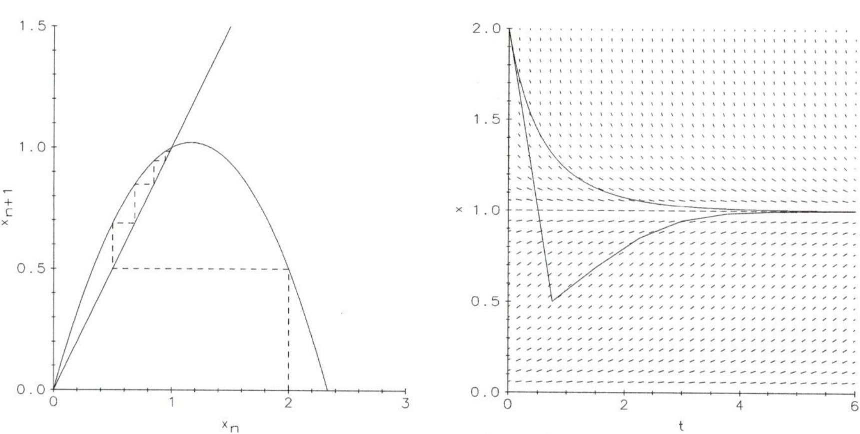


*Figure 2.8.2* *Generation of Euler solution:* $0 < a < 1$, $a = rh = \frac{3}{4}$.

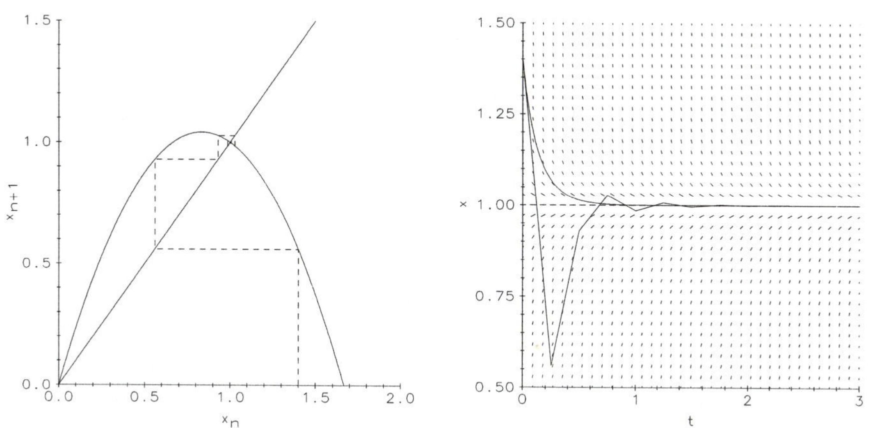


*Figure 2.8.3* *Generation of Euler solution:* $1 < a < 2$, $a = rh = \frac{3}{2}$.

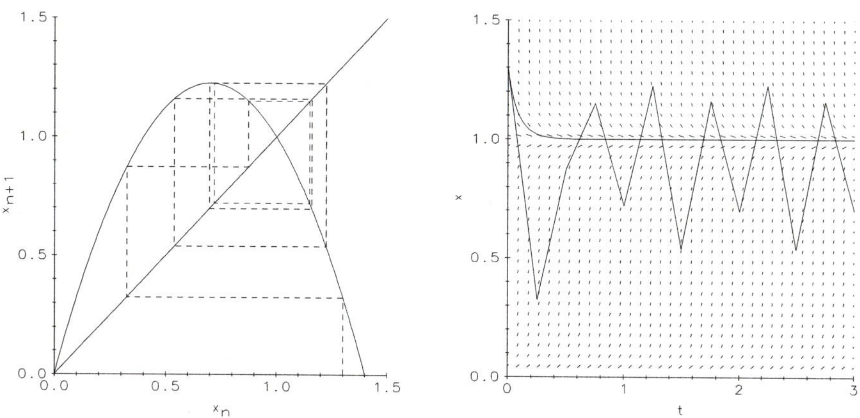


*Figure 2.8.4* *Generation of Euler solution:* $2 < a < 3$, $a = rh = \frac{5}{2}$.

____________ / ____________
hardware software

____________ / ____________
name date

Answer questions in the space provided,
or on attached sheets or carefully labeled graphs.

____________ / ____________
course section

## *2.8 Convergent Euler Sequences*

***Abstract*** When $a = rh$ is small enough, the Euler sequences defined by (3) converge to 1 as $t \to \infty$. This behavior is simulated in this experiment.

**1.** Solving IVP (1) using Euler's method, show that the Euler sequence $x_0, x_1, x_2, \ldots$ generated by the initial choice $x_0$ is a constant sequence only for $x_0 = 0$ and $x_0 = 1$. Put $a = rh$ and show that if $0 < a \leq 3$, then for any initial choice $x_0$ in the interval $0 \leq x_0 \leq (1 + a)/a$ it follows that the entire Euler sequence $x_1, x_2, \ldots$ is in the same interval.

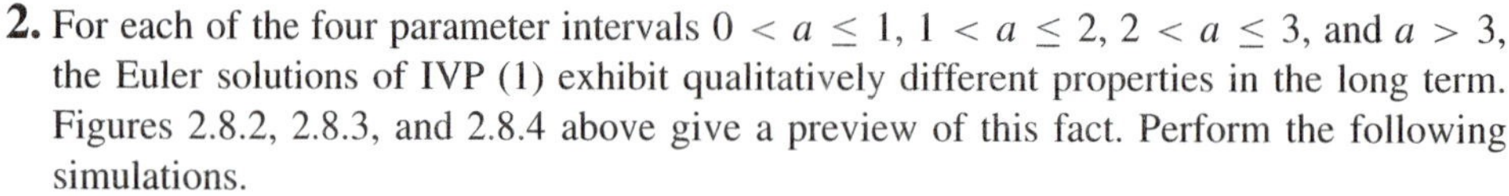

**2.** For each of the four parameter intervals $0 < a \leq 1$, $1 < a \leq 2$, $2 < a \leq 3$, and $a > 3$, the Euler solutions of IVP (1) exhibit qualitatively different properties in the long term. Figures 2.8.2, 2.8.3, and 2.8.4 above give a preview of this fact. Perform the following simulations.

**(a)** Let $r = 10$, $h = 0.05$ (so $a = rh = 0.5$). Construct Euler solutions of IVP (1) with $N = 10$ for each of the initial values $x_0 = 0.3, 0.8, 1.3, 2.0$. Graph the Euler solution for $x_0 = 2.0$ and compare it to the exact solution curve for IVP (1) and describe qualitative differences.

**(b)** Let $r = 100$, $h = 0.015$ (so $a = rh = 1.5$). Construct Euler solutions of IVP (1) with $N = 10$ for each of the initial values $x_0 = 0.5, 0.75, 1.25$, and $1.5$. Graph the Euler solution for $x_0 = 1.5$ and compare it to the exact solution curve for IVP (1). Describe qualitative differences.

_____/_____ hardware software

_____/_____ name date

Answer questions in the space provided,
or on attached sheets or carefully labeled graphs.

_____/_____ course section

## *2.9 Period Doubling and Chaos: Graphical Evidence*

***Abstract*** The long-term behavior of Euler solutions of IVP (1) depends very strongly on the value of the parameter $a = rh$. For "small" values of $a$ and $x_0$ near 1, the Euler solution tends to the constant solution $x \equiv 1$ as the number $N$ of Euler points increases. On the other hand, for "large" values of $a$ and $x_0$ near (but not equal to) 1 the Euler solution seemingly wanders chaotically about the constant $x \equiv 1$ as $N$ increases, and this behavior will be examined in this experiment.

* **1.** Solving IVP (1) using Euler's method, put $a = rh$ and use the graph in Figure 2.8.1 to show in a convincing way that for $1 < a < 2$ the Euler sequence (2) generated by any $x_0$ in the interval $0 \leq x_0 \leq (1+a)/a$ converges to 1 as $n \to \infty$. (Hint: First show that for any initial point $x_0$ the corresponding Euler sequence eventually is trapped in the interval $(1+a)^3(3-a)/16a \leq x_n \leq (1+a)^2/4a$ and that $1/a < (1+a)^3(3-a)/16a < 1 < (1+a)^2/4a < (1+a)/a$).

**2. (Period doubling)** Let $r = 100$, $h = 0.023$ (so $a = rh = 2.3$). Construct the Euler solution for IVP (1) with $N = 50$ and the initial value $x_0 = 1.3$. Graph this Euler solution and compare it to the exact solution curve of IVP (1). Repeat this simulation with $r = 100$, $h = 0.025$ (hence $a = rh = 2.5$). Do the values of these Euler solutions eventually settle down to any pattern? If so how would you describe it?

**3. (Chaotic wandering)** Let $r = 100$, $h = 0.04$ (so $a = rh = 4$). Construct the Euler solutions of IVP (1) with $N = 50$ for initial values $x_0 = 0.25, 0.50, 0.75, 1.2$. Graph each of these Euler solutions and compare them to the exact solutions of IVP (1). Can you detect any pattern in the values of any of these Euler solutions?

* **4.** Using the iteration (2), create a 2D plot as follows: For each value $a_0 = rh$ with $0 < a_0 < 4$ on the vertical axis, place a dot at each $x_n$, $n \geq 100$, on the horizontal line $a = a_0$. What information emerges? Describe the plot in words.

_______________/_______________ hardware software

_______________/_______________ name date

Answer questions in the space provided, or on attached sheets or carefully labeled graphs.

_______________/_______________ course section

## 2.10 *Period Doubling and Chaos: Theory*

***Abstract*** Let $a = rh$ in IVP (1). The behavior of Euler solutions of IVP (1) is quite different on the parameter sets $0 < a \leq 1$, $1 < a \leq 2$, $2 < a < 3$, and $a \geq 3$. This was illustrated for specific data sets in Experiments 2.8–2.9. The theoretical basis for this behavior will be explored in this experiment.

***Special instructions*** The open-ended nature of this experiment makes it a candidate for a team effort. Follow the instructions in Appendix A for writing the laboratory report.

---

Give a convincing justification for the following general description of the behavior of Euler solutions for IVP (1) in the indicated parameter regimes. (Hint: Use the recursion relation (3) and its geometric implementations in Figures 2.8.1, 2.8.2, 2.8.3, and 2.8.4). What does this behavior imply about the choice of step size $h$?

- For $0 < a \leq 1$ the Euler solution $\{x_n\}$ generated by any choice of initial point $x_0$ in the interval $0 < x_0 < (1 + a)/a$ behaves as follows:
  - ☐ If $0 < x_0 < 1$, then $\{x_n\}$ rises steadily toward 1, as $n \to \infty$.
  - ☐ If $x_0 = 1$ or $1/a$, then $x_n \equiv 1$, for $n \geq 1$.
  - ☐ If $1 < x_0 < 1/a$, then $\{x_n\}$ decreases steadily toward 1, as $n \to \infty$.
  - ☐ If $1/a < x_0 < (1 + a)/a$, then $0 < x_1 < 1$ and $\{x_n\}$ rises steadily toward 1, as $n \to \infty$.
- For $1 < a \leq 2$ the Euler sequence $\{x_n\}$ generated by the initial choice of $x_0$ in the interval $0 < x_0 < (1 + a)/a$ behaves as follows:
  - ☐ If $0 < x_0 < 1/a$, the Euler sequence $\{x_n\}$ rises steadily until for some $N$, $x_N > 1$, and for $n > N$, $\{x_n\}$ alternates about 1 while steadily approaching 1, as $n \to \infty$.
  - ☐ If $x_0 = 1/a$ or 1, then $x_n = 1$ for all $n \geq 1$.
  - ☐ If $1/a < x_0 < 1$, then $\{x_n\}$ alternates about 1 while steadily approaching 1 as $n \to \infty$.
  - ☐ If $1 < x_0 < (1 + a)/a$, then $0 < x_1 < 1$, and thereafter $\{x_n\}$ behaves as described above.
- **(Chaotic wandering)** For $a > 2$ the Euler solution $\{x_n\}$ generated by any initial value $x_0$ in the interval $0 < x_0 < (1 + a)/a$ does not approach 1 as $n \to \infty$. How would you describe the behavior of $\{x_n\}$?

Initial value problems (abbreviated **IVPs**) arise naturally in real-world applications and are basic to the study of ODEs. Some theoretical details on IVPs appear below, along with some background on numerical methods for solving IVPs.

Let $f(t, x)$ be defined in some region $R$ of the $tx$-plane and suppose that $(t_0, x_0)$ is a point in $R$. Consider again the IVP

$$\begin{aligned} x' &= f(t, x) \\ x(t_0) &= x_0 \end{aligned} \tag{1}$$

### Existence and Uniqueness

**Uniqueness Principle:** *In IVP (1) let $f$ and $\partial f/\partial x$ be continuous on a region $R$ in the $tx$-plane containing $(t_0, x_0)$. Then on any $t$-interval $I$ containing $t_0$ there is at most one solution of IVP (1).*

An immediate consequence of the Uniqueness Principle is that no two solution curves can intersect anywhere in $R$.

*Example 1* **(IVP with many solutions)** The IVP $x' = 3x^{2/3}$, $x(t_0) = 0$, has infinitely many solutions on the whole $t$-axis, for any $t_0$. These solutions to the IVP can be "constructed" from the special solutions $x \equiv 0$ and $x = (t + C)^3$, for any constant $C$. Note that the Uniqueness Principle does not apply, since for any $t_0$ there is no region in the $tx$-plane containing $(t_0, 0)$ for which $\partial f/\partial x = 2x^{-1/3}$ is continuous.

**Existence Principle:** *In IVP (1) let $f$ and $\partial f/\partial x$ be continuous in a region $R$ of the $tx$-plane containing $(t_0, x_0)$. Then IVP (1) has a (unique) solution on an interval containing $t_0$ in its interior.*

Note that the Existence Principle does not say how big the solution interval is (it merely asserts its existence).

### Extension of a Solution

First observe that solvers are set up to solve IVP (1) for forward time and for backward time, but not for the "two-sided" problem (1). Consider the **forward initial value problem** based on (1) and let $S$ be the closed rectangle $t_0 \leq t \leq t_0 + a$, $x_0 - b \leq x \leq x_0 + b$, where $a$ and $b$ are positive constants. See Figure 2.N.1. In the course of proving the Existence Principle it is shown that a solution of the forward IVP "lives" at least on the

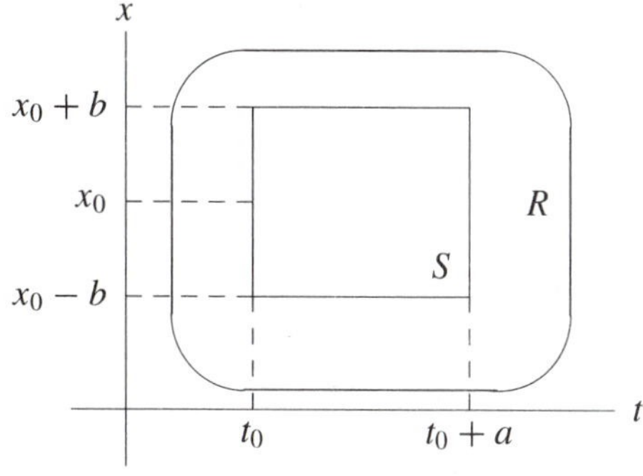


*Figure 2.N.1 Geometry for a forward initial value problem.*

interval $t_0 \le t \le t_0 + c$, where $c = \min\{a, b/M\}$ and $M > 0$ is any constant such that $|f(t,x)| \le M$ for all $(t,x)$ in $S$. A similar result holds for the backward IVP based on (1). Note that the forward and backward solutions "fit" together smoothly at $t = t_0$ to form the "two-sided" solution for IVP (1). Now suppose that IVP (1) has a solution on the closed interval $I = [c, d]$ and that the point $(d, x(d))$ lies in $R$. From the above observation, we know that IVP (1) with $t_0 = d$, $x_0 = x(d)$ has a forward solution that "lives" on an interval $d \le t \le e$, where $e$ is calculated by construction of a rectangle analogous to $S$ in Figure 2.N.1. This forward solution and the original solution "fit" together smoothly to form an **extended solution** over the interval $c \le t \le e$. Solutions can also be extended backward in a similar fashion if the point $(c, x(c))$ is in $R$. Using these techniques, solution curves over specific time intervals can be extended forward and backward in time. Now we state the

**Extension Principle:** *In IVP (1) let $f$ and $\partial f/\partial x$ be continuous in a region $R$ containing $(t_0, x_0)$. Then for any closed, bounded region $D$ in $R$ with $(t_0, x_0)$ in $D$, the forward and backward extended solution curve of IVP (1) exits $D$ through its boundary.*

***Example 2*** **(Extension interval of an IVP)** Consider the IVP

$$\begin{aligned} x' &= x^2 \\ x(0) &= 1 \end{aligned} \tag{2}$$

where $R$ is the entire $tx$-plane. Observe that IVP (2) has the maximally extended solution $x = (1-t)^{-1}$, for $t < 1$. Pretending not to know this, let us take a rectangle $S$ as in Figure 2.N.1, defined by $0 \le t \le a$, $1 - b \le x \le 1 + b$, for some positive constants $a$ and $b$. Since $|x^2| \le (1+b)^2$ for all $(t,x)$ in $S$, we may take $M = (1+b)^2$. Thus we conclude that the forward problem for (2) has a solution that lives at least on $0 \le t \le c$ where $c = \min\{a, b/(1+b)^2\}$. Now since the maximum value of $b/(1+b)^2$ for $b > 0$ is $\frac{1}{4}$ when $b = 1$, we see that the largest possible value for $c$ is also $\frac{1}{4}$. Hence, taking $a = 2$, $b = 1$ we see that the forward solution of IVP (2) is at least defined for $0 \le t \le \frac{1}{4}$. Notice that the solution curve of the forward IVP (2) exits the vertical side of the rectangle $S_1\colon 0 \le t \le \frac{1}{4}$, $0 \le x \le 2$, not through its top side. Using the point $t_0 = \frac{1}{4}$, $x_0 = \frac{4}{3}$ as a new initial point, we can extend the solution curve constructed above, that lives on $0 \le t \le \frac{1}{4}$, to a solution that lives on a larger interval. Placing the rectangle $S_2$: $\frac{1}{4} \le t \le \frac{1}{4} + a$, $|x - \frac{4}{3}| \le b$, in the standard position about the initial point $(\frac{1}{4}, \frac{4}{3})$ and noting that $|x^2| \le (\frac{4}{3} + b)^2$ for $(t,x)$ in $S_2$, we may take $M = (\frac{4}{3} + b)^2$. Thus the extended solution is at least defined on the interval $\frac{1}{4} \le t \le \frac{1}{4} + c$, where $c = \min\{a, b/(\frac{4}{3} + b)^2\}$. The maximum value that can be obtained for $c$ is when $b = \frac{4}{3}$. In that case $(\frac{4}{3})/(\frac{4}{3} + \frac{4}{3})^2 = \frac{3}{16}$, and if $a = 1$ (for instance), then $c = \frac{3}{16}$ We now have an extended solution of the forward IVP (2) that lives at least on the interval $0 \le \frac{1}{4} + \frac{3}{16}$. We could continue this process indefinitely; the process never terminates. Notice that, quite correctly, the solution cannot be extended beyond $t = 1$.

### *Sensitivity*

The following result gives an answer to the question of how the solution of IVP (1) changes as the data change.

**Sensitivity Principle:** *In IVP (1) let $f$, $\partial f/\partial x$ be continuous on the region $R$ in $tx$-space containing $(t_0, x_0)$. Then given any tolerance $E > 0$, there exist positive constants $c$, $H$, $K$ such that:*

1. *For any function $g(t,x)$ with $g$, $\partial g/\partial x$ continuous on $R$, and $|g(t,x)| \le H$ for all $(t,x)$ in $R$, and*

2. *For any $y_0$ with $|x_0 - y_0| \le K$, the following estimate holds for all $|t - t_0| < c$:*

$$|x(t) - y(t)| \le E$$

*where $x = y(t)$ is the solution of the IVP*

$$x' = f(t, x) + g(t, x)$$
$$x(t_0) = y_0$$

***Example 3*** **(Sensitivity to data changes)** For any constant $k$ the problem $x' + kx = 0$, $x(0) = x_0$, has the solution $x = x_0 e^{-kt}$ for all $t$. Denote by $x(t)$ and $\bar{x}(t)$ the respective solutions of this problem for the data pair $(k, x_0)$ and $(\bar{k}, \bar{x}_0)$. Using the Triangle Inequality we have (subtracting and adding $e^{-\bar{k}t}x_0$)

$$\begin{aligned} |x(t) - \bar{x}(t)| &= \left| e^{-kt}x_0 - e^{-\bar{k}t}x_0 + e^{-\bar{k}t}x_0 - e^{-\bar{k}t}\bar{x}_0 \right| \\ &\le |x_0| \left| e^{-kt} - e^{-\bar{k}t} \right| + e^{-\bar{k}t} |x_0 - \bar{x}_0| \end{aligned} \qquad (3)$$

To make computation a bit easier suppose that $k, \bar{k} > 0$, and that $0 \le t \le T$, for some fixed $T$. The Mean Value Theorem applied to the function $g(k) \equiv e^{-kt}$ implies that $\left| e^{-kt} - e^{-\bar{k}t} \right| \le T \left| k - \bar{k} \right|$, which when inserted into (3) yields

$$|x(t) - \bar{x}(t)| \le |x_0| \, T \left| k - \bar{k} \right| + |x_0 - \bar{x}_0|, \qquad 0 \le t \le T \qquad (4)$$

From (4) we see that the closer $\bar{k}$ and $\bar{x}_0$ are to $k$ and $x_0$, respectively, the more $\bar{x}(t)$ looks like $x(t)$ on the interval $0 \le t \le T$. Thus, small perturbations in the data (which may be due to experimental error) cause only small perturbations in the solution to the problem of this example over a finite time interval.

### *Solvability, Numerical Solutions*

Euler's method for solving IVPs is an example of a class of algorithms called **one-step methods** because $x_n$ is calculated at each step using data only from the preceding step. In general, there are three components included in a one-step method. Finding an approximation $x(t)$ of the a solution of an IVP for some $T > t_0$ requires the following tasks:

1. Partition the interval $t_0 \le t \le T$ into $N$ subintervals of equal length by the points $t_n = t_0 + nh$, $n = 1, 2, \ldots, N$, where $h = (T - t_0)/N$.
2. Choose an **approximating rate function** $A(t, x, h)$ and define the sequence

$$x_n = x_{n-1} + hA(t_{n-1}, x_{n-1}, h), \qquad n = 1, 2, \ldots, N \qquad (5)$$

3. Find a method for estimating the error $|x(T) - x_N|$ in terms of the step size $h$.

For Euler's method, note that $A(t_n, x_n, h) = f(t_n, x_n)$. One-step algorithms for which the approximating rate function uses averages of the rate function $f(t, x)$ at two or more points over the interval $t_n \le t \le t_{n+1}$ are said to be **Runge-Kutta methods**. The so-called **classical fourth order Runge-Kutta method** is given by

$$A(t_n, x_n, h) = \tfrac{1}{6}(k_1 + 2k_2 + 2k_3 + k_4)$$

where

$$\begin{aligned} k_1 &= f(t_n, x_n) \\ k_2 &= f(t_n + h/2, x_n + k_1 h/2) \\ k_3 &= f(t_n + h/2, x_n + k_2 h/2) \\ k_4 &= f(t_{n+1}, x_n + k_3 h) \end{aligned}$$

### *Errors*

A major source of errors in calculating numerical approximations to the solution of IVP (1) lies in function evaluations and the chopping of long decimal strings. These **round-off errors** are individually small, but they can be devastating in their cumulative effect.

A quite different kind of error arises from the algorithm (5) itself. First of all, for each $n = 1, 2, \ldots, N$ there is the **local discretization error** (or **formula error** or **truncation error**),

$$e_n = |y_n(t_n) - x_n|, \qquad n = 1, 2, \ldots, N$$

where $x_n$ is generated by the algorithm (5) and $y_n(t)$ is the solution of the IVP

$$\begin{aligned} y' &= f(t, y) \\ y(t_{n-1}) &= x_{n-1} \end{aligned}$$

The number $e_n$ measures how much $x_n$ deviates from the true solution of IVP (1) in the hypothetical case that $x(t_{n-1})$ happens to be exactly the number $x_{n-1}$. The **global** (or **accumulated**) **discretization error at step** $n$ is

$$E_n = |x(t_n) - x_n|, \qquad n = 1, 2, \ldots, N$$

where $x(t)$ solves IVP (1) and $x_n$ is generated by the algorithm (5). The error $E_N = |x(T) - x_N|$ is the most important since our goal is to approximate $x(T)$.

The approximate rate function $A(t, x, h)$ of a one-step method involves more and more function evaluations as the discretization error of the method diminishes. Hence the better the approximation of the method at each step becomes, the more function evaluations and arithmetic operations to perform, and thus, the more opportunities for round-off errors. Thus designing an algorithm for solving IVP (1) that is extremely accurate *and* very efficient is not a trivial task.

The analysis of discretization and round-off errors is central if one is to have any confidence that an approximate solution to an IVP is reasonably close to the exact solution. In recent years this analysis has focused on the question of determining the best all-around numerical algorithms for general computer solution of IVPs. There are two such algorithms in wide use: the fourth order Runge-Kutta method and the Adams-Bashforth-Moulton multistep method (not described here). Users of either method may be reasonably confident that solutions of most IVPs will be reasonably accurate. However, no method will always work well for all problems.

### *Some General Advice*

The user is well advised to have done some preliminary "pencil-and-paper" work on an IVP before using a computer. One should know what to expect, as well as which answers are plausible and which are not. Such knowledge is often related to whether the right questions have been asked. For example, it would be meaningless to ask the computer to estimate $x(1)$ if $x' = x^2$, $x(0) = 1$, since the exact solution is $x = (1 - t)^{-1}$. Expecting to find $x(T)$ for $T > 1$ would not be meaningful either. Thus, it may not always be possible to spot potential difficulties in using a numerical algorithm solely from the form of the IVP. Discretion, common sense, and preliminary study, as well as skepticism about the numerical results, are always well advised.

CHAPTER 3

# First Order Equations

In the three hundred years of study of differential equations, many techniques have been invented to solve IVPs for the ODE $x' = f(t, x)$ if the rate function $f$ has a special form. This chapter explores how some of these techniques can be enhanced with numerical solutions and graphics.

## *Properties of Solutions*

A **first order IVP in normal form** may be written as

$$x' = f(t, x), \quad x(t_0) = x_0 \tag{1}$$

If $f$ and $\partial f/\partial x$ are continuous in a region $R$ in the $tx$-plane containing $(t_0, x_0)$, then (1) has at most one solution on any $t$-interval $I$ containing $t_0$. Moreover, (1) does have a solution on some interval $I$ and this solution can be extended both forward and backward in time, ultimately exiting any closed bounded region $D$ in $R$, $D$ containing $(t_0, x_0)$. Thus, solutions of (1) cannot "disappear" inside $R$. The region $R$ may be pictured as being filled with the solution curves of $x' = f$, curves that rise and fall, but never turn back, never touch, and never just "end." Figure 3.1 portrays a solution curve for a linear first order ODE. Chapters 1–2 have set the vocabulary for discussing (1) and its solution, have taken up computing and numerical techniques, and have pointed out computing pitfalls.

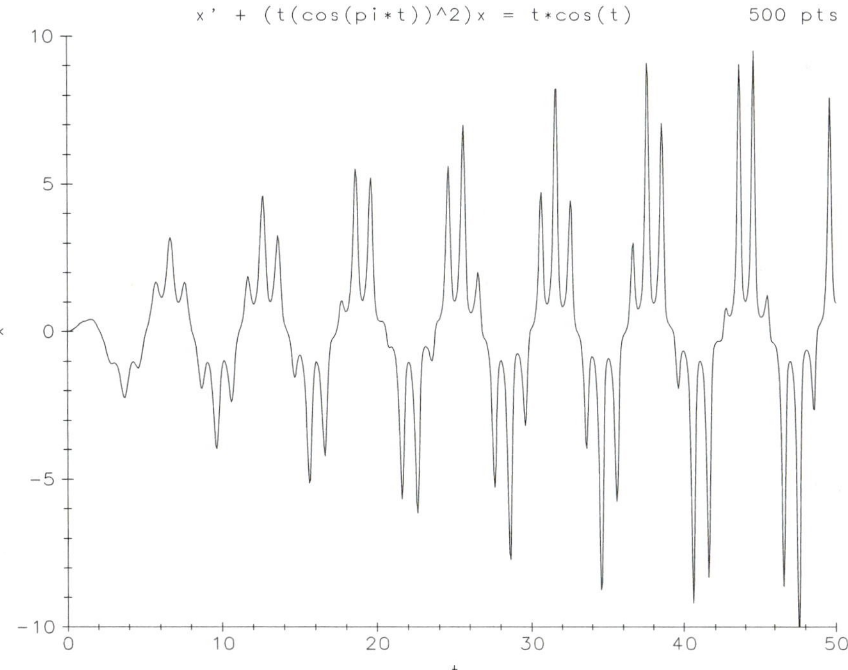


*Figure 3.1 The solution curve of a first order linear IVP with $x(0) = 0$.*

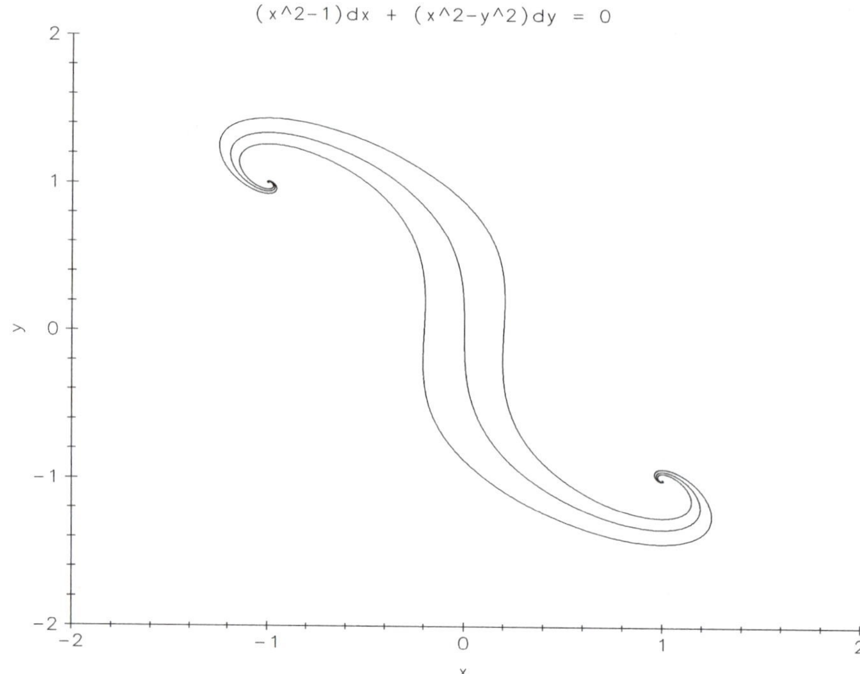


*Figure 3.2* *Three integral curves of a nonlinear first order ODE in differential form.*

## *Technique Based on Planar Autonomous Systems*

Solvers/graphers present a useful technique for finding approximate solutions to

$$N(x, y)\frac{dy}{dx} + M(x, y) = 0 \tag{2}$$

by considering the orbits of the "equivalent" autonomous planar system

$$\frac{dx}{dt} = -N(x, y), \qquad \frac{dy}{dt} = M(x, y) \tag{3}$$

The orbits of the system (3) are called **integral curves** for the ODE (2). **Solution curves** of ODE (2) are just certain connected "pieces" of these integral curves described as follows. Let $x = x(t)$, $y = y(t)$ be a portion $\Gamma$ of an orbit of system (3) over a $t$-interval on which $x'$ never vanishes. Then $\Gamma$ can be parameterized by $x$ instead of $t$ by solving the equation $x = x(t)$ for $t$ in terms of $x$ to obtain $t = t(x)$ and then replacing $t$ in $y(t)$ by $t(x)$. Thus $y$ is a function of $x$ through $y = y(t(x))$. Using the Chain Rule, $dy/dx = (dy/dt)(dt/dx) = (dy/dt)/(dx/dt) = -M(x, y)/N(x, y)$. Thus, $\Gamma$ is a solution curve for ODE (2). Note that the ODE (2) is often written in the differential form $M\,dx + N\,dy = 0$.

***Example 1*** **(Solution curves via integral curves)** The three integral curves in Figure 3.2 are orbits of the planar autonomous system

$$x' = -x^2 + y^2, \qquad y' = x^2 - 1$$

Thus, solution curves of $(x^2 - y^2)(dy/dx) + x^2 - 1 = 0$ are just "pieces" of these integral curves that have no vertical tangents. This approach shows how far a solution curve can be extended. Note that one integral curve may be "composed" of many solution curves.

# Linear First Order ODEs: Properties of Solutions

***Purpose*** To study the behavior of solutions of first order linear ODEs, to see how the solutions respond to inputs, and to determine what happens in the neighborhood of a singularity.

***Keywords*** First order linear ODE, superposition, singularity, input, driving term, output, response

***See also*** Figure 3.1; Atlas Plates `First Order D-F`

***Background*** A **linear first order IVP** has the form

$$\begin{aligned} a(t)x' + b(t)x &= c(t) \\ x(t_0) &= x_0 \end{aligned} \tag{1}$$

where the functions $a(t)$, $b(t)$, and $c(t)$ are continuous on a $t$-interval $I$ that contains $t_0$. If $a(t)$ does not vanish anywhere in $I$, the ODE in (1) may be divided by $a(t)$ to obtain the **normal linear first order IVP**

$$\begin{aligned} x' + p(t)x &= f(t) \\ x(t_0) &= x_0 \end{aligned} \tag{2}$$

The **integrating factor technique** may be used to construct all solutions of the ODE in (2):

- Find the **integrating factor** $\mu = e^{P(t)}$, where $P(t)$ is any antiderivative of $p(t)$.
- Multiply each side of the ODE in (2) by $\mu$ to obtain $\mu(x' + px) \equiv (\mu x)' = \mu f$.
- Antidifferentiate each side of the above ODE and divide by $\mu$, to obtain

$$x(t) = e^{-P(t)}C + e^{-P(t)} \int^t e^{P(s)} f(s)\,ds \tag{3}$$

where $C$ is any constant.

If the specific antiderivative $P_0(t) = \int_{t_0}^t p(s)\,ds$ is used, then from (3) we see that the solution of IVP (2) is the sum of two terms:

- The **zero-input response**: $e^{-P_0(t)}x_0$
- The **zero-initial-data response**: $e^{-P_0(t)} \int_{t_0}^t e^{P_0(s)} f(s)\,ds$

This may be formulated as a basic principle:

**Superposition Principle I:** *The solution of IVP (2) is the superposition (sum) of the solution of the undriven IVP,*

$$x' + p(t)x = 0, \quad x(t_0) = x_0 \qquad \textbf{(zero-input response)}$$

*and the solution of the driven ODE with zero initial data,*

$$x' + p(t)x = f(t), \quad x(t_0) = 0 \qquad \textbf{(zero-initial-data response)}$$

This principle is a particular instance of another more general principle:

**Superposition Principle II:** *Suppose that $x_i(t)$, $i = 1, 2$, are the respective solutions of the IVPs $x_i' + p(t)x_i = f_i(t)$, $x_i(t_0) = \alpha_i$, $t$ in $I$. Then $x = C_1x_1 + C_2x_2$ solves the IVP $x' + p(t)x = C_1f_1 + C_2f_2$, $x(t_0) = C_1\alpha_1 + C_2\alpha_2$, $t$ in $I$, for any constants $C_1$ and $C_2$.*

Thus, an IVP with input or initial data that is a sum of other inputs or initial data can be solved piecemeal and the respective solutions superposed (see Figure 3.1.1).

The integrating factor technique produces solution formulas. Computer solvers and graphics allow the user to find and graph the solution of IVP (2) without having to carry

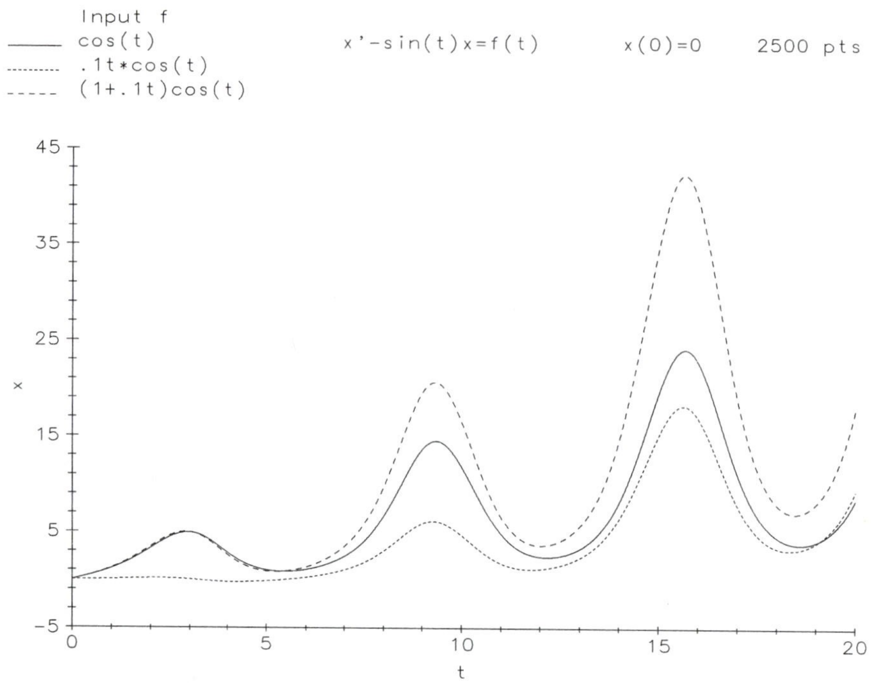

*Figure 3.1.1 Superposition of inputs implies superposition of outputs.*

out the explicit evaluation of the two integrals required to implement solution formula (3). See Atlas Plate `First Order D` for a portrait of the solution curves of a particular IVP.

***Observation*** **(Singularities)** IVP (1) is not in normal form, but most of the discussion above assumes the normalized form (2). Now suppose we want to solve (1) over a time domain where normalization is not possible (e.g., over a region in which $a(t)$ vanishes at least once). A number $t_1$ such that $a(t_1) = 0$ is a **singularity** of (1). Assuming that it is not removable (i.e., that $b(t)/a(t)$ and $c(t)/a(t)$ cannot be defined to be continuous through $t_1$), then the Existence and Uniqueness Principles cannot be applied to IVP (2). Just how solutions of an ODE behave as they approach a singularity depends upon the nature of the coefficients of the ODE. This topic is taken up in Experiment 3.2. Atlas Plates `First Order E, F` show two kinds of behavior near singularities.

***Observation*** **(Piecewise continuity)** The methods for solving IVP (2) apply even if $p(t)$ and $f(t)$ are only piecewise continuous. A function is **piecewise continuous** if it is continuous on each bounded interval of its domain except, possibly, at a finite number of points at which the function has finite jumps.

_______________/_______________ hardware software

_______________/_______________ name date

Answer questions in the space provided, or on attached sheets or carefully labeled graphs.

_______________/_______________ course section

## *3.1 Superposition*

***Abstract*** The superposition properties of solutions of first order linear ODEs are studied.

**1.** Select one of the IVPs below and solve the following problems.

☐ **A.** $x' + (1 + t + t^2)x = f(t), x(0) = x_0$

☐ **B.** $x' + (t \sin t)x = f(t), x(0) = x_0$

☐ **C.** $x' + x = f(t), x(0) = x_0$

☐ **D.** $x' + x/(1 + t^2) = f(t), x(0) = x_0$

**(a)** Set $f(t) = 0$ and solve the undriven IVP with $x_0 = 1$ over the interval $0 \le t \le 10$.

**(b)** Set $f(t) = 3 \cos t^2$ and solve the driven IVP with $x_0 = 0$.

**(c)** Now solve the full IVP with $f(t) = 3 \cos t^2$ and $x_0 = 1$. If possible, overlay in a single graph the graphs of the solutions in (a) and (b) and verify that the solution to your IVP is the superposition of those of (a) and (b). Otherwise, hold the three graphs up to a strong light and observe the superposition, or else use a ruler and measure the distances on the graphs from the $t$-axis.

**2.** Consider the IVP $(1 + t^2)x' + 2tx = e^{-t}, x(0) = x_0$.

**(a)** Normalize the ODE, and use an integrating factor to solve.

**(b)** Use a solver and plot the graphs of solutions for various values of $x_0$, $|x_0| \le 10$ in the rectangle $|t| \le 10$, $|x| \le 10$.

**3.** Plot the graphs for each of the following IVPs for various values of $x_0$ on as large a time interval as possible.

**(a)** $tx' + 2x = \sin t, t > 0, x(1) = x_0$

**(b)** $x' + 2t^{-1}x = t^{-2} \cos t, t > 0, x(\pi) = x_0$

**(c)** $tx' + 2x = t^2 + t - 2, x(1) = x_0$

_______________ / _______________ 
hardware software

_______________ / _______________ 
name date

Answer questions in the space provided,
or on attached sheets or carefully labeled graphs.

_______________ / _______________ 
course section

## *3.2 Singularities*

***Abstract*** The varieties of solution behavior near a singularity of a first order linear ODE are studied.

**1.** The ODE $tx' + 2x = 4t^2$ has a singularity at $t = 0$.

**(a)** Normalize the ODE and use the integrating factor technique to find a formula for all solutions, $t > 0$. Repeat for $t < 0$. Explain why there is exactly one solution for which $x(0) = 0$, but there are no solutions at all if $x(0) = \alpha \neq 0$. Explain why the Existence and Uniqueness Principles do not apply on any interval containing $t = 0$.

**(b)** Sketch the graphs of the solutions for several initial conditions $x(t_0) = x_0$. Sketch the graph of the unique solution passing through the origin. What happens to all other solutions as $t \to 0$?

**2.** Find all the solutions of the ODEs in Atlas Plates `First Order E, F` by normalizing and constructing integrating factors. For which values of $t$ does the ODE have a singularity? Explain. Duplicate the graphs of the two plates.

**3.** Select one of the IVPs that follow. For which values of $t$ does the ODE have a singularity? Explain. Solve explicitly on a $t$-interval containing a singularity if possible; if not possible, explain why not. Graph solutions. In (B)–(C) use several initial conditions.

☐ **A.** $x' + (\cos^2 t)x = 2\csc^2 t,\ x(\pi/2) = 1$

☐ **B.** $t^2x' + 2tx = 1,\ x(t_0) = x_0,\ t_0 \neq 0$

☐ **C.** $tx' - x = t^2,\ x(t_0) = x_0,\ t_0 \neq 0$

☐ **D.** $tx' + 2tx = \sin t,\ x(0) = 1$

# Linear First Order ODEs: Data

***Purpose*** To show how the solution of a first order linear IVP changes if the data (initial conditions, input) are changed, and to study the conditions for bounded output.

***Keywords*** Initial data, input, driving force, response, output, sensitivity, bounded functions

***See also*** Experiments 3.1–3.2; Atlas Plate `First Order D`

***Background*** The solution of the normal first order linear IVP

$$\begin{aligned} x' + p(t)x &= f(t) \\ x(t_0) &= x_0 \end{aligned} \tag{1}$$

is given by

$$x(t) = e^{-P_0(t)}x_0 + e^{-P_0(t)} \int_{t_0}^{t} e^{P_0(s)} f(s)\, ds \tag{2}$$

where $P_0(t) = \int_{t_0}^{t} p(s)\, ds$ and $p(t)$ and $f(t)$ are assumed to be at least piecewise continuous on a $t$-interval containing $t_0$. The solution $x(t)$ of (1) is the **output**. The **data of the problem** are the numbers $t_0$ and $x_0$ of the initial condition, and the **input** or **driving function** $f(t)$. There are other data elements, of course: for example, $p(t)$, the time span of the solution, the number of points plotted on the screen, and so on. For our purposes, however, the data consist of the initial data set and the input.

How sensitive is the solution (2) of IVP (1) to changes in the data? This is a question of considerable importance in the design of a stable operating system. It is critical that the small changes and inevitable uncertainties in data not lead to unmanageable changes in output. Fortunately, there is a crude estimate on the magnitude of changes in the output in response to changes in the input. First note that a function $g(t)$ is bounded on an interval $I$ if there is a constant $M$ such that $|g(t)| \leq M$ for all $t$ in $I$.

**Bounded Input/Bounded Output Principle:** *Suppose that there are positive constants $p_0$ and $M$ such that for all $t$ in $I$, $p(t) \geq p_0$ and $|f(t)| \leq M$. Then for all $t$ in $I$ with $t \geq t_0$, the solution $x(t)$ of (1) satisfies the estimate*

$$|x(t)| \leq e^{-p_0(t-t_0)}|x_0| + \frac{M}{p_0}\left(1 - e^{-p_0(t-t_0)}\right) \leq \max\left(|x_0|, \frac{M}{p_0}\right) \tag{3}$$

One immediate consequence of (3) is the following result: Suppose that $x(t)$ solves the IVP (1) and that $y(t)$ solves the IVP

$$\begin{aligned} y' + p(t)y &= g(t) \\ y(t_0) &= y_0 \end{aligned}$$

where the input $g(t)$ is at least piecewise continuous on the same $t$-interval $I$ as $f(t)$. Then for $t$ in $I$, $t \geq t_0$, we have

$$|x(t) - y(t)| \leq e^{-p_0(t-t_0)}\,|x_0 - y_0| + \frac{M^*}{p_0}\left(1 - e^{-p_0(t-t_0)}\right) \tag{4}$$

if $p(t) \geq p_0 > 0$ and $|f(t) - g(t)| \leq M^*$ for all $t$ in $I$.

The estimates of (3) and (4) are rough and for many specific systems they can be improved considerably. Using computer solvers it is often easier to plot the solutions directly and use your eyes to detect the changes in the solutions as the data are changed.

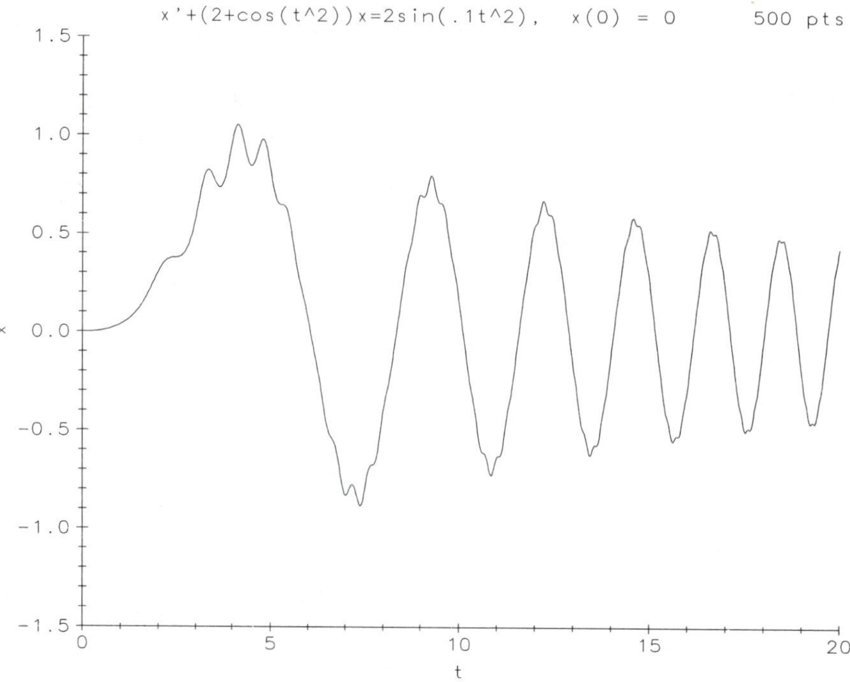


*Figure 3.3.1 Graph of bounded output.*

***Example 1*** **(Input/output bounds)** Consider the IVP

$$\begin{aligned} x' + (2 + \cos(t^2))x &= 2\sin(0.1t^2) \\ x(0) &= x_0 \end{aligned} \qquad t \geq 0$$

with input $f(t) = 2\sin(0.1t^2)$. There is no possibility of applying formula (2) to express the solution explicitly in terms of familiar functions. However, we have that for $t \geq 0$

$$p(t) = 2 + \cos(t^2) \geq 1$$

$$\left|2\sin(0.1t^2)\right| \leq 2$$

Hence, with $p_0 = 1$ and $M = 2$ in (3) we have

$$|x(t)| \leq e^{-t}\,|x_0| + 2(1 - e^{-p_0 t}) \leq \max(|x_0|, 2)$$

See Figure 3.3.1 for the graph of the output with the initial condition $x(0) = 0$. Observe that $|x(t)| \leq 2$ as the inequality above requires.

***Observation*** **(Sensitivity)** The coefficient $p(t)$ could be changed as well, and the effect of the change on the solution calculated or computed. In the Chapter 2 Notes see the Sensitivity Principle and Example 3.

_______________/_______________ hardware / software

_______________/_______________ name / date

Answer questions in the space provided, or on attached sheets or carefully labeled graphs.

_______________/_______________ course / section

## *3.3 Dependence on Data*

***Abstract*** The effects of changes in the data on the solution of an IVP are studied.

**1.** Consider the IVP $x' + kx = 2\cos 3t$, $x(0) = 1$, $0 \leq t \leq 10$.

**(a)** Graph the solution of the IVP for $k = -2, -1, 0, 1, 2$. Describe what you see and make a conjecture about the behavior of the solution as a function of $k$ as $k$ increases.

**(b)** Prove or disprove your conjecture by finding an explicit formula for the solution of the IVP and determining what happens as $k$ increases. (Hint: Use the integrating factor $e^{kt}$ to solve the ODE.)

**2.** Consider the IVP $x' + (\cos t)x = 2\cos(\omega t)$, $x(0) = 1$. Graph the solution of this IVP for $0 \leq t \leq 50$ and $\omega = 1, 2, 3, \pi$. Describe what you see, and make a conjecture about the way the solution changes as $\omega$ increases.

* **3.** Find continuous functions $p(t)$ and $f(t)$ such that the solution $x(t)$ of the IVP $x' + p(t)x = f(t)$, $x(0) = 1$, $0 \leq t \leq 20$, has the property indicated below. Graph that solution.

**(a)** $x(t)$ is periodic with period $2\pi$ and attracts nearby solutions.

**(b)** $x(t)$ is periodic with period $2\pi$ and repels nearby solutions.

**(c)** $x(t)$ grows faster than any exponential of the form $e^{at}$, $a > 0$.

## 3.4 *Bounded Input/Bounded Output*

***Abstract*** Examples of the Bounded Input/Bounded Output Principle are examined. The hypotheses of that principle are shown to be sharp.

**1.** Duplicate Atlas Plate `First Order D`. Explain why the Bounded Input/Bounded Output Principle does *not* apply. Are all solutions of the ODE bounded as $t \to \infty$? Explain.

**2.** Select one of the IVPs below and do the following problems.

☐ **A.** $(1+t)x' + (1+2t)x = 1+t,\ t \geq 0,\ x(0) = x_0$

☐ **B.** $(1+t^2)x' + t^2 x = (1+t^2)\sin(e^t),\ t \geq 1,\ x(1) = x_0$

☐ **C.** $\tan(t)x' + x = \sin(t^2+1),\ \pi/4 \leq t \leq \pi/3,\ x(\pi/4) = x_0$

**(a)** The ODE can be written in the form $x' + p(t)x = f(t)$. Find the largest $p_0$ and the smallest positive $M$ such that for all $t$ in the given interval $p(t) \geq p_0$ and $|f(t)| \leq M$. Then use (3) to find an upper bound for $|x(t)|$.

**(b)** For various values of $x_0$ plot the solution $x(t)$ of the IVP over a sufficiently long $t$-interval that the long-term solution can be described. Verify from the graphs that the magnitude of the solutions remains less that the upper bound found in (a). Write on the graph the values of max $|x(t)|$ over the solution $t$-range. If possible, plot both the input $f(t)$ and the output $x(t)$ on the same graph.

**3.** The following problems show that the hypotheses $p(t) \geq p_0 > 0$ and $|f(t)| \leq M$ cannot be weakened in the Bounded Input/Bounded Output Principle.

**(a)** Show that the ODE $x' + t^{-2}x = e^{1/t},\ t \geq 1$, has unbounded solutions. (Hint: Show that $te^{1/t}$ is a solution.) Graph solutions for several values of $x(1)$. Solve over the $t$-range $1 \leq t \leq 10$. Which hypothesis of the principle is not satisfied?

**(b)** Show that the ODE $x' + x = t$ has unbounded solutions on the interval $t \geq 0$. Graph several solutions. Which hypothesis of the principle is not satisfied?

# Separable ODEs: Implicit Solutions

***Purpose*** To study the behavior of solution curves of separable first order ODEs.

***Keywords*** Separable ODEs, implicit solutions, integral curves

***See also*** Introductory text of Chapter 3; Experiment 3.7

***Background*** **Separable** first order ODEs may be written in any of the forms:

$$M(x) + N(y)dy/dx = 0 \tag{1}$$

$$M(x)\,dx + N(y)\,dy = 0 \tag{2}$$

$$dy/dx = -M(x)/N(y) \tag{3}$$

$$dx/dy = -N(y)/M(x) \tag{4}$$

where $M(x)$ and $N(y)$ are continuously differentiable functions. A separable ODE may be solved by "separating" the variables to write the ODE in the form (2) and antidifferentiating with respect to each variable separately. Indeed, note that if $F(x)$ and $G(y)$ denote antiderivatives of $M(x)$ and $N(y)$ respectively, then any solution $y(x)$ of ODE (1) must satisfy the condition that $(d/dx)(F(x) + G(y(x)) = 0$ for all $x$ values where $y(x)$ is defined. Thus, the Mean Value Theorem implies that the solution $y(x)$ is defined implicitly by the relation $F(x) + G(y) = C$, where $C$ is a constant. Conversely, for any constant $C$, a differentiable function $y(x)$ that satisfies the relation $F(x) + G(y(x)) = C$ must be a solution of ODE (1). Thus, the relation $F(x) + G(y) = C$ is said to define solutions of ODE (1) **implicitly**. The solution curve passing through $(x_0, y_0)$ is defined implicitly by $F(x) + G(y) = F(x_0) + G(y_0)$. The function $F(x) + G(y)$ is an **integral** of the ODE, and the collection of points $(x, y)$ for which $F(x) + G(y) = F(x_0) + G(y_0)$ defines the **integral curve** through $(x_0, y_0)$. Solving the equation $F(x)+G(y) = F(x_0)+G(y_0)$ for $y$ in terms of $x$ on some interval then gives the solution of the separable ODE (1) that satisfies the condition $y(x_0) = y_0$.

***Example 1*** **(Explicitly defined solution curves)** The first order ODE

$$\frac{dy}{dx} = (1 + y^2)(1 + \sin x) \tag{5}$$

separates to

$$\frac{1}{1 + y^2}dy = (1 + \sin x)dx$$

and integrates to

$$\arctan y = x - \cos x + C$$

where $C$ is a constant. Each value of $C$ determines an integral curve of (5). If a solution curve through a particular point in the $xy$-plane is wanted, then $C$ is chosen accordingly. For example, if the curve is to pass through the origin, then $C$ must be chosen so that $\arctan 0 = 0 - \cos 0 + C$ (i.e., $C = 1$). The solution curve desired is thus defined by

$$y = \tan(x - \cos x + 1)$$

which is explicit, but not exactly easy to analyze.

***Example 2*** **(Implicitly defined solution curves)** The first order ODE

$$(4x^3 + 1)dx + (1 + 2\sin y)dy = 0 \tag{6}$$

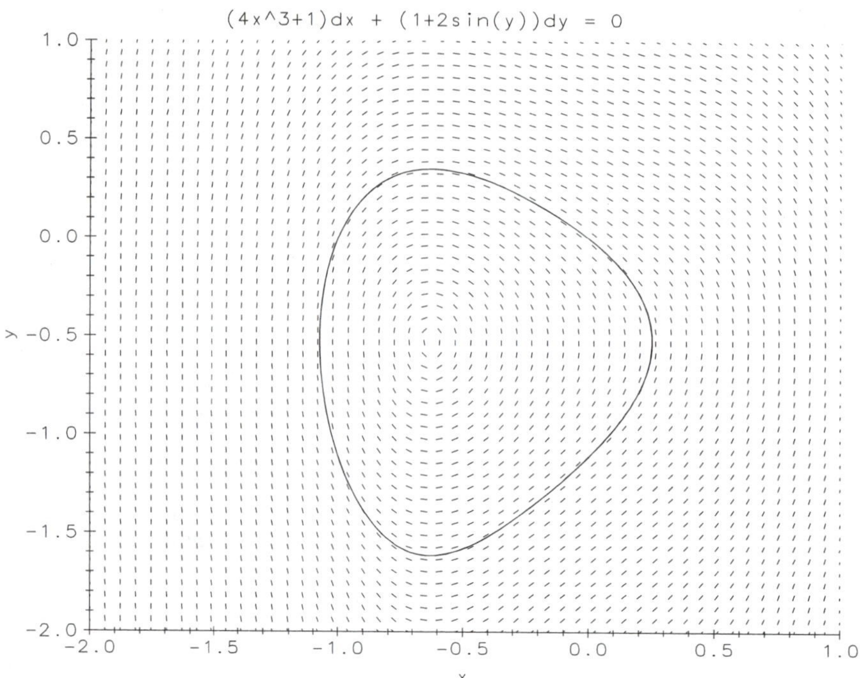


*Figure 3.5.1 An integral curve passing through the origin.*

is in separated form. Integrating, we obtain

$$x^4 + x + y - 2\cos y = C$$

where $C$ is a constant. The integral curve through the origin must satisfy the condition $-2\cos(0) = C$; hence, $C = -2$ and the desired integral curve is defined implicitly by

$$x^4 + x + y - 2\cos y = -2 \qquad (7)$$

Figure 3.5.1 shows this integral curve. Equation (7) cannot be explicitly solved for $y$ in terms of $x$ but the solution curve can easily be identified from the integral curve in Figure 3.5.1. "Cut" the integral curve at the two points where the tangent line to the integral curve is vertical. The upper and lower arcs thus formed are both solution curves for the ODE (6), but only the upper arc satisfies the initial condition.

***Observation*** **(Solution curves via integral curves)** A problem that could arise in dealing with (3) or (4) directly is that there may be singularities (for example, (3) has problems at the roots of $N(y)$). Problems with singularities can be avoided by considering the autonomous system of first order ODEs

$$\begin{aligned} dx/dt &= -N(y) \\ dy/dt &= M(x) \end{aligned} \qquad (8)$$

If $x(t)$, $y(t)$ solves the system (8), then dividing the second ODE by the first we see that if $x = x(t)$ is solved for $t$ in terms of $x$ to obtain $t = t(x)$ and substituted for $t$ in $y(t)$ to obtain $y = y(x) = y(t(x))$ then $y(x)$ solves ODE (3). In other words, the orbit $x = x(t)$, $y = y(t)$ of the planar system (8) defines parametrically a solution curve for the ODE (3) if the orbit has no vertical tangents. The advantage of using (8) is that the graph of a parametrically defined curve may double back on itself, have a vertical tangent, and in general behave in a more versatile way than solution curves of (3). It is not hard to see that the orbits of the

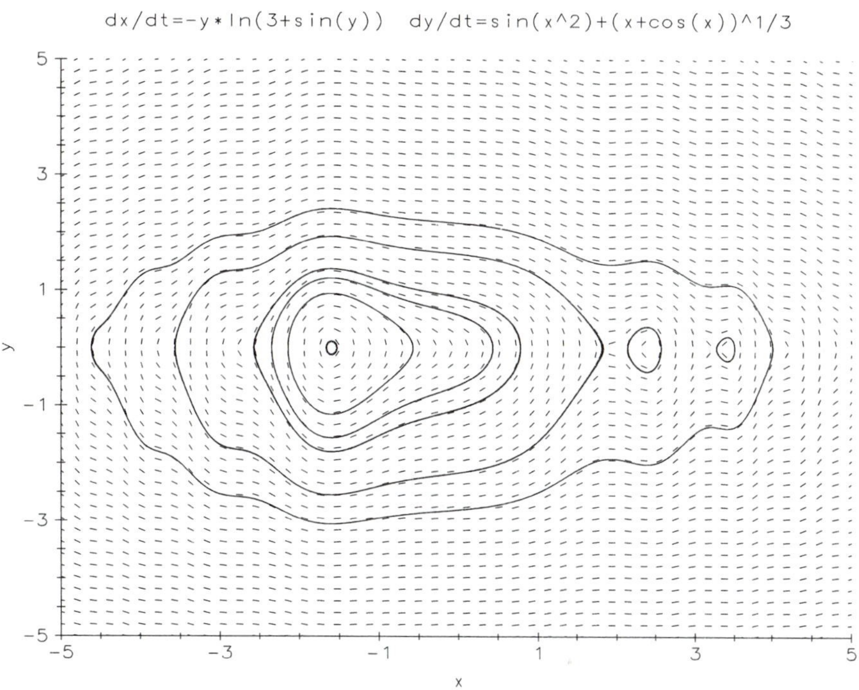


*Figure 3.5.2 Integral curves for* $[\sin(x^2) + (x + \cos x)^{1/3}]dx + [y \ln(3 + \sin y)]dy = 0$.

planar system (8) are integral curves of the ODE (1). Solvers can handle a system like (8) very easily. One simply identifies the functions $M(x)$ and $N(y)$, initial data $x_0$ and $y_0$, and a time interval $t_0 \le t \le t_1$ for solving.

***Example 3*** **(Example 2 revisited)** Replace ODE (6) in Example 2 with the system

$$\begin{aligned} dx/dt &= -1 - 2\sin y \\ dy/dt &= 4x^3 + 1 \end{aligned}$$

Imposing the initial data $x(0) = 0$, $y(0) = 0$ and solving this planar autonomous system with a solver/grapher results in the integral curve in the $xy$-plane displayed in Figure 3.5.1.

***Example 4*** **(When there are no solution formulas)** The ODE

$$\left[\sin(x^2) + (x + \cos x)^{1/3}\right] dx + [y \ln(3 + \sin y)]\, dy = 0$$

is in separated form. In theory, each term can be integrated. However, the antiderivatives of the bracketed terms are not expressible in terms of currently known functions. Figure 3.5.2 displays several of the integral curves for this ODE. Note that the system form (8) of this ODE appears in Figure 3.5.2.

***Observation*** The examples treated above suggest that computer solvers and graphers may be needed to determine the geometric nature of integral curves of separable equations even if some or all of the terms can be integrated in terms of elementary functions.

_______________/_______________
hardware software

_______________/_______________
name date

Answer questions in the space provided,
or on attached sheets or carefully labeled graphs.

_______________/_______________
course section

## 3.5 *Separable ODEs: Implicit Solutions*

***Abstract*** Integral curves of separable equations are graphed.

**1.** Duplicate Figure 3.5.1 and 3.5.2. Add other integral curves. Construct the integral curve through $(0, -1)$. With a highlighter, identify the solution curve of the ODE through $(0, -1)$ that is defined over the largest possible $x$-interval.

**2.** Use a solver/grapher to plot a solution curve of the IVP

$$(1 + 2\sin y)\frac{dy}{dx} = -4x^3 - 1, \qquad y(0) = 0$$

Use your solver/grapher to extend the solution curve of this IVP to the largest possible $x$-interval. (Hint: See Figure 3.5.1.)

**3.** Duplicate Atlas Plate `Planar System D`, and draw arrowheads to show the direction of increasing time. Write out the corresponding first order ODE in the form $M\,dx + N\,dy = 0$, antidifferentiate, and find the implicit formula for the curves viewed in the Altas Plate. From the formula, find the four special solution curves each consisting of a single point. Which curves are simple and closed? How can you use the formula to distinguish the unbounded from the bounded integral curves?

**4.** Select one of the separable equations below and perform the tasks that follow.

- ☐ **A.** $\tan^2 y\,dy = \sin^3 x\,dx$
- ☐ **B.** $dy/dx = (\csc y)(4 - x^2)$
- ☐ **C.** $dy/dx = (1 + 2x + x^2)/(1 + y^3)$
- ☐ **D.** $y^2 dy/dx = x^2 \sin^2 x$
- ☐ **E.** $\sec(x^2)\sin(1 - e^y)(dy/dx) = 1$

**(a)** Convert the ODE to a system of the form (8) and graph several integral curves on the screen $|x| \leq 3$, $|y| \leq 3$. Mark the integral curve through the point $x = 1$, $y = 1$. Extend that integral curve as far as you can.

**(b)** Find all special orbits, that is, point orbits where both $M(x)$ and $N(y)$ vanish. Focus on one of these points and graph as many of the nearby integral curves as possible.

# Nonlinear ODEs: Homogeneous Functions

***Purpose*** To show how a change of variables transforms a first order ODE involving a function that is homogeneous of order 0 into a separable ODE and to observe the geometric similarities in the family of solution curves.

***Keywords*** Homogeneous functions and their orders, homothety

***See also*** Chapter 3 Introduction; Experiment 3.7

***Background*** A function $f(x, y)$ is **homogeneous of order** $K$ if for every positive constant $c$, $f(cx, cy) = c^K f(x, y)$ for all $x$ and $y$. For example, $x^2 + 3xy$ is homogeneous of order 2 since $(cx)^2 + 3(cx)(cy) = c^2(x^2+3xy)$; $x^2+y$ is not homogeneous since $(cx)^2+cy \neq c^K(x^2+y)$ for any number $K$; $(y^2/x^2)\sin(y/x)$ is homogeneous of order 0 since $[(cy)^2/(cx)^2]\sin(cy/(cx)) = (y^2/x^2)\sin(y/x)$. Consider the ODE

$$\frac{dy}{dx} = f(x, y) \tag{1}$$

where $f$ is homogeneous of order 0. Introduce the new variable $v = y/x$. Differentiating with respect to $x$ we have

$$\frac{dv}{dx} = \frac{1}{x}\frac{dy}{dx} - \frac{y}{x^2} = \frac{1}{x}f(x, xv) - \frac{v}{x} = \frac{1}{x}(f(1, v) - v) \tag{2}$$

where the $0^{\text{th}}$ order homogeneity of $f$ has been used. Equation (2) is separable; integration yields $\ln|x| = g(v) + C_1$, where $g(v)$ is any antiderivative of $(f(1, v) - v)^{-1}$ and $C_1$ is any constant. Exponentiating and removing the absolute value, we have that

$$x = Ce^{g(v)} = Ce^{g(y/x)} \tag{3}$$

where $C$ is any constant. Thus, solutions of (1) satisfy (3). Multiplying each side of (3) by any nonzero constant $M$, we have

$$Mx = MCe^{g(y/x)} = MCe^{g(My/Mx)}$$

Thus, if the point $(x, y)$ traces out a solution curve of (1) the point$(Mx, My)$ also traces out a solution curve. This is the **homothetic property**. Homothety implies that the uniform contraction ($0 < M < 1$), dilation ($M > 1$), or reflection ($M = -1$) of a solution curve of (1) along radial lines through the origin generates another solution curve. Figure 3.6.1 shows homothetic solution curves.

***Observation*** **(Solution curves via integral curves)** Consider the ODE

$$\frac{dy}{dx} = \frac{y^2 + 2xy}{2x^2 + 2xy}$$

To find solutions that pass close to points where the denominator vanishes, it is simpler to find the integral curves of this ODE by plotting the orbits of the equivalent system (see Figure 3.6.1)

$$\frac{dx}{dt} = 2x^2 + 2xy$$
$$\frac{dy}{dt} = y^2 + 2xy$$

See Experiment 3.5 or 3.7 for a discussion of this approach to solving first order ODEs.

Answer questions in the space provided,
or on attached sheets or carefully labeled graphs.

## *3.6 Nonlinear ODEs: Homogeneous Functions*

***Abstract*** The ODE $dy/dx = f$, where $f$ is homogeneous of order 0, is solved as an equivalent planar system.

**1.** Duplicate Figure 3.6.1. Find $f$ and show that it is homogeneous of order 0.

**2.** Select an ODE listed below and do the following problems.

☐ **A.** $\dfrac{dy}{dx} = \dfrac{y^2 + 2xy}{x^2}$

☐ **B.** $\dfrac{dy}{dx} = \dfrac{-4x - 3y}{2x + y}$

☐ **C.** $\dfrac{dy}{dx} = \dfrac{4x^2 + 7xy + 2y^2}{x^2}$

☐ **D.** $\dfrac{dy}{dx} = \dfrac{8xy^3}{-3x^4 + 6x^2y^2 + y^4}$

**(a)** Verify that the function on the right is homogeneous of order 0.

**(b)** Graph orbits in the $xy$-plane. Observe the homothetic property by graphing the orbit through a point $(x_0, y_0)$, then the orbit through points $(Mx_0, My_0)$ for $M = \frac{1}{2}, 2, -1$. Locate all the straight line orbits (if any) and plot orbits in the sectors bounded by these lines.

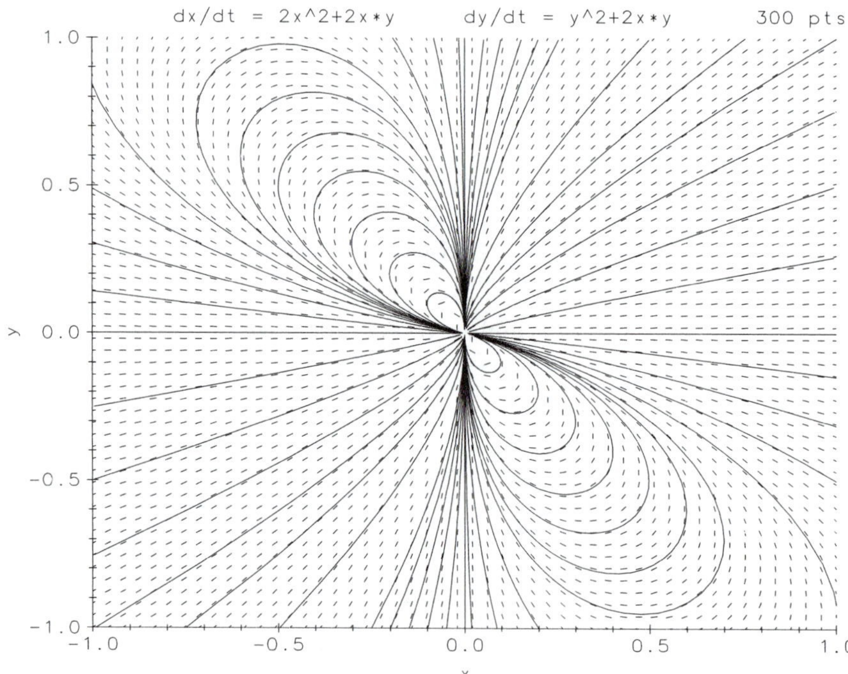


*Figure 3.6.1 Homothetic solution curves.*

# The ODE: $M(x, y)dx + N(x, y)dy = 0$

***Purpose*** To compute and plot integral curves of $M(x, y)dx + N(x, y)dy = 0$.

***Keywords*** First order ODEs, integral curves for first order ODEs, integrals, exact equations, planar systems, separable equations

***See also*** Experiments 3.5, 3.6; Atlas Plates `First Order G-I`

***Background*** The first order ODE $M(x, y) + N(x, y)y' = 0$ is written in the differential form

$$M(x, y)dx + N(x, y)dy = 0 \tag{1}$$

in order to take advantage of the exactness and integrating factor techniques for finding a solution formula. Notice that the coefficients $M(x, y)$ and $N(x, y)$ in equation (1) are not unique since the multiples $\alpha(x, y)M(x, y)$ and $\alpha(x, y)N(x, y)$ would work equally well in any region of the $xy$-plane where $\alpha(x, y)$ is continuous and does not vanish.

• **Solution curves via integrals**. Equation (1) is **exact** in some simply connected region of the $xy$-plane if there is a function $K(x, y)$ such that $\partial K/\partial x \equiv M$ and $\partial K/\partial y \equiv N$ in that region. For example, the separable equation $N(y)y' + M(x) = 0$ is exact, and $K(x, y)$ can be taken to be $F(y) + G(x)$, where $F$ and $G$ are any antiderivatives of $N$ and $M$, respectively. Since any solution $y(x)$ of an exact equation (1) satisfies the condition $d(K(x, y(x)))/dx \equiv 0$, it follows that

$$K(x, y(x)) = K(x_0, y_0), \text{ where } y_0 = y(x_0) \tag{2}$$

The function $K(x, y)$ is an **integral** of equation (1), and the curve defined implicitly by $K(x, y) = K(x_0, y_0)$ is the **integral curve through** $(x_0, y_0)$. From equation (2) we see that the solution curve of an exact ODE through $(x_0, y_0)$ must be confined to the integral curve (2) through $(x_0, y_0)$. Not all equations of the form (1) are exact. Most can be made exact by multiplying by a nonvanishing **integrating factor** $\mu(x, y)$ so that $(\mu M)dx+(\mu N)dy = 0$ is exact. (Recall the **exactness test**: $\partial(\mu M)/\partial y \equiv \partial(\mu N)/\partial x$.) An integral $K$ or an integrating factor $\mu$ may be hard to find for a given ODE $Mdx + Ndy = 0$.

• **Solution curves via orbits**. Even if an integral $K(x, y)$ of equation (1) is known, the nature of the integral curves $K(x, y) =$ constant may be hard to decipher. ODE solver/graphers allow the user to plot integral curves directly without having to find integrating factors or integrals. This approach to solving equation (1) is described as follows: Before using the computer, write the ODE $Mdx + Ndy = 0$ in the form $dy/dx = -M(x, y)/N(x, y)$, and then as the first order system:

$$\begin{aligned} dx/dt &= -N(x, y) \\ dy/dt &= M(x, y) \end{aligned} \tag{3}$$

If $x = x(t)$, $y = y(t)$ is the solution of (3) for which $x_0 = x(0)$ and $y_0 = y(0)$ and if $K(x, y)$ is an integral of (1) that is defined on a region containing $(x_0, y_0)$, then $K(x(t), y(t)) = K(x_0, y_0)$ for all $t$ in some interval containing $t_0 = 0$. Thus an orbit of (3) is an **integral curve** of the ODE $M(x, y) + N(x, y)y' = 0$ when the equation $M(x, y)dx + N(x, y)dy = 0$ is exact. Orbits of system (3) are called **integral curves** of the ODE $M(x, y) + N(x, y)y' = 0$ even when the equation $M(x, y)dx + N(x, y)dy = 0$ is not exact. It is not hard to show that any "piece" of an integral curve in the $xy$-plane without a vertical tangent is a solution curve for the ODE $M(x, y) + N(x, y)y' = 0$.

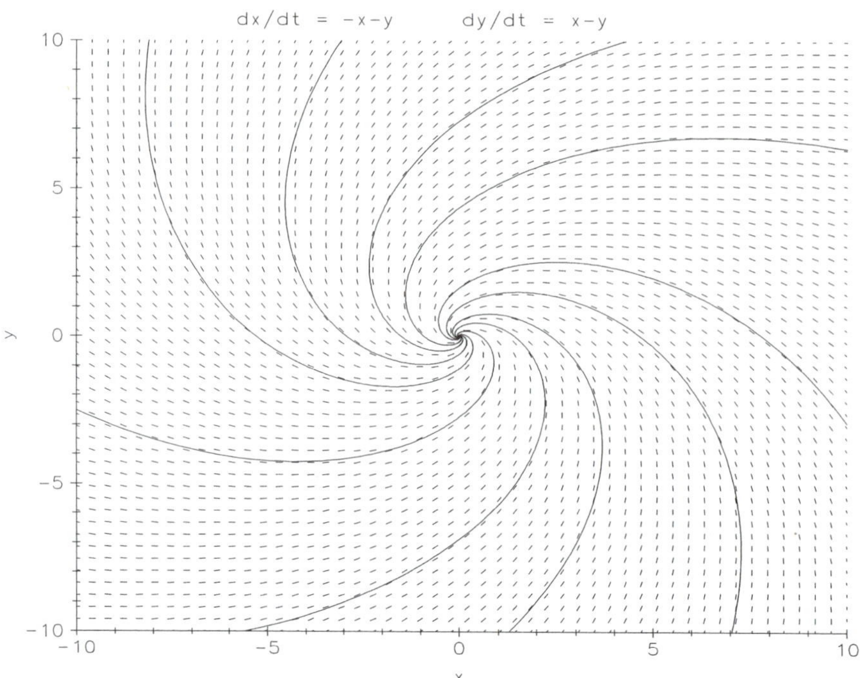


*Figure 3.7.1 Integral curves of the planar system (4).*

***Observation*** **(Common zeros of $M$ and $N$)** At points where $N$ vanishes, $dy/dx$ is undefined, hence a solver for $dy/dx = -M/N$ would have difficulties. Observe that system (3) does not suffer from this problem. If $M(x, y) \neq 0$ at a point where $N(x, y) = 0$, the integral curve has a vertical tangent. If both $M(x, y)$ and $N(x, y)$ are 0 at a point $(x, y)$, the point is an **equilibrium (stationary, rest,** or **critical) point.** In the example below, $M(x, y)$ and $N(x, y)$ vanish simultaneously at the origin (0, 0), which is the sole equilibrium point. Equilibrium points should be suspected where integral curves behave "strangely."

***Example 1*** The ODE $(x - y)dx + (x + y)dy = 0$ is not exact. Although an integrating factor can be found, we instead replace the ODE by the system

$$\begin{aligned} dx/dt &= -x - y \\ dy/dt &= x - y \end{aligned} \tag{4}$$

Solvers will plot a direction field as well as solution curves for system (4), hence also the integral curves for the ODE. See Figure 3.7.1. The integral curves seem to meet at the origin, and so (0, 0) is probably a singular point for system (4). This guess is easily verified. It is instructive to compare this plot with the one in Figure 3.9.2.

_______________/_______________
hardware software

_______________/_______________
name date

Answer questions in the space provided,
or on attached sheets or carefully labeled graphs.

_______________/_______________
course section

## 3.7 Planar Systems and $M\,dx + N\,dy = 0$

***Abstract*** Integral curves provide the key to solving $M\,dx + N\,dy = 0$. This experiment explores the relation between integral curves and solution curves for this first order ODE.

**1.** The ODE $2x\,dx + 4y\,dy = 0$ is both exact and separable.

**(a)** Find an integral $K(x, y)$ for the ODE. Find the equation of each of the three integral curves passing, respectively, through the points $(1, 1)$, $(-1, 2)$, and $(-2, 0)$.

**(b)** Solve and graph the equivalent system $dx/dt = -4y$, $dy/dt = 2x$, for the three sets of initial data $x_0 = 1$, $y_0 = 1$; $x_0 = -1$, $y_0 = 2$; and $x_0 = -2$, $y_0 = 0$. What is the connection between this graph and your answer to (a)? Indicate with a highlighter some solution curves $y = y(x)$.

**2.** Reproduce Atlas Plate `First Order G`. With a highlighter indicate some solution curves $y = y(x)$.

**3.** Select one of the ODEs listed below and do the problems that follow.

- ☐ **A.** $-y\,dx + (y\cos y + x)dy = 0$
- ☐ **B.** $(y - 2x)dx + (x + 1)dy = 0$
- ☐ **C.** $(1 + 3x^2)dx + (3y^2 - 1)dy = 0$
- ☐ **D.** $y\,dx + (2xy - e^{-2y})dy = 0$

**(a)** Replace the ODE by a system, $dx/dt = -N(x, y)$, $dy/dt = M(x, y)$, and find all equilibrium points of the system.

**(b)** Use your answer to (a) and a solver for planar systems to graph the four integral curves passing, respectively, though $(1, 1)$, $(0, 2)$, $(-1, 1)$, and $(-3, 0)$.

**(c)** Find as many integral curves of the ODE in the square $S: -5 \le x \le 5, -5 \le y \le 5$, as you can. The resulting "picture" is a partial portrait of the integral curves of the ODE. Include enough integral curves to identify the distinctive features of the portrait, but not so many that the features blur in a maze of curves.

**(d)** On the graph produced in (c) indicate with a highlighter some solution curves of the ODE. Mark all equilibrium points.

_______________/_______________
hardware software

_______________/_______________
name date

Answer questions in the space provided,
or on attached sheets or carefully labeled graphs.

_______________/_______________
course section

## 3.8 *Construction of Integral Curves: The Cat*

***Abstract*** Construction of solution curves for first order ODEs can be carried out in many ways with the computer. A solver for planar systems can be used to find the integral curves of a first order ODE, and then these integral curves can be used to find the solution curves. This experiment extends this technique to more complex systems.

***Special instructions*** The open-ended nature of this experiment makes it a appropriate for a team effort. Follow the instructions in Appendix A for writing the laboratory report.

Find some functions $M(x, y)$ and $N(x, y)$ such that the face of a cat appears in the portrait of the integral curves of the ODE $M(x, y)dx + N(x, y)dy = 0$. The functions $M$ and $N$ need not necessarily be continuous in the region of the $xy$-plane in which the cat is portrayed. Some suggestions:

- First, try selecting your own functions $M(x, y)$ and $N(x, y)$ and a screen, $a \leq x \leq b$, $c \leq y \leq d$, and then constructing a portrait of the integral curves of your own ODE, $M\,dx + N\,dy = 0$.
- Use your solver to reproduce the Atlas Plates `Planar System C` and `D`. What first order ODEs of the form $M\,dx + N\,dy = 0$ have the curves in these plates as integral curves? Do these integral curves remind you of anything? Are the functions $M(x, y)$ and $N(x, y)$ continuous over the $xy$-region of interest? What are some equilibrium points of the planar systems, and how do the integral curves of the related first order ODE behave near these "singular" points?
- Use your solver to reproduce the Atlas Plates `First Order G` and `H`. Do the curves in these plates remind you of anything? Are the functions $M(x, y)$ and $N(x, y)$ continuous? How do the integral curves behave near points where both $M$ and $N$ vanish?
- Try defining ODEs of the form $M\,dx + N\,dy = 0$ where the expressions used to define $M$ and $N$ can change from one part of the region to the next. For example, use an "if ... then ... else ..." construction as is done in the Atlas Plate `Planar System F`. What first order ODE of the form $M\,dx + N\,dy = 0$ has the curves in `Planar System F` as integral curves? Are the functions $M$ and $N$ continuous? Do the curves in this plate remind you of anything? Try reproducing this plate using your own solver. Fill in some of the blank spaces. Do you run into any difficulties?
- Begin with the equation of a curve that you want for an integral curve and differentiate each side of the equation to construct an ODE, reversing the usual process. For example, the ellipse defined by $x^2 + 10y^2 = 5$ is an integral curve of the equation, $2x\,dx + 20y\,dy = 0$, as can be shown by taking differentials.
- Go "shopping" in the Atlas for pieces of the cat. For example, consider Atlas Plate `Two-cycled System A` for the eyes of the cat.
- Consider arcs of the integral curves of $y\,dx - x\,dy = 0$ for the cat's whiskers.

# ODEs in Polar Coordinates

***Purpose*** To transform ODEs in rectangular coordinates to equivalent ODEs in polar coordinates.

***Keywords*** Polar coordinates, rectangular coordinates, Chain Rule

***Background*** Writing the general normal first order ODE in terms of the rectangular coordinate variables $x$ and $y$ instead of the variables $t$ and $x$, we have

$$\frac{dy}{dx} = f(x, y) \tag{1}$$

ODE (1) can be transformed into polar coordinates $r$ and $\theta$ by the relationships shown in Figure 3.9.1. Polar variables are a good choice if elliptical or spiraling integral curves are present.

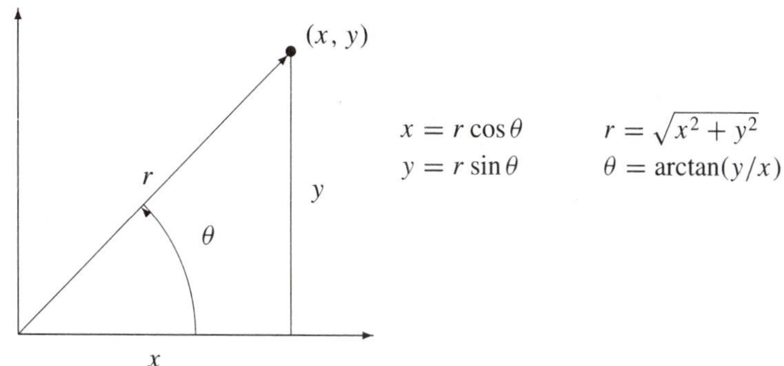


*Figure 3.9.1 Polar and rectangular coordinates.*

Suppose that $y = g(x)$ is a solution of (1); in polar coordinates this relationship becomes

$$r \sin\theta = g(r\cos\theta) \tag{2}$$

which implicitly defines $r$ as a function of $\theta$. Differentiating each side of (2) with respect to $\theta$ and using the Chain Rule and (1), we have

$$\begin{aligned} \frac{dr}{d\theta}\sin\theta + r\cos\theta &= g'(r\cos\theta)\left(\frac{dr}{d\theta}\cos\theta - r\sin\theta\right) \\ &= f(r\cos\theta, r\sin\theta)\left(\frac{dr}{d\theta}\cos\theta - r\sin\theta\right) \end{aligned} \tag{3}$$

Equation (3) is a first order ODE in $r$ and $\theta$ (not in normal form) and is equivalent to (1). Equation (3) seems more complicated than (1), but it is actually simpler for certain functions.

***Example 1*** **(Changing ODEs to polar form)** Applying (3) to the ODE

$$\frac{dy}{dx} = \frac{y - x}{y + x} \tag{4}$$

we have

$$\frac{dr}{d\theta}\sin\theta + r\cos\theta = \frac{r\sin\theta - r\cos\theta}{r\sin\theta + r\cos\theta}\left(\frac{dr}{d\theta}\cos\theta - r\sin\theta\right) \tag{5}$$

After canceling the common factor $r$ in the quotient, multiplying each side by $\sin\theta + \cos\theta$, simplifying, and again canceling a common factor, ODE (5) reduces to

$$\frac{dr}{d\theta} = -r \tag{6}$$

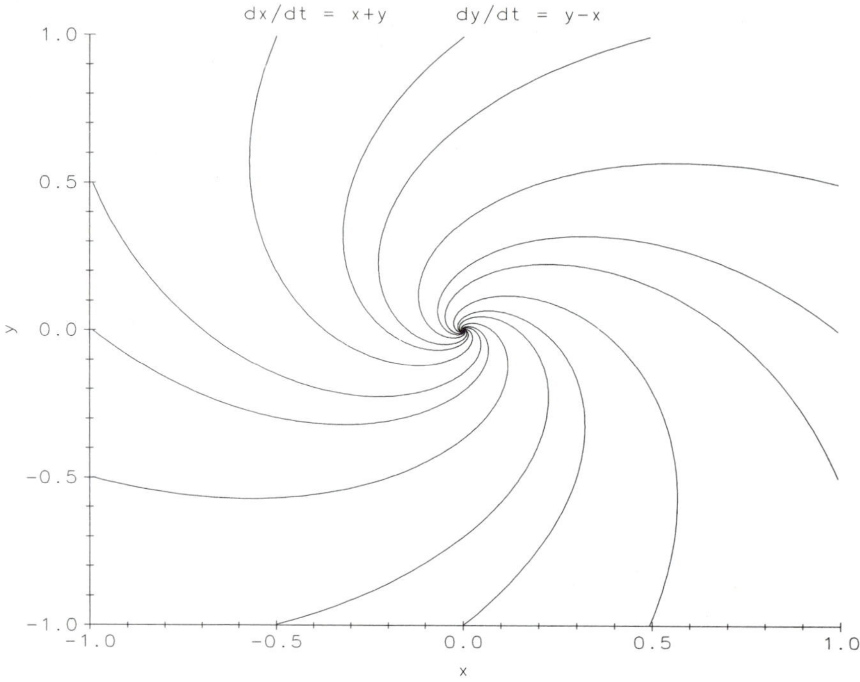


*Figure 3.9.2 The spiral solutions of (4).*

The solutions of (6) are given by $r = Ce^{-\theta}$, where $C$ is any nonnegative constant. Taking logarithms, and returning to $xy$-variables, we have the (implicit) solutions of (4)

$$\ln\left(x^2 + y^2\right)^{1/2} = \ln C - \arctan(y/x) \tag{7}$$

Figure 3.9.2 shows the spirals defined by (7). Alternatively, the plot in Figure 3.9.2 could have been produced by taking solutions $r(\theta)$ of (6) and plotting $r(\theta)\cos\theta$ against $r(\theta)\sin\theta$ on the Cartesian axes.

***Observation*** **(Polar form of an autonomous planar system).** Let $(x(t)), y(t))$ solve the system

$$\begin{aligned} x' &= f(x, y) \\ y' &= g(x, y) \end{aligned} \tag{8}$$

Then the point $(x(t), y(t))$ on the orbit of (8) can be identified by the polar coordinates $r(t), \theta(t)$, where $x(t) = r(t)\cos\theta(t)$, $y(t) = r(t)\sin\theta(t)$. Differentiating these relations with respect to $t$ (using the Chain Rule) and solving the derived system for $r'$ and $\theta'$ yields the equivalent system

$$\begin{aligned} r' &= \cos\theta f(r\cos\theta, r\sin\theta) + \cos\theta g(r\cos\theta, r\sin\theta) \\ \theta' &= \frac{1}{r}\{-\sin\theta f(r\cos\theta, r\sin\theta) + \cos\theta g(r\cos\theta, r\sin\theta)\} \end{aligned} \tag{9}$$

__________/__________ __________/__________
hardware software name date

Answer questions in the space provided,
or on attached sheets or carefully labeled graphs.

__________/__________
course section

## *3.9 ODEs in Polar Coordinates*

***Abstract*** First order ODEs in rectangular coordinates $x$ and $y$ are rewritten and solved in polar coordinates.

**1.** Select one of the ODEs listed below, and peform the following tasks.

☐ **A.** $\dfrac{dy}{dx} = \dfrac{x + 2y}{2x - y}$

☐ **B.** $\dfrac{dy}{dx} = \dfrac{x + y - y(x^2 + y^2)}{x - y - x(x^2 + y^2)}$

☐ **C.** $\dfrac{dy}{dx} = \dfrac{-x + 5y - y(x^2 + y^2)}{5x + y - x(x^2 + y^2)}$

**(a)** Convert the ODE to a the form $dr/d\theta = f(r, \theta)$ and solve that ODE for $r$ as a function of $\theta$. Finally, write the solution in rectangular coordinates.

**(b)** Plot solution curves of the ODE in the $xy$-plane.

**2.** Consider the autonomous planar system

$$x' = y(x^2 + y^2)^{-1/2}$$
$$y' = x(x^2 + y^2)^{-1/2}$$

**(a)** Find an equivalent autonomous planar system in the polar coordinates $r = r(t)$ and $\theta = \theta(t)$. (Hint: See the system in (9).)

**(b)** Using the result in (a), find orbits of the original system. Verify your calculations by use of an ODE solver to plot orbits of the original system.

**3.** Four bugs are at the corners of a square of side $a$. Each bug crawls counterclockwise at a constant speed $b$ directly toward a neighbor.

* **(a)** Find the path of a bug. (Hint: Use the position vector approach to track the position of one bug (see Appendix B.2). Expressing the position vector in polar form, explain why $dr/dt = -r d\theta/dt$ for a bug located at the polar point $(r, \theta)$, where the center of the square is the origin of the coordinate system. Thinking of $r = r(\theta)$, find an expression for $dr/d\theta$ and then find the path in polar coordinates.)

* **(b)** Graph the path of a bug. Explain why the four bugs meet at the origin in finite time even though each has spiraled around the origin infinitely many times (i.e., $\theta \to \infty$ along each path).

# Comparison of Solutions of Two ODEs

***Purpose*** To determine how much the solutions of two first order ODEs can differ as time goes on if the rate functions differ by a known amount.

***Keywords*** Perturbations, growth estimates, sensitivity

***See also*** Experiments 3.3–3.4; Problem 2b in Experiment 2.4

***Background*** An ODE model of a physical system is rarely precise. The effect of this imprecision can be estimated by finding outer bounds for solutions of an IVP whose data set is known only to lie within certain bounds. The Sensitivity Principle of Chapter 2 is one way to approach the question. Experiments 3.3–3.4 provide an answer in the setting of first order linear ODEs. Here we shall take up the general case of a nonlinear (i.e., not necessarily linear) first order ODE in normal form.

Suppose that $f(t, x)$, $g(t, x)$, $\partial f/\partial x$ and $\partial g/\partial x$ are continuous in a region $R$ of $tx$-space and that $(t_0, x_0)$ is a point inside $R$. Think of $f$ as the rate function and $g$ as a perturbation of $f$ or, perhaps, a measure of the uncertainty in $f$ itself. Suppose that a bound on the uncertainty is known: $|g(t, x)| \leq M$ for all $(t, x)$ in $R$. Suppose, moreover, that a bound on the variation in $f$ as $x$ changes is also known: $|\partial f(t, x)/\partial x| \leq L$ for all $(t, x)$ in $R$. Then we have the following principle concerning the difference $|x(t) - \tilde{x}(t)|$ between the solutions of the two IVPs,

$$x' = f(t, x), \quad x(t_0) = x_0 \tag{1}$$

$$\tilde{x}' = f(t, \tilde{x}) + g(t, \tilde{x}), \quad \tilde{x}(t_0) = \tilde{x}_0 \tag{2}$$

over some interval $t_0 \leq t \leq t_0 + c$.

**Perturbation Principle:** *Let $x(t)$, $\tilde{x}(t)$ be the solutions of (1), (2) respectively. There are positive numbers $b$ and $c$ such that if $|x_0 - \tilde{x}_0| < b$ and $t_0 \leq t \leq t_0 + c$ then*

$$|x(t) - \tilde{x}(t)| \leq |x_0 - \tilde{x}_0| \, e^{L(t-t_0)} + \frac{M}{L}(e^{L(t-t_0)} - 1) \tag{3}$$

*where the constants $M$ and $L$ are as defined above.*

The estimate given by (3), although often much too rough, leads to one somewhat surprising result: the differences between the solutions grow at most like $e^{L(t-t_0)}$. That is typical of the growth of a linear rate equation even though neither (1) nor (2) needs to be linear at all. In many cases computer calculation suggests that the effect of a perturbation is much smaller than the upper bound given by (3).

***Example 1*** **(Sensitivity to changes in rates and data)** Consider the IVPs for $t \geq 0$:

$$x' = x/10, \quad x(0) = a$$

$$\tilde{x}' = \tilde{x}/10 + e^{-\tilde{x}^2}, \quad \tilde{x}(0) = \tilde{a}$$

Here $f(x) = x/10$ and $g(\tilde{x}) = e^{-\tilde{x}^2}$. Observe that the bounds $L$ and $M$ are easily calculated: $|\partial f/\partial x| = 1/10 = L$ and $\left|e^{-\tilde{x}^2}\right| \leq e^0 = 1 = M$. Thus, (3) implies that the solutions $x(t)$ and $\tilde{x}(t)$ of the two IVPs satisfy

$$|x(t) - \tilde{x}(t)| \leq |a - \tilde{a}| \, e^{t/10} + 10(e^{t/10} - 1) \tag{4}$$

______________/______________ hardware software

______________/______________ name date

Answer questions in the space provided, or on attached sheets or carefully labeled graphs.

______________/______________ course section

## *3.10 Comparison of Solutions of Two ODEs*

***Abstract*** The effects of uncertainties in the rate function of a first order ODE are studied.

**1.** Duplicate Figure 3.10.1. Find $M$ and $L$ and verify that the Perturbation Principle holds.

**2.** Select one of the IVPs below and answer the following questions. The terms in curly braces are the perturbations in the rate functions and initial values. The region $R$ is the entire $tx$-plane except as indicated.

$x' = x\sin t + \{(1+x^2)^{-1}\}, \qquad x(0) = 1 + \{0.1\}$
$x' = e^{-x^2}\cos(t^3) + \{0.001e^{-x^2}\}, \qquad x(0) = 1 + \{0.001\}$
$x' = x(10-x) + \{100\cos t\}, \quad x(0) = 5 + \{0.1\}, \quad x \geq 0$

* **(a)** Use (3) to find an upper bound on $|x(t) - \tilde{x}(t)|$ where $x$ and $\tilde{x}$ are the respective solutions of the IVP and the perturbed IVP.

* **(b)** Solve the IVP and the perturbed IVP over the $t$-range $0 \leq t \leq 10$ and graph the solutions. Compare the actual differences $|x(t) - \tilde{x}(t)|$ over the $t$-range to the upper bound for these differences found in (a).

* **(c)** Extend the $t$-range as far as possible toward $\infty$ and use the corresponding graphs to make a plausible estimate of $|x(\infty) - \tilde{x}(\infty)|$.

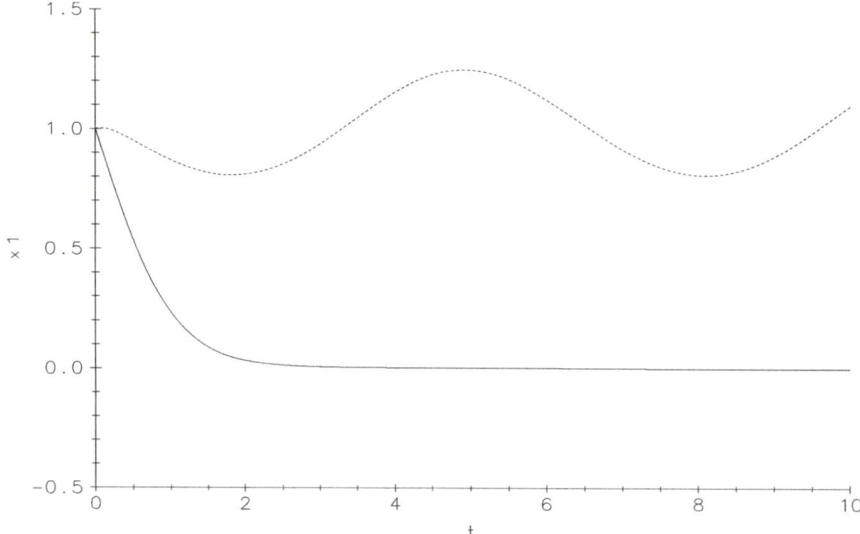


*Figure 3.10.1 The effect of a perturbation on a linear ODE.*

# Harvesting a Species

***Purpose*** To show the effect of harvesting on a single species whose growth is logistic.

***Keywords*** Population models, logistic model for populations, harvesting, first order ODEs, Balance Law

***See also*** Experiments 1.6–1.8, 2.2, 2.8–2.10, them 3.16–3.17; Appendix B.1

***Background*** These experiments treat the accumulation of a "substance" (a species) in an isolated environment. Left to its own, $P(t)$, the population of the species at time $t$, follows a rate law (usually not of first order). The population may be subject to external agents (e.g., hunters) which "drive" the population. When an agent acts to diminish the population the action is called **harvesting**. An action which increases the population is called **stocking**. The rate of change of the population, $dP(t)/dt$, can be found in the following way. Suppose, for example, that in the absence of external agents the population represented by $P(t)$ evolves with the rate law given by the **logistic model** (see Appendix B.1 for a derivation). If the **harvesting rate** is given by a nonnegative function $H(t)$ then the Balance Law yields the ODE

$$P'(t) = r\left(1 - \frac{P(t)}{K}\right)P(t) - H(t) \tag{1}$$

where the **intrinsic rate coefficient** $r$ and the **carrying capacity** (or **saturation level**) $K$ are both positive constants.

In the absence of harvesting ($H \equiv 0$), ODE (1) is autonomous and can be analyzed by sign analysis (Experiment 2.2). In particular it follows that the population rises or falls toward the carrying capacity $K$ as time advances (see also Appendix B.1).

Harvesting diminishes the growth rate and augments a decay rate, and so one might expect that the population curves of a harvested species would tend toward a diminished carrying capacity. This is indeed true if the harvesting rate is a positive constant that is not too large and if the initial population lies above some survival threshold. However, if the harvesting rate exceeds a critical value, the population becomes extinct regardless of its initial numbers. The experiments here explore models of constant, periodic, and seasonal harvesting rates.

***Observation*** **(Explicit solution)** If $H$ is a constant in ODE (1) then the variables separate. An explicit formula can be obtained for the solution $P(t)$, but this fact is not used in the experiments. See the Observation on the cover sheet of Experiments 2.8–2.10.

***Example 1*** **(Light harvesting)** Figure 3.11.1 is typical for species whose population grows logistically and is harvested at a relatively low constant rate. Note that if harvesting is "light" then initial population values below a survival threshold lead to extinction, but initial population values above that threshold lead to saturation. In Figure 3.11.2 note that "heavy" harvesting always leads to extinction regardless of the initial value of the population. These matters will be addressed in the experiments.

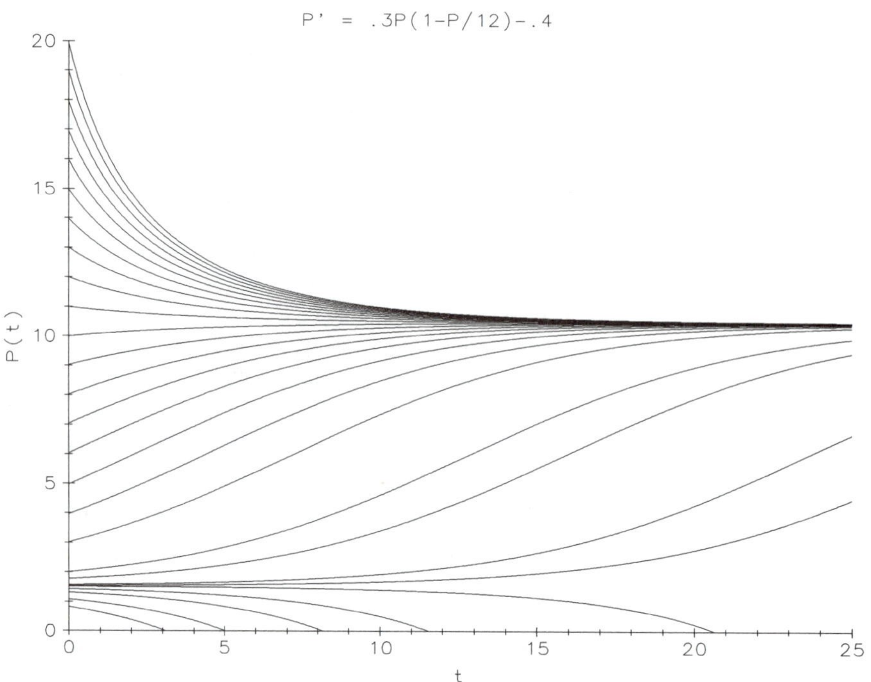


*Figure 3.11.1* *Solution curves for logistic growth with light harvesting.*

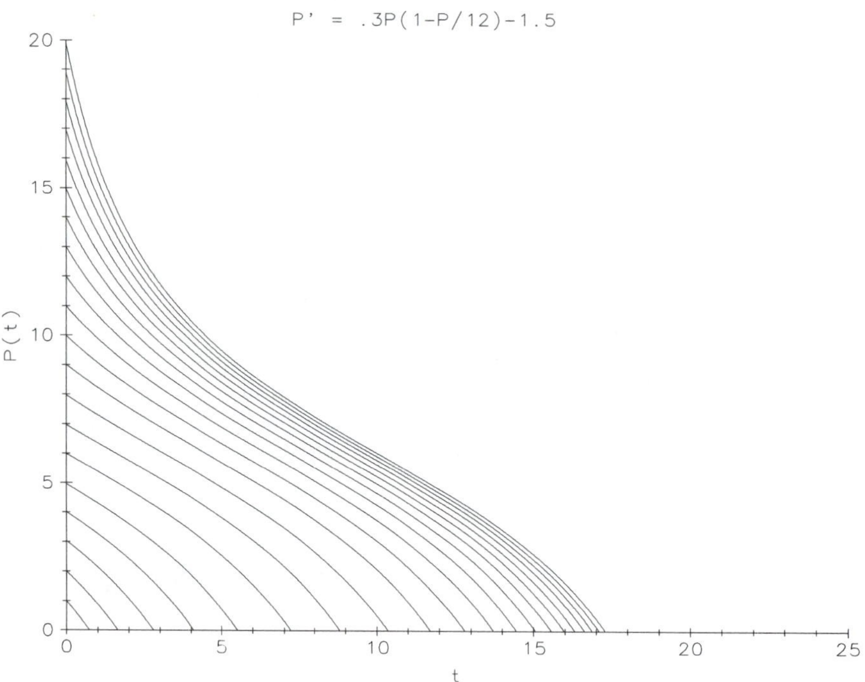


*Figure 3.11.2* *Solution curves for logistic growth with heavy harvesting.*

_______________/_______________ 
hardware software

_______________/_______________ 
name date

Answer questions in the space provided,
or on attached sheets or carefully labeled graphs.

_______________/_______________ 
course section

## *3.11 Constant Rate Harvesting*

***Abstract*** This experiment examines the effect of constant rate harvesting on a population described by ODE (1).

**1.** **(No harvesting)** Suppose that $H = 0$ in ODE (1).

**(a)** Explain why $P(t) \equiv 0$ and $P(t) \equiv K$ are solutions of the ODE. These are the **equilibrium** or **steady state populations**. If the population is larger than the environment can support (i.e., if $P(t)$ exceeds the carrying capacity), you might expect the population to decline. Does this happen with this rate equation? What happens if the population is less than the carrying capacity? An equilibrium level is an **attractor** if nearby population curves tend to that level as time advances; it is a **repeller** if the nearby curves move away. Identify each of the two equilibrium levels as an attractor or repeller.

**(b)** Suppose that $r = 1\ (\text{year})^{-1}$ (corresponding to a natural percentage growth rate of 100% per year) and that the environment will support a population of $10^6$ individuals. Solve and graph the solution of the initial value problem

$$P' = r\,(1 - P/K)\,P, \quad P(0) = P_0$$

over a time span of 20 years with six values of $P_0$ varying from $4 \times 10^5$ to $1.5 \times 10^6$. Include $P_0 = 10^6$. Plot $P(t)$ against $t$ for each value of $P_0$, putting all six plots on the same graph, if possible.

**(c)** From (a) and (b), give a general description of how the population changes with time's advance. Explain why $K$ is called the carrying capacity or the saturation level of the population.

**2.** **(Constant rate harvesting)** Suppose that the population is harvested at a constant rate of $H$ individuals per year. The model ODE is (1) with $H$ a positive constant.

**(a)** Suppose $r = 1$ (year)$^{-1}$, $K = 10^6$ individuals, and $H = 10^5$ individuals per year. Solve and graph ODE (1) with $P(0) = P_0$, where $P_0$ ranges in increments of $2 \times 10^5$ from $10^5$ to $10^6$. Let the span of time be 20 years in each case. Plot $P(t)$ against $t$ for each value of $P_0$, putting all six graphs on the same plot. From these graphical results, discuss the future of the population in terms of the initial values $P_0$. Can extinction occur? How many steady states are there now? Is there a stable steady state?

* **(b)** Use the quadratic formula to find the roots $P_1$ and $P_2$ of the rate function $r(1 - P/K)P - H$. Explain why $P_1$ and $P_2$ are both positive and real if $H < rK/4$, but are complex conjugates if $H > rK/4$. The value $H = rK/4$ is the **critical harvesting rate.** Explain why extinction is inevitable if the harvesting rate is set above the critical value.

**(c)** Suppose $r = 1$ (year)$^{-1}$, $K = 10^6$ individuals, and $H = 4 \times 10^5$ individuals per year. Plot $P(t)$ against $t$ for values of $P_0$ ranging from $10^5$ to $10^6$ in increments of $10^5$. Let the span of time be 20 years in each case. From the results, what happens to the population as time advances? Does $H = 4 \times 10^5$ exceed the critical harvesting rate given in (b)?

_______________/_______________
hardware software

_______________/_______________
name date

Answer questions in the space provided,
or on attached sheets or carefully labeled graphs.

_______________/_______________
course section

## 3.12 *Variable Rate Harvesting*

***Abstract*** This experiment examines the effect of variable rate harvesting on a population described by the model ODE (1).

**1.** Suppose a population is harvested at sinusoidally varying harvesting rates. A corresponding ODE is

$$P' = r(P/K)P - H_0(1 + \sin \pi t)$$

where $H_0$ is a positive constant.

**(a)** Plot the solutions on a single graph of $P(t)$ against $t$ for a range of values of $P_0$ from $5 \times 10^5$ to $1.5 \times 10^6$ individuals, where $r = 0.1$ (year)$^{-1}$, $K = 10^6$ individuals and $H_0 = 10^5$ individuals/year.

**(b)** From the results in (a) discuss the long-term behavior of this population.

* **(c)** Plot the solutions with the data of (a), but for a sequence of values of $H_0$ ranging from $10^4$ to $10^6$. Discuss the effects on the population of moving from a low amplitude to a high amplitude sinusoidal harvesting policy.

* **2.** Choose your own variable rate harvesting model. Explain what happens to the population given your harvesting rate funciton.

_______________/_______________
hardware software

_______________/_______________
name date

Answer questions in the space provided, or on attached sheets or carefully labeled graphs.

_______________/_______________
course section

## 3.13 *Intermittent Harvesting*

***Abstract*** This experiment examines the effect of intermittent harvesting on a population described by the model ODE (1).

***Special instructions*** If your solver does not accept the "square wave" $H(t)$ of Problem 1, then do Problem 2 instead.

**1.** Suppose the population is harvested, but only every other year. In this setting, the harvesting rate $H(t)$ in ODE (1) is piecewise constant:

$$H(t) = \begin{cases} H_0 & \text{if} \quad 0 \le t \le 1, \quad 2 \le t \le 3, \ldots \\ 0, & \text{if} \quad 1 < t < 2, \quad 3 < t < 4, \ldots \end{cases}$$

* **(a)** With an intermittent harvest, the population can be harvested at a relatively high rate for one year, then allowed to grow back toward the carrying capacity during the following year. What factors should be considered in designing a harvesting policy for a long-term sustainable yield?

* **(b)** Set $r = 0.01$ (year)$^{-1}$, $K = 10^6$ individuals, $H_0 = 4000$ individuals/year. With several values of $P_0$ ranging from $10^5$ to $10^6$ use the computer to solve ODE (1) with intermittent harvesting. Solve over a 20-year span of time. Place all your graphs on the same plot.

* **(c)** From the results in (b), outline a specific intermittent harvesting policy for a long-term sustainable yield of 4000 individuals every other year. To avoid population extinction, the policy should provide a lower bound below which the population must not be allowed to fall.

**2.** Repeat Problem 1 for the harvesting rate function

$$H_1(t) = H_0 \left\{ 1 + \frac{2}{\pi} \sum_{k=0}^{4} \frac{\sin(2k+1)\pi t}{2k+1} \right\}$$

Graph $H_1(t)$ and compare it to the graph of $H(t)$ in Problem 1. If you have solved Problem 1, compare those results with the results of this problem. Conclusions?

# Salt Levels in a Brine

***Purpose*** To model the changing amount of salt dissolved in brine as an initial value problem, and to use computer solvers and graphics to analyze the behavior of the solutions of the model equations.

***Keywords*** Brines, balance law, transient, steady state, Law of Mass Action, saturated solution, linear, separable, and Bernoulli/Riccati ODEs;nonlinear systems

***See also*** Experiments 1.6–1.8, 3.1–3.4, 3.11, 3.16

***Background*** A brine is a solution of salt in water. If the brine is in a tank equipped with fill and drain pipes, then the total amount of dissolved salt in the tank varies as the salt concentration in the inflow stream changes and the inflow and outflow rates are adjusted. The amount of salt in the tank can be modeled by appeal to the

**Balance Law:** *Net rate of change = rate in − rate out.*

The "rate in" term is the rate at which salt is added to the brine by means of the inflow stream, and the "rate out" term is the rate at which salt leaves the tank through the outflow pipe. It is assumed throughout that the inflow stream is instantaneously mixed with the brine in the tank so that at any given time the concentration of salt in the tank is uniform. Each of these terms (i.e., the "rate in" and "rate out") is the product of an appropriate brine flow rate with a corresponding salt concentration.

***Example 1*** **(Brine flow)** A tank with a capacity of 4000 liters holds 2000 liters of brine that contains 50 kg of dissolved salt. Brine with a salt concentration of 0.2 kg per liter is piped into the tank at a rate of 40 liters per minute. Well-mixed brine is drawn off at $a$ liters per minute. The model IVP for the amount $x(t)$ of salt in the tank at time $t$ is given by

$$x' = (0.2)(40) - \frac{x}{2000 + (40 - a)t}a, \qquad x(0) = 50$$

Figure 3.14.1 shows $x(t)$ for three values of the outflow rate $a$. The model is valid only for times $t$ such that $0 \leq 2000 + (40 - a)t \leq 4000$.

The salt is already dissolved in the brine for the phenomenon described above. Now suppose that a tank contains a brine into which bulk salt is suddenly dumped. The **saturation level** (or **carrying capacity**) of a given volume of brine is the largest amount of dissolved salt it can carry. The Law of Mass Action applies to the rate of change of the amount of dissolved salt:

**Law of Mass Action:** *The rate of change of the amount of dissolved salt is proportional to the amount of salt undissolved and the amount that can still be dissolved.*

***Example 2*** **(Bulk salt)** Suppose that 200 kg of bulk salt is placed in a tank holding 2000 liters of brine already containing $x_0$ kg of dissolved salt. Suppose that the saturation level of the brine is 300 kg of salt. Figure 3.14.2 displays solution curves of the separable, nonlinear initial value problem. Curves representing the amount of dissolved salt $x(t)$ are drawn for various values of $x_0$.

***Observation*** **(Composite model)** Bulk salt is placed in a tank holding a brine and brine also runs into and out of the tank. The corresponding mathematical model is a pair of coupled nonlinear rate equations for the amount of bulk salt that has dissolved and for the total amount of dissolved salt (from all sources) in the tank.

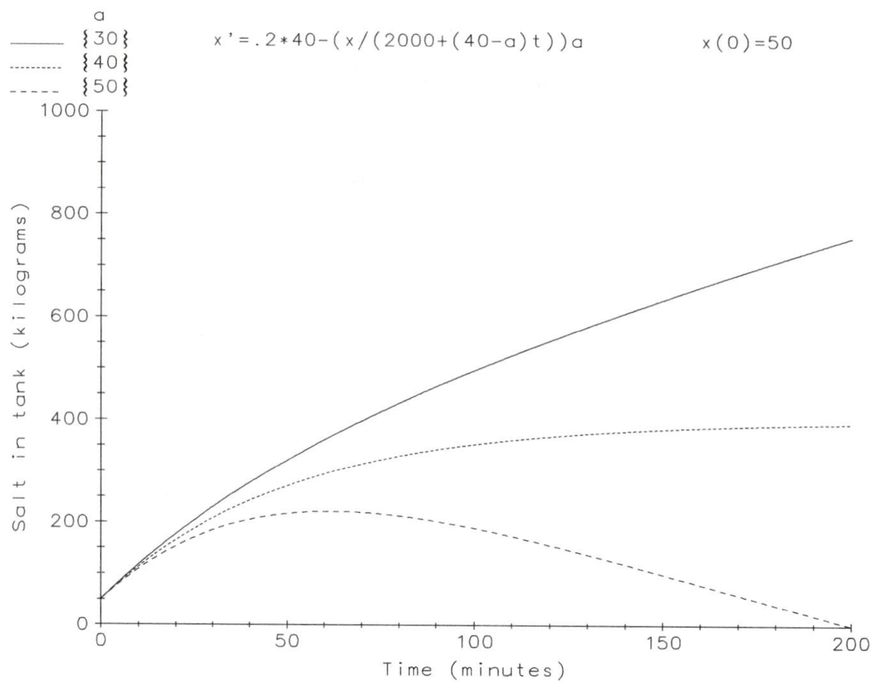


*Figure 3.14.1 Amount of dissolved salt for three outflow rates (see Example 1).*

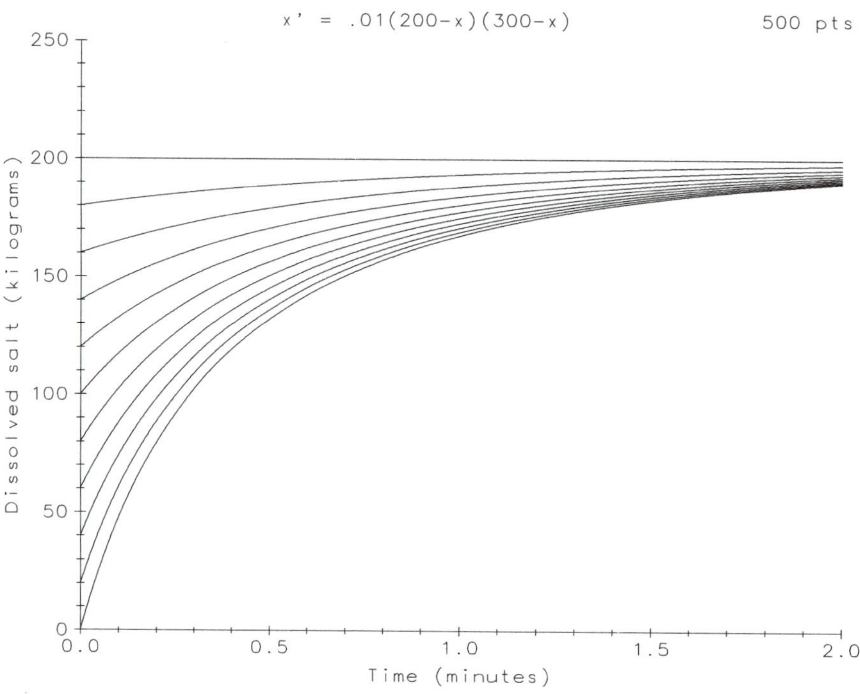


*Figure 3.14.2 Amount of dissolved bulk salt (see Example 2).*

______/______ ______/______
hardware software name date

Answer questions in the space provided, or on attached sheets or carefully labeled graphs.

______/______
course section

## *3.14 Linear Brine Models*

***Abstract*** The phenomenon of brine flowing in and out of a tank is modeled by linear ODEs. The amount of dissolved salt in the tank is graphed as a function of time.

**1.** This problem treats the simple constant rate brine flow model of Example 1.

**(a)** Explain why the ODE is valid only over some maximal finite time interval, $0 \leq t < T_{\max}$. Find $T_{\max}$ for $a < 40$, and for $a > 40$.

**(b)** Solve the ODE for six values of the outflow rate $a$ ranging from 0 to 100 liters/min. Use the largest time intervals that are physically and computationally realistic. Mark on the solution curves the points at which the tank has emptied or overflowed and the values of $a$, $T$, and $x(T)$ for these points.

**(c)** Rewrite the ODE model in linear form, find an integrating factor $\mu$, and find a formula for the amount $x(t)$ of dissolved salt as a function of time $t$ and outflow rate $a$. For the six values of $a$ and $T$ used in (b) use the formula to calculate six values of $x(T)$. Explain any discrepancies between your results in (b) and (c).

**2.** This problem treats inflow and outflow rates and concentrations that may vary over time.

**(a)** A 4000 liter tank is half filled with brine containing 50 kg of dissolved salt. Brine with a salt concentration of $0.01(1 + .5 \sin t)$ kg/liter flows into the tank at a rate of 40 liters/min. The well-mixed brine flows out at the same rate. After one hour the inflow is shut off and the tank drains. Plot the amount of salt in the tank over a 2 hour time span. Identify the points where the amount of salt in the tank reaches a local or an absolute maximum or minimum.

* **(b)** Data for real devices are only approximately known. Suppose that the data in (a) are known to lie between 90 and 110% of the given values. Suppose moreover that the allowable concentration of salt in the vat must always lie between 0.015 and 0.028 kg/liter whenever both inflow and outflow pipes are open. Does the system operate within these limits? Write comments on the graphs.

* **(c)** Construct a general ODE model for the amount of salt in the tank for a brine inflow/outflow problem that allows variable inflow rates and concentrations and a variable outflow rate. (Suggestion: Use the Balance Law after expressing the volume $V(t)$ of brine in the tank at time $t$ in terms of initial volume and an integral of the difference between inflow and outflow rates. Be sure to limit the time span by consideration of the tank capacity. You may want to set up a pair of rate equations to be solved simultaneously, one for $V(t)$ and another for the amount of salt.) Attach graphs of the varying amount of salt for your own choice of varying flow rates and input concentration. Explain why the model equations are of Bernoulli or Riccati type.

__________ / __________
hardware software

Answer questions in the space provided,
or on attached sheets or carefully labeled graphs.

__________ / __________
name date

__________ / __________
course section

## *3.15 Nonlinear Brine Models*

***Abstract*** Bulk salt slowly dissolves in a closed tank of brine. Inflow and outflow pipes are opened and brine flows into and out of the tank. The model ODEs for this composite phenomenon of dissolution and mixing form a coupled nonlinear system.

***Special instructions*** Do Problem 1 before Problem 2.

**1.** Bulk salt is placed in a tank containing 1000 liters of brine and dissolves at a rate determined by the Law of Mass Action.

**(a)** Write out an IVP for the amount $x(t)$ of dissolved salt in the tank at time $t$. Use $x_0$ for the initial amount of dissolved salt in the brine, $A$ for the initial amount of bulk salt, $C$ for the carrying capacity of the brine, and $k$ for the rate constant.

**(b)** Suppose that $x_0 = 50$ kg, $A = 20$ kg, $C = 60$ kg, and $x = 55$ kg 10 minutes after the bulk salt starts to dissolve. Use a computer plot of $x(t)$ to estimate the rate constant $k$. Attach a graph.

* **(c)** Plot the amount $x(t)$ of dissolved salt for your choice of time-varying data: $A(t)$ (corresponding to continual resupply of bulk salt), $C(t)$, and $k(t)$. Explain your model.

* **2.** Bulk salt is placed in a tank of brine that is equipped with inflow and outflow pipes. The bulk salt dissolves at a rate determined by the Law of Mass Action. Brine enters the tank, mixes instantaneously with the brine already in the tank, and exits through the outflow pipe. Create a general model for determining the amount of salt in the tank at time $t$. Plot solutions of the model equations for four different data sets, at least two of which involve time-varying data. Explain your model and your data sets. (Suggestion: Let $x(t)$ and $y(t)$ denote, respectively, the amount of dissolved bulk salt and the total amount of dissolved salt (including salt from the inflow stream). Use the Law of Mass Action and the Balance Law to write a pair of coupled nonlinear rate equations for $x(t)$ and $y(t)$. Be sure to identify tank overflow and emptying constraints for your model equations.)

# Bimolecular Chemical Reactions

***Purpose*** To model concentrations of species in a bimolecular chemical reaction using first order ODEs and to show how the concentrations tend to equilibrium levels.

***Keywords*** Chemical reactions, Balance Law, Conservation Law, Law of Mass Action, equilibria, separable ODEs

***See also*** Experiments 2.2, 3.11, 3.14–3.15; Appendix B.3

***Background*** In a bimolecular chemical reaction the molecules of two species interact and create one or more products. For example, in the elementary bimolecular reaction

$$A + B \to C + D \tag{1}$$

the interaction of one molecule of reactant $A$ with one of $B$ results in the transformation of the pair into one molecule each of products $C$ and $D$. If the reaction occurs in a constant-volume chemical reactor, the quantities may be specified by concentrations. $[X(t)]$ denotes the concentration of species $X$ in the reactor at time $t$. Since material is conserved, reactants $A$ and $B$ disappear at the same rate that products $C$ and $D$ are generated. The equality of rates may be expressed by the

**Law of Conservation:** $\dfrac{d[C]}{dt} = \dfrac{d[D]}{dt} = -\dfrac{d[A]}{dt} = -\dfrac{d[B]}{dt}$

Integration of these equalities from initial time $t = 0$ gives an alternate form of the law,

$$[C(t)] - c = [D(t)] - d = -[A(t)] + a = -[B(t)] + b \tag{2}$$

where $a, b, c, d$ are the respective initial concentrations. From (2) we see that the initial data and one of the concentrations determine the other three concentrations. To find one of the concentrations, $[C(t)]$ for example, a second chemical principle is used:

**Law of Mass Action:** *The rate of an elementary reaction is proportional to the product of the concentrations of the reactants.*

For example, in reaction (1) we have from mass action and from (2)

$$\frac{d[C]}{dt} = k_1[A][B] = k_1(a + c - [C])(b + c - [C])$$

where $k_1$ is the positive constant of proportionality (the **rate constant** of the reaction). Denoting $[C]$ by $x$ and the time derivative by $x'$, we have the model IVP based on a separable ODE,

$$x' = k_1(a + c - x)(b + c - x), \quad x(0) = c \tag{3}$$

The constant solutions of (3) are $x(t) \equiv a + c$ and $x(t) \equiv b + c$. They represent possible equilibrium levels of the concentration of species $C$. Only the lower of the two levels is attainable chemically. By (2) the higher level corresponds to the impossible case of a negative concentration of reactant $B$ (if $a > b$) or $A$ (if $a < b$).

In reality, most chemical reactions are reversible,

$$A + B \underset{k_{-1}}{\overset{k_1}{\rightleftharpoons}} C + D \tag{4}$$

where $k_{-1}$ is the rate constant of the backward reaction. Suppose that both reactions are elementary and bimolecular. Then an IVP that models the concentration of species $C$, say, is obtained using the Law of Mass Action for each reaction and applying the

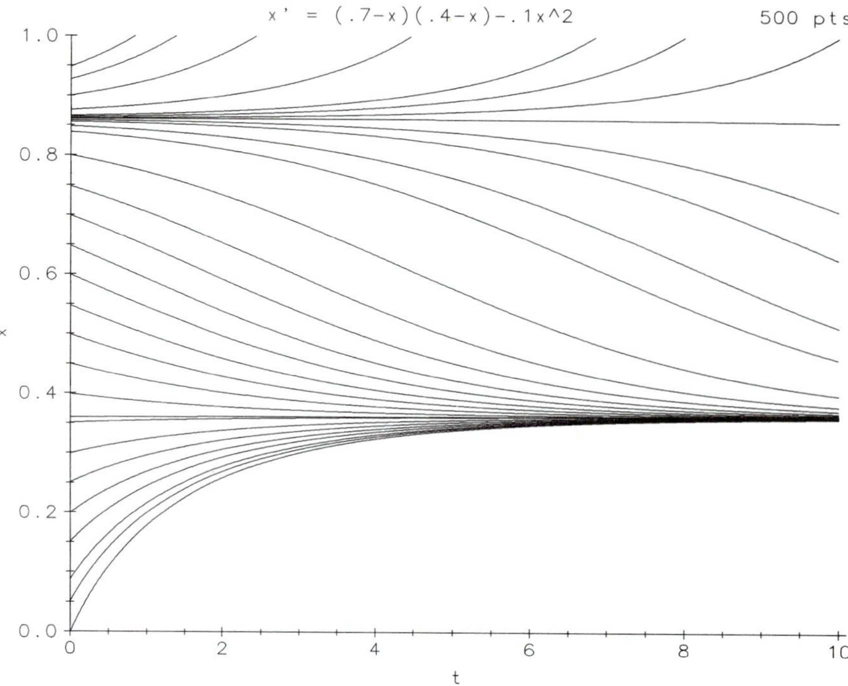


*Figure 3.16.1 Solution curves for the ODE of Example 1.*

**Balance Law:** *Net rate = rate of creation − rate of destruction*
to obtain $[C]' = k_1[A][B] - k_{-1}[C][D]$, $[C(0)] = c$. Using (2) and denoting $[C(t)]$ by $x(t)$, again we have an IVP with a separable ODE,

$$x' = k_1(a + c - x)(b + c - x) - k_{-1}x(d - c + x), \quad x(0) = c \tag{5}$$

Equilibrium levels for $x$ are calculated by equating the quadratic rate function in (5) to zero.

***Example 1*** **(A bimolecular reaction)** The IVP

$$x' = (.7 - x)(.4 - x) - 0.1x^2, \qquad x(0) = c \tag{6}$$

models a reversible reaction with $k_1 = 1$, $k_{-1} = 0.1$, $a + c = 0.7$, $b + c = 0.4$, and $d = c$ in (5). The equilibrium levels are found by setting the right-hand side of (6) to zero and solving for $x$ to obtain $x_1 \approx 0.36$ and $x_2 \approx 0.86$. Figure 3.16.1 shows solution curves for various values of $x(0) = c$. Only those curves with $0 \le c \le 0.4$ correspond to real chemical reactions. If $c > 0.4$, then the initial concentration of species $B$ is negative (since $b + c = 0.4$) and has no physical meaning. Observe that the chemically realistic concentrations tend, as time goes on, to the lower equilibrium level.

***Observation*** **(Different phenomena, but a common model)** The ODEs that model a bimolecular chemical reaction, constant rate harvesting (Experiment 3.11), and the dissolving of bulk salt (Experiments 3.14–3.15) all have the same form; hence one ODE models several physical phenomena.

_______________/_______________ hardware software

_______________/_______________ name date

Answer questions in the space provided, or on attached sheets or carefully labeled graphs.

_______________/_______________ course section

## 3.16 *Quadratic Rates as Bimolecular Models*

***Abstract*** Rate equations are interpreted as models of bimolecular chemical reactions; rate constants and equilibrium levels are determined; solution curves are computed and graphed, and the chemically realistic curves are identified.

**1.** Select one of ODEs below and do the following problems.

☐ **A.** $x' = (0.10 - x)(0.30 - x)$

☐ **B.** $x' = (0.10 - x)(0.30 - x) - 0.001x^2$

☐ **C.** $x' = 0.1(0.70 - x)(0.30 - x)$

☐ **D.** $x' = 0.1(0.70 - x)(0.30 - x) - 0.01x(0.1 - x)$

**(a)** The ODE models chemical reaction (1) or (4). Is it a reversible reaction? Draw a suitable reaction diagram for the ODE in terms of species $A$, $B$, $C$, $D$ in such a way that $x(t)$ denotes the concentration of product $C$. What are the values of the rate constants? What are the equilibrium values $x_1$ and $x_2$, $x_1 \leq x_2$?

$k_1 =$ _______ $k_{-1} =$ _______. $x_1 =$ _______ $x_2 =$ _______.

Why do solution curves rise as time increases if $x < x_1$ or $x > x_2$, but fall if $x_1 < x < x_2$?

**(b)** Express the initial values $a$, $b$, and $d$, of $[A]$ and $[B]$ (and $[D]$ if appropriate) in terms of the initial value $c$ of $[C]$, and the constants of the given ODE (see (5)). Find the largest value of the constant $c$ that is chemically realistic (see Example 1).

$a =$ _______ $b =$ _______ $d =$ _______. $c_{\max} =$ _______.

**(c)** Plot solution curves of the ODE for $0 \leq t \leq 20$, where $x(0) = c$ has the values $0$, $0.5x_1$, $x_1$, $c_{\max}$, $0.5(x_1 + x_2)$, $x_2$, $2x_2$, and several other positive values. Label all chemically realistic curves and the equilibrium curve they approach as $t$ increases.

**(d)** Plot the concentrations $[A(t)]$, $[B(t)]$, and $[D(t)]$, $0 \leq t \leq 20$, for the initial values $a, b, c, d$ given in (b) and (c). Label all chemically realistic curves and the equilibrium curves they approach as $t$ increases.

__________ / __________
hardware software

__________ / __________
name date

Answer questions in the space provided,
or on attached sheets or carefully labeled graphs.

__________ / __________
course section

## 3.17 *Modeling a Bimolecular Reaction*

***Abstract*** A specific bimolecular reaction is modeled by an ODE. Computer graphics are used to estimate a rate constant.

**1.** Acetic acid and an alcohol react to produce an acetate and water, and this reaction is reversible. Initially, acid and alcohol are present in the same amounts.

**(a)** Explain why $A + B \underset{k_{-1}}{\overset{k_1}{\rightleftharpoons}} C + D$ depicts the chemical reaction. Give the chemical names of $A$, $B$, $C$, and $D$. Write the four rate equations for the concentrations $[A]$, $[B]$, $[C]$, $[D]$. For example, $[A]' = -k_1[A][B] + k_{-1}[C][D]$.

**(b)** Explain why the rate equation $x' = k_1(a - x)^2 - k_{-1}x(d - c - x)$ models the production of acetate. Give the chemical meanings of the constants in the rate equation and of the function $x(t)$.

**(c)** Suppose acetic acid and alcohol each have an initial concentration of 0.9. The initial concentrations of acetate and water are 0. Suppose that $k_{-1} = 0.25k_1$. After 64 days, measurements show that the concentration of the acetate product is 0.25. Use a computer solver and graphics to estimate the value of the forward rate constant $k_1$ that is consistent with the data. Use your solver to estimate the time $T$ when the concentration of acetate has reached 0.5. Estimate the equilibrium level of acetate.

**(d)** Repeat (c) with initial concentrations of acid and alcohol equal to 0.7, while those of acetate and water are respectively 0.1 and 0.3.

CHAPTER 4

# Second Order Equations

Most of this chapter is devoted to linear second order ODEs. Experiments 4.7–4.9 treat some nonlinear models of oscillating masses connected by springs. There are other nonlinear second order models of considerable importance, but these appear in the next chapter, where the ODEs are converted to planar systems.

## *IVPs for Second Order ODEs*

The **general second order IVP in normal form** is

$$\begin{aligned} x'' &= F(t, x, x') \\ x(t_0) &= r \\ x'(t_0) &= s \end{aligned} \tag{1}$$

If $x(t)$ solves the ODE in (1), then the 3D plot of the points $(t, x(t), x'(t))$ is called a **solution curve** for the ODE. If the function $F$ and its partial derivatives $\partial F/\partial x$ and $\partial F/\partial x'$ are continuous in a 3D region $R$ of $txx'$-space that contains the initial data point $(t_0, r, s)$, then the IVP (1) has a unique solution on some $t$-interval containing $t_0$. Moreover, this solution curve can be extended both forward and backward in time until it approaches the boundary of $R$. A consequence of uniqueness is that different solution curves never touch or cross one another. However, note that the graphs of two solutions $x_1(t)$ and $x_2(t)$ in $tx$-space *may* intersect at a point $(t^*, x^*)$, but that $x_1'(t^*) \neq x_2'(t^*)$ by uniqueness.

## *Linear Second Order ODEs*

A widely used form of (1) is that for which $F$ is linear in $x$ and $x'$. The most general (but nonnormal) linear form of (1) is

$$a(t)x'' + b(t)x' + c(t)x = f(t) \tag{2}$$

where $a$, $b$, $c$, and $f$ are continuous on a common $t$-interval. The driving term $f(t)$ is often called the **input** of the ODE (2) and the solution $x(t)$ is called the **output** (or **response**). The solution sets of linear equations such as (2) are generated from a handful of solutions, and this makes them particularly easy to treat once a pair of generating solutions has been found. The main reason that (2) has been so useful in science and mathematics is that many of the core phenomena of mechanics, electric circuits, automatic control, physical chemistry, and celestial mechanics can be modeled by such ODEs.

On any interval in which $a(t)$ does not vanish, ODE (2) may be normalized by dividing by $a(t)$, and we shall do so wherever convenient. However, some of the most intriguing behavior occurs near the roots of $a(t)$, the **singularities** of the equation.

## *Some Examples*

The four plots in Figure 4.1 display various output modes of a linear ODE driven by a periodic input. The growing nature of the output is an example of **resonance**, a response triggered when the input excites the so-called natural frequency of the ODE. See Experiment 4.3. The graphs in Figure 4.2 illustrate one response mode of an undriven linear ODE that models an aging spring—the spring stretches without bound (see Experiment 4.9).

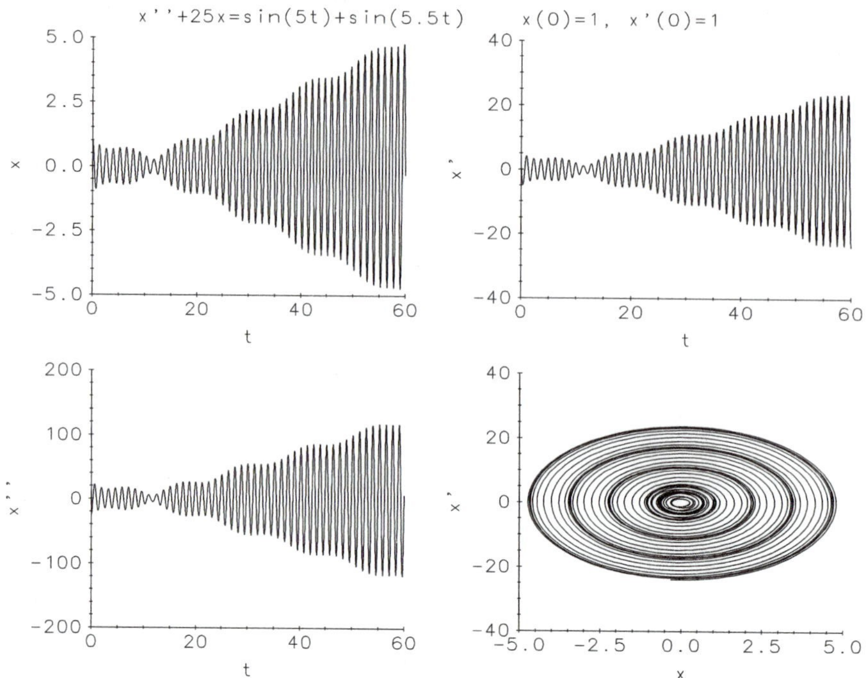


*Figure 4.1 A driven second order linear ODE.*

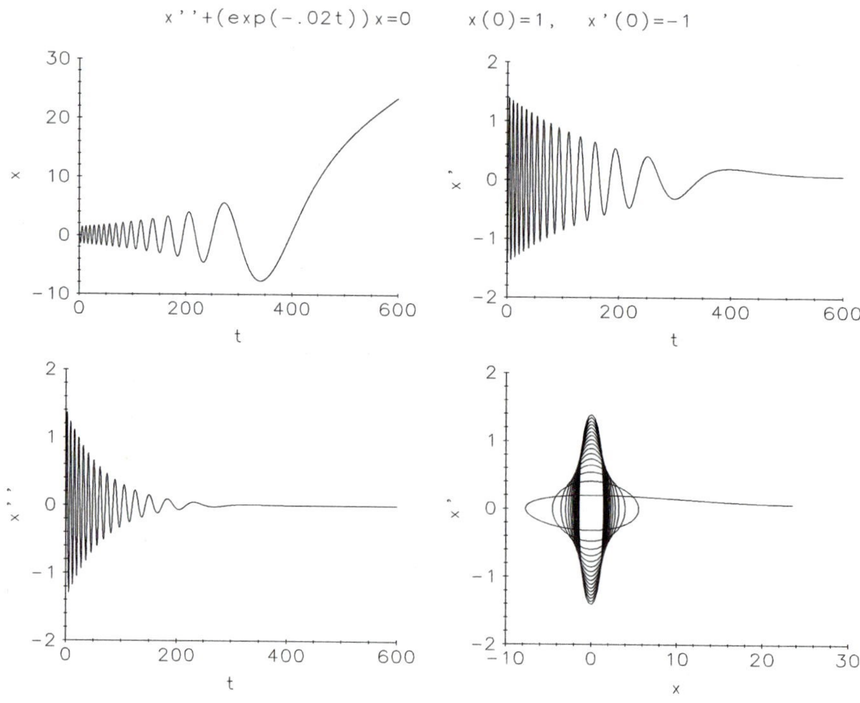


*Figure 4.2 Stretching of an aging spring.*

# Properties of Solutions

***Purpose*** To examine the properties of the solutions of general second order linear ODEs.

***Keywords*** Second order linear ODE, superposition, input, driving term, homogeneous ODE, particular solution, general solution, basis, spanning set, independence

***See also*** Experiments 3.1–3.4

***Background*** The **general second order linear ODE** is

$$a(t)x'' + b(t)x' + c(t)x = f(t) \tag{1}$$

where the **coefficients** $a$, $b$, and $c$, and the **input** or **driving term** $f(t)$ are continuous on a $t$-interval $I$. To include the on/off inputs frequently used by engineers, $I$ is often specialized to $0 \le t < \infty$, and continuity generalized to piecewise continuity. In this experiment, however, the coefficients and the input are assumed to be continuous on $I$ and $a(t)$ is assumed not to vanish on $I$.

• **Finding all solutions of ODE (1)**. The theoretical process of finding all solutions of (1) proceeds as follows:

☐ Find a pair of linearly independent solutions $x_1(t)$ and $x_2(t)$ on $I$ of the **undriven** or **homogeneous** ODE

$$a(t)x'' + b(t)x' + c(t)x = 0 \tag{2}$$

Solutions $x_1(t)$ and $x_2(t)$ are **linearly independent** on $I$ if there are no constants $K_1$ and $K_2$ (not both 0) such that $K_1x_1(t) + K_2x_2(t) \equiv 0$ for all $t$ in $I$. In that case the set $\{x_1(t), x_2(t)\}$ is called a **basis** or an **independent spanning set** for the solution space of (2). Every solution of the undriven ODE (2) can be written in the form $C_1x_1(t) + C_2x_2(t)$, where $C_1$ and $C_2$ are arbitrary constants.

☐ Find one solution $x_p(t)$ of the driven ODE (1)—any solution will do. This is called a **particular solution**; hence the subscript $p$.

☐ Superpose these solutions to construct the **general solution** of (1):

$$x(t) = C_1x_1(t) + C_2x_2(t) + x_p(t)$$

☐ To find the solution of ODE (1) which satisfies the initial conditions $x(t_0) = r$, $x'(t_0) = s$, find constants $C_1$ and $C_2$ from the two linear algebraic equations

$$\begin{aligned} r &= C_1x_1(t_0) + C_2x_2(t_0) + x_p(t_0) \\ s &= C_1x_1'(t_0) + C_2x_2'(t_0) + x_p'(t_0) \end{aligned} \tag{3}$$

The independence of $x_1(t)$ and $x_2(t)$ ensures unique solvability for $C_1$ and $C_2$.

• **Constant Coefficients Case**. If the coefficients $a$, $b$, and $c$ are constants, then a pair of independent solutions of the undriven ODE (2) may be constructed using the roots of the characteristic polynomial (see Experiment 4.2). The variable coefficient case is harder; indeed, there is no general method for constructing an explicit spanning set that applies to every variable coefficient ODE (2). In some notable special cases, however, an explicit construction is possible. Nevertheless, spanning sets always exist (in fact, there are infinitely many), and a computer solver/graphics package may be used to graph the elements of such a spanning set. A particular solution $x_p(t)$ of (1) may (sometimes) be constructed by the **method of undetermined coefficients**, or (always) by the **method of variation of parameters**. The first method is straightforward (if it works at all), and the second requires

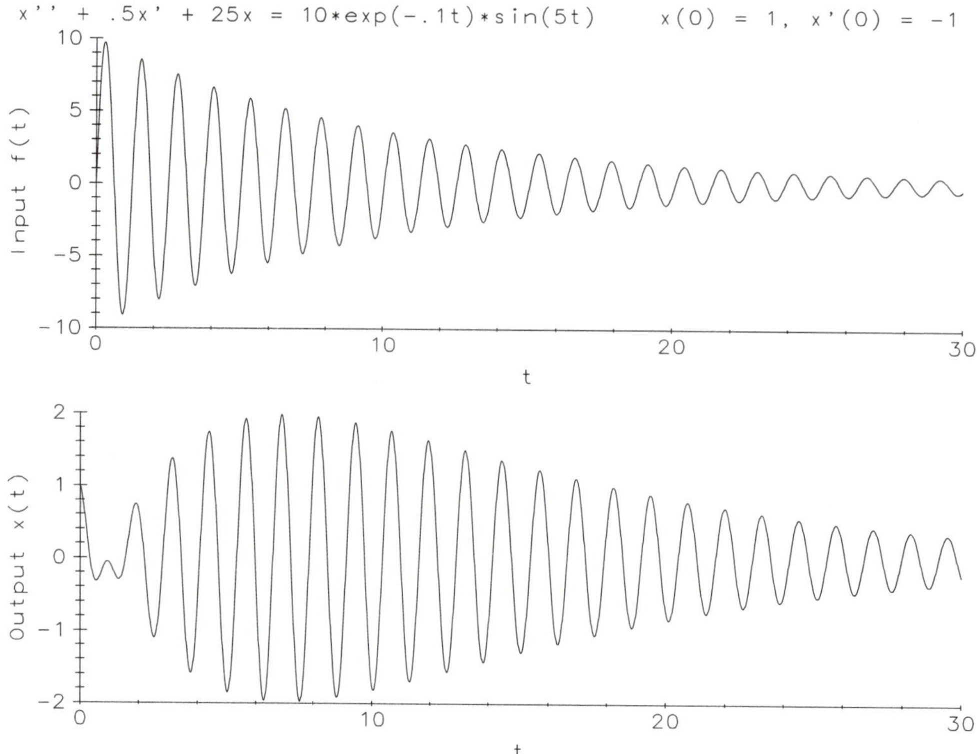


*Figure 4.1.1 Input/output graphs.*

a basis and some integrations. Both methods are described in textbooks on ODEs. Neither method is given here, since computer solvers do not rely on explicit solution formulas.

• **Computer generated solutions**. With a solver/graphics package available, the IVP (with $t_0$ in $I$)

$$\begin{aligned} &a(t)x'' + b(t)x' + c(t)x = f(t) \\ &x(t_0) = r, \quad x'(t_0) = s \end{aligned} \tag{4}$$

can be solved either directly or by first converting it to the equivalent first order system

$$\begin{aligned} x' &= y & x(t_0) &= r \\ y' &= \frac{1}{a}(f - cx - by) & y(t_0) &= s \end{aligned}$$

Recall the assumption that $a(t) \neq 0$, for $t$ in $I$. The solution $x(t)$ of the IVP (4) is the **output** corresponding to the **input** $f$ and the initial data (see Figure 4.1.1). Using the superposition properties of solutions of linear ODEs, the process for solving IVP (4) may be split into the following steps:

1. Find $x_1(t)$, the solution of the IVP $a(t)x'' + b(t)x' + c(t)x = 0$, $x(t_0) = r$, $x'(t_0) = 0$.
2. Find $x_2(t)$, the solution of the IVP $a(t)x'' + b(t)x' + c(t)x = 0$, $x(t_0) = 0$, $x'(t_0) = s$.
3. Find $x_p(t)$, the solution of the IVP $a(t)x''+b(t)x'+c(t)x = f(t)$, $x(t_0) = 0$, $x'(t_0) = 0$.
4. Superpose $x_1$, $x_2$, and $x_p$ to obtain the solution $x = x_1 + x_2 + x_p$ of the IVP (4).

__________ / __________
hardware software

__________ / __________
name date

Answer questions in the space provided,
or on attached sheets or carefully labeled graphs.

__________ / __________
course section

## 4.1 *Properties of Solutions*

***Abstract*** The solution of a second order linear IVP is constructed by superposing the solutions of related IVPs.

***Special instructions*** When solving each IVP in Problems 2 and 3, follow the steps 1–4 on the cover sheet. If your graphics package supports overlays, the four graphs may be drawn on one screen.

**1.** Duplicate Figure 4.1.1. Repeat with initial data $x(0) = 0$, $x'(0) = 0$; $x(0) = 10$, $x'(0) = -5$; and $x(0) = -5$, $x'(0) = -10$. On the basis of your work, make a conjecture about $\lim_{t\to\infty} x(t)$ and $\lim_{t\to\infty} x'(t)$ for all solutions $x(t)$ of the ODE.

**2.** Show that steps 1–4 from the cover sheet lead to a solution of IVP (4).

**3.** Select one of the IVPs below and perform the tasks that follow.

☐ **A.** $x'' + 5x' + 4x = \sin t$, $x(0) = 2$, $x'(0) = 3$, $0 \le t \le 10$
☐ **B.** $x'' + 2x' - 3x = 1 + e^{-t}$, $x(0) = -1$, $x'(0) = 5$, $0 \le t \le 5$
☐ **C.** $x'' + tx = \cos t$, $x(0) = 2$, $x'(0) = -3$, $0 \le t \le 10$
☐ **D.** $t^2x'' - tx' + 2x = 1$, $x(1) = -2$, $x'(1) = 2$, $0 < t \le 2$

**(a)** Graph the solutions of the related IVPs of steps 1 and 2 from the cover sheet.

**(b)** Graph the solution of the related IVP of step 3.

**(c)** Graph the solution $x(t)$ of the IVP and use a ruler to verify that step 4 holds. Write directly on the graph.

**(d)** Replace the given time interval by $0 \le t \le T$ for various "large" values of $T$. What happens to the solution as $T \to \infty$?

**4.** Select one of the IVPs below and perform the tasks that follow.

☐ **A.** $(1 - t^2)x'' - 2tx' + 12x = \cos t$, $x(0) = 2$, $x'(0) = -2$, $-1 < t < 1$
☐ **B.** $t^2x'' + tx' + (t^2 - 1)x = 1$, $x(1) = 1$, $x'(1) = 1$, $0 < t \le 2$
☐ **C.** $tx'' + (1 - t)x' + 10x = 3$, $x(1) = 1$, $x'(1) = 2$, $0 < t \le 3$
☐ **D.** $(1 - t^2)x'' - tx' + 9x = e^{-t}$, $x(0) = 2$, $x'(0) = -1$, $-1 < t < 1$

**(a)** Identify the singularities.

**(b)** Solve the IVP, following steps 1–4 from the cover sheet.

**(c)** What happens to the solution as $t$ approaches a singularity?

# Constant Coefficient Linear ODEs: Undriven

***Purpose*** To find all real-valued solutions of constant coefficient undriven linear ODEs and explore the behavior and stability of those solutions.

***Keywords*** Constant coefficient linear ODEs, characteristic polynomial, transients, harmonic motion, free vibrations, harmonic oscillator, oscillators, oscillating period

***See also*** Experiment 4.1

***Background*** The ODE in normal form is

$$x'' + ax' + bx = 0 \tag{1}$$

where $a$ and $b$ are real constants. The first step in constructing the general solution $x(t)$ of (1) is to find the roots of the **characteristic polynomial** $r^2 + ar + b$. There are three cases depending upon the nature of the roots $r_1$ and $r_2$. Note that $r_1, r_2 = (-a \pm \sqrt{a^2 - 4b})/2$:

- $r_1 \neq r_2$, both real: $x(t) = C_1 e^{r_1 t} + C_2 e^{r_2 t}$.
- $r_1 = r_2$, real: $x(t) = e^{r_1 t}(C_1 + C_2 t)$.
- $r_1 = \alpha + i\beta$, $r_2 = \alpha - i\beta$, $\alpha$, $\beta$ real, $\beta \neq 0$: $x(t) = e^{\alpha t}(C_1 \cos \beta t + C_2 \sin \beta t)$.

$C_1$ and $C_2$ are arbitrary real constants. If initial conditions $x(0) = u$, $x'(0) = v$ are imposed, $C_1$ and $C_2$ must be particularized. If the coefficients $a$ and $b$ are changed, the roots $r_1$ and $r_2$ change, and the behavior of the solutions changes accordingly. It is often convenient to rewrite (1) in equivalent system form

$$\begin{aligned} x' &= y \\ y' &= -bx - ay \end{aligned} \tag{2}$$

Initial conditions for system (2) have the form $x(0) = u$, $y(0) = v$ where $u$ and $v$ are specified constants. Figures 4.2.1 and 4.2.2 show how solutions are affected by changes in the initial data.

Of particular interest in the applications are conditions that characterize the behavior of solutions of ODE (1) starting "near" the equilibrium solution $x \equiv 0$. This is done by examining the stability (at the equilibrium point (0, 0)) of the equivalent planar system (2). Informally, stability means that for any solution $(x(t), y(t))$ for which $|x(0)|$ and $|y(0)|$ are small, $|x(t)|$ and $|y(t)|$ remain small as $t \to \infty$. More precisely, the definitions below give the range of possibilities.

- **Stability**. ODE (1) (or system (2)) is **stable** if for any tolerance $E > 0$ there is an allowable deviation $D$ such that if $(x^2(0) + y^2(0))^{1/2} < D$, then $(x^2(t) + y^2(t))^{1/2} < E$ for all $t \geq 0$. It can be shown that if the real parts of the roots of the characteristic polynomial are nonpositive, and if zero is not a double root, then (1) is stable. From the formula for the roots, we see that the coefficient criterion for stability is: $a$ and $b$ must be nonnegative and not both zero.
- **Asymptotic stability**. ODE (1) (or system (2)) is **asymptotically stable** if it is stable and $|x(t)| \to 0$ and $|y(t)| \to 0$ as $t \to \infty$. That is, for an asymptotically stable ODE both $x(t)$ and $y(t)$ are **transients** (i.e., decay to zero as time advances). In this case the real parts of the characteristic roots $r_1$ and $r_2$ must be negative. The solutions in the case of distinct negative real roots are said to be **overdamped**; solutions are **critically damped** if $r_1 = r_2 < 0$; solutions are **underdamped** if $r_1 = \bar{r}_2 = \alpha + i\beta$, $\alpha < 0$, $\beta \neq 0$. Asymptotic stability is even stronger than stability. Hence, the coefficient criterion for asymptotic stability is a little stronger than that for stability: Both $a$ and $b$ must be positive.

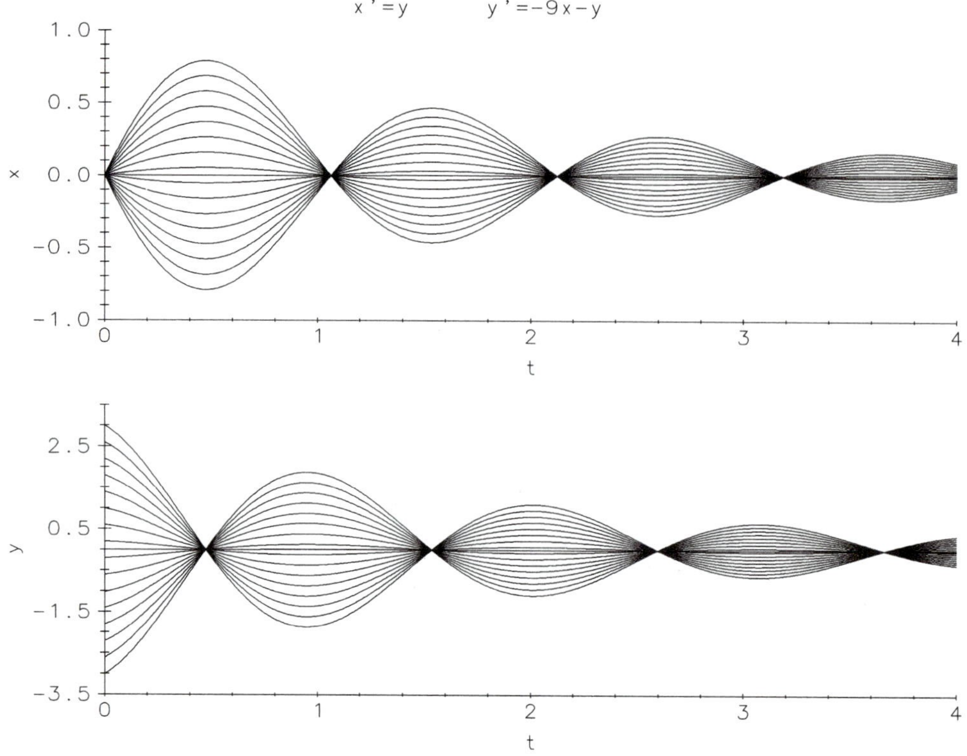


*Figure 4.2.1 The effect of changing the initial velocity $y_0$ in an asymptotically stable ODE which has complex characteristic roots.*

- **Neutral stability**. ODE (1) (or system (2)) is **neutrally stable** if it is stable but not asymptotically stable. The criterion for neutral stability is: One of $a$, $b$ must be positive, the other must be zero.
- **Instability**. ODE (1) (or system (2)) is **unstable** if it is not stable. The criterion for instability is the negativity of $a$ or $b$ (or both).

***Observation*** **(Periodic solutions, free oscillations)** A nonconstant solution $x(t)$ of an ODE is **periodic** of **period** $T$ if $x(t+T) = x(t)$ for all $t$, where $T$ is the smallest positive number for which this is true. A periodic solution of an autonomous ODE such as (1) is called a **free oscillation** because it arises freely and not in response to a periodic coefficient or to a periodic driving force. The nonconstant solutions of ODE (1) are free oscillations of period $2\pi/\beta$ if and only if the roots of the characteristic polynomial are pure imaginary, $r_1 = \bar{r}_2 = i\beta$, $\beta > 0$ This happens if and only if $a = 0$ and $b > 0$ (note that $\beta^2 = b$). Observe that ODE (1) and system (2) are neutrally stable in this case. Sometimes the periodic motion of ODE (1) is said to be **harmonic** and ODE (1) is said to be a **harmonic oscillator**.

***Observation*** **(Oscillatory solutions)** If the roots $\alpha \pm i\beta$ of the characteristic polynomial are such that $\alpha \neq 0$, $\beta > 0$ then the solutions are **oscillatory** with **quasiperiod** $2\pi/\beta$, but they are not actually periodic since no solution curve repeats itself. If $\alpha > 0$ the amplitude of the oscillation grows without bound; if $\alpha < 0$ the amplitude decays to 0.

## *4.2 Constant Coefficient Linear ODEs: Undriven*

***Abstract*** The stability properties of a constant coefficient undriven linear ODE are explored as the coefficients are changed.

***Special instructions*** Use the overlay feature (if your hardware/software package has one) to plot solution curves or orbits corresponding to different coefficients or initial data on the same graph. If this is not possible, plot separate graphs.

**1.** Reproduce Figures 4.2.1 and 4.2.2. Write the second order ODE involved. Find the roots of the characteristic polynomial. Is the ODE asymptotically stable, neutrally stable, or unstable? Explain.

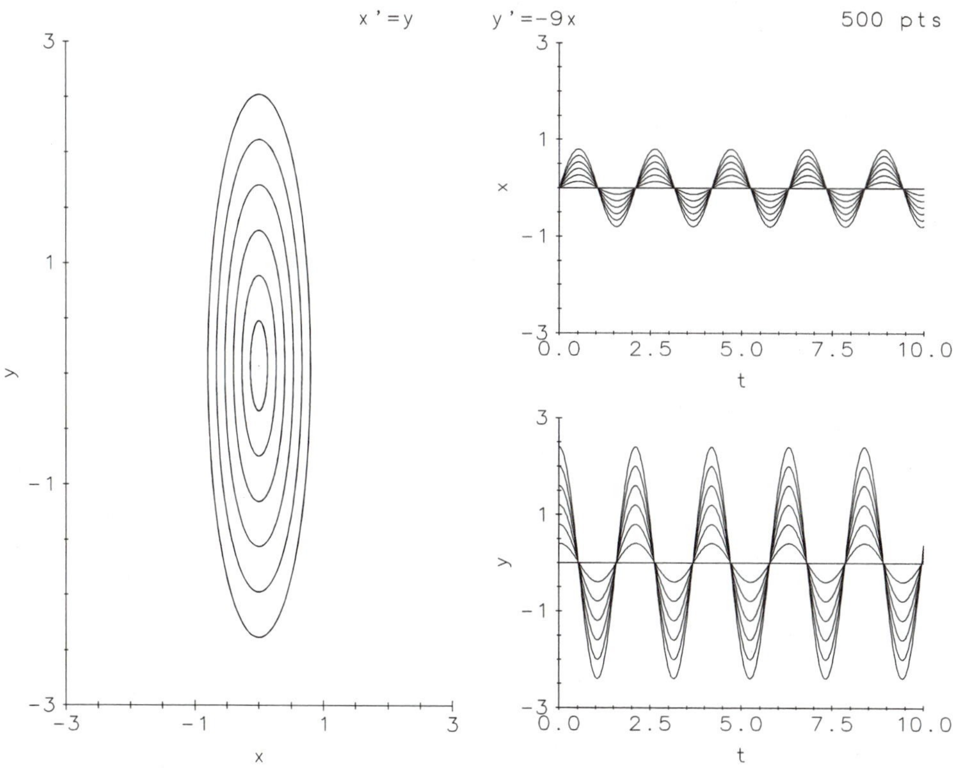


*Figure 4.2.2 The periodic position curves, velocity curves, and orbits of a harmonic oscillator: effect of changing the initial velocity $y_0$.*

**2.** Select one of the ODEs listed below and perform the tasks that follow.

☐ **A.** $x'' + ax' + x = 0$

☐ **B.** $x'' + x' + bx = 0$

**(a)** Find all values of the unspecified coefficient for which the ODE is stable, asymptotically stable, neutrally stable, or unstable.

**(b)** Find all values of the unspecified coefficient for which the nonconstant solutions are periodic (specify the period), oscillatory but not periodic (specify the quasiperiod), overdamped, underdamped, or critically damped.

**(c)** Set $x(0) = x'(0) = 1$ and graph solutions $x(t)$ of the IVP for several values of the unspecified coefficient chosen to illustrate a variety of behaviors and stability types. Tell which values correspond to what behaviors. Plot orbits in the $xx'$-plane as well.

**3.** Consider the IVP $x'' - x' + bx = 0$, $x(0) = 1$, $x'(0) = -1$. Graph $x(t)$ for several values of $b$ (positive, negative, and zero). Are there any values of $b$ for which the motion is oscillatory? Will some choice of $b$ lead to stability?

**4.** Repeat Problem 3 for the ODE $x'' + ax' - x = 0$, but with the coefficient $a$ as the "tuning" parameter.

# Constant Coefficient Linear ODEs: Driven

***Purpose*** To determine the response of a constant coefficient linear ODE to a variety of driving terms.

***Keywords*** Resonance, beats, Bounded Input/Bounded Output Principle

***See also*** Experiments 3.4, 4.1–4.2, 4.5–4.6; Atlas Plates `Second Order A-C`

***Background*** The driven second order constant coefficient linear ODE is

$$\begin{aligned} x'' + ax' + bx &= f(t) \\ x(0) = r, \quad x'(0) &= s \end{aligned} \tag{1}$$

where the driving term $f(t)$ is at least piecewise continuous on the $t$-range, $t \geq 0$. The solution of the ODE in (1) has the general form

$$x(t) = C_1 x_1(t) + C_2 x_2(t) + x_p(t) \tag{2}$$

where $x_1$ and $x_2$ are independent solutions of the undriven ODE, $x'' + ax' + bx = 0$, and $x_p(t)$ is any particular solution of the ODE in (1). There are unique values for the constants $C_1$ and $C_2$ so that $x(t)$ as given by (2) satisfies the initial data.

• **No damping**. First suppose that there is no damping in the ODE in (1) (i.e., $a = 0$) and that the coefficient $b$ is positive. For convenience set $b = k^2$. The solutions of the undriven ODE in (1) are free oscillations and have the form

$$x = C_1 \sin kt + C_2 \cos kt$$

where $k$ is called the **natural frequency**. Next suppose that the driving force is a sinusoid $F_0 \cos \omega t$ whose **amplitude** $F_0$ is positive and whose frequency (the **driving frequency**) is $\omega$. Assume that $\omega \neq k$. Solving the ODE

$$x'' + k^2 x = F_0 \cos \omega t \tag{3}$$

by the method of undetermined coefficients we obtain a particular solution

$$x_p(t) = \frac{F_0}{k^2 - \omega^2} \cos \omega t$$

which is called a **forced oscillation** (of frequency $\omega$). For arbitrary constants $C_1$ and $C_2$ the general solution of ODE (3) is

$$x(t) = C_1 \sin kt + C_2 \cos kt + \frac{F_0}{k^2 - \omega^2} \cos \omega t \tag{4}$$

• **Beats**. Suppose that we seek the solution of ODE (3) that satisfies $x(0) = 0, x'(0) = 0$. From (4) we see that $C_1 = 0$, $C_2 = -F_0(k^2 - \omega^2)^{-1}$. Hence, we have

$$\begin{aligned} x(t) &= \frac{F_0}{k^2 - \omega^2}(\cos \omega t - \cos kt) \\ &= \left[\frac{2F_0}{k^2 - \omega^2} \sin\left(\frac{k - \omega}{2} t\right)\right] \sin\left(\frac{k + \omega}{2} t\right) \end{aligned} \tag{5}$$

where we have used the trigonometric identity

$$\cos \alpha - \cos \beta = 2 \sin\left(\frac{\beta - \alpha}{2}\right) \sin\left(\frac{\beta + \alpha}{2}\right)$$

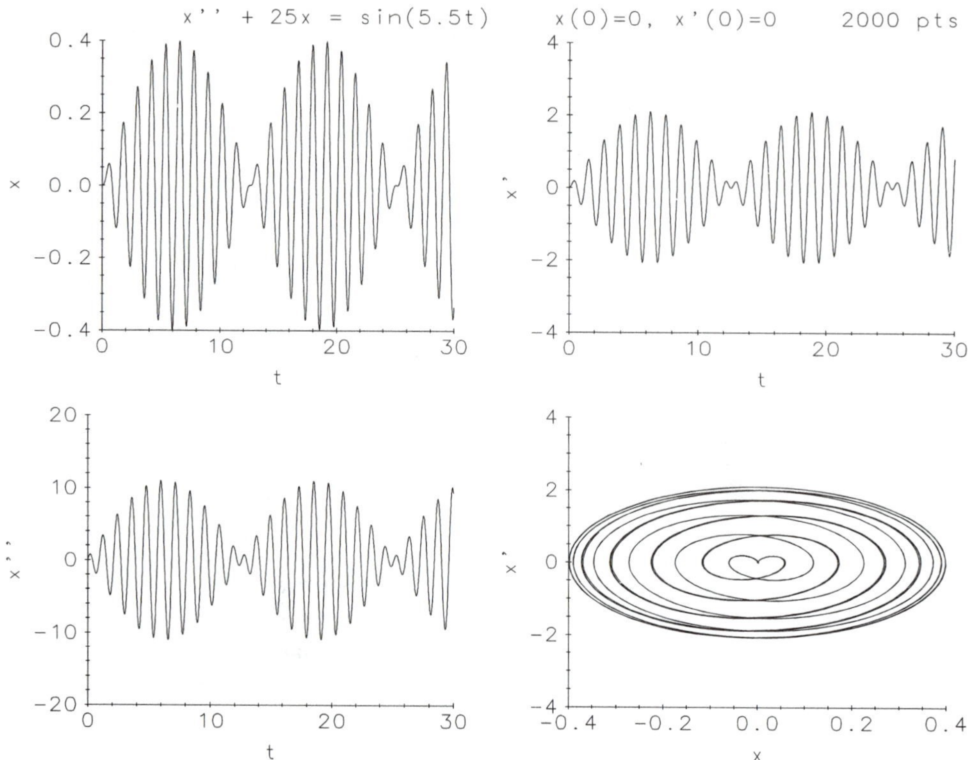


*Figure 4.3.1 The phenomenon of beats.*

If $|k - \omega|$ is small compared to $|k + \omega|$, then (5) shows that the solution displays the phenomenon of **beats** in which the amplitude (i.e., the term in brackets in (5)) of a high frequency sinusoid (the term $\sin(\frac{k+\omega}{2}t)$) slowly and periodically varies. Another term for the phenomenon is **amplitude modulation**. See Figure 4.3.1. Beats and amplitude modulation occur when free and forced oscillations of different frequencies appear in connection with some linear second order ODE with constant coefficients.

• **Resonance**. Suppose the frequency $\omega$ of the driving force exactly matches the natural frequency $k$ of the ODE. In this case no function of the form $A \cos kt$ or $A \sin kt$ could possibly be a solution of

$$x'' + k^2 x = F_0 \cos kt \tag{6}$$

since these functions are already solutions of the undriven equation. In fact, one particular solution of (6) has sinusoidal form with a steadily growing amplitude,

$$x_p(t) = \left(\frac{F_0}{2k} t\right) \sin kt$$

Thus, every solution of (6) contains a sinusoid with unbounded amplitude. This phenomenon is called **resonance**. See Atlas Plate `Second Order C` for an example.

***Observation*** **(Effect of damping)** If damping takes place (i.e., if $a \neq 0$ in (1)), then the solutions discussed above change again. The details of exactly what happens under those circumstances, as well as the case of general driving terms, are left to the experiments.

_______________/_______________
hardware software

_______________/_______________
name date

Answer questions in the space provided,
or on attached sheets or carefully labeled graphs.

_______________/_______________
course section

## *4.3 Beats and Resonance*

***Abstract*** The phenomena of beats and resonance are explored in this experiment.

**1.** Duplicate Atlas Plate `Second Order C` and Figure 4.1. Explain in your own words why the amplitude of the solution appears to be growing without bound. Explain the "bumps" in the $xt$-graph of Figure 4.1 in the Chapter 2 Introduction.

**2.** Select one of the IVPs below and do the problems that follow.

☐ **A.** $x'' + 25x = \sin \omega t,$ $x(0) = 0, x'(0) = 0$

☐ **B.** $x'' + 9x = 10 \cos \omega t,$ $x(0) = 1, x'(0) = 1$

☐ **C.** $x'' + x = \sin \omega t + \cos \omega t,$ $x(0) = 0, x'(0) = 1$

☐ **D.** $x'' + \pi^2 x = 3 \sin \omega t,$ $x(0) = 0, x'(0) = 0$

**(a)** Choose three values of $\omega$ so that corresponding solutions display (in turn) resonance, beats, and the superposition of two periodic functions, the ratio of whose periods is irrational. In the latter case, is the solution periodic? What is the period of the beats in the second case? Identify these properties on the graphs.

**(b)** Choose a variety of initial conditions for the beat case in (a). Is there much change in the beat phenomenon as the initial conditions are changed? Explain.

**3.** **(Damped resonance)** By graphing the solutions for different values of $\omega$, which value of $\omega$ maximizes the amplitude of the solution of the IVP, $x'' + x' + 25x = 10 \sin \omega t$, $x(0) = 0$, $x'(0) = 0$? Explain.

## *4.4 General Driving Forces*

***Abstract*** The response of the IVP, $x''+ax'+bx = f(t)$, to various bounded and unbounded periodic and nonperiodic driving forces is studied.

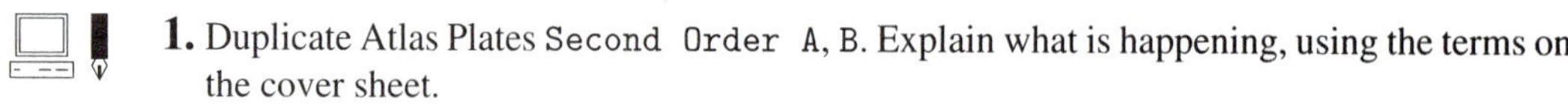

**1.** Duplicate Atlas Plates `Second Order A, B`. Explain what is happening, using the terms on the cover sheet.

**2.** Draw the graphs of the solutions of each of the following IVPs for $0 \le t \le 30$. Explain what is happening in terms of the phenomena described on the cover sheet.

**(a)** $x'' + 0.5x' + 25x = \sin(5t)$, $x(0) = 1$, $x'(0) = 1$

**(b)** $x'' + 0.5x' + 25x = 10 \sin t$, $x(0) = 1$, $x'(0) = 1$

**(c)** $x'' + 0.5x' + 25x = 10e^{-t/10} \sin(5t)$, $x(0) = 1$, $x'(0) = 1$

**(d)** $x'' + 25x = 10 \sin^5(5t) + e^{-t} \cos(5t)$, $x(0) = 1$, $x'(0) = 1$

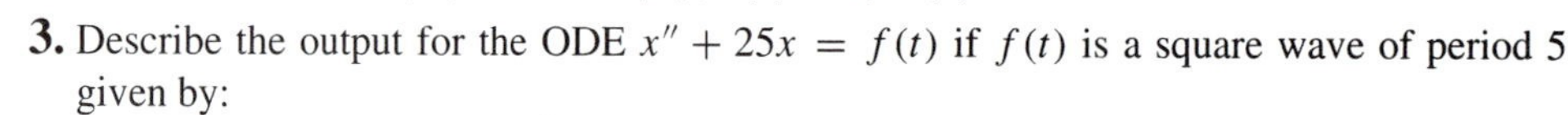

* **3.** Describe the output for the ODE $x'' + 25x = f(t)$ if $f(t)$ is a square wave of period 5 given by:

$$f(t) = \begin{cases} 1, & 0 \le t < 1 \\ -1, & 1 \le t < 5 \\ f \text{ extended periodically beyond } t = 5 \end{cases}$$

Attach graphs of solutions for various initial conditions.

* **4.** Look at Experiments 3.3–3.4 and then formulate a Bounded Input/Bounded Output Principle for the ODE $x'' + ax' + bx = f(t)$. Show by example that if $a \le 0$ or $b \le 0$, then the principle does not hold. State which ODEs in Problem 2 satisfy the hypotheses of the principle and explain why.

# Frequency Response Modeling

***Purpose*** To examine the frequency response approach to solving a second order linear constant coefficient ODE and its connection with system parameter identification.

***Keywords*** Characteristic roots, natural frequencies, sinusoidal input, transfer function

***See also*** Experiments 4.3–4.4; Atlas Plate `Second Order A`

***Background*** The second order constant coefficient ODE $x'' + ax' + bx = f(t)$ arises in modeling systems that involve oscillatory motion (see experiments on springs, pendulums, circuits). The constants $a$ and $b$ are determined by the system, and the driving term $f(t)$, which is externally controlled, is called the **input** to the system. Presuming a bit on the applications, we will write the ODE as

$$x'' + 2cx' + k^2x = f(t) \tag{1}$$

where $c$ and $k$ are nonnegative constants with $c < k$, and the input function has the sinusoidal form $F_0 \sin \omega t$, where the positive constants $F_0$ and $\omega$ are the **input amplitude** and **driving frequency**, respectively. The system response, $x(t)$, called the **output**, is characterized in terms of a so-called **steady state solution** (to be defined below) to the driven equation and solutions of the corresponding undriven ODE, called **transients**. The characteristic polynomial for (1) is

$$P(r) = r^2 + 2cr + k^2$$

which has the two complex conjugate roots $r_1 = -c + i\sqrt{k^2 - c^2}$, $r_2 = -c - i\sqrt{k^2 - c^2}$. Thus all transient solutions of ODE (1) have the form

$$x_{tr} = Ae^{-ct}\cos(k^2 - c^2)^{1/2}t + Be^{-ct}\sin(k^2 - c^2)^{1/2}t \tag{2}$$

where $A$, $B$ are arbitrary constants. All transients decay to zero exponentially as $t \to \infty$.

• **Sinusoidal inputs**. Now we turn to finding a particular solution of the driven ODE (1) with driving term $f(t) = F_0 \sin \omega t$, that is:

$$x'' + 2cx' + k^2x = F_0 \sin \omega t \tag{3}$$

Toward that end we first replace the driving term in (3) by a complex exponential function whose imaginary part is $F_0 \sin \omega t$, and then find a response to this new input function. The imaginary part of this response will then solve ODE (3) because the ODE has real coefficients. Note that $\text{Im}(F_0 e^{i\omega t}) = F_0 \sin \omega t$ and that ODE (3) assumes the complex form

$$P(D)[x] = F_0 e^{i\omega t} \tag{4}$$

where $D = d/dt$ and $P(D) = D^2 + 2cD + k^2$ is the second order linear operator which, when applied to a function $x(t)$, produces the function $x'' + 2cx' + k^2x$. Now since $P(D)[e^{rt}] \equiv P(r)e^{rt}$ for any real (or complex) constant $r$, we are led to look for a particular solution of ODE (4) in the form $x_p(t) = Ae^{i\omega t}$, where $A$ is a constant to be determined. Observe that $P(D)[Ae^{i\omega t}] = AP(i\omega)e^{i\omega t}$ and that $P(i\omega) = (i\omega)^2 + 2c(i\omega) + k^2 = k^2 - \omega^2 + i2c\omega$ does not vanish unless $k = \omega$ and $c = 0$. Hence, except for the special case $k = \omega$, $c = 0$, we see that if $H(r) = 1/P(r)$ then

$$x_p(t) = H(i\omega)F_0 e^{i\omega t} \tag{5}$$

solves the ODE (4). $H(r)$ is called the **transfer function** for ODE (1). To extract the imaginary part of the solution in (5) we first write $H(i\omega)$ in Cartesian form

$$H(i\omega) = \frac{1}{k^2 - \omega^2 + 2ic\omega} = \frac{k^2 - \omega^2}{(k^2 - \omega^2)^2 + 4c^2\omega^2} + i\frac{-2c\omega}{(k^2 - \omega^2)^2 + 4c^2\omega^2} \tag{6}$$

and then in polar form as follows:

$$H(i\omega) = M(\omega)e^{i\theta(\omega)} \tag{7}$$

$$M(\omega) = \frac{1}{\sqrt{(k^2-\omega^2)^2 + 4c^2\omega^2}} \qquad \theta(\omega) = \cot^{-1}\left(\frac{\omega^2 - k^2}{2c\omega}\right) \tag{8}$$

with $-\pi \le \theta(\omega) \le 0$ (Note: From (6) the complex number $H(i\omega)$ always "points" downward, and so the polar angle $\theta(\omega)$ is in the indicated range.) Now using (7) in (5) we see that

$$\begin{aligned} H(i\omega)F_0 e^{i\omega t} &= F_0 M(\omega) e^{i(\omega t + \theta(\omega))} \\ &= F_0 M(\omega)[\cos(\omega t + \theta(\omega)) + i\sin(\omega t + \theta(\omega))] \end{aligned}$$

and extracting the imaginary part of this function we find that

$$M(\omega)F_0 \sin(\omega t + \theta(\omega)) \tag{9}$$

is a real-valued solution of ODE (3). Note from (2) that all other solutions of ODE (3) decay as $t \to +\infty$; hence the solution in (9) is known as the **steady state solution**.

• **Gain and phase shift**. Notice that $M(\omega)$ and $\theta(\omega)$ do not depend on the initial data. Comparing (9) with the input $F_0 \sin \omega t$, we see that the steady state solution has the same sinusoidal form, but a different amplitude and phase. The ratio of the steady state amplitude to the input amplitude is $M(\omega)$ and is called the **gain**. The steady state solution is shifted in phase by $|\theta(\omega)|/\omega$ radians to the right, so it is common to refer to $\theta(\omega)$ as the **phase shift**. The graphs of gain and phase shift against $\omega$ (using a $\log_{10}$ scale on the $\omega$-axis) are called **Bode plots** and give valuable information about the system. Engineers call the quantity $20\log_{10} M(\omega)$ the gain in **decibels** (dB), and use decibel units on the gain axis. Degrees are used on the $\theta$-axis instead of radians. See Figure 4.5.1. These two graphs constitute the **frequency response curves** for the system.

***Observation*** **(Parameter identification)** The graph of the phase-shifted sinusoid $\sin(\omega t + \theta)$ against $t$ can be obtained from the graph of $\sin \omega t$ against $t$ as follows: If $\theta$ is positive (resp., negative) then shifting the graph of $\sin \omega t$ to the left (resp., right) by $\frac{\theta}{\omega}$ radians results in the graph of $\sin(\omega t + \theta)$. This clearly follows from writing $\sin(\omega t + \theta) = \sin(\omega(t + \frac{\theta}{\omega}))$. Sometimes it is known that a system can be modeled by an ODE of the type given in (1) but the system parameters cannot be accurately measured (i.e., the constants $c$ and $k$ are not accurately known). If a sinusoid $\sin \omega t$ is used as input in ODE (1) and if the gain $M(\omega)$ and phase shift $\theta(\omega)$ in the steady state output (9) can be experimentally determined, then formulas (8) form a basis for determining the system constants $c$ and $k$.

***Example 1*** Consider the ODE $x'' + 2x' + 4x = \sin 2t$ and initial data $x(0) = 0.5$, $x'(0) = 1$. We use a solver to produce the input-output graphs in Figure 4.5.2 and use these graphs to compute $M(2)$ and $\theta(2)$ and compare these values to the theoretical values given by (8). Observe that the output quickly settles down to a steady state. Using the vertical dashed lines we estimate that the output is shifted to the right by about 0.75 units, hence $\theta(2)$ is about 1.5 radians. From (8) we have that $\cot(\theta(2)) = 0$, and since $-\pi \le \theta(\omega) \le 0$, it follows that $\theta(2) = -\frac{\pi}{2}$. Since $\frac{\pi}{2} \approx 1.57$, our graphical procedure has done well. Using the horizontal dashed lines, we estimate $M(2) \approx 0.26$. From (8) we see that $M(2) = 0.25$, exactly.

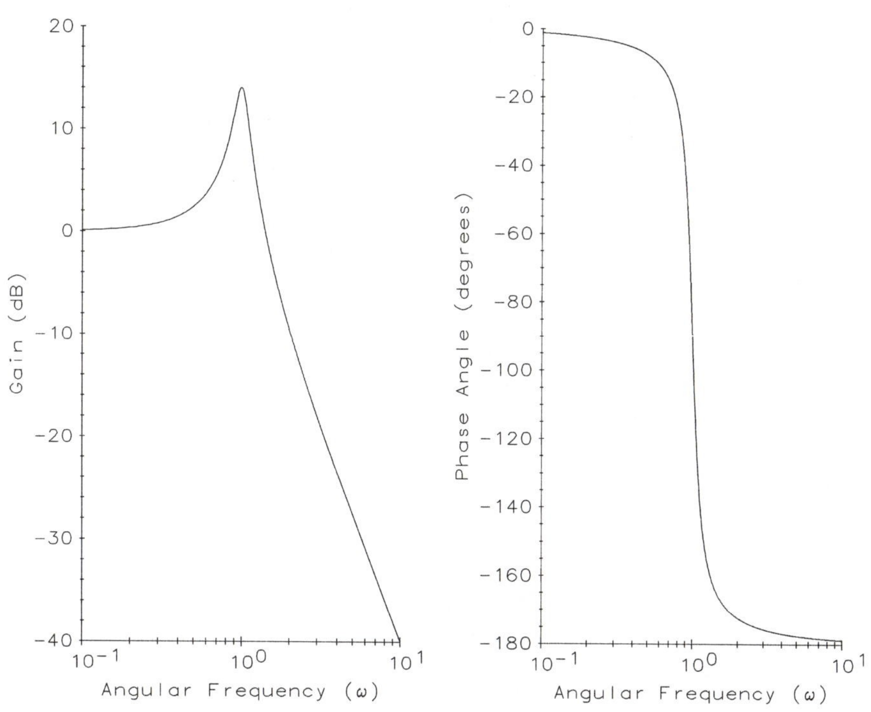


*Figure 4.5.1 Typical Bode plot.*

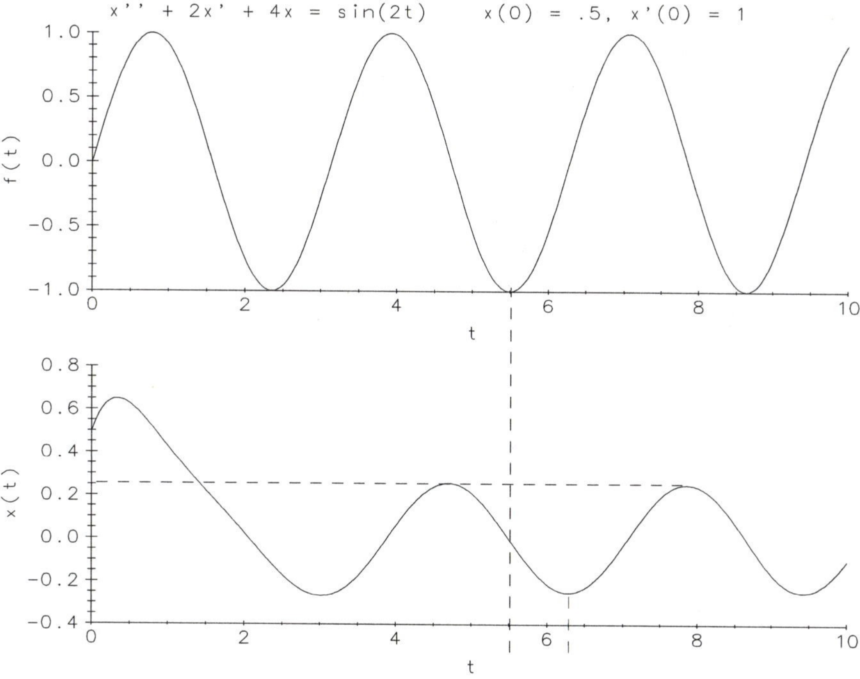


*Figure 4.5.2 Graphical determination of steady state response.*

_______________/_______________
hardware software

_______________/_______________
name date

Answer questions in the space provided, or on attached sheets or carefully labeled graphs.

_______________/_______________
course section

## 4.5 Parameter Identification

***Abstract*** This experiment explores the frequency response modeling technique to estimate system parameters.

**1.** Consider the ODE $x'' + x' + 2x = F_0 \sin \omega t$.

**(a)** Use an ODE solver to find enough values of the gain $M(\omega)$ and the phase shift $\theta(\omega)$ to make rough sketches of $M$ and $\theta$ versus $\omega$.

**(b)** Use the formulas in (8) to plot the graphs of $M(\omega)$ and $\theta(\omega)$. Compare these plots with the rough sketches drawn in (a).

**2.** Find the parameters $c$ and $k$ of the system whose Bode plots are shown in Figure 4.5.1.

**3.** Consider the ODE, $x'' + 2cx' + k^2x = f(t)$ with system parameters $c$ and $k$ and input $f(t)$.

**(a)** Assume that $k > c > 0$, but their exact values are not known. Design an experiment that will ultimately lead to the (approximate) identification of the parameters $c$ and $k$.

**(b)** Choose values for $c$ and $k$ and use the procedure developed in (a) to verify your chosen values of the system parameters.

**4.** Graph the solution of the IVP $x'' + 4x = \sin(3t)$, $x(0) = 0$, $x'(0) = 0$.

* **(a)** Does the forced oscillation have a nonzero phase shift? Can the forced oscillation be detected by examining the graphical output? Is the solution of IVP periodic? If so, why, and what is its period?

* **(b)** Repeat with input $\sin(\pi t)$ instead of $\sin(3t)$.

hardware / software    name / date

Answer questions in the space provided, or on attached sheets or carefully labeled graphs.    course / section

## 4.6 *Gain and Phase Shift*

***Abstract*** The steady state response of the ODE $x'' + 2cx' + k^2x = f(t)$, with $0 < c \leq k$, can be characterized in terms of an amplitude magnification and a phase shift applied to the sinusoidal driving term $f(t)$. This experiment explores this connection.

**1.** Use the solution curve for the IVP

$$x'' + 2x' + 4x = 3\sin 2t$$
$$x(0) = 0.5, \quad x'(0) = 0.5$$

to estimate the gain $M(2)$ and phase shift $\theta(2)$ for the steady state forced response. Compare these estimates with the exact values calculated from (8). Find the system transfer function.

**2.** Find the steady state solutions and the transient solutions of each of the following systems.

**(a)** $x'' + 8x' + 5x = 4\sin 3t$
**(b)** $x'' + 24x' + 25x = 2\sin(13t) - \sin(12t)$

**3.** Perform the tasks below:

**(a)** Verify using (8) that the general features of the graphs of $M(\omega)$ and $\theta(\omega)$, for fixed positive constants $k$ and $c$ with $k^2 > 2c^2$, are similar to the plots in the cover sheet. What is the value $\omega_r$ where $M(\omega)$ achieves a maximum ($\omega_r$ is called the **resonant frequency** of the system)? What are $M(0)$, $M(\omega_r)$? What is the value $\omega_0$ where $\theta(\omega)$ changes inflection?

**(b)** What happens to the frequency $\omega_r$ and the maximum value $M(\omega_r)$ determined in (a), for fixed $k > 0$, as $c \to 0$? Interpret this observation for a vibrating mechanical system generated by ODE (1) with very small damping constant $c$ and a sinusoidal driving term with frequency close to a natural (i.e., undamped) frequency of the system.

**4.** Consider the IVP based on the ODE (1) with $f(t) = \cos \omega t$:

$$\begin{aligned} x'' + 0.2x' + x &= \cos \omega t \\ x(0) = 0, \quad x'(0) &= 0 \end{aligned} \tag{10}$$

**(a)** Let $\omega = 0.5$. Plot the solution of (10) over a time sufficiently long that the transient part of the solution has become negligible and the steady state response is clearly visible. From the graph, determine the amplitude, $M$, of the steady state response, and compare it to the amplitude of the driving function.

**(b)** Now determine the phase shift, $\theta$, between the driving function and the steady state response. (Hint: See Figure 4.5.2.)

**(c)** The amplitude, $M$, and the phase shift, $\theta$, do not depend on initial conditions. Explain why this is so. Confirm this fact by repeating (a) and (b) for several sets of initial conditions.

**(d)** Determine the amplitude, $M$, and the phase shift, $\theta$, for various other values of $\omega$ by repeating (a) and (b) for the IVP (10). For instance, choose the values $\omega = 0.25, 0.75, 1, 1.5, 2$, and 3. If you have time, try additional values near $\omega = 1$. Sketch the graphs of $M$ versus $\omega$, and $\theta$ versus $\omega$. .

**(e)** Solve the IVP (10) analytically and determine the steady state response. Find $M$ and $\theta$ as functions of $\omega$. Compare these results with those of (a)–(d).

# Springs

***Purpose*** To model the motion of a mass attached to a spring, and to study the behavior of the solutions of the resulting second order ODE.

***Keywords*** Springs, oscillations, Hooke's Law; linear, soft, hard, and aging springs; Bessel functions

***See also*** Experiments 4.2, 4.3–4.4; Atlas Plates `Spring A-C`; Appendix B.2, B.5

***Background*** A mass $m$ is attached to the end of a (massless) spring, the other end of which is fastened to a wall. If the spring is compressed or extended along its axis and then released, the mass moves back and forth on its horizontal supporting surface as the spring vibrates. The nature of the motion depends on the nature and strength of the spring force, on the frictional forces that tend to dampen motion, and on any other external forces that act on the mass. The system is at equilibrium when the mass is at rest and the spring is neither stretched nor compressed. Let $x$ measure the displacement of the mass along the spring's axis from the equilibrium position; $x > 0$ corresponds to a stretched spring, $x < 0$ a compressed spring. Figure 4.7.1 displays the spring-mass system stretched beyond equilibrium, subject to an **external force** $G(t)$ acting parallel to the $x$-axis, and with frictional forces represented by a **dashpot** or **damper**, which is shown as a cylinder of fluid through which a piston moves. The gravitational force on the mass acts in a vertical direction and is assumed to have no effect on horizontal motion.

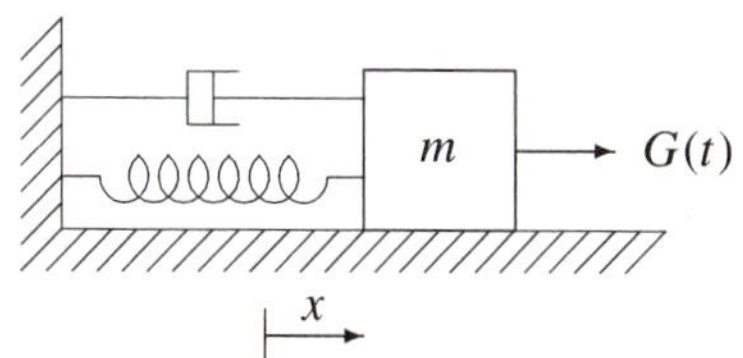


*Figure 4.7.1 A mass-spring system with dashpot and external force.*

The problem now is to describe the changing displacement $x(t)$ of the mass under the action of the forces. Newton's Second Law may be applied in the $x$-direction, but assumptions are needed about the spring and damping forces. Common assumptions based on observation of real springs are:

**Spring Force:** *The force exerted by the spring acts parallel to the $x$-axis and has magnitude that is a function $F(x)$ of the distance of the mass from the position of equilibrium.*

**Viscous Damping Force:** *The damping force on the mass has magnitude proportional to the velocity $x'$ of the mass and acts in the direction opposite to the motion.*

Under these assumptions, Newton's Second Law implies that

$$mx''(t) = F(x(t)) - cx'(t) + G(t) \tag{1}$$

where $c > 0$ is the **damping constant**. Careful measurements suggest that if the displacement from equilibrium is small, then the spring force $F(x)$ is an odd function (i.e., $F(-x) = -F(x)$), and points in the opposite direction to the displacement $x$. Four commonly used models with these properties are:

$$\begin{aligned}
&\textbf{Hooke's Law (linear spring)} \quad & F(x) &= -kx & (2)\\
&\textbf{Hard spring} \quad & F(x) &= -kx - bx^3 & (3)\\
&\textbf{Soft spring} \quad & F(x) &= -kx + bx^3 & (4)\\
&\textbf{Aging spring} \quad & F(x,t) &= -k(t)x & (5)
\end{aligned}$$

where $k$ and $b$ are positive (See Figure 4.7.2). Observe that in comparison to the Hooke's Law force, the hard spring force strengthens with extension, the soft spring force weakens with extension, and the aging spring force weakens with time.

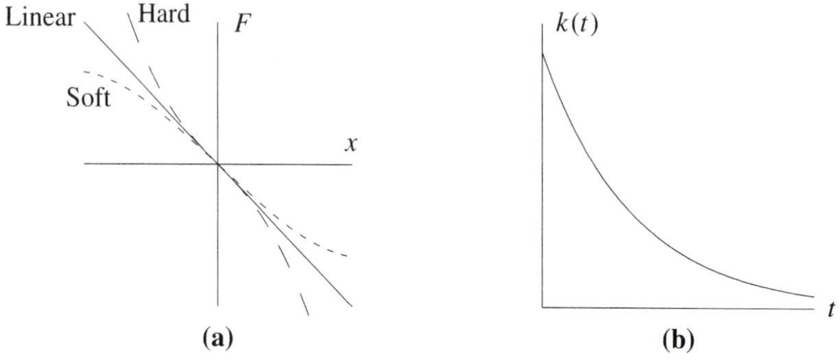


*Figure 4.7.2 (a) Time-invariant spring forces. (b) Decaying coefficient in an aging spring.*

The equations of motion of the linear spring and of the aging spring are the linear equations (6) and (7), but the hard and soft springs have the nonlinear models (8) and (9):

$$mx'' + cx' + kx = G(t) \tag{6}$$
$$mx'' + cx' + k(t)x = G(t) \tag{7}$$
$$mx'' + cx' + kx + bx^3 = G(t) \tag{8}$$
$$mx'' + cx' + kx - bx^3 = G(t) \tag{9}$$

For convenience, the external force term $G(t)$ appears on the right hand sides of these equations, the other force terms on the left. Atlas Plates `Spring A-C` display the motions of undriven, undamped, constant coefficient linear, hard, and soft spring model ODEs, respectively. Orbits are sketched in the $xy$-plane where $y = x'$; in two cases, plots of $x, x'$ versus $t$ are also shown. Periodic oscillations appear in the $xy$-plane as simple closed curves. The other orbits in Atlas Plate `Spring C` correspond to unbounded stretching.

***Example 1*** Figure 4.7.3 shows orbits of an undriven, damped soft spring ($c = .1, k = 1, b = .1$).

***Example 2*** Figure 4.7.4 shows the motion of an undamped linear spring driven by the square wave

$$G(t) = \begin{cases} 2, & 0 \le t \le 1 \\ 0, & 1 < t < 2 \\ \text{extended periodically with period 2} \end{cases}$$

$G(t)$ is denoted by $2\mathrm{SQW}(t, 50, 2)$ in Figure 4.7.4, where SQW is shorthand for "square wave." The number 50 indicates a 50% duty cycle (the function "on" for the first 50% of the period), and the number 2 is the period. In this case the amplitude of the square wave is 2. The IVP for this setting of a driven, undamped linear spring with unit mass attached is

$$x'' + 5x = G(t), \qquad x(0) = 0, \qquad x'(0) = 0 \tag{10}$$

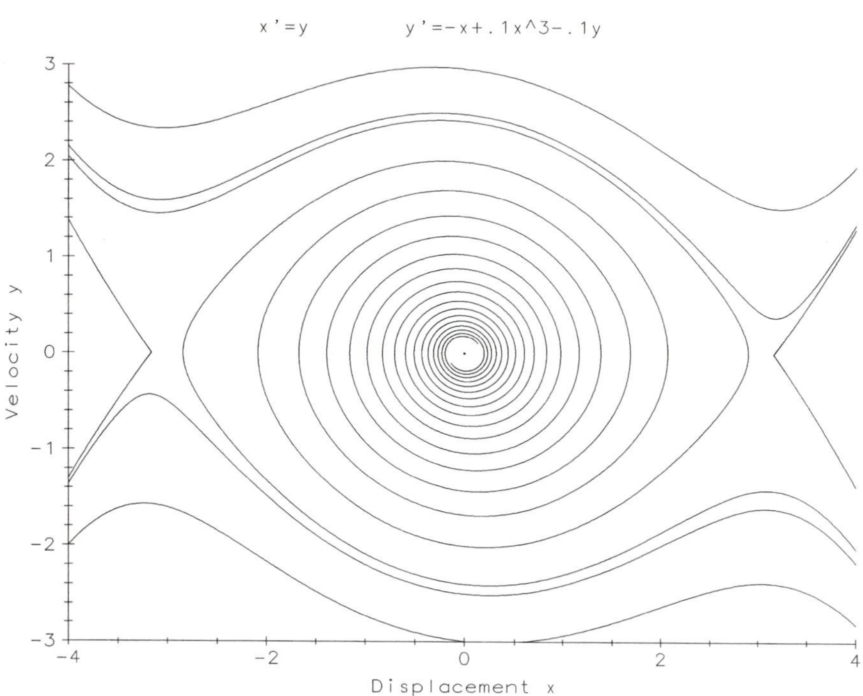


*Figure 4.7.3* *Motions in xy-state space of a damped, undriven soft spring.*

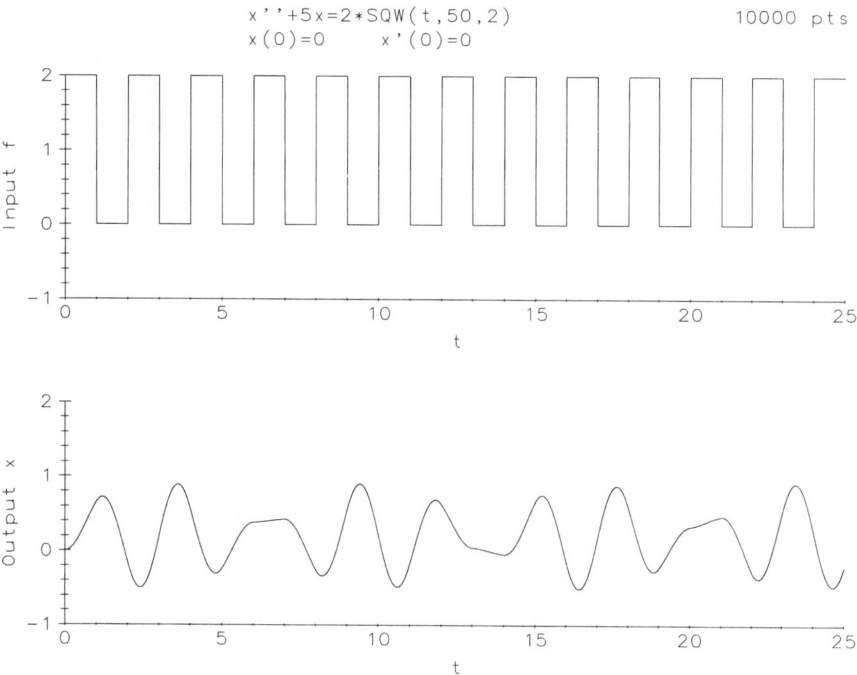


*Figure 4.7.4* *Input/output graphs of a linear spring system driven by a square wave.*

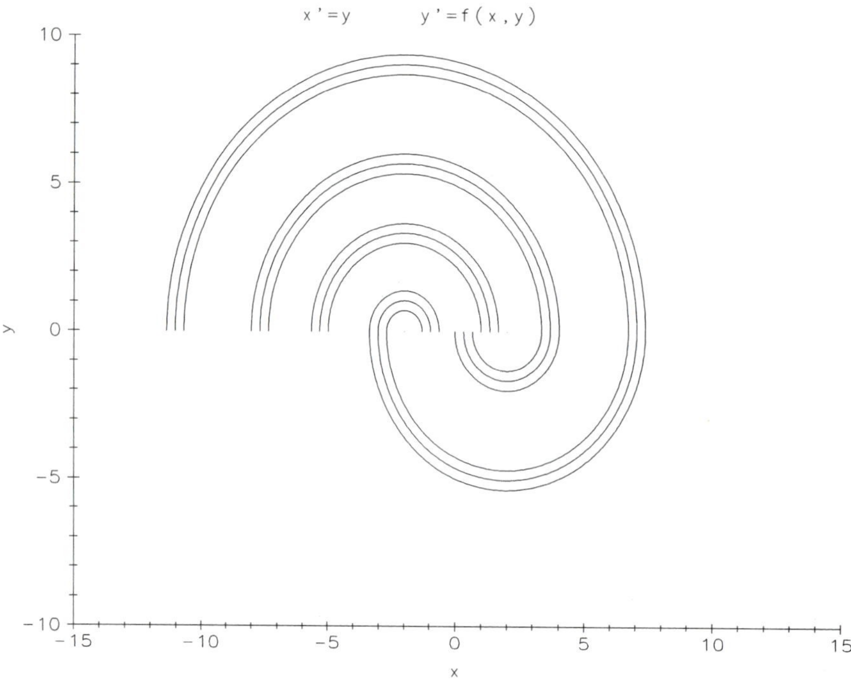


*Figure 4.7.5 Spring-mass motion with dry friction (see Example 3 for a definition of $f$).*

***Example 3*** **(Dry friction)** The above models differ in the nature of the spring force, but the friction force is the same, $-cx'$, in all models, although $c = 0$ in the undamped cases. Suppose now that the dashpot is removed from the spring-mass system and that the sliding surface is roughened. If the mass is displaced slightly from equilibrium and released from rest, it remains where it is, held in place by the **dry friction** of the surface, which is strong enough to cancel the Hooke's Law spring force. However, if the displacement is large enough, the spring force exceeds the force of dry friction and motion of the mass toward equilibrium begins. While the mass is moving, the dry frictional force is constant in magnitude and opposite to the direction of motion. A model of dry friction is the piecewise-linear model

$$mx'' = f(x, x') = \begin{cases} -kx + \mu & \text{if } x' < 0 \\ -kx - \mu & \text{if } x' > 0 \\ -kx & \text{if } x' = 0 \text{ and } k|x| > \mu \\ 0 & \text{if } x' = 0 \text{ and } k|x| \le \mu \end{cases} \tag{11}$$

where $k$ and $\mu$ are positive constants. Orbits of the motion are successive arcs of semicircles centered alternately at $(-\mu/k, 0)$ and $(\mu/k, 0)$ in the $xy$-plane according as $y = x' > 0$ or $y = x' < 0$. Motion continues until an arc terminates in the **dead zone**, the interval $|x| \le \mu/k$ on the $x$-axis. In the dead zone, frictional and spring forces cancel out. Figure 4.7.5 displays orbits of system (11) where $m = 1, k = 1, \mu = 2$.

***Observation*** **(Linear springs)** The most widely used model for spring-mass motion is that of Hooke's Law, with or without linear damping $-cx'$, and with or without an external driving force $G(t)$. The ODE in this case is linear, and all the results of Experiments 4.1–4.6 apply.

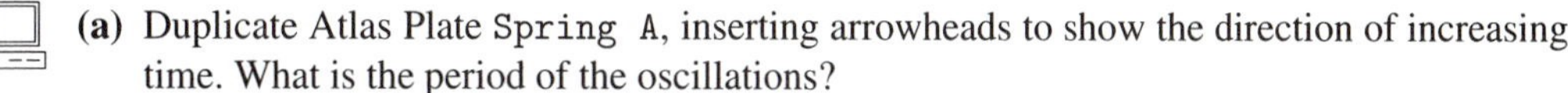


Answer questions in the space provided, or on attached sheets or carefully labeled graphs.

## *4.7 Linear Springs*

***Abstract*** The motions of a linear spring are studied; the restoring force is assumed to obey Hooke's Law with a constant coefficient. The motion of a spring-mass system hanging vertically is considered. In addition, the piecewise linear model with dry frictional damping is treated.

**1.** Consider the constant coefficient linear spring model (6).

**(a)** Duplicate Atlas Plate `Spring A`, inserting arrowheads to show the direction of increasing time. What is the period of the oscillations?

**(b)** Insert damping and driving terms so that $x'' + 0.01x' + 0.2x = 2\cos\omega t$. If $x(0) = 0$, $x'(0) = 0$, what value of $\omega$ produces a response of maximal amplitude? Sketch graphs like those in Atlas Plate `Spring A` for several values of $\omega$ including that value yielding maximum amplitude.

**(c)** Springs have elastic limits of stretching/compression. If the positive numbers $L_1$ and $L_2$ are the **elastic limits** of the spring, the model ODE is valid only if the displacement $x(t)$ satisfies $-L_1 \le x(t) \le L_2$. If the limits are exceeded, the spring breaks, stretches without bound, or for some other reason cannot be modeled by a simple ODE. Suppose that for the model ODE in (b) the elastic limit is reached at $|x| = 1$, that $\omega = 1.6$, and that $x(0) = 0$, $x'(0) = v_0 > 0$. Sketch graphs like those in Atlas Plate `Spring A` for various values of $v_0$. Estimate the critical value of $v_0$ for which the spring reaches its elastic limit.

**2.** Duplicate Figure 4.7.4 for the driven undamped spring modeled by $x'' + 5x = G(t)$, $x(0) = 0$, $x'(0) = 0$, where $G(t)$ is the square wave defined on the cover sheet. Replace the Hooke's Law constant 5 in the ODE by the parameter $k$, $1 \le k \le 5$. What value of $k$ maximizes the amplitude of the response on $0 \le t \le 10$? Attach a graph and explain your work on the graph.

* **3. (Backlash model).** Consider the "backlash" model sketched below, where $x$ measures the displacement of the mass $m$ from a point midway between the plates of the massless springs in the equilibrium position.

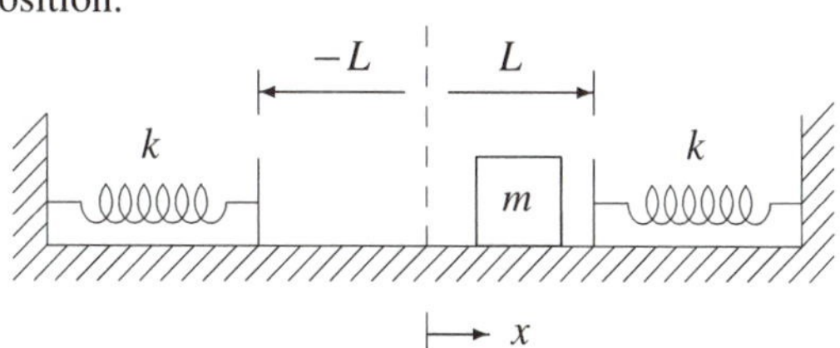


The mass moves freely between the springs until it hits one of them and compresses it; that spring then rebounds, releasing the mass when the equilibrium position is regained, and the massless spring abruptly stops. The motion of the mass repeats but with the mass now moving toward the other spring. Neither spring stretches beyond its equilibrium position. The surface is assumed to be frictionless.

**(a)** Model the motion of the mass up to the time it hits the second spring. (Suggestion: Try $m(x - L)'' + k(x - L) = 0$ for the initial stage of the motion.) Explain the equation, and write down appropriate initial data. Explain why the period $T$ is 4 times the amount of time it takes that mass to move from an initial position of maximal spring compression to the position $x = 0$. Find $T$ as a function of $m$, $k$, $L$, and $a$. (Suggestion: Divide the motion into two phases: (i) mass moves with a spring, and (ii) mass moves freely at constant speed between the springs.)

**(b)** Use a solver/grapher to plot the motion of the mass over a time span of three periods. Compare the period obtained graphically with that determined in (a).

**4. (Dry friction)**. Consider the piecewise linear dry friction model of system (11).

* **(a)** Explain the model in your own words.

* **(b)** Create orbits like those shown in Figure 4.7.5, using $m = 1, k = 1, \mu = 2$. Observe that the orbit consists of a union of circular arcs (of successively smaller radii) that continue until the mass reaches the dead zone, $|x| \leq 2$, $y = 0$. Explain the motion. Attach orbital graphs and graphs of $x$ and $x'$ plotted against $t$, $0 \leq t \leq 20$. Explain why every orbit eventually reaches the dead zone regardless of initial data.

**5. (Vertical spring)** Suppose that a Hooke's Law linear spring is fastened to the ceiling and hangs vertically. Suppose that a mass $m$ is attached to the spring and the spring is allowed to stretch a distance $h$, where the spring force balances the gravitational force. Explain why $kh = mg$. If $x$ denotes the displacement from the equilibrium position, explain why (6) is a reasonable model for the vertical motion of the mass. Then let $m = 1$, $k = 0.1$, $G(t) = 10 \cos \omega t$. Vary $\omega$ over the interval $0 \leq \omega \leq 5$ and find the value of $\omega$ that maximizes the amplitude of $x(t)$ if $x(0) = 0$, $x'(0) = 0$, $0 \leq t \leq 20$. Attach graphs.

_____/_____ hardware software

_____/_____ name date

Answer questions in the space provided, or on attached sheets or carefully labeled graphs.

_____/_____ course section

## *4.8 Hard and Soft Springs*

***Abstract*** The motions of a mass suspended by a hard or a soft spring are considered. A revised soft spring model is proposed.

**1.** Duplicate Atlas Plate `Spring B` for the motions of an undamped, undriven hard spring. Insert arrowheads on orbits indicating increasing time. Estimate the periods, and explain why the period decreases as the cycle amplitude increases. What are the minimal and the maximal periods?

**2.** Model the motion of a damped, undriven hard spring by $x'' + cx' + x + x^3 = 0$. Graph orbits in the $xx'$-state space for a variety of positive values of $c$. What happens to all orbits as $t \to \infty$? Justify your answer. Find the critical value of $c$ that separates underdamping (oscillating and decaying motion about equilibrium) from overdamping.

**3.** The hard spring equation with a periodic sinusoid driving term is $mx'' + cx' + kx + bx^3 = A_0 \cos \omega t$. Let $x(0) = 0$, $x'(0) = 0$.

**(a)** Rescale the equation by setting $x = Au$, and $t = Bs$, and choosing $A$ and $B$ so that the rescaled equation can be written as $d^2u/ds^2 + c_1 du/ds + u + b_1 u^3 = \cos(\omega_1 s)$ with parameters, $c_1, b_1, \omega_1$. (Hint: See Appendix B.5 for a rescaling of the soft spring.)

**(b)** Vary $c_1$, $b_1$, and $\omega_1$ (one at a time) from their respective "base" values of 1, 1, 1 and see if you can "break" the spring, if its elastic limit is given by $|x| = 100$. Explain your work and graph the orbits in the $xx'$-plane in each case.

**4.** Duplicate Atlas Plate `Spring C` for the motions of an undamped and undriven soft spring, inserting arrowheads of time. Explain why the model loses physical significance if $|x| \geq \sqrt{10}$. What do you think happens to the spring if $x(t)$ increases to $\sqrt{10}$ as $t \to t_1 > 0$, where $x'(t_1) > 0$? Explain. If possible, plot the orbits as $t$ increases through $t_1$.

**5.** Consider the damped, undriven soft spring model $x'' + cx' + x - x^3 = 0$, with elastic limit $|x| = 1$.

**(a)** Let $x(0) = 0$, $x'(0) = 100$, and graph orbits for various values of $c$. Estimate the minimal value of $c$ for which motion stays within the elastic limit. Explain your work on your graphs.

**(b)** Delete the damping term and add a driving term $\cos \omega t$. Graph orbits for $x(0) = 0$, $x'(0) = 100$, and various values of $\omega$. Is there a value of $\omega$ for which the motion stays within the elastic limit?

* **6.** The soft spring model given by (9) is seriously flawed since it treats compression and extension as equivalent. Explain what is wrong with this model. The model

$$F(x) = \begin{cases} -kx & \text{if } x < 0 \\ -kx + bx^3 & \text{if } 0 \leq x \leq \sqrt{k/b} \\ 0 & \text{if } x > \sqrt{k/b} \end{cases}$$

is more realistic. Explain. Using this new soft spring model, carry out a graphical analysis of the motions for various values of $k$ and $b$.

_______________/_______________
hardware software

_______________/_______________
name date

_______________/_______________
course section

Answer questions in the space provided, or on attached sheets or carefully labeled graphs.

## *4.9 Aging Springs*

***Abstract*** Various models of a linear spring with a decaying Hooke's Law coefficient are considered. One of these leads to the Bessel equation of order 0.

***Special instructions*** This experiment should be done by a team. Consult Appendix A before writing the group report. Do the following in considering the motion of an aging spring.

- Duplicate Figure 4.2 in the Chapter 4 Introduction. Repeat with $x(0) = 0, x'(0) = A$ and construct graphs for various values of $A$. Extend your analysis to $0 \le t \le 1000$. What happens to the velocity and the acceleration as $t \to 1000$? Explain. Estimate the minimum value of $A$ for which the spring stops oscillating at time $T = 100$ and begins to stretch without bound.
- Consider the general model $mx'' + k_0 e^{-at} x = 0$ for an aging spring, where $m$, $k_0$, and $a$ are positive constants. Rescale time by $s = t/a$ so that the model becomes $x'' + k_1 e^{-t} x = 0$, where $k_1$ is a positive constant, and the rescaled time $s$ is again denoted by $t$. Now set $\tau = 2\sqrt{k_1} e^{-t/2}$ and show that this second (and nonlinear) rescaling of time changes the equation of the aging spring into Bessel's equation of order 0:

$$\tau^2 \frac{d^2x}{d\tau^2} + \tau \frac{dx}{d\tau} + \tau^2 x = 0$$

  Explain your earlier graphs in terms of properties of Bessel functions of order 0.
- Create your own model of an aging spring. Analyze it graphically, and justify your model. Use reasonable elastic limits.
- Create a model of an aging hard spring. Explain your model and draw orbital and component graphs for various values of the parameters. Use reasonable elastic limits.
- Insert a linear damping term, $-cx'$, in all the models considered. Is it possible to choose $c$ large enough to prevent the spring from stretching without bound as $t \to \infty$? Explain.

# Circuits

***Purpose*** To study the ODE model of the flow of electrical energy in a simple $RLC$ circuit, and to experiment with the use of the circuit as a radio tuner.

***Keywords*** Circuit, voltage, current, charge, resistance, capacitance, inductance, tuner, filter, engineering functions, frequency, input, output, superposition, amplitude

***See also*** Experiments 4.2–4.7; Appendix B.4

***Background*** As outlined in Appendix B.4, the charge $q(t)$ (in coulombs) on the capacitor at time $t$ (in seconds) and the current $I(t)$ (in amperes) in the simple $RLC$ circuit of Figure 4.10.1 satisfy the respective IVPs (1) and (2). The IVP for charge is

$$Lq'' + Rq' + \frac{1}{C}q = \mathcal{E}(t), \qquad q(0) = q_0, \qquad q'(0) = I_0 \tag{1}$$

where $L$ is the inductance (in henries), $R$ the resistance (in ohms), $C$ the capacitance (in farads), and $\mathcal{E}(t)$ the voltage (in volts) of the external power source. If $\mathcal{E}$ is continuously differentiable, then the IVP (1) has the equivalent formulation in terms of current

$$LI'' + RI' + \frac{1}{C}I = \mathcal{E}'(t), \qquad I(0) = I_0, \qquad I'(0) = \frac{\mathcal{E}(t_0) - RI_0 - q_0/C}{L} \tag{2}$$

If $q(t)$ is the solution of IVP (1), then $I = q'(t)$ is the solution of IVP (2).

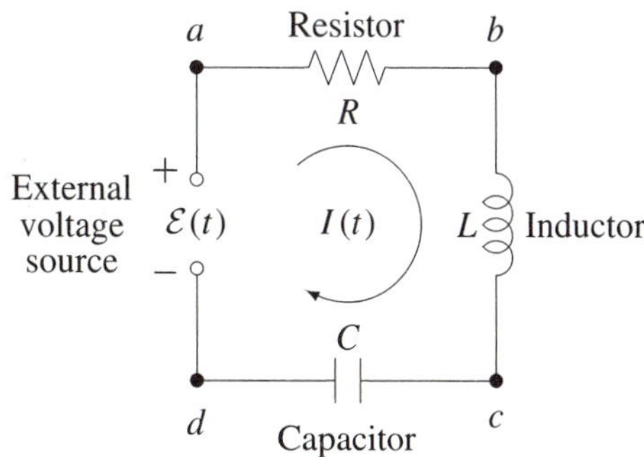


*Figure 4.10.1 The $RLC$ circuit.*

The notable feature of these ODEs for charge and current is that they have the same form as the ODE for the displacement $x$ from equilibrium of a linear spring-mass system (Experiment 4.7):

$$mx'' + cx' + kx = G(t) \tag{3}$$

The table below shows this formal correspondence in detail.

| **Circuit Element** | **Spring Element** |
|---|---|
| Inductance $L$ | Mass $m$ |
| Resistance $R$ | Damping coefficient $c$ |
| Reciprocal capacitance $1/C$ | Spring constant $k$ |
| Charge $q$, current $I$ | Displacement $x$ |
| External voltage $\mathcal{E}(t)$, voltage rate $\mathcal{E}'(t)$ | External force $G(t)$ |

We shall focus on two problems: the response of the $RLC$ circuit to piecewise-linear input voltages, and the tuning of the circuit so that it filters out (by attenuation) all but a prescribed narrow band of input frequencies. Although the discussion here is in terms of voltages, currents, and charges, it applies as well to any ODE of the same form.

• **The engineering functions**. In engineering and scientific applications inputs are often piecewise-linear functions whose graphs consist of line segments joined at "corners," the so-called **engineering functions**. Four of these functions and some extensions are listed below. Note that in three cases the extensions are periodic; $p$ denotes the period. The **duty cycle** $d$ of a periodic function is the percentage of the period when the function is "on." By convention, the "on" interval is always at the start of a period. The list of engineering functions includes: (a) square pulses and waves, (b) triangular pulses and waves, (c) sawtooth pulses and waves, and (d) step and stair.

(a)
$$\text{SQP}\,(t,a) = \begin{cases} 1, & 0 \le t \le a \\ 0, & \text{otherwise} \end{cases}$$
$$\text{SQW}\,(t,d,p) = \begin{cases} 1, & 0 \le t \le p(d/100) \\ 0, & p(d/100) < t \le p \\ \text{extended with period } p \end{cases}$$

(b)
$$\text{TRP}\,(t,a) = \begin{cases} 2t/a, & 0 \le t \le a/2 \\ 2-2t/a, & a/2 < t \le a \\ 0, & \text{otherwise} \end{cases}$$
$$\text{TRW}\,(t,d,p) = \begin{cases} 2t/p(d/100), & 0 \le t \le p(d/100)/2 \\ 2-2t/p(d/100), & p(d/100)/2 < t \le p(d/100) \\ 0, & p(d/100) < t \le p \\ \text{extended with period } p \end{cases}$$

(c)
$$\text{SWP}\,(t,a) = \begin{cases} t/a, & 0 \le t \le a \\ 0, & \text{otherwise} \end{cases}$$
$$\text{SWW}\,(t,d,p) = \begin{cases} t/p(d/100), & 0 \le t \le p(d/100) \\ 0, & p(d/100) < t \le p \\ \text{extended with period } p \end{cases}$$

(d)
$$\text{Step}(t) = \begin{cases} 1, & t \ge 0 \\ 0, & \text{otherwise} \end{cases}$$
$$\text{Stair}(t,a) = \begin{cases} 1, & 0 \le t \le a \\ 2, & a < t \le 2a \\ \vdots & \\ n, & (n-1)a < t \le na \end{cases}$$

If $\mathcal{E}(t)$ (or $\mathcal{E}'(t)$) is one of these functions, then it is not easy to find an explicit formula for the charge $q(t)$ on the capacitor (or the current $I(t)$ in the circuit) as the solution of IVP (1) (or IVP (2)). However, if the engineering functions are predefined in the solver package or if they can be defined by the user, the solver/graphics package will draw the graph of the output $q(t)$ (or $I(t)$) for given initial data. Figures 4.10.2–4.10.4 display input/output graphs when some of the engineering functions are used as inputs.

• **The $RLC$ circuit as a frequency tuner**. The $RLC$ circuit of Figure 4.10.1 can be used as a tuner to select one frequency out of a band of frequencies in the input voltage $\mathcal{E}(t)$. The algebra is simple if we follow the common engineering practice of using the complex exponential $F_0e^{i\omega t}$, where $F_0$ is a positive real number, as a complex input voltage rather than using its real part $F_0 \cos \omega t$ or its imaginary component $F_0 \sin \omega t$ (recall that $F_0e^{i\omega t} = F_0 \cos \omega t + i F_0 \sin \omega t$). This approach has also been used in Experiments 4.5–4.6. We shall consider the ODEs

$$Lq'' + Rq' + \frac{1}{C}q = F_0e^{i\omega t} \tag{4}$$

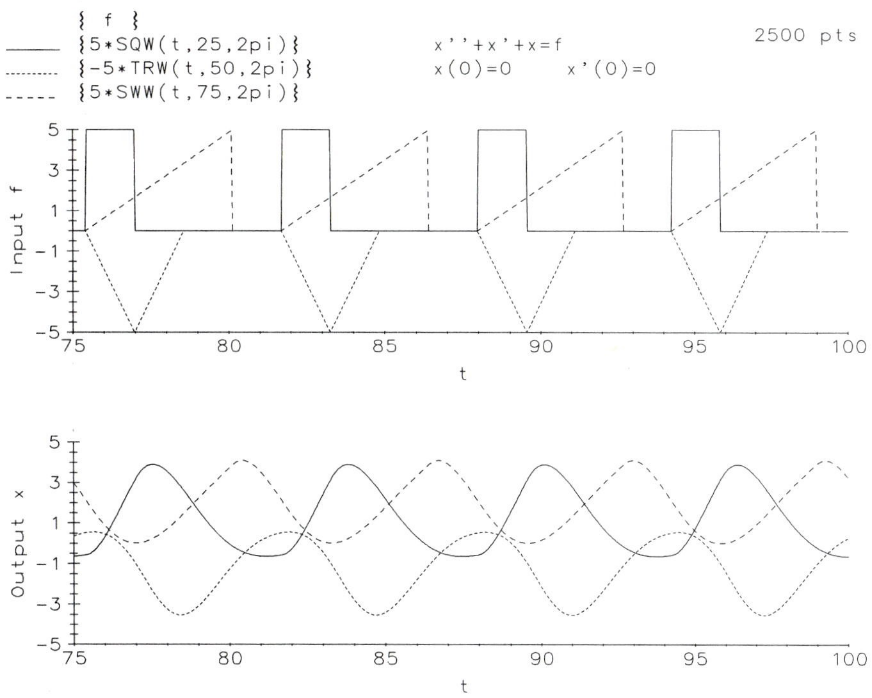


*Figure 4.10.2 Responses of an RLC circuit ($R = 1$, $L = 1$, $C = 1$) to three periodic engineering functions with duty cycles of 25, 50, and 75%; output plotted for $75 \leq t \leq 100$ after transients have decayed.*

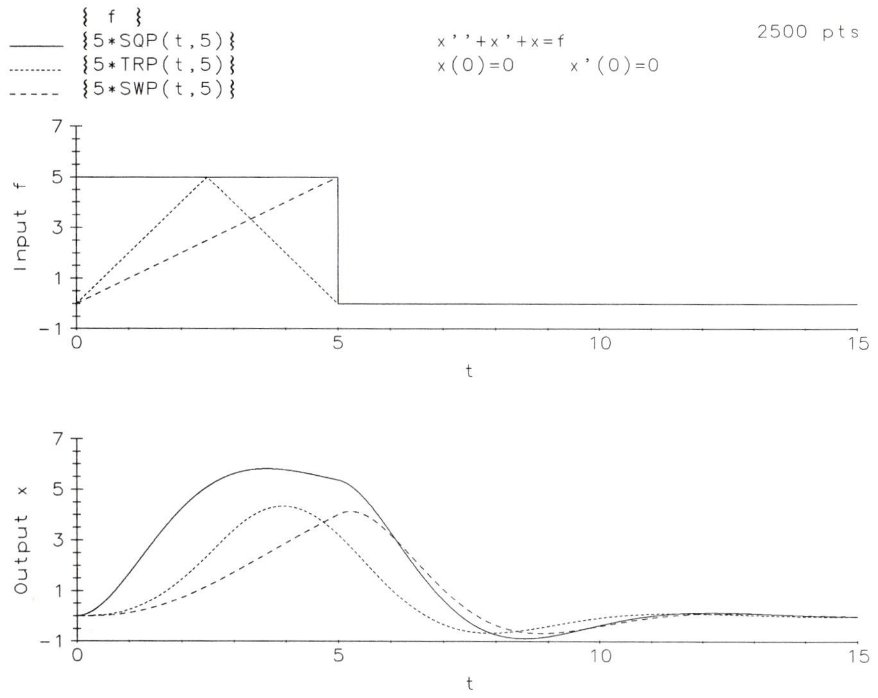


*Figure 4.10.3 Responses of an RLC circuit ($R = 1$, $L = 1$, $C = 1$) to three engineering functions.*

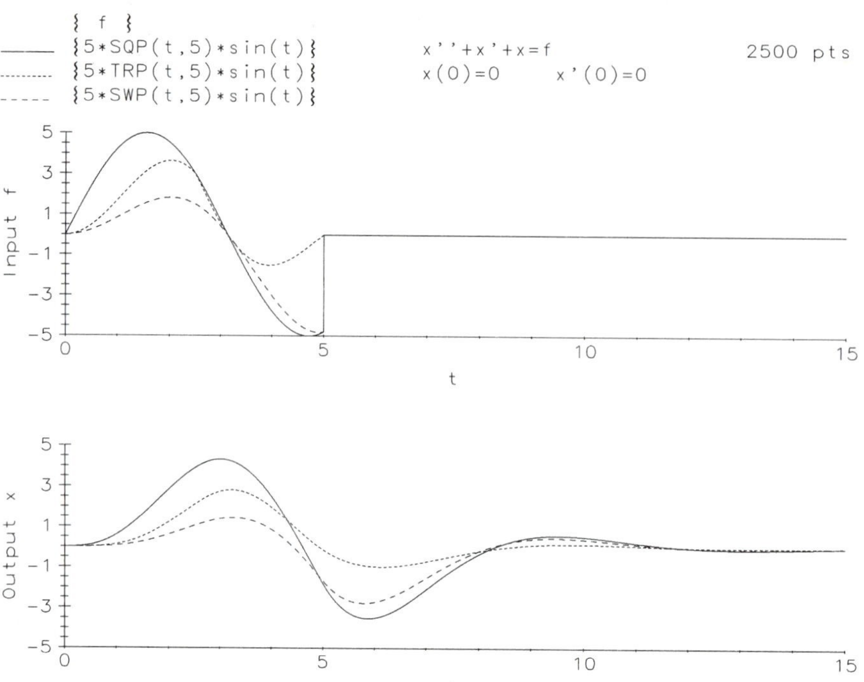


*Figure 4.10.4 Responses of an RLC circuit ($R = 1$, $L = 1$, $C = 1$) to the products of engineering functions and a sinusoid.*

$$LI'' + RI' + \frac{1}{C}I = F_0 i\omega e^{i\omega t} \tag{5}$$

Consider the steady state solution $q_p(t)$ of (4) obtained by the method of undetermined coefficients. Because $L$, $R$, and $C$ are assumed to be positive, all solutions of the undriven equation, $Lq'' + Rq' + q/C = 0$, tend to 0 as $t \to \infty$; hence, the steady state solution $q_p(t)$ represents the long-term behavior of the (complex) charge on the capacitor in the circuit. Given the relationship between the ODE for the charge and that for the current, the steady state (complex) current is easily found from $I_p(t) \equiv q_p'(t)$. Using undetermined coefficients and the relation between $q_p$ and $I_p$, we find

$$q_p = \frac{F_0}{\left(1/C - \omega^2 L\right) + i\omega R} e^{i\omega t} \tag{6}$$

$$I_p = \frac{F_0}{R + i\,(\omega L - 1/\omega C)} e^{i\omega t} \tag{7}$$

The denominator in (7) is called the **complex impedance**.

The amplitude of the input voltage $F_0 e^{i\omega t}$ is $F_0$ since $|F_0 \cos\omega t + i\,F_o \sin\omega t| = F_0\sqrt{\cos^2\omega t + \sin^2\omega t} = F_0$. The amplitude of the steady state output current $I_p(t)$ is

$$\left|I_p(t)\right| = \frac{F_0}{|R + i\,(\omega L - 1/\omega C)|} = \frac{F_0}{\left[R^2 + (\omega L - 1/\omega C)^2\right]^{1/2}} \tag{8}$$

Suppose that we want to "tune" the circuit by changing the capacitance (or inductance) to maximize the amplitude of $I_p(t)$. From (8) we see that the maximal response amplitude is attained by selecting $C$ (or $L$) so that

$$\omega L - \frac{1}{\omega C} = 0 \tag{9}$$

or

$$\frac{1}{LC} = \omega^2 \tag{10}$$

The ratio of the steady state amplitude from (8) to the input amplitude $F_0$ is the **gain**; thus, the gain is largest if (10) is satisfied.

Now suppose that the $RLC$-oscillator of a radio circuit receives periodic inputs from a number of broadcasting stations. In this setting the input voltage and its derivative have the forms

$$\mathcal{E}(t) = \sum_{k=1}^{N} F_k e^{i\omega_k t}, \qquad \mathcal{E}'(t) = \sum_{k=1}^{N} F_k i\omega_k e^{i\omega_k t} \tag{11}$$

Since the ODE for current is linear, the steady state response to an input $\mathcal{E}'(t)$ of mixed amplitude and frequencies is the superposition of the steady state responses to the individual amplitudes and frequencies. From (7) we have

$$I_p(t) = \sum_{k=1}^{N} \frac{F_k}{R + i\,(\omega_k L - 1/\omega_k C)} e^{i\omega_k t} \tag{12}$$

For good reception of a particular station, say station 12, the amplitude

$$\frac{F_{12}}{\left(R^2 + (\omega_{12} L - 1/\omega_{12} C)^2\right)^{1/2}} \tag{13}$$

should be much larger than the amplitudes of all the other stations. The amplitude of the radio's response to station 12 can be magnified with a good directional antenna or by placing the radio near the transmitter of the station. A better strategy is to tune the radio to station 12 by adjusting the capacitance so that $\omega_{12} L = 1/\omega_{12} C$. Of course, nothing will help much if a nearby station broadcasts a strong signal at a frequency close to $\omega_{12}$. However, under the right conditions tuning the circuit will filter out all the frequencies except $\omega_{12}$ in the sense that the amplitude given by (13) with $\omega_{12} L = 1/\omega_{12} C$ dominates the output amplitudes of the other stations.

***Example 1*** **(Tuning a circuit)** These results apply without change if real sines and cosines are used as inputs in place of complex exponentials. Figure 4.10.5 shows the effect of tuning a circuit with $L = 1$ and $R = 0.1$ and ODE

$$I'' + 0.1 I' + \frac{1}{C} I = \cos t + 2\cos 2t = (-\sin t - \sin 2t)'$$

The circuit is first tuned to the input voltage, $-\sin t$, by setting $1/C$ equal to the square of the input frequency $\omega_1 = 1$ (hence $1/C = 1$). Observe that the circuit responds essentially with a sinusoidal wave with frequency $\omega_1 = 1$. Thus, although the input is a superposition of two frequencies, $\omega_1 = 1$ and $\omega_2 = 2$, the tuned circuit assigns such a low value to the amplitude of the component of the output with frequency $\omega_2 = 2$ that it essentially "disappears." The dotted curve in the lower graph of Figure 4.10.5 results when the circuit is tuned to the input voltage, $-\sin 2t$, by setting $1/C = 2^2$ with similar consequences. The dashed curve in the lower graph illustrates what happens when $1/C = 100$ and the circuit is tuned neither to $\omega_1$ nor to $\omega_2$.

Figure 4.10.6 displays the effect of a slight "detuning" of the circuit. The coefficient $1/C$ is first set to the tuning value of 1 to permit selection of the input signal with frequency 1. Changing $1/C$ to 0.9 or 1.1 results in a reponse signal that has a phase shift and a smaller amplitude.

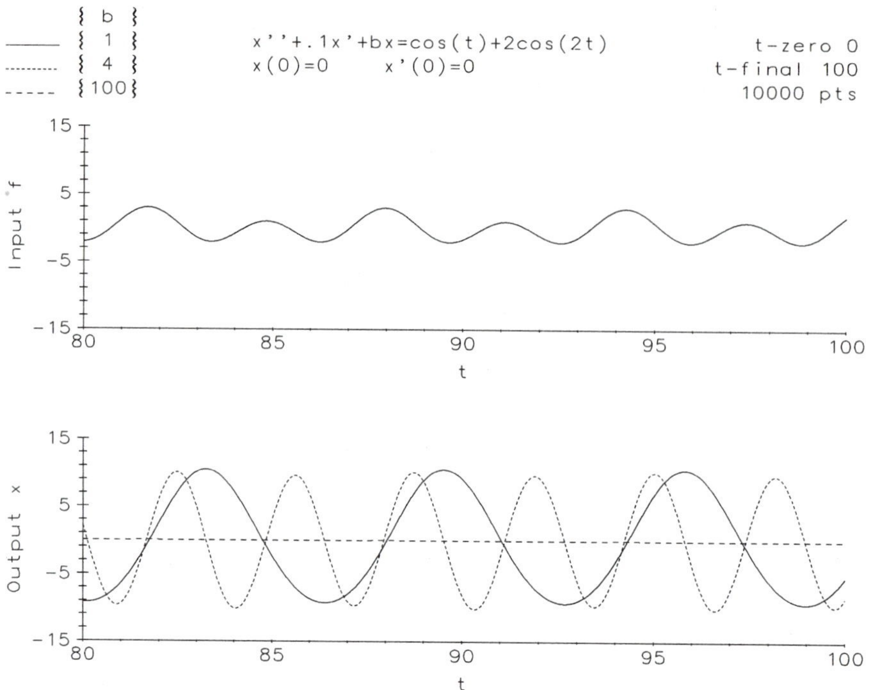


*Figure 4.10.5 Tuning an RLC circuit to an input frequency by changing parameter b (see Example 1).*

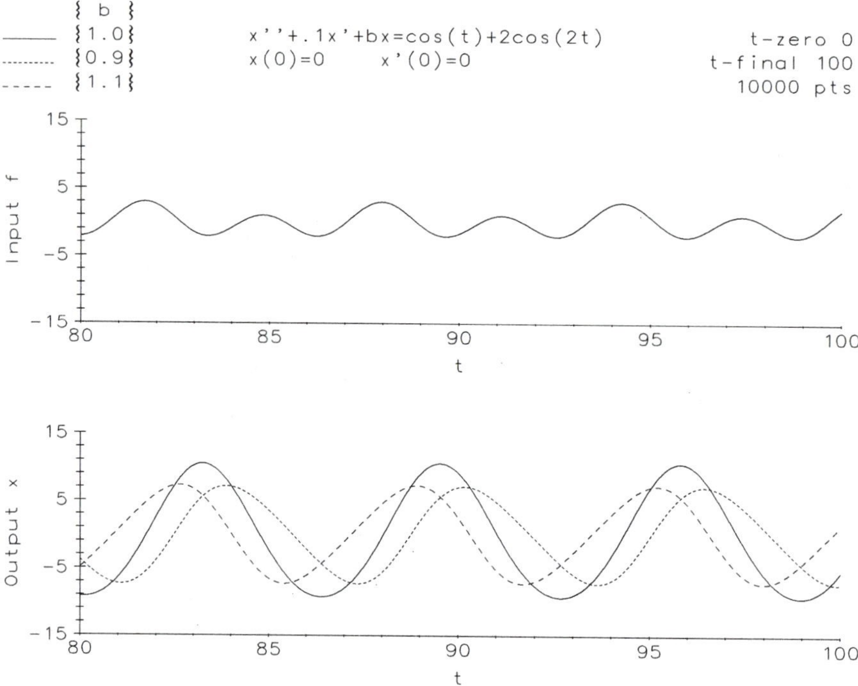


*Figure 4.10.6 The response of a poorly tuned RLC circuit (see Example 1).*

_______________/_______________
hardware software

_______________/_______________
name date

Answer questions in the space provided,
or on attached sheets or carefully labeled graphs.

_______________/_______________
course section

## 4.10 *Simple RLC Circuit*

***Abstract*** The responses of an $RLC$ circuit to a variety of inputs are studied.

***Special instructions*** The inputs in some of the ODEs are engineering functions. If the local solver package does not include these functions, replace each by a reasonable approximation, using sums and products of the functions $\exp(-at^2)$, $at(1+t^2)^{-1}$, $\cos(at)$, and $\sin(at)$.

**1.** Select one of the IVPs below and do the problems that follow.

☐ **A.** $q'' + \frac{R}{L}q' + \frac{1}{LC}q = \frac{1}{L}\cos t,\quad q(0) = 0,\quad q'(0) = 0$

☐ **B.** $q'' + \frac{R}{L}q' + \frac{1}{LC}q = \frac{t/L}{1+t^2},\quad q(0) = 0,\quad q'(0) = 0$

☐ **C.** $I'' + \frac{R}{L}I' + \frac{1}{LC}I = \frac{t^2/L}{1+t^2},\quad I(0) = 0,\quad I'(0) = 0$

☐ **D.** $I'' + \frac{R}{L}I' + \frac{1}{LC}I = \frac{t/L}{1+t^2}\sin t,\quad I(0) = 0,\quad I'(0) = 0$

☐ **(a)** If $R = 1$, $1 \le L \le 2$, $0.1 \le C \le 0.2$, use your solver to find the values of $L$ and $C$ that maximize the maximal amplitude of the charge on the capacitor (or current in the circuit) over the time span $0 \le t \le 10$. Then find the values that minimize the maximal amplitude. Explain how the optimal values of $L$ and $C$ were found.

☐ **(b)** Repeat (a), now allowing $R$ to vary as well, $1 \le R, L, C \le 2$, and find the appropriate values of $R$, $L$, and $C$ for maximization and minimization of the output amplitude.

☐ **2.** Solve the ODEs of Figures 4.10.2–4.10.4 but plot for $0 \le t \le 100$. Repeat for $75 \le t \le 100$.

* ☐ **3.** Select one of the $RLC$ models below. Let $q_0 = 0$, $I_0 = 0$, $L = 1$, $R = 0.5$, $C = 1$, and $0 \le t \le 20$.

☐ **A.** $\mathcal{E}(t) = A\cdot\text{TRW}(t, 20, 3)$ in equation (1)

☐ **B.** $\mathcal{E}(t) = A\cdot\text{SWW}(t, 30, 5)$ in equation (1)

☐ **C.** $\mathcal{E}'(t) = A\cdot\text{TRP}(t, 2)$ in equation (2)

☐ **D.** $\mathcal{E}'(t) = A\cdot\text{SQP}(t, 1)$ in equation (2)

☐ **E.** $\mathcal{E}'(t) = A\cdot\text{stair}(t, 5)$ in equation (2)

Graph the solution for $A = 1$ and for $A = \cos(\omega t)$ using different values of $\omega$, where $0 \le \omega \le 2\pi$. Find the values for $\omega$ for which the solution has maximal amplitude.

**4.** The engineering function $\text{SQW}(t, d, p)$, where $d$ is small, acts as a **sampling function**. For example, $\text{SQW}(t, 1, 2\pi)\cos t$ displays the values of the cosine function over the short intervals $2n\pi \le t \le 2\pi(n + 0.01)$ and is zero elsewhere. Select one of the IVPs below and do the problems that follow. Use the interval $0 \le t \le 20$.

☐ **A.** $I'' + 0.1I' + 10I = \text{SQW}(t, 1, 2\pi)\cos(\pi t)$, $I(0) = 0$, $I'(0) = 0$

☐ **B.** $10I'' + I' + 20I = \text{SQW}(t, 2, \pi)\cos(\frac{t}{2})$, $I(0) = 0$, $I'(0) = 0$

☐ **C.** $I'' + 25I = \text{SQW}(t, 1, \frac{2\pi}{5})\cos(5t)$, $I(0) = 0$, $I'(0) = 0$

* ☐ **(a)** Graph $I(t)$ and $I'(t)$ against $t$ and $I'$ against $I$. Does the solution approach a periodic state?

* ☐ **(b)** Replace the square wave function in the input by the triangular wave function $\text{TRW}(t, 50, p)$. Find the value of $p$ that yields a solution of maximal amplitude.

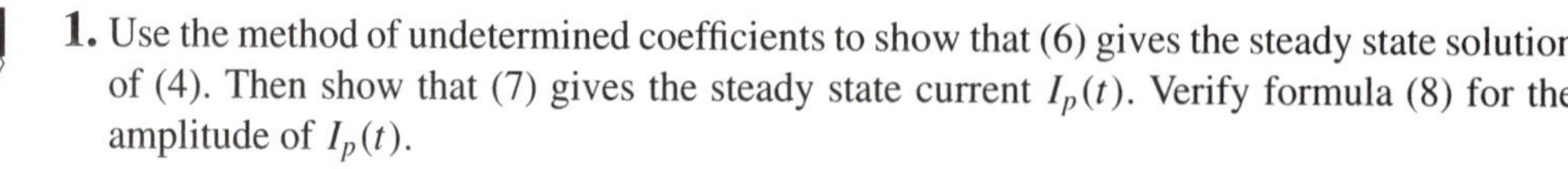


Answer questions in the space provided,
or on attached sheets or carefully labeled graphs.

## *4.11 Tuning a Circuit*

***Abstract*** The simple $RLC$ circuit is used as a filter for amplifying one input frequency while tuning out the others.

**1.** Use the method of undetermined coefficients to show that (6) gives the steady state solution of (4). Then show that (7) gives the steady state current $I_p(t)$. Verify formula (8) for the amplitude of $I_p(t)$.

**2.** Duplicate Figure 4.10.5. Choose $1/C$ to tune for station 1 ($f(t) = \cos t$), then for station 2 ($f(t) = \cos(2t)$). Choose $1/C$ close to but not equal to the tuning value for a station (as in Figure 4.10.6). What happens? What happens if $1/C$ is chosen halfway between the tuning values for stations 1 and 2? Attach the corresponding graphs.

* **3.** Let $R = 1$, $L = 1$ in an $RLC$ circuit. Select one of the following superpositions of mixed amplitudes and frequencies as the input voltage to the circuit and perform the tasks that follow concerning tuning the circuit to maximize the amplitude of the response current corresponding to an input frequency.

☐ **A.** $\mathcal{E}(t) = 5\cos t + 5\cos 1.1t + 10\cos(0.1t)$

☐ **B.** $\mathcal{E}(t) = 5\cos t + 5\cos 1.1t + 5\cos(0.9t)$

☐ **C.** $\mathcal{E}(t) = \sum_{k=1}^{10} k\cos(kt)$

Use your computer to find the values of $C$ that best tune the circuit to each of the input frequencies. Attach graphs of the outputs and use a marker to highlight the oscillations corresponding to the tuned frequency. (Reminder: The input to the ODE for $I(t)$ is $\mathcal{E}'(t)$.)

**4.** Two transmitters are to be located on the same mast, broadcasting on distinct frequencies, but at the same amplitude: $\mathcal{E}(t) = A\cos t + A\cos\omega t$, $\omega \neq 1$. By experimenting with different values of $\omega$, find the smallest $\omega > 1$ that permits an $RLC$ tuner with $R = 1$ and $L = 1$ to tune to either station and produce an output current for the tuned station that is at least 10 times that for the other. (Reminder: The input to the ODE for $I(t)$ is $\mathcal{E}'(t)$.)

CHAPTER 5

# Planar Systems

Equilibrium points and cycles are the dominant features of planar autonomous systems. Many of the experiments in this chapter are organized around the study of such points and cycles. Stability of motion is another important idea that is introduced in this chapter.

## *Initial Value Problems*

**Planar IVPs in normal form** may be written as

$$\begin{aligned} x' &= f(t, x, y), \qquad x(t_0) = x_0 \\ y' &= g(t, x, y), \qquad y(t_0) = y_0 \end{aligned} \tag{1}$$

If $f$ and $g$ do not explicitly depend on $t$, the system is **autonomous**. Many of the systems considered in this chapter are autonomous. If $f$, $g$, $\partial f/\partial x$, $\partial f/\partial y$, $\partial g/\partial x$, and $\partial g/\partial y$ are continuous in a region $R$ of $txy$-space that contains $(t_0, x_0, y_0)$, then (1) has a unique solution $x(t)$, $y(t)$, on some $t$-interval $I$ containing $t_0$, and at most one solution on any $t$-interval. The solution on $I$ can be extended in time beyond $I$ until the **solution curve** of points $((t, x(t), y(t))$ nears the boundary of $R$. Each solution defines a solution curve in $R$, and $R$ is filled with these curves, one through every point $(t_0, x_0, y_0)$. Although the collection of curves may appear to be a tangle, the curves never turn back in time, never touch, and never "end" inside $R$. Figure 5.1 shows an example of a solution curve.

## *Orbits of Planar Systems*

The solution curves of ODE (1) may be projected onto the $xy$-plane. These projected curves are called the **orbits**. As noted in Experiments 2.1–2.2, time may be translated forward or backward in an autonomous system without changing the orbits. For this reason, $t_0$ is usually taken to be 0 in solving an autonomous ODE. One consequence of this time invariance is that distinct orbits of an autonomous system can never touch or cross one another; thus, orbits in $xy$-space share this property with solution curves in $txy$-space. Orbits, unlike solution curves, may "end" inside the projection of the $txy$-region $R$ onto the $xy$-plane. For example, if an orbit tends to an equilibrium point as $t \to \infty$, the orbit appears to terminate at that point. Of course, in the full $txy$-space the solution curve actually approaches the boundary of $R$ as $t \to +\infty$. The following notation is used for orbits: if $x = x(t)$, $y = y(t)$, $t$ in $I$, is the solution of an autonomous IVP with $t_0 = 0$, then the corresponding orbit in the $xy$-plane is denoted by $\phi_t(p)$, where $p = (x_0, y_0)$ and $\phi_0(p) = p$.

## *Linear and Nonlinear Systems*

Linear systems have particularly simple features. If the coefficients of a linear system are constants, there is a striking and visual categorization of possible orbital structures (Experiment 5.2). The collection of distinct portraits of nonlinear autonomous planar systems is much more varied, but the alternatives of Poincaré and Bendixson imply that there are definite limits to what orbits can do (Experiment 5.16). Figures 5.1 and 5.2 show some of the features of nonautonomous ODEs. Observe that the orbit in Figure 5.2 intersects itself, but this does not violate the Uniqueness Principle, since the orbit passes through an intersection point at different times with different values for $x'$ and $y'$.

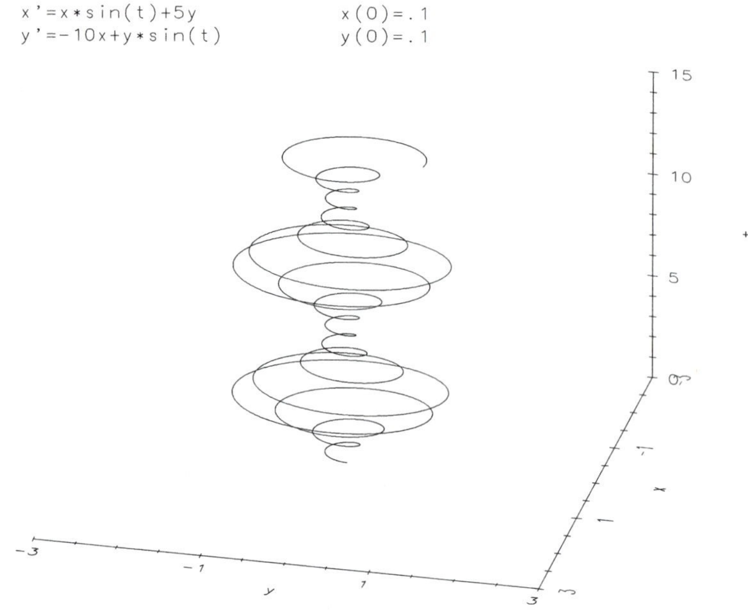


*Figure 5.1 Solution curve of a planar IVP.*

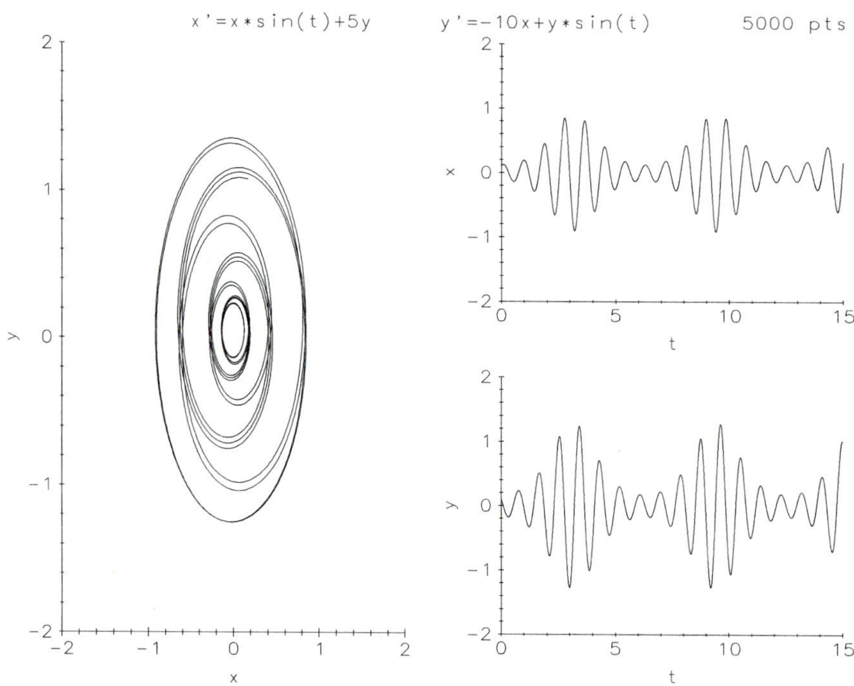


*Figure 5.2 Orbit and component curves of the solution in Figure 5.1.*

# Autonomous Linear Systems

***Purpose*** To determine the nature and the stability properties of the orbits of a homogeneous planar autonomous linear system with constant coefficients.

***Keywords*** Linear system, planar system, system matrix, triangular system, constant coefficients, equilibrium point, eigenvalue, eigenvector, stability, asymptotic stability, neutral stability, instability, node, center, saddle, focus

***See also*** Experiments 3.1, 4.2; Atlas Plate `Planar System A`; Chapter 5 Notes

***Background*** The general **homogeneous planar autonomous linear system with constant coefficients** has the form

$$\begin{aligned} x' &= ax + by \\ y' &= cx + dy \end{aligned} \tag{1a}$$

or

$$u' = Au \qquad u = \begin{bmatrix} x \\ y \end{bmatrix}, \quad A = \begin{bmatrix} a & b \\ c & d \end{bmatrix} \tag{1b}$$

where the matrix $A$ of real constants is called the **system matrix**. A solution of (1a) is a column vector $[x(t)\ y(t)]^T$, where $^T$ denotes **transpose**, with components $x(t)$ and $y(t)$ which together satisfy the equations of (1a). One of the sources of a system such as (1a) is any scalar ODE of order 2. For example, the constant coefficient undriven linear second order ODE $x'' + p_0x' + q_0x = 0$ leads directly to

$$\begin{aligned} x' &= y \\ y' &= -q_0x - p_0y \end{aligned} \tag{2}$$

by introducing $y = x'$ as a second state variable.

The solutions of system (2) are easily found by solving $x'' + p_0x' + q_0x = 0$ to obtain $x(t)$. Then construct $y(t)$ as $x'(t)$. If system (1a) is triangular then it may also be solved piecemeal. A **lower triangular system** (or **cascade**) has the form

$$\begin{aligned} x' &= ax \\ y' &= cx + dy \end{aligned} \qquad A = \begin{bmatrix} a & 0 \\ c & d \end{bmatrix} \tag{3}$$

This system can be solved by starting at the top with the uncoupled equation $x' = ax$, solving it to obtain $x = C_1e^{at}$ and inserting that function in the rate equation for $y$, which is a first order linear rate equation in $y$ solvable by the integrating factor technique of Experiment 3.1. **Upper triangular systems** are solved similarly, but from the bottom up.

The linear system (1a) is solved in a different way if it is neither a cascade nor of form (2). Eigenelement methods are appropriate in this case. The **eigenelements** of the system matrix $A$ include the **spectrum**, i.e., the set of **eigenvalues** of $A$, each repeated according to its multiplicity as a root of the **characteristic polynomial** $p(\lambda) = \det[A - \lambda I] = \lambda^2 - (a + d)\lambda + ad - bc$, where det denotes the determinant and $I$ is the $2\times2$ identity matrix (a diagonal matrix with 1's on the principal diagonal). The other eigenelements aside from the eigenvalues are the **eigenvectors** which are defined for each eigenvalue $\lambda_j$ as the vectors $v \neq 0$ for which $Av = \lambda_j v$. After adjoining the vector 0, the set of eigenvectors corresponding to a given eigenvalue $\lambda$ form a (linear) **eigenspace** $V_\lambda$. A maximal set of independent eigenvectors (i.e., a basis for $V_\lambda$) is included among the eigenelements for each eigenvalue. If $\lambda$ is a simple eigenvalue then it has a one-dimensional eigenspace, and a single eigenvector is included among the eigenelements in that case. A double eigenvalue (i.e., a double root of the characteristic polynomial) may have a two-dimensional eigenspace in which case a pair of independent eigenvectors is included. If the dimension of the eigenspace is 1, while $\lambda$ is a double eigenvalue only one eigenvector is put into the set of eigenelements. This is the **deficiency** case, and it causes a problem when eigenelements are used to find all of the solutions of (2). See Example 3 in the Chapter 5 Notes.

Suppose now that a set of eigenelements of $A$ (i.e., the spectrum of $A$ and a collection of independent eigenvectors) has been calculated algebraically. Suppose that $\lambda$ is an eigenvalue of $A$ and $v$ is a corresponding eigenvector. Then $u = e^{\lambda t}v$ is a solution of (1a) since

$$u' = (e^{\lambda t}v)' = \lambda e^{\lambda t}v = e^{\lambda t}(\lambda v) = e^{\lambda t}Av = A(e^{\lambda t}v) = Au$$

If $\lambda_1$ and $\lambda_2$ are the eigenvalues of $A$ and if $\{v^1, v^2\}$ is a corresponding set of independent eigenvectors, then the set of all solutions of (1a) is given by the **general solution**

$$u = C_1 e^{\lambda_1 t}v^1 + C_2 e^{\lambda_2 t}v^2$$

where $C_1$ and $C_2$ are arbitrary constants (which are real or complex as appropriate). The constants $C_1$ and $C_2$ must be particularized to determine the solution passing through a given point at a given time. Examples 1–3 of the Chapter 5 Notes illustrate how all of this works out in practice, Example 3 taking up the deficiency case mentioned above.

***Observation*** **(Orbital portraits of linear systems)** Orbital portraits of systems such as (1a) are dominated by the straight line orbits generated by the real eigenvectors of $A$ as seen in Figures 5.2.1–5.2.3. These line orbits are also evident in the **saddle** and **node** figures in Atlas Plate `Planar System A`. Complex eigenvalues $\lambda = \alpha \pm \beta i$ are associated with orbits that rotate around the equilibrium point at the origin, spiraling inward if $\alpha < 0$, outward if $\alpha > 0$, or closing to form a circle (or an ellipse) if $\alpha = 0$. See the **center** and **focus** figures in Atlas Plate `Planar System A`.

***Observation*** **(Stability)** Finally, there is the important question of stability (in this connection see Experiment 4.2). Note that system (1a) always has the origin as an equilibrium point. System (1a) is **stable (at the origin)** if for every number $E > 0$ there is a number $D > 0$ such that if $({x_0}^2 + {y_0}^2)^{1/2} < D$ then the solution $[x(t)\ y(t)]^T$ of the IVP

$$\begin{aligned} x' &= ax + by \\ y' &= cx + dy \\ x(0) &= x_0, \quad y(0) = y_0 \end{aligned}$$

satisfies $(x^2(t) + y^2(t))^{1/2} < E$, all $t \geq 0$. The test for stability is that the eigenvalues of $A$ have nonpositive real parts and that 0 not be a double eigenvalue with deficient eigenspace. Asymptotic stability is a stronger kind of stability. System (1a) is **(globally) asymptotically stable** if it is stable and if every solution of (1a) tends to the origin as $t \to \infty$. This will be true if and only if the eigenvalues of $A$ have negative real parts. System (1a) is **neutrally stable** at the origin if it is stable, but not asymptotically stable. A system is **unstable** at the origin if it is not stable. The various kinds of spectra, types of stability, and linear orbital portraits lead to the following table:

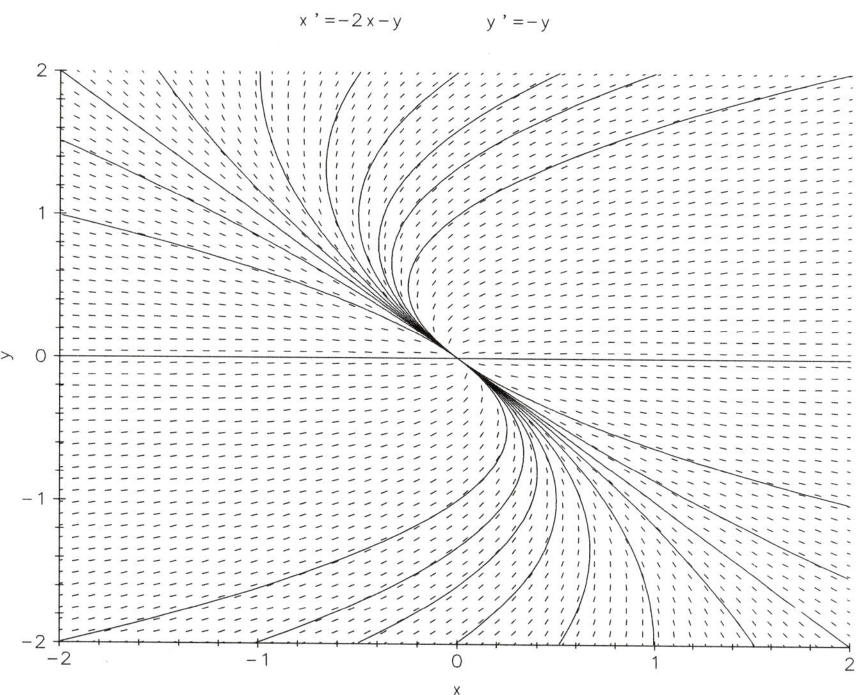


*Figure 5.2.1* *Asymptotically stable improper node: eigenspaces along the* $x$*-axis and the line* $y = -x$.

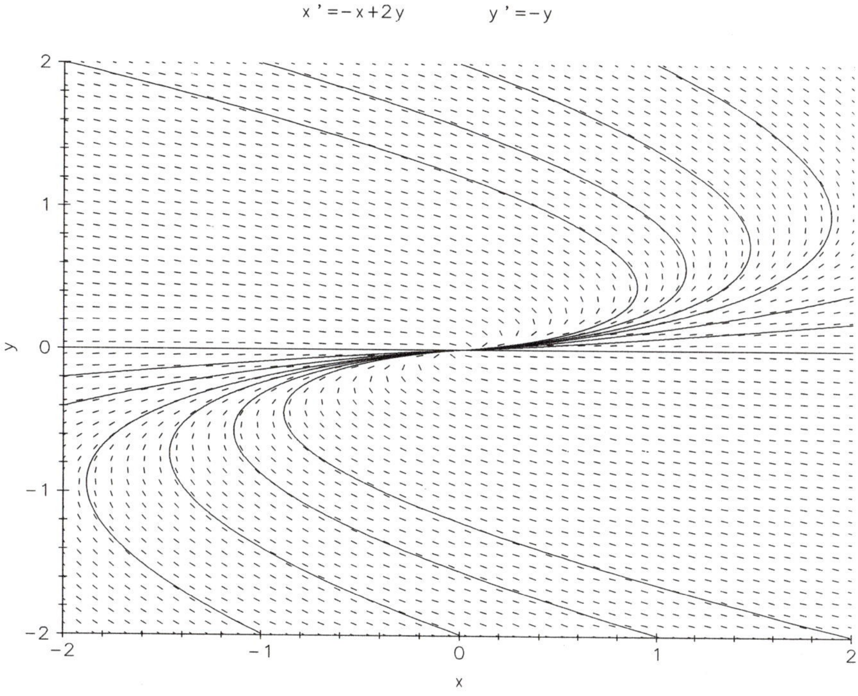


*Figure 5.2.2* *Asymptotically stable proper node: eigenspace along the* $x$*-axis.*

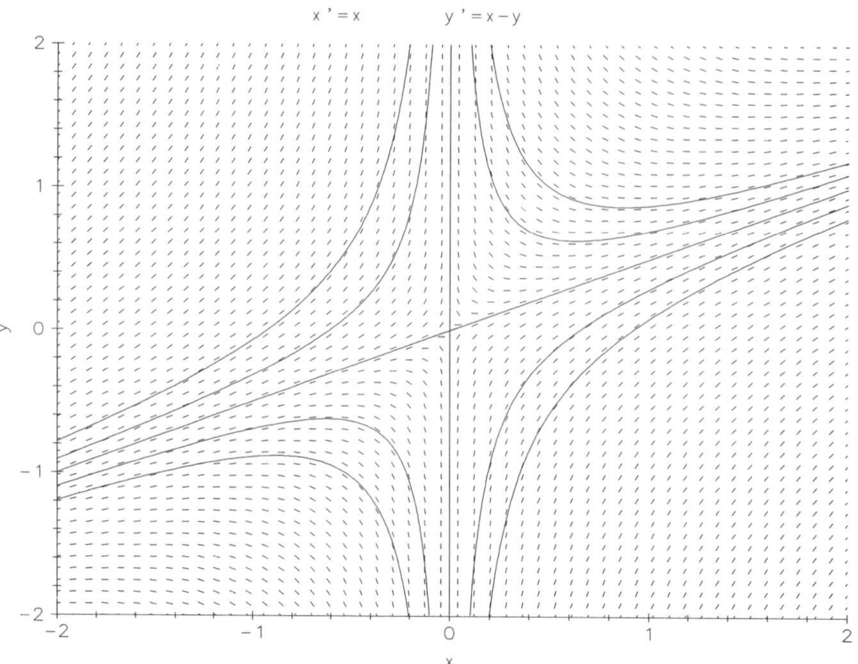


*Figure 5.2.3 Unstable saddle: eigenspaces along the y-axis and the line $y = x$.*

| Spectrum | Stability Type | Name |
|---|---|---|
| $\lambda_1 < \lambda_2 < 0$ | Asymptotically stable | Improper node |
| $\lambda_1 = \lambda_2 < 0$ (2D eigenspace) | Asymptotically stable | Star node |
| $\lambda_1 = \lambda_2 < 0$ (1D eigenspace) | Asymptotically stable | Proper node |
| $\lambda_1 = \alpha + \beta i = \bar{\lambda}_2, \alpha < 0, \beta \neq 0$ | Asymptotically stable | Focus |
| $\lambda_1 = \beta i = \bar{\lambda}_2, \beta \neq 0$ | Neutrally stable | Center |
| $\lambda_1 < 0 < \lambda_2$ | Unstable | Saddle |
| $0 < \lambda_1 = \lambda_2$ (1D eigenspace) | Unstable | Proper node |
| $0 < \lambda_1 = \lambda_2$ (2D eigenspace) | Unstable | Star node |
| $0 < \lambda_2 < \lambda_1$ | Unstable | Improper node |
| $\lambda_1 = \alpha + \beta i = \bar{\lambda}_2, \alpha > 0, \beta \neq 0$ | Unstable | Focus |

***Observation*** **(Other cases)** The list above omits the case of a zero eigenvalue, but this case is explored in Experiment 5.2. The examples treated here all have an equilibrium point at the origin, but it could just as easily be placed at any other position by a translation of coordinates. The system for this case has the form

$$x' = ax + by + e$$
$$y' = cx + dy + f$$

where $e$ and $f$ are any constants. The stability type and the graphs of the orbits are unchanged by the translation.

_______________/_______________
hardware software

_______________/_______________
name date

Answer questions in the space provided,
or on attached sheets or carefully labeled graphs.

_______________/_______________
course section

## *5.2 Gallery of Pictures*

***Abstract*** A collection of orbital portraits of planar autonomous linear systems is constructed. Representative component curves and solution curves are also constructed.

***Special instructions*** Solution curves may be omitted if your platform does not support 3D graphics.

**1.** Reproduce Figures 5.2.1–5.2.3. Draw arrowheads on the orbits to indicate the direction of increasing time. Highlight the eigenspaces.

**2.** Reproduce Atlas Plate `Planar System A`, using four separate graphs if necessary. Draw arrowheads on the orbits to show the direction of increasing time. Highlight eigenspaces.

**3.** For each system below, draw a portrait of the orbits on the screen, $|x| \le 2$, $|y| \le 2$, using a suitable $t$-range. On the graph write the spectrum of the system matrix, identify any visible eigenspaces and the corresponding eigenvalue, and give the portrait its name (from the listing on the cover sheet) and its stability type. Indicate the advance of time by arrowheads on the orbits. Highlight eigenspaces.

**(a)** $x' = -x,\ y' = -2y$
**(b)** $x' = -x,\ y' = -y$
**(c)** $x' = -x,\ y' = x - y$
**(d)** $x' = x,\ y' = 2y$
**(e)** $x' = x,\ y' = y$
**(f)** $x' = x,\ y' = x + y$
**(g)** $x' = x,\ y' = -2y$
**(h)** $x' = -x + 10y,\ y' = -10x - y$
**(i)** $x' = 10y,\ y' = -10x$
**(j)** $x' = x + 10y,\ y' = -10x + y$
**(k)** $x' = 0,\ y' = 0$
**(l)** $x' = 0,\ y' = x$
**(m)** $x' = 0,\ y' = -y$
**(n)** $x' = 0,\ y' = y$

**4.** For each system in Problem 3, plot component graphs and solution curves for "typical" orbits.

**5.** The cover sheet lists ten stability types, but Problem 3 gives fourteen. For each of the extra four, construct orbital portraits, mark equilibrium points, and identify stability type.

**6.** Show that the origin is a saddle point for the system $x' = -x + 2y$, $y' = y$ and draw an orbital portrait on the screen $|x| \le 5$, $|y| \le 15$. Explain the striking similarity with the portrait in Atlas Plate `First Order F`.

**7.** Select a pair of nonorthogonal lines passing through the origin. Construct a linear system whose system matrix has eigenvalues $\lambda_1 = -1$, $\lambda_2 = -2$ with these lines as corresponding eigenspaces. Graph orbits of your system between the lines. What is the "name" of the system (refer to the listing on the cover sheet).

______________/______________ hardware / software

______________/______________ name / date

Answer questions in the space provided, or on attached sheets or carefully labeled graphs.

______________/______________ course / section

## 5.3 *Stability*

***Abstract*** The stability character of a planar autonomous linear system is studied.

**1.** For each of the following systems, find the eigenvalues, determine the stability type and name (from the list on the cover sheet), and construct an orbital portrait. Draw arrowheads on the orbits to indicate the direction of increasing time. Write the eigenvalues and stability type on the graph. Identify any visible eigenspaces and circle all equilibrium points.

**(a)** $x' = x + 3y,\ y' = 4x - 6y$
**(b)** $x' = -x + 4y,\ y' = -3x - 2y$
**(c)** $x' = 2y,\ y' = -8x$
**(d)** $x' = 3x - 2y,\ y' = 4x - y$
**(e)** $x' = 3x - 2y,\ y' = 2x - 2y$
**(f)** $x' = -4x - 3y,\ y' = 4x + 3y$
**(g)** $x' = x + y,\ y' = -x - y$
**(h)** $x' = x + y,\ y' = 4x + y$
**(i)** $x' = 4x - 3y,\ y' = 8x - 6y$
**(j)** $x' = -x - 4y,\ y' = x - y$
**(k)** $x' = x + 2y,\ y' = -5x - y$
**(l)** $x' = -6x + 4y,\ y' = -x - 2y$
**(m)** $x' = 4x - 2y,\ y' = 8x - 4y$

**2.** Consider the planar autonomous linear system

$$x' = -2x - y + 1$$
$$y' = -y - 1$$

The system has only one equilibrium point. Where is it? Plot an orbital portrait of this system near the equilibrium point. Indicate the direction of increasing time with arrowheads on the orbits. How does this portrait compare to Figure 5.2.1. Write all your comments directly on the graph.

**3.** Consider the system $x' = -x + ay$ and $y' = 2x - 3y$. Attach graphs that show how the orbital portraits change as the parameter $a$ is changed from $-10$ to $+10$. For which values of $a$ is there a sudden change in the nature of the orbital portrait? Label each graph with a suitable name chosen from the list on the cover sheet.

# Driven Linear Systems

***Purpose*** To examine the response of a planar linear constant coefficient system to a driving term.

***Keywords*** Driving forces, driven linear system, response, Bounded Input/Bounded Output Principle

***See also*** Experiments 3.4, 4.3–4.4, 5.2–5.3; Chapter 5 Notes

***Background*** A **driven planar linear system** with constant system matrix A has the form

$$\begin{aligned} x' &= ax + by + F_1(t) \\ y' &= cx + dy + F_2(t) \end{aligned} \tag{1}$$

where the system matrix

$$A = \begin{bmatrix} a & b \\ c & d \end{bmatrix}$$

has constant real entries and the **driving** or **input** vector $F(t) = [F_1(t)\ \ F_2(t)]^T$ has components that are at least piecewise continuous on a $t$-range $I$, usually taken to be $t \geq 0$. Let $u(t) = [x(t)\ y(t)]^T$ denote a solution of the corresponding system $u' = Au + F$. If the solution vectors $u^1(t)$ and $u^2(t)$ generate the two-dimensional solution space of the undriven system

$$\begin{aligned} x' &= ax + by \\ y' &= cx + dy \end{aligned} \tag{2}$$

and if $u_p(t)$ is a particular solution vector of the driven system (1), then the general solution $u(t)$ of (1) is given by

$$u(t) = C_1 u^1(t) + C_2 u^2(t) + u_p(t) \tag{3}$$

where $C_1$ and $C_2$ are arbitrary constants. If initial data are available (i.e., $x(0) = x_0$, $y(0) = y_0$), then $C_1$ and $C_2$ can be particularized to ensure that the solution given by (3) meets the initial conditions.

The construction of the vector functions $u^1$ and $u^2$ through the use of the eigenelements of $A$ is described in the Experiment 5.2 and in the Chapter 5 Notes. A particular solution $u_p(t)$ of (1) may be explicitly constructed (sometimes) by the method of undetermined coefficients, or (always) by a vector version of the method of variation of parameters. However, since we are not concerned with formulas here, we shall only describe a three-step process in solving the IVP

$$\begin{aligned} x' &= ax + by + F_1(t) \\ y' &= cx + dy + F_2(t) \\ x(0) &= x_0, \quad y(0) = y_0 \end{aligned} \tag{4}$$

The steps are

- Solve the undriven IVP

$$\begin{aligned} x' &= ax + by \\ y' &= cx + dy \\ x(0) &= x_0, \quad y(0) = y_0 \end{aligned}$$

  to obtain a vector solution $u(t) = [x(t)\ y(t)]^T$. This can either be done analytically (see Experiment 5.2–5.3 and Chapter 5 Notes) or with a computer solver.

- Analytically or with a computer solver find the solution $u_p(t)$ of (4) with $x_0$, $y_0$ both replaced by 0.

- Superpose to obtain the solution $u(t) + u_p(t)$ of (4).

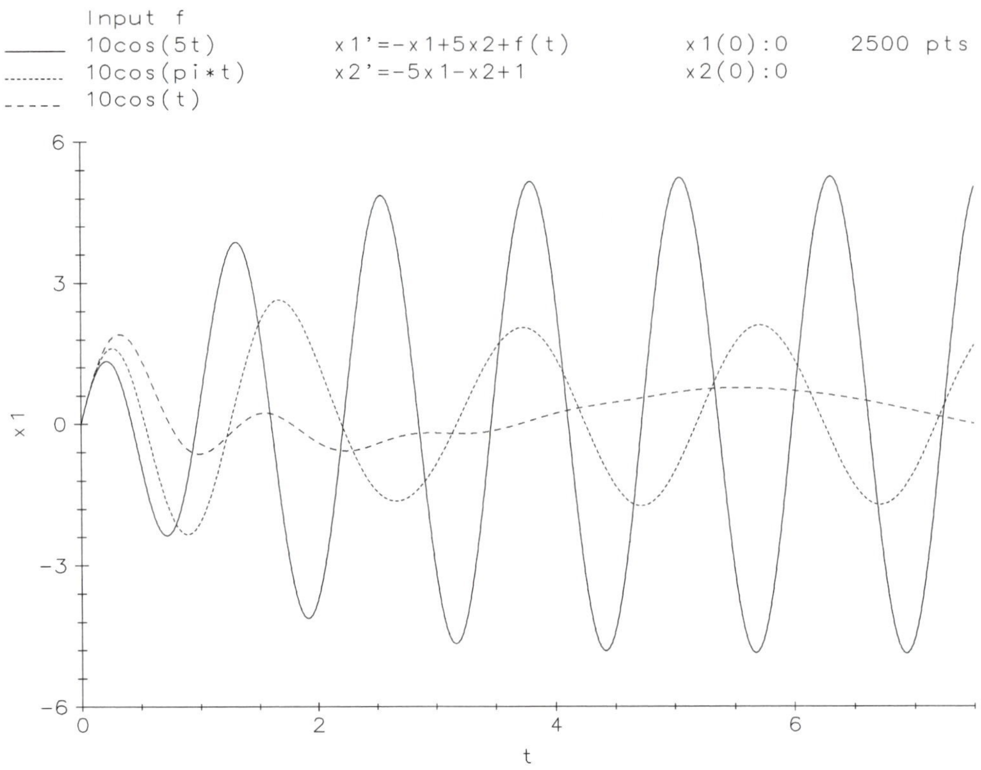

*Figure 5.4.1 Approach to steady state: responses to three bounded input functions.*

This three-step process shows how the solution of IVP (4) is the sum of the solutions of two subsidiary IVPs; the first summand is the response to the initial data in the absence of a driving force, while the second is the response to the driving force when the initial data are trivial (i.e., $x_0 = 0$, $y_0 = 0$).

If the eigenvalues of the system matrix $A$ have negative real parts, then all solutions of the undriven system (2) decay to the origin, $x = 0$, $y = 0$ as $t \to \infty$, leaving active only a particular solution $u_p(t)$ in (3). The latter solution is called the **steady state** of the system, and the decaying solutions to the undriven system are called **transients**. See Figure 5.4.1 for an example of approach to steady state.

***Observation*** **(Relation between input and output)** The Bounded Input/Bounded Output Principle mentioned in Experiment 3.4 is valid here as well. If the eigenvalues of the system matrix $A$ have negative real parts and if the components of the driving vectors are bounded for $t \geq 0$ (i.e., if there is a constant $M$ such that $(F_1^2(t) + F_2^2(t))^{1/2} \leq M$ for $t \geq 0$), then every solution $[x(t)\ y(t)]^T$ of (1) is bounded for $t \geq 0$ (i.e., there is a constant $K$ such that $(x^2(t) + y^2(t))^{1/2} \leq K$, for $t \geq 0$).

_______________/_______________
hardware software

_______________/_______________
name date

Answer questions in the space provided,
or on attached sheets or carefully labeled graphs.

_______________/_______________
course section

## 5.4 *Driven Linear Systems*

***Abstract*** The response of a linear system to a vector driving force is studied in a variety of particular cases.

**1.** Select one of the following Atlas plates and answer the questions that follow.

☐ **A.** `Second Order A`
☐ **B.** `Second Order B`
☐ **C.** `Second Order C`

**(a)** Convert the ODE into a planar system. Write the system matrix and the (vector) driving force.

**(b)** Find the eigenvalues of the system matrix. Find a bound for the magnitudes of the components of the vector driving forces over the $t$-range $t \geq 0$. Does the Bounded Input/Bounded Output Principle apply? Explain.

**(c)** Duplicate the Atlas Plates `Second Order A-C`. Replace the vector driving force of the corresponding system first by $[0\ t]^T$ and then by $[0\ e^{-t}\cos t]^T$. Draw the corresponding component plots and explain what you see. Write on the graphs.

**2.** Select one of the following IVPs and perform the tasks that follow.

☐ **A.** $x' = y,\ y' = -10x - 5y + 10\cos 3t;\ x(0) = 1,\ y(0) = 1$
☐ **B.** $x' = 5x + 3y + \cos 2t,\ y' = -6x - 4y + \sin 2t;\ x(0) = 2,\ y(0) = 3$
☐ **C.** $x' = 2x - 5y + \cos t,\ y' = x - 2y;\ x(0) = 1,\ y(0) = 1$
☐ **D.** $x' = -x - 4y + e^{-t}\cos t,\ y' = x - y + \sin 2t;\ x(0) = 1,\ y(0) = 1$

**(a)** Graph the orbit on the screen $|x| \leq 2$, $|y| \leq 2$ over the $t$-range $0 \leq t \leq 10$.

**(b)** Does the Bounded Input/Bounded Output Principle apply? Explain.

**(c)** Build up the solution in stages by first graphing the components of the solution of the IVP without the driving force, then the driven IVP but with the trivial initial data $x(0) = 0$, $y(0) = 0$. Finally, add the two solutions and plot the orbit of the given IVP. Verify by measuring with a ruler that the components do indeed add as claimed.

**3.** For the following systems, plot component graphs and orbits of the steady state and several other solutions. Use a highlighter to distinguish the steady state solution.

**(a)** $x' = y,\ y' = -x - 2y + \cos t$

**(b)** $x' = -x + 10y + \cos 10t,\ y' = -10x - y + \sin 10t$

**4.** Consider the driven planar system $u' = Au + F(t)$, where $A$ is a 2×2 matrix of real constants, and $u$ and $F(t)$ are the transposes of the vectors $[x\ y]$ and $[F_1\ F_2]$, respectively.

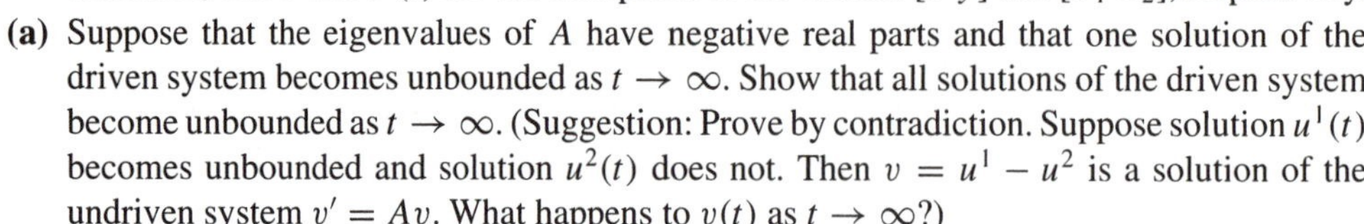

* **(a)** Suppose that the eigenvalues of $A$ have negative real parts and that one solution of the driven system becomes unbounded as $t \to \infty$. Show that all solutions of the driven system become unbounded as $t \to \infty$. (Suggestion: Prove by contradiction. Suppose solution $u^1(t)$ becomes unbounded and solution $u^2(t)$ does not. Then $v = u^1 - u^2$ is a solution of the undriven system $v' = Av$. What happens to $v(t)$ as $t \to \infty$?)

**(b)** Illustrate (a) by considering the system

$$x' = -x - 5y + 6 + t$$
$$y' = 5x - y + 1 - 5t$$

Show that $x = t$, $y = 1$ is an unbounded solution. Then plot enough other component curves and orbits to convince yourself that other orbits are unbounded as well.

* **(c)** Suppose that the driving term $F(t)$ of the system $u' = Au + F(t)$ is periodic of period $T$. It is known that the driven system must always possess a forced oscillation if the undriven system has no free oscillation of period $T$. (A **forced oscillation** is a periodic solution $u(t)$ of period $T$ of the driven system; a **free oscillation** is a periodic solution of the undriven system.) Show that the driven system cannot have more than one forced oscillation. (Hint: Suppose that $u^1(t)$ and $u^2(t)$ are forced oscillations. Show that $u^1(t) - u^2(t)$ either vanishes for all $t$ or is a free oscillation of period $T$.)

**(d)** Illustrate (c) by plotting component curves and orbits of $x' = -3x + 4y + \cos(\pi t)$, $y' = 4x - 3y + 7\sin(\pi t)$. Highlight the forced oscillation.

# Interacting Species

***Purpose*** To examine ODE models of the interactions of species, to make predictions about changes in the populations, and to study sensitivity to changes in rate constants.

***Keywords*** Interacting species, populations, predator, prey, satiation, competing species, equilibrium point, cycle

***See also*** Experiments 1.6, 3.11; Appendix B.1; Atlas Plates `Predator-Prey A-C`

***Background*** Appendix B.1 presents a simple ODE model of the interactions between species. Suppose that $x(t)$ and $y(t)$ denote the respective populations of two species at time $t$. One model for the changing rates of the populations is the IVP

$$\begin{aligned} x' &= R_1 x \qquad x(0) = x_0 \\ y' &= R_2 y \qquad y(0) = y_0 \end{aligned} \tag{1}$$

where $R_1$ and $R_2$ are the **intrinsic rate coefficients**. Each coefficient measures the contribution of the average individual of the species to the overall growth rate of the population. These coefficients generally are not constants, but functions of time, populations, individual ages, and a host of other parameters. We shall average out the time variation in the coefficients and assume that the two species form a community isolated from all other influences. In this setting $R_1$ and $R_2$ are functions only of the populations $x$ and $y$. The rate constants $a, b, \ldots,$ in the models below are taken to be positive.

• **Predator-prey interaction.** Vito Volterra introduced the system

$$\begin{aligned} x' &= (-a + by)x \\ y' &= (c - dx)y \end{aligned} \tag{2}$$

where $x(t)$ and $y(t)$ are the respective populations of the predator species and its prey; $a$ and $c$ are the decay and growth coefficients of each species as if the other were absent. Volterra assumed that the number of predator-prey encounters is proportional to the population of each (**Law of Mass Action**); $b$ and $d$ measure, respectively, predator efficiency in converting food into fertility and the probability that an encounter removes one of the prey. See Atlas Plate `Predator-Prey A`. It can be shown that the equilibrium populations $x = c/d$ and $y = a/b$ are also the time averages of the populations over each cycle.

• **Overcrowding.** Excess population may diminish the intrinsic rate coefficients. This intraspecies form of mass action may be included in (2) as follows

$$\begin{aligned} x' &= (-a - \alpha x + by)x \\ y' &= (c - \gamma y - dx)y \end{aligned} \tag{3}$$

where $-\alpha x^2$ and $-\gamma y^2$ model intraspecies overcrowding.

• **Constant effort harvesting.** If both predator and prey are harvested at rates proportional to the populations, then the following ODEs are an alternate form of (2):

$$\begin{aligned} x' &= (-a + by)x - H_1 x \\ y' &= (c - dx)y - H_2 y \end{aligned} \tag{4}$$

where $H_1$ and $H_2$ measure harvesting efficiency. In Figure 5.5.1, $H = H_1 = H_2$.

• **Competition.** Not all two-species interactions are of the predator-prey type. For example, two species may compete for a common resource in short supply. A model for this kind of interaction is

$$\begin{aligned} x' &= (a - bx - cy)x \\ y' &= (d - ex - fy)y \end{aligned} \tag{5}$$

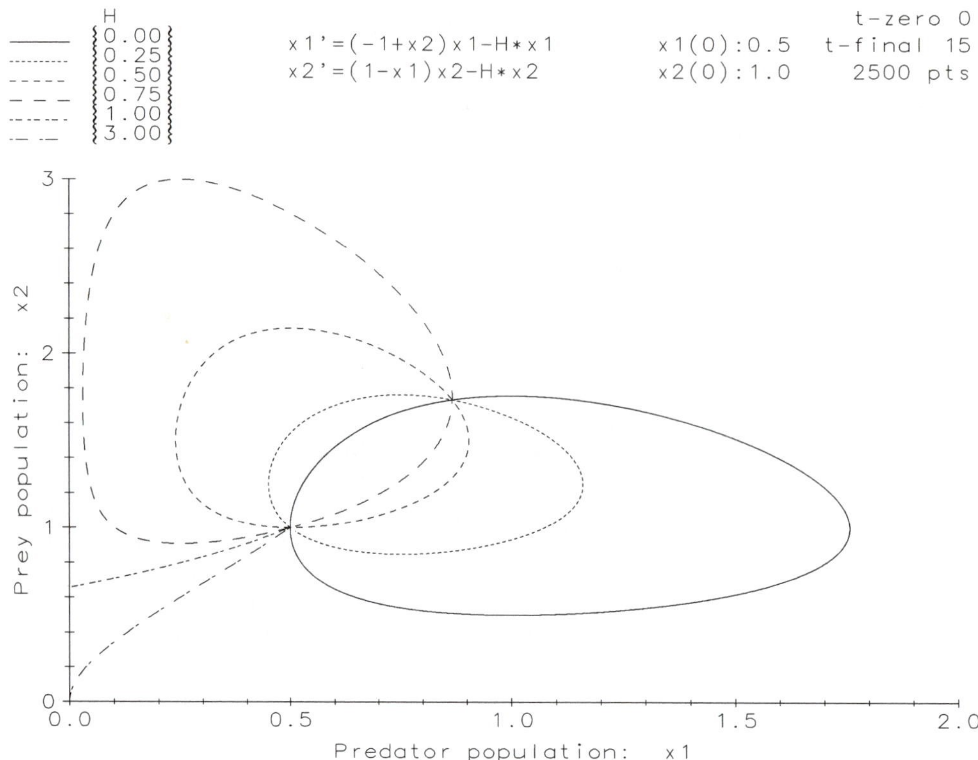


*Figure 5.5.1 Overharvesting (large H) destroys the predator-prey community.*

Both interspecies and intraspecies competition are modeled in system (5). Population curves for a competition model are displayed in Atlas Plate `Predator-Prey C`.

• **Satiable predation.** When food is plentiful, the predator's appetite is satiated, and an increase in prey population has little effect on the interaction terms in the rate equations. One model is

$$\begin{aligned} x' &= (-a + by/(k+y))x \\ y' &= (c - \gamma y - dx/(k+y))y \end{aligned} \tag{6}$$

where prey overcrowding is also present. See Atlas Plate `Predator-Prey B` and Experiment 5.18 for an interesting property of this system.

***Observation*** **(Population quadrant)** These models make sense only in the **population quadrant,** $x \geq 0$, $y \geq 0$. An orbit starting in that quadrant always stays there since the positive axes are unions of orbits; the Uniqueness Principle prevents any other orbit from touching the axial orbits.

***Observation*** **(Model validity)** The models have their flaws. No account is taken of the time delay between an action and its effect on population rates. Averaging the rates of change over all categories of age, sex, fertility, and health is of dubious validity. More intricate models could be (and have been) constructed, but most of these have their origins in the models presented above.

_______________/_______________ hardware software

_______________/_______________ name date

Answer questions in the space provided, or on attached sheets or carefully labeled graphs.

_______________/_______________ course section

## *5.5 Predator-Prey Models: Harvesting, Overcrowding*

***Abstract*** Volterra's models of a two-species predator-prey interactions are explored. The effects of constant effort harvesting and of overcrowding are investigated.

**1.** Duplicate Atlas Plate `Predator-Prey A`, which depicts the population cycles of a particular instance of Volterra's equations (2). There are no harvesting or overcrowding terms in this model. Insert arrowheads on orbits to show the direction of increasing time. Plot component curves for $0 \leq t \leq 20$ and estimate the period $T$ of each. Does $T$ diminish, increase, or stay fixed as the starting points $(x_0, y_0)$ of distinct orbits are chosen closer to the equilibrium point (25, 100)? Plot enough component curves near equilibrium to estimate the limiting value of the period as $(x_0, y_0) \to (25, 100)$. Write your answers on the graphs.

**2.** The system of ODEs in Atlas Plate `Predator-Prey A` may be linearized in a region about the equilibrium point (25, 100) by rewriting the ODEs in new variables $u = x - 25$ and $v = y - 100$ and discarding all nonlinear terms. Show that the new system is $u' = v/4$, $v' = -8u$. Graph orbits and component curves for the linear system, and calculate the (constant) period. Explain why the period should be the same as the limiting period estimated in Problem 1. Write explanations on your graphs.

**3.** The system of Figure 5.5.1 shows what happens to the orbits of a predator-prey community as the harvesting coefficient $H$ is increased. Duplicate that figure, and insert arrowheads to indicate the direction of increasing time. For each of the six values of the harvesting coefficient $H$, mark the locations of all equilibrium points inside or on the edge of the population quadrant. Explain what happens to the time-average populations as the harvesting coefficient is increased from 0 to 1 and beyond.

**4.** On the basis of Problem 3 explain why limited harvesting of both species is a boon to the prey, but not to the predator. A species of ladybug once kept the cottony cushion scale pest of California orange trees under control. During the 1950s the broad-range insecticide DDT was applied to further control the scale. Why did that turn out to be a bad idea?

**5.** Overcrowding further limits the growth of a predator-prey community. Set $a = d = \alpha = \gamma = 1$, $b = c = 2$ in system (3) and plot several orbits inside the population quadrant and the corresponding component curves. Explain what is happening. Mark all the equilibrium populations inside or on the edge of the population quadrant. Replot with $\gamma = 2, 3, 4, 5$. Explain any significant changes in the nature of the orbits, the component curves, and the periods of cycles.

_______________/_______________
hardware software

_______________/_______________
name date

Answer questions in the space provided,
or on attached sheets or carefully labeled graphs.

_______________/_______________
course section

## *5.6 Competing Species*

***Abstract*** Can competing systems coexist? This experiment answers the question.

**1.** System (5) models the dynamics of a pair of species that compete for a common resource that is in limited supply. Explain the meaning of each term in the $x$-rate equation.

**2.** Show that by letting $x = A\tilde{x}$, $y = B\tilde{y}$, $t = C\tilde{t}$, where $A = a/b$, $B = a/c$, $C = 1/a$, changing to $\tilde{x}$, $\tilde{y}$, $\tilde{t}$ variables, and then using the old variable names, system (5) can be written as $x' = (1 - x - y)x$, $y' = (a_1 - a_2 x - a_3 y)$, where $a_1 = d/a$, $a_2 = e/b$, $a_3 = f/c$, using three parameters instead of the original six. Duplicate Atlas Plate `Predator-Prey C` ($x$ survives, $y$ does not) and Figure 5.6.1 ($x$ and $y$ coexist), marking all equilibrium points inside and on the edge of the population quadrant and inserting arrowheads on the orbits to show the direction of increasing time. Beginning with the values of $a_1$, $a_2$, $a_3$ in the Atlas Plate, change the parameters by small amounts, plotting orbits after each change, until Figure 5.6.1 is obtained. Continue making small changes until a system is obtained in which $y$ survives, but $x$ does not. On your last graph write your conclusions.

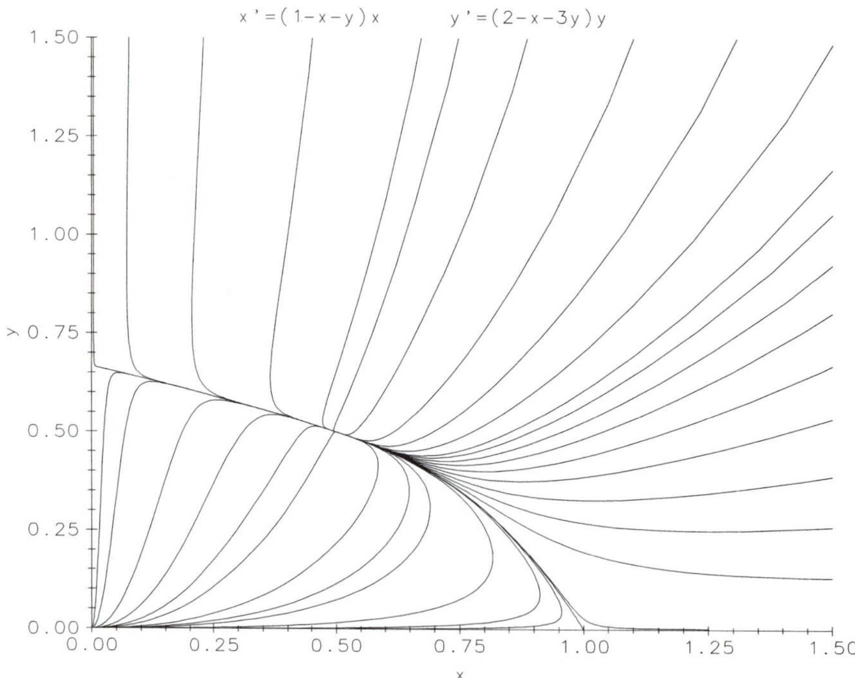


*Figure 5.6.1 Coexistence of competing species.*

# The Pendulum

***Purpose*** To investigate the motion of a simple pendulum. Solutions of the corresponding nonlinear ODE and its linearization are compared. The motion of a driven upended pendulum is also studied.

***Keywords*** Newton's Laws of Motion, simple pendulum, separatrix, periodic motion, driven pendulum

***See also*** Experiments 5.2–5.3, 4.7–4.9, 4.10–4.11; Appendix B.2; Atlas Plates `Pendulum A-C`, `Pendulum (Upended) A-C`

***Background*** The **simple pendulum** consists of a bob of mass $m$ hanging on a (presumed massless) rigid rod of fixed length $L$ firmly attached to a horizontal support. The pendulum is in equilibrium when the bob and rod are aligned with the local vertical and at rest. It is presumed that all motions of the bob take place in a plane. If the pendulum moves in a resistive medium, we assume that the force of resistance is proportional to the velocity of the bob. To derive the equation of motion consider the position vector $\mathbf{R}$ from the support of the pendulum to the center of the bob. Note that $\mathbf{R}(t)$ tracks the motion of the (center of the) bob and that $\mathbf{R}'(t)$, $\mathbf{R}''(t)$ are the velocity and acceleration vectors for this motion. Introducing polar coordinates in the plane of oscillation with the origin at the support and $\theta$ measured positively in the counterclockwise direction from the downward position, we can write $\mathbf{R}(t) = r(t)\hat{\mathbf{r}}(t)$, where $r(t)$ is the length of $\mathbf{R}(t)$ and $\hat{\mathbf{r}}(t)$ is a unit vector directed along $\mathbf{R}(t)$. Note that $r(t) \equiv L$ since the rod has fixed length. Denoting by $\hat{\boldsymbol{\theta}}(t)$ the unit vector orthogonal to $\mathbf{R}(t)$ and pointing in the direction of increasing $\theta$ we can calculate $\mathbf{R}'(t)$ and $\mathbf{R}''(t)$ in terms of $\hat{\mathbf{r}}(t)$ and $\hat{\boldsymbol{\theta}}(t)$ as follows. From Figure 5.7.1 we see that

$$\hat{\mathbf{r}} = \sin\theta\hat{\mathbf{\imath}} - \cos\theta\hat{\mathbf{\jmath}}, \qquad \hat{\boldsymbol{\theta}} = \cos\theta\hat{\mathbf{\imath}} + \sin\theta\hat{\mathbf{\jmath}}$$

where $\hat{\mathbf{\imath}}$, $\hat{\mathbf{\jmath}}$ are (fixed) Cartesian unit vectors along the positive $x$ and $y$-axes, respectively. Observe that

$$\hat{\mathbf{r}}' = (\cos\theta\hat{\mathbf{\imath}} + \sin\theta\hat{\mathbf{\jmath}})\theta' = \theta'\hat{\boldsymbol{\theta}}, \qquad \hat{\boldsymbol{\theta}}' = (-\sin\theta\hat{\mathbf{\imath}} + \cos\theta\hat{\mathbf{\jmath}})\theta' = -\theta'\hat{\mathbf{r}} \tag{1}$$

hence using (1) and the fact that $r(t) \equiv L$ we see

$$\mathbf{R}' = (L\hat{\mathbf{r}})' = L\theta'\hat{\boldsymbol{\theta}}, \qquad \mathbf{R}'' = (L\theta'\hat{\boldsymbol{\theta}})' = L\theta''\hat{\boldsymbol{\theta}} - L(\theta')^2\hat{\mathbf{r}} \tag{2}$$

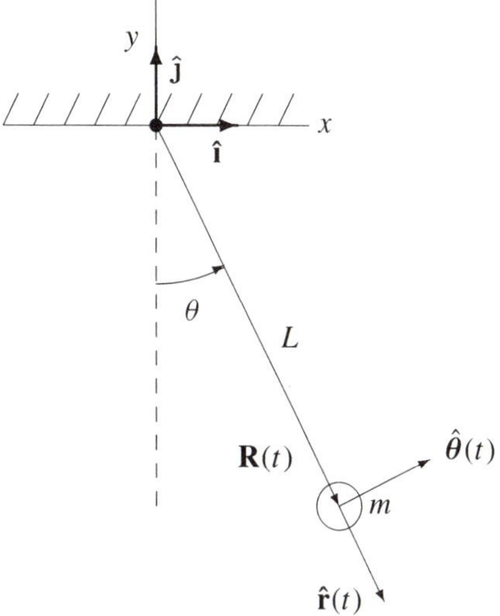


*Figure 5.7.1 Geometry for motion of a pendulum bob.*

Newton's Second Law states that $m\mathbf{R}''(t) = \mathbf{F}(t)$, where $\mathbf{F}$ is the resultant force acting on the bob at time $t$. $\mathbf{F}$ is the composition of three individual forces: (1) the force of gravity (acting downward), (2) the resistive force, and (3) the tension in the rod (acting in the direction of $-\hat{\mathbf{r}}$) that keeps the bob moving in a circular orbit. The force of gravity can be expressed as $(mg\cos\theta)\hat{\mathbf{r}} - (mg\sin\theta)\hat{\boldsymbol{\theta}}$, where $g$ is the gravitational constant. If the coefficient of proportionality for the resistive force is $c$, then that force can be expressed as $-c\mathbf{R}'(t) = -cL\theta'\hat{\boldsymbol{\theta}}$. Note that $c > 0$ if damping is present and $c = 0$ if not. Equating coefficients of $\hat{\boldsymbol{\theta}}$ in $m\mathbf{R}'' = \mathbf{F}$ and rearranging terms, we obtain the ODE for $\theta(t)$:

$$mL\theta'' + cL\theta' + mg\sin\theta = 0 \tag{3}$$

Equating the coefficients of $\hat{\mathbf{r}}$ only determines the magnitude of the tension in the rod, hence will be ignored.

The ODE for $\theta(t)$ is clearly nonlinear due to the presence of the $\sin\theta$ term. Putting $x = \theta$, $y = \theta'$ converts the second order ODE (3) into the nonlinear first order system

$$\begin{aligned} x' &= y \\ y' &= -\frac{g}{L}\sin x - \frac{c}{m}y \end{aligned} \tag{4}$$

The equilibrium points for this autonomous system are located at $x = n\pi$, $y = 0$, where $n = 0, \pm1, \pm2, \ldots$. Atlas Plate `Pendulum A` displays the orbits of the undamped simple pendulum. The wavy orbits at the top and bottom depict roll-over motions for which the pendulum has enough energy to swing all the way around its pivot. The top orbits correspond to counterclockwise motion with $\theta$ monotonically increasing in time, clockwise motion for the bottom orbits; these orbits are oscillatory but not periodic. The periodic orbits are the ovals centered at the equilibrium point $(2n\pi, 0)$; they correspond to the typical back-and-forth swings of the pendulum. Figure 5.7.2 shows a spray of orbits corresponding to the pendulum starting out in the vertically downward position with a spread of positive angular velocities. The orbits are plotted over two units of time, the spray dividing into those orbits with enough energy to go over the top and those with lower energy. Atlas Plate `Pendulum B` shows the oscillatory, non-periodic orbits of the damped simple pendulum.

• **Linearized pendulum**. System (4) can be linearized about the equilibrium point $(0, 0)$ by replacing $\sin x$ by the first term in its Taylor series about $x = 0$ to obtain

$$\begin{aligned} x' &= y \\ y' &= -\frac{g}{L}x - \frac{c}{m}y \end{aligned} \tag{5}$$

Because $\sin x$ is approximately $x$ for small values of $x$, it is reasonable to expect solutions of linear system (5) for small $x$ to approximate solutions of the nonlinear system (4). The linear ODE that reduces to (5) if $\theta = x$ is

$$\theta'' + \frac{c}{m}\theta' + \frac{g}{L}\theta = 0 \tag{6}$$

The eigenvalues of the matrix of the linear system (5) are $-c/2m \pm (c^2/4m^2 - g/L)^{1/2}$. Hence if $c > 0$, both eigenvalues have negative real parts and so the origin is an asymptotically stable focus for the linear system (5). If $c^2/4m^2 > g/L$, then both eigenvalues are negative and all solutions of (5) decay exponentially to $(0, 0)$. The system (5) is said to be **overdamped** in this case. If $c^2/4m^2 < g/L$, then all solutions of (5) are decaying sinusoids with angular frequency $\omega = (g/L - c^2/4m^2)^{1/2}$. In this case (5) is said to be **underdamped**. When $c^2/4m^2 = g/L$, the system (5) is said to be **critically damped**. Figure 5.7.3 displays the inward spiralling orbits of an underdamped linearized pendulum. Near the origin these orbits resemble those of Atlas Plate `Pendulum B`, but at a distance they are very different. Which are more realistic?

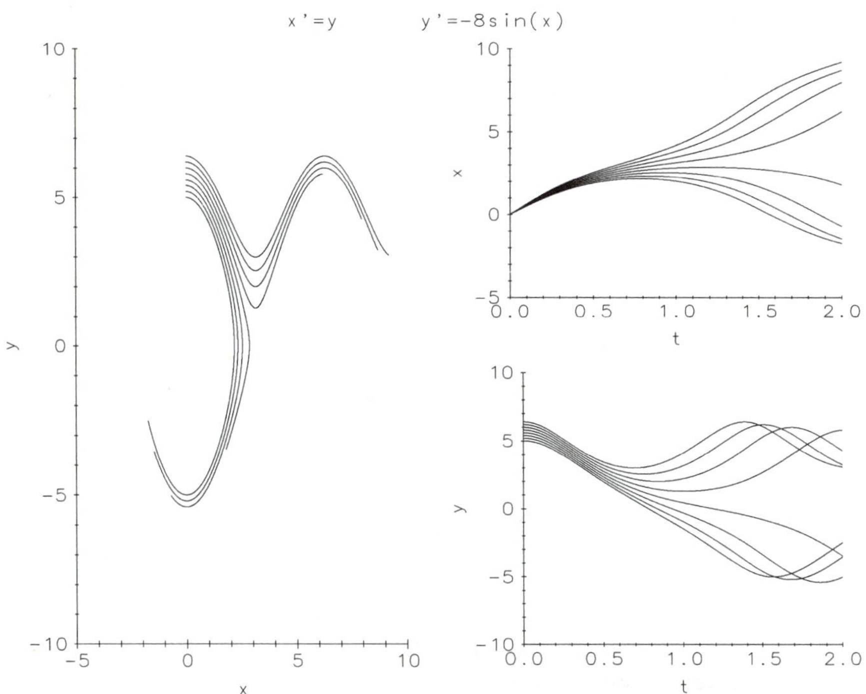


*Figure 5.7.2* *Orbital segments of the undamped simple pendulum: $x(0) = 0$, varying $y(0)$.*

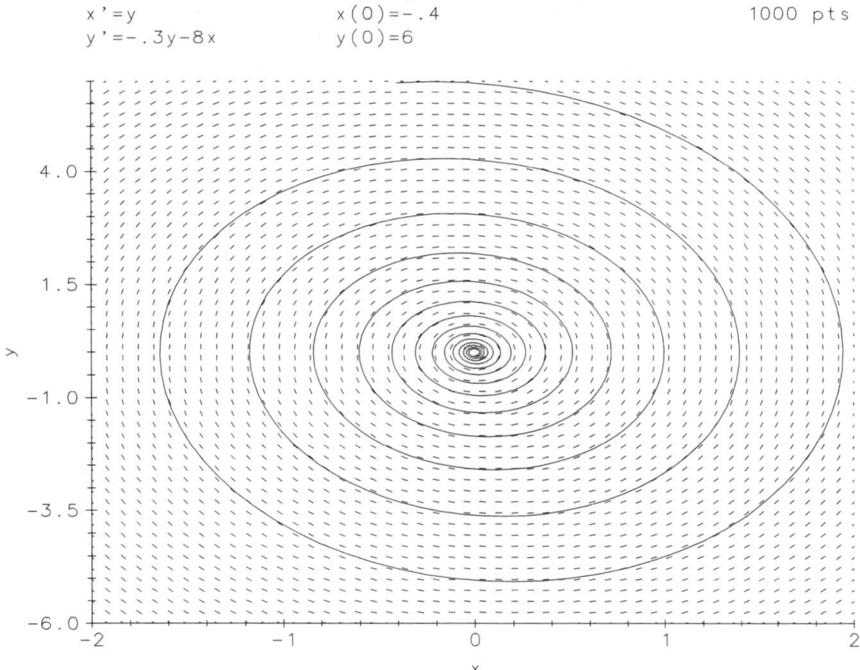


*Figure 5.7.3* *An orbit for an underdamped linearized pendulum.*

• **Stability for nonlinear system (4)**. It will be seen in Experiments 5.12–5.13 that because the origin is an asymptotically stable focus for the linear system (5) when $c > 0$, it follows that nonlinear system (4) has essentially the same orbital portrait near the origin. The equilibrium point (0, 0) for (4) corresponds physically to the pendulum hanging straight down and at rest, and the asymptotic stability property just means that if the pendulum undergoes a small perturbation from this rest position (0, 0), the subsequent motion "seeks" to return the pendulum to this rest position.

• **Integral of the motion**. When no damping is present ($c = 0$) an interesting fact can be extracted from (3). Let $\theta(t)$ be any solution of (3). Multiplying (3) through by $\theta'(t)$, the equation can be rewritten as

$$\left(\tfrac{1}{2}mL(\theta')^2 - mg\cos\theta\right)' = 0$$

Antidifferentiate this equation, multiply by $L$ and arrange constants of integration to obtain

$$\tfrac{1}{2}m(L\theta')^2 + mgL(1 - \cos\theta) = C \tag{7}$$

Equation (7) expresses the following law

**Law of Energy Conservation:** *Kinetic + potential energy remains constant.*

The kinetic energy is the energy of angular motion, and the potential energy is the gravitational energy. This implies that any orbit $(x(t), y(t))$ of system (4), with $c = 0$, must lie on a level curve of the function $I(x, y) \equiv y^2 + (2g/L)(1 - \cos x)$. Such a function is called an **integral** for system (4). To find all orbits it is merely necessary to plot all the level curves of $I(x, y)$. The Atlas Plate `Pendulum A` gives a good idea of what these level curves look like. Much can be inferred about the behavior of orbits from these level curves.

• **Driven upended pendulum**. Now suppose that the pendulum is "upended"; that is, $\theta = 0$ corresponds to the bob standing straight up. The equation of motion can be obtained from (3) by replacing $\theta$ by $\theta + \pi$ and using the fact that $\sin(\theta + \pi) = -\sin\theta$. Suppose, moreover, that a force $F(t)$ (e.g., a magnetic force, or pulsating shocks to the horizontal pivot that are transmitted to the bob by the rod) acts on the pendulum bob. The new equation of motion is

$$mL\theta'' + cL\theta' - mg\sin\theta = F(t) \tag{8}$$

As in (5), the Taylor series expansion about $\theta = 0$ of $\sin\theta$

$$\sin\theta = \theta - \frac{\theta^3}{3!} + \frac{\theta^5}{5!} - \frac{\theta^7}{7!} + \cdots$$

can be truncated and used to replace the function $\sin\theta$. The purpose of Experiment 5.9 is to examine the relation between the solutions of (8) and solutions of equations like (9) or (10)

$$mL\theta'' + cL\theta' - mg\theta = F(t) \tag{9}$$

$$mL\theta'' + cL\theta' - mg\left(\theta - \frac{\theta^3}{3!}\right) = F(t) \tag{10}$$

in which $\sin\theta$ is replaced by a Taylor truncation.

_______________/_______________
hardware software

_______________/_______________
name date

Answer questions in the space provided,
or on attached sheets or carefully labeled graphs.

_______________/_______________
course section

## 5.7 *The Undriven Pendulum: Linear Model*

***Abstract*** To explore the influence of the damping coefficient on the motion of a damped (but undriven) linear oscillator such as a simple linear pendulum (or a mass on a spring).

**1.** In (6) take $g/L = 1$ and $c = 0$.

**(a)** Plot $\theta$ versus $t$, for $0 \leq t \leq 50$, where $\theta(t)$ solves (6) and satisfies the initial conditions $\theta(0) = 2$, $\theta'(0) = 0$. Determine the period $T$ of $\theta(t)$ from your plot. Highlight a single period of $\theta(t)$ directly on your graph.

**(b)** Repeat (a) for other initial conditions. Do all initial conditions give rise to periodic solutions? (Roll-over motions are not considered to be periodic.) Describe how the period and amplitude of periodic motions of the pendulum depend on the initial data.

**2.** In (6) take $g/L = 1$ and define $\gamma = c/m$.

**(a)** Solve (6) with $\gamma = 0.15$ and initial conditions $\theta(0) = 2$, $\theta'(0) = 0$, and plot $\theta$ versus $t$ over $0 \leq t \leq 50$. Estimate the time $\tau$ at which the solution becomes negligibly small. Use whatever definition you wish of "negligibly small," but use the same definition in later parts of this experiment. Also determine the **quasiperiod** $T$ of the motion, that is, the interval between successive times when $\theta = 0$ and $\theta' > 0$. Write directly on graphs.

**(b)** Repeat (a) for $\gamma = 1, 1.5, 1.9, 1.95$.

**(c)** Using results of (a) and (b), form a table showing the values of $\tau$ and $T$ for each value of $\gamma$. Describe the qualitative dependence of $\tau$ on $\gamma$; of $T$ on $\gamma$. Construct a simple function expressing the relation between $\tau$ and $\gamma$ that is qualitatively consistent with your tabulated values and quantitatively as accurate as possible. Do the same for the relation between $T$ and $\gamma$. Using the function you constructed, estimate the value of $\gamma$ for which $T \to \infty$.

**(d)** Find the solution formula for the ODE $\theta'' + \gamma\theta' + \theta = 0$ subject to the initial conditions $\theta(0) = 2, \theta'(0) = 0$, or other initial conditions of your choice. Determine formulas showing how $\tau$ and $T$ depend on $\gamma$. Compare these results with those of (c).

**(e)** The value of $\gamma$ for which $T \to \infty$ is called the critical value of $\gamma$. It separates those values of $\gamma$ for which solutions are damped oscillations from those for which the solutions are decaying exponentials. Determine this value experimentally by plotting solutions for values of $\gamma$ near the critical value found in (c) and (d). Then check your answer by determining the critical value of $\gamma$ from the analytical solution found in (d).

_______________ / _______________
hardware software

_______________ / _______________
name date

Answer questions in the space provided,
or on attached sheets or carefully labeled graphs.

_______________ / _______________
course section

## 5.8 *The Undriven Nonlinear Pendulum*

***Abstract*** The nonlinear ODE for the simple pendulum is examined in some detail.

**1.** Determine the equilibrium points of system (4) in the $xy$-plane. Describe the configuration of the pendulum corresponding to each such point. Do the equilibrium points depend on $c$?

**2.** Take $m = L = g = 1$, and $c = 0$.

**(a)** Plot the orbits of system (4) that originate at the points $(-12, 1)$, $(-12, 1.5)$, $(-12, 2)$, and $(-12, 3)$. Explain the different kinds of physical motion of the pendulum that correspond to closed and nonclosed orbits. Confirm your explanation by plotting $x$ versus $t$ for each of the given initial conditions. (Remember that $x$ stands for $\theta$, so that a plot of $x$ versus $t$ describes the angular position of the pendulum as a function of time.)

**(b)** Repeat (a) for orbits that originate at $(-6, y_0)$, $(0, y_0)$, and $(6, y_0)$, for some value $y_0$.

**(c)** Consider the periodic motions of the pendulum described by the orbits of system (4) originating at $(1, 0)$, $(1.5, 0)$, $(2, 0)$, and $(3, 0)$. Plot $x$ versus $t$ and determine the period $T$ of the motion from the graph in each of these cases. How does $T$ depend on the initial position of the pendulum?

* **(d)** The orbit of system (4) originating at $(\alpha, 0)$, $0 < |\alpha| < \pi$, has period

$$T = 4\sqrt{\frac{L}{g}} \int_0^{\pi/2} \frac{d\phi}{\sqrt{1 - k^2 \sin^2 \phi}}, \quad \text{where} \quad k = \sin\left(\frac{\alpha}{2}\right)$$

Evaluate this integral (numerically) with $L/g = 1$ and with $\alpha = 1, 1.5, 2$, and $3$. Compare the results with those in (c).

**3.** Take $m = L = g = 1$.

**(a)** With $c = 0.1$ repeat the calculations described in Problems 2a and 2b. Compare these phase portraits with those above. What would be the effect of increasing $c$?

**(b)** When $c > 0$, orbits of system (4) starting in a certain zone approach the origin as $t \to \infty$, while orbits starting in other zones approach one of the other equilibrium points. A curve that separates two adjacent zones is called a **separatrix**. By plotting orbits that start at various initial points, try to locate the separatrix between the zone containing the origin and the zone containing the equilibrium point $(2\pi, 0)$.

_______________ / _______________
hardware software

_______________ / _______________
name date

Answer questions in the space provided,
or on attached sheets or carefully labeled graphs.

_______________ / _______________
course section

## *5.9 The Driven Upended Pendulum*

***Abstract*** Solution curves of the driven upended pendulum are examined after approximating the $\sin\theta$ term with various truncations of its Taylor series.

***Special instructions*** In the problems below, set $g/L = 6/11$ and $G(t) = F(t)/mL$ in (8) to obtain the IVP

$$\theta'' + \frac{c}{m}\theta' - \frac{6}{11}\sin\theta = G(t), \qquad \theta(0) = a, \quad \theta'(0) = b$$

**1.** Let $c = 0$, $G = 0$, $a \neq 0$, $b = 0$ in the IVP above and graph the solution of the IVP over $0 \leq t \leq 100$. Then replace $\sin\theta$ by $\theta - \theta^3/6$ and repeat. In each case graph orbit and component curves. What causes the differences in the graphs?

**2.** In the IVP above set $c/m = 0.07$ and $G = 0$, and in the $\theta\theta'$-plane graph solutions of this modified IVP for several values of $a$ and $b$, including $a = -1.4$, $b = 1$.

**3.** In the IVP above set $c/m = 0.07$ and $G(t) = 0.3\cos t$, and graph the solution in the $\theta\theta'$-plane for $a = -1.4$, $b = 1$. Compare with Atlas Plate `Pendulum (Upended) A`.

**4.** For the tasks below, take $c/m = 0.07$ and $G(t) = 0.3\cos t$, $a = -1.4$, $b = 1$ in the IVP above.

**(a)** In the IVP above replace $\sin\theta$ by the first term in its Taylor series. In the $\theta\theta'$-plane graph the long-term behavior of the solution of this modified IVP. How would you describe this behavior in words? Compare this solution to the solution of Problem 3.

**(b)** Repeat (a) for a two-term Taylor approximation for $\sin\theta$ in the IVP above. Compare this solution to the solution in (a). Conclusions? Compare with Atlas Plates `Pendulum (Upended) B`.

**(c)** Repeat (a) for a three-term Taylor approximation for $\sin\theta$ in the IVP above. Compare this solution to the solution in (a) and (b). Conclusions?

**(d)** Repeat (a) for a four-term Taylor approximation for $\sin\theta$ in the IVP above. Compare this solution to the solution in (a), (b), and (c). Conclusions?

# Duffing's Equation

***Purpose*** To observe the oscillations and the chaotic wanderings of solutions of Duffing's equation and its equivalent planar system.

***Keywords*** Duffing's equation, chaotic wandering, sensitivity

***See also*** Experiment 5.9; Atlas Plates `Duffing A-C`; Chapter 6 Notes

***Background*** The motion of a long, thin, elastic steel beam with its top end embedded in a pulsating frame and the bottom end hanging just above two magnets can be modeled by

$$x'' + cx' - ax + bx^3 = A\cos\omega t \tag{1}$$

where $a, b, c, A$, and $\omega$ are constants. This is known as **Duffing's equation**. The equivalent first order system is

$$\begin{aligned} x' &= y \\ y' &= ax - bx^3 - cy + A\cos\omega t \end{aligned} \tag{2}$$

Observe that (1) resembles the ODE for a driven upended pendulum (Experiment 5.9) if $\sin\theta$ is replaced by the first two terms of its Taylor expansion. Orbits of (2) have a remarkable variety of behaviors, ranging from oscillatory, to periodic, to chaotic. For some sets of values of the parameters and some sets of initial data, orbits appear to approach a strange attractor.

• **Poincaré time sections**. One of the best ways to display the orbits of a driven system in which the driving force is periodic is to take "snapshots" of the orbit as it is being traced out in time, snapshots that show the orbital point $(x(t), y(t))$ only when $t$ is an integer multiple of the period. The resulting picture of "dots" is called a **Poincaré time section** of the orbit. If the orbit is periodic with the same period as the driving function, a single dot will show. If the orbit does not have that period, or is not periodic at all, a "dust" of points appears on the plot. In fact, Poincaré time sections can be taken of any orbit of any system at integer multiples of any fixed time span chosen by the user. See Figure 5.10.4 and Atlas Plates `Duffing C` and `Poincaré A-C` for examples of Poincaré time sections.

• **Sensitivity**. Some orbits of (2) are sensitive to changes in initial data even if the driving term is absent ($A = 0$). Figure 5.10.1 shows that the long-term behavior of one of the orbits completely changes if its starting point is shifted from the point $(1, 0.54)$ to the point $(1, 0.56)$. The first orbit spirals toward the equilibrium point $(-1, 0)$, the second toward the equilibrium point $(1, 0)$. Thus, the eventual behavior of this orbit as time increases is sensitive to a change in the initial data.

***References*** Consult the material below for additional background and details.


1. Guckenheimer, J., Holmes, P. *Nonlinear Oscillations, Dynamical Systems, and Bifurcations of Vector Fields*. Springer-Verlag, New York, 1983.
2. Holmes, P., Moon, F.C. *J. of Appl. Mech.*, (**108**, 1021–1032 (1983)).
3. Thompson, J.M.T., Stewart, H.B. *Nonlinear Dynamics and Chaos*. John Wiley & Sons, New York, 1986.

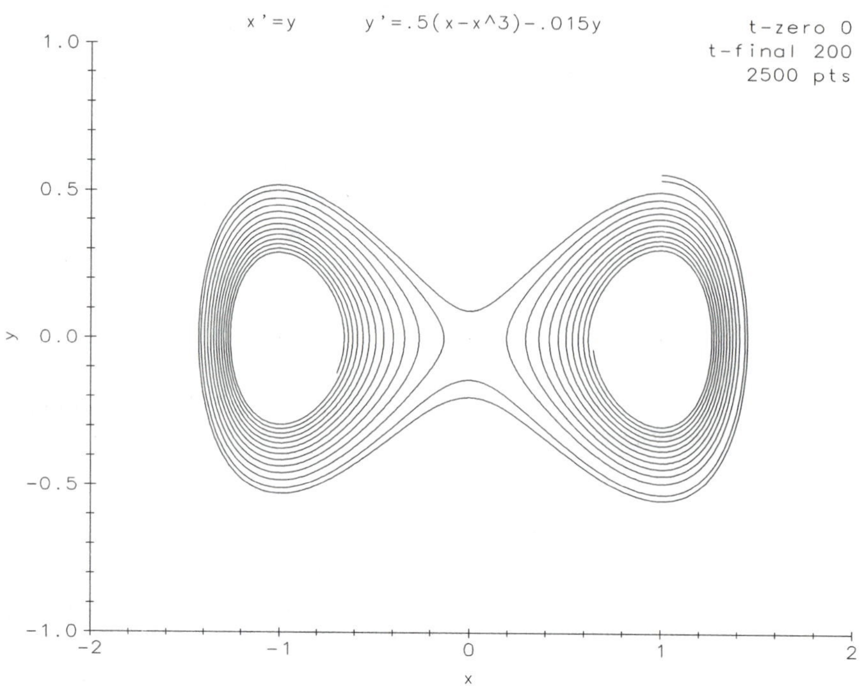


*Figure 5.10.1 Damped, undriven Duffing system: sensitivity to small changes in initial data.*

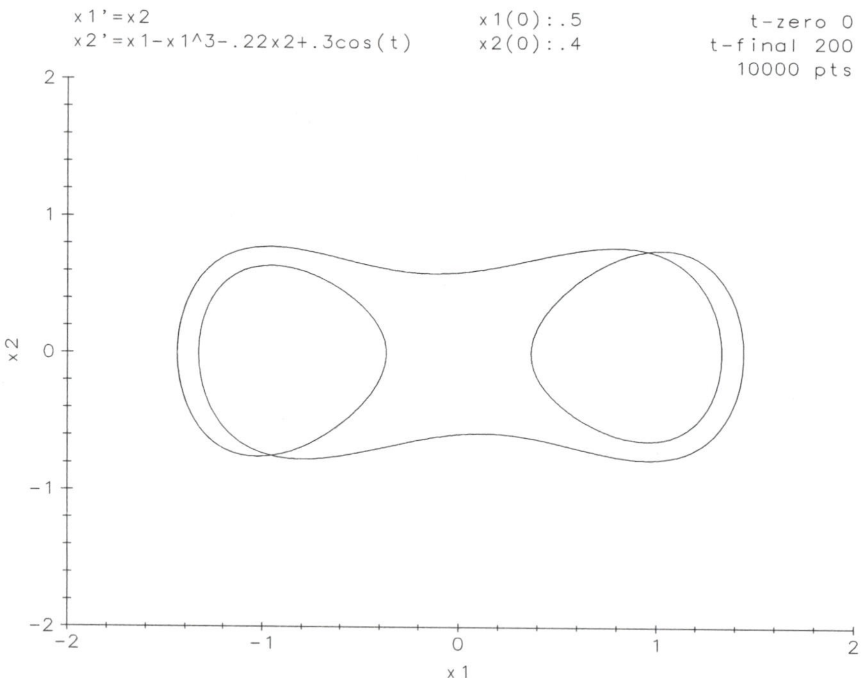


*Figure 5.10.2 A stable periodic orbit plotted for* $180 \leq t \leq 200$.

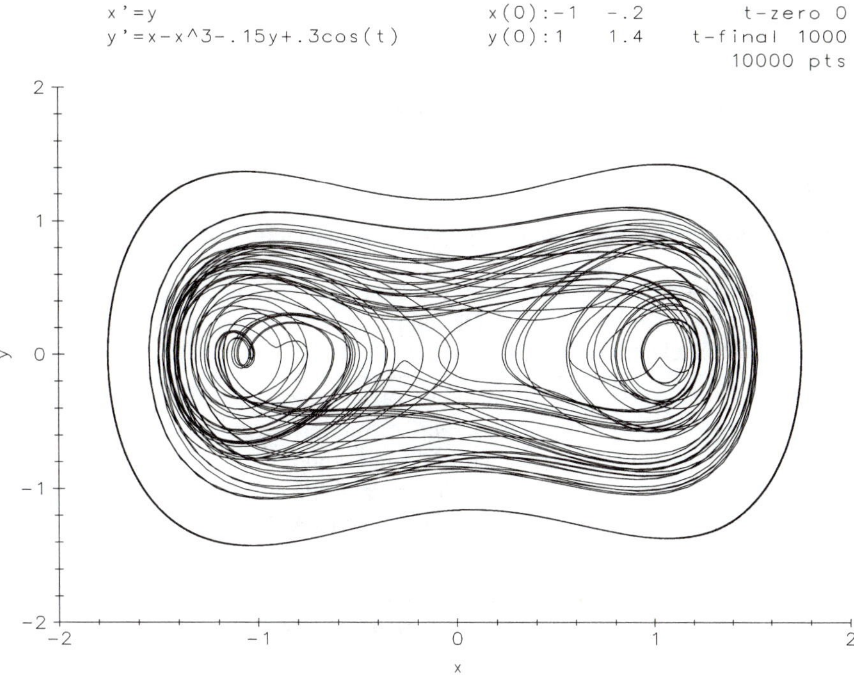


*Figure 5.10.3 A stable periodic orbit, and an orbit of chaotic wandering (both orbits plotted for* $500 \leq t \leq 1000$).

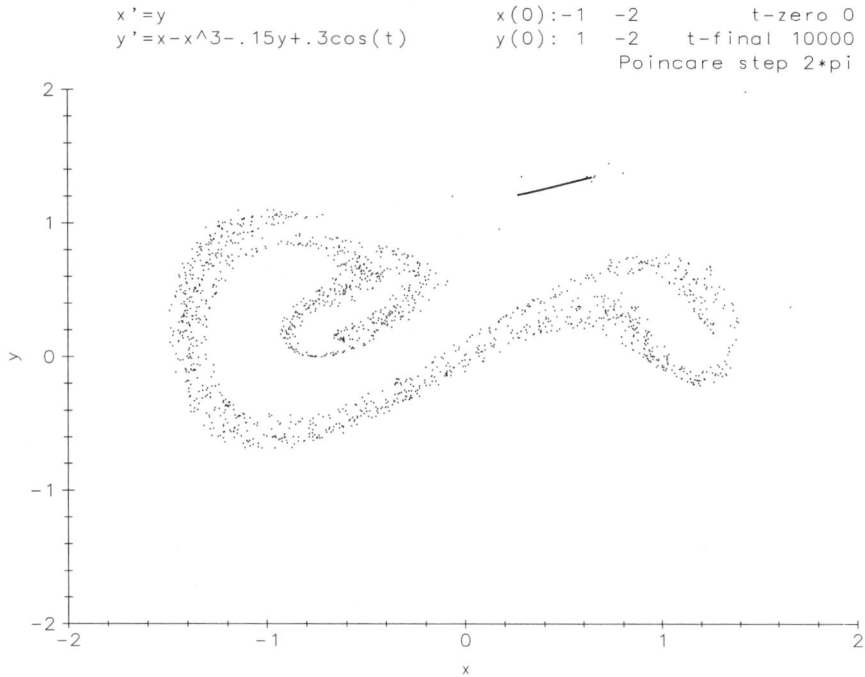


*Figure 5.10.4 The Poincaré time section* ($\Delta t = 2\pi$) *of two orbits.*

_______________/_______________
hardware software

_______________/_______________
name date

Answer questions in the space provided,
or on attached sheets or carefully labeled graphs.

_______________/_______________
course section

## 5.10 *Duffing's Equation*

***Abstract*** Orbits and Poincaré time sections of various Duffing systems are plotted and explored.

***Special instructions*** This experiment is best done by a team. At least one of the references listed on the cover sheet should be consulted for additional background and suggestions. The team's report should address the points listed below.

- Analyze and graph the orbits of an undriven, undamped Duffing system: $x' = y$, $y' = .5(x - x^3)$. Find the three equilibrium points. Find an orbital figure-eight that consists of the origin and two other orbits. Why are all but five orbits periodic? What happens to the periods of the orbits inside a lobe of the figure-eight as the amplitudes diminish to zero? As the periodic orbits approach the boundary of the lobe? Analyze the periods of the orbits outside the lobes.
- Add damping and obtain the system of Figure 5.10.1, in which orbits are graphed for $x_0 = 1$, $y_0 = .54$ and $x_0 = 1$, $y_0 = .56$. One orbit tends to $(-1, 0)$, the other to $(1, 0)$ as $t \to \infty$. Fix $x_0$ at 1 and plot orbits for several values of $y_0$ from 0 to 5.0. For each $y_0$ solve for $0 \leq t \leq 200$. Do you see any pattern? Construct a table having the values of $y_0$ in the first column, and the number of times the orbit starting at $(1, y_0)$ crosses the $y$-axis as $t$ increases from 0 to 200 in the second. From the table, estimate the length of each $y_0$ interval that generates orbits with a given number of crossings. Explain why all orbits starting inside any one of these $y_0$ intervals approach one of the equilibrium points $(-1, 0)$, $(1, 0)$, while the orbits starting in adjacent intervals approach the other equilibrium point. What happens to the lengths of these $y_0$ intervals as $y_0$ increases? Any explanation? How does `Duffing A` fit in the setting here?
- Change the damping a little and add a periodic driving force to obtain the Duffing system of Figure 5.10.2. What is the period of the periodic orbit shown? Is it a stable attractor (i.e., do nearby orbits approach it as $t$ increases)?
- Change the damping back to that of Figure 5.10.1 but keep the periodic driving force to obtain the system in Figure 5.10.3. What is the period of the periodic orbit visible in the picture? What is the other orbit doing? These orbits are solved for $0 \leq t \leq 1000$ but plotted in Figure 5.10.3 only for $500 \leq t \leq 1000$. What happens if the driving and plotting are done over other time spans? Duplicate Atlas Plates `Duffing B`, `C` and explain the information they contain in the context of Figure 5.10.3.
- Plot Poincaré time sections over $0 \leq t \leq 10000$ for several initial points near $(-1, 1)$. Use the Duffing system of Figure 5.10.4. Explain what you see.
- Figure 5.10.4 shows Poincaré time sections of two orbits. The one with all the dots is the section of the chaotic orbit in Figure 5.10.3. What is the other orbit doing?
- Rescale $x$ and $t$ and reduce the number of parameters in (1) from five $(c, a, b, A, \omega)$ to three. (Hint: See Appendix B.5.) Then carry out a study of what happens as one or more of the three parameters is varied. Use Poincaré time sections if appropriate. Explain what you see. Do you see a strange attractor?

# Planetary Motion

***Purpose*** To examine **Kepler's** and **Newton's Laws** for planetary motion.

***Keywords*** Newton's Laws of Motion, two-body problem, Newton's Equations of Planetary Motion, conic sections in polar form, vectors, precession

***See also*** Appendix B.2

***Background*** It took Kepler[1] twenty years of study of observational data to formulate the results derived below. By the end of the 17th century it would take but a few hours to derive the same results mathematically. Newton's concepts of force and gravitation and the new techniques of integral and differential calculus made the difference.

In Appendix B.2 Newton's Laws of Planetary Motion are derived by applying Newton's Laws of Motion to a two-body system in a central force field created by their mutual gravitational attraction. It is shown that the vector $\mathbf{r}(t)$ from the body of mass $M$ (the "sun") to the body of mass $m$ (the "planet") satisfies the second order vector differential equation

$$\mu\mathbf{r}'' = \frac{-GMm}{r^3}\mathbf{r} \tag{1}$$

where $\mu = mM(m+M)^{-1}$ is the **reduced mass**, $G$ is the universal gravitational constant,[2] and $r = \|\mathbf{r}\|$ is the length of the vector $\mathbf{r}$. Using (1) it follows that $(\mathbf{r}\times\mathbf{r}')' \equiv 0$ and therefore the motion of $m$ about $M$ takes place in a fixed plane. Taking the coordinate frame $\{\hat{\mathbf{i}}, \hat{\mathbf{j}}, \hat{\mathbf{k}}\}$ with $\hat{\mathbf{k}}$ orthogonal to this plane, and $(r, \theta)$ as polar coordinates in this plane (as indicated in Figure 5.11.1), it is shown in Appendix B.2 that $r(t)$ and $\theta(t)$ satisfy the ODEs

$$\begin{aligned} r'' - (\theta')^2 r &= -\frac{G(M+m)}{r^2} \\ \theta'' r + 2\theta' r' &= 0 \end{aligned} \tag{2}$$

which are **Newton's Equations of Planetary Motion**.

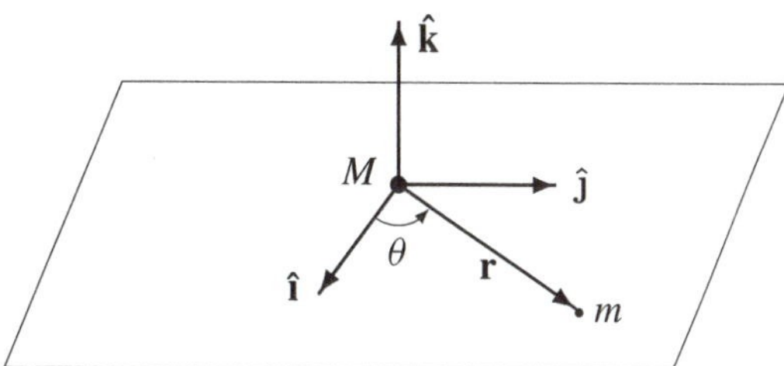


*Figure 5.11.1 Geometry of planetary motion.*

The parametrically defined curve

$$\mathbf{r}(t) = r(t)\cos\theta(t)\hat{\mathbf{i}} + r(t)\sin\theta(t)\hat{\mathbf{j}}$$

in the plane is called an **orbit** of the "planet"; hence the plane is called the orbital plane (this notion of orbit is slightly different from that of an orbit for a system of ODEs).

Whatever the shape of the orbital curve, it has the following surprising property (also called **Kepler's Second Law**).

1. Johannes Kepler (1571-1630) was a German astronomer who discovered the three laws of planetary motion given in this experiment. Although the laws are deduced here from Newton's laws using calculus, Kepler, who died 12 years before Newton was born, came to his conclusions after a twenty-year study of the observational data obtained by the Danish astronomer, Tycho Brahe.
2. $G$ is approximately $6.67 \times 10^{-11}$ N$\cdot$ m$^2$/kg$^2$

**Equal Areas-Equal Times:** *The vector* $\mathbf{r}(t)$ *of the orbit sweeps out equal areas in equal times.*

To show this, first set up a moving reference frame of orthogonal unit vectors $\hat{\mathbf{r}}$, $\hat{\boldsymbol{\theta}}$ in the orbital plane as follows (see also Appendix B.2):

$$\hat{\mathbf{r}} = \cos\theta\hat{\mathbf{i}} + \sin\theta\hat{\mathbf{j}}, \qquad \hat{\boldsymbol{\theta}} = -\sin\theta\hat{\mathbf{i}} + \cos\theta\hat{\mathbf{j}}$$

and notice that

$$\hat{\mathbf{r}}' = \theta'\hat{\boldsymbol{\theta}}, \qquad \hat{\boldsymbol{\theta}}' = -\theta'\hat{\mathbf{r}}$$

Hence, it follows that

$$\mathbf{r} = r\hat{\mathbf{r}}, \qquad \mathbf{r}' = r'\hat{\mathbf{r}} + r\theta'\hat{\boldsymbol{\theta}} \tag{3}$$

Since $(\mathbf{r}\times\mathbf{r}')' \equiv 0$, $\mathbf{r}\times\mathbf{r}'$ is a constant vector (parallel to $\hat{\mathbf{k}}$ in Figure 5.11.1). Using (3) we see that $\mathbf{r}\times\mathbf{r}' = r^2\theta'\hat{\mathbf{r}}\times\hat{\boldsymbol{\theta}}$ and so for all $t$

$$r^2(t)\theta'(t) = h \tag{4}$$

where $h$ is a constant. We may also think of $\mathbf{r}$ as a function of the polar angle $\theta$. Hence the area of the region swept out by $\mathbf{r}(t)$ as $t$ increases from $t_0$ to $t_1$ is

$$A = \frac{1}{2}\int_{\theta(t_0)}^{\theta(t_1)} r^2(\theta)\,d\theta = \frac{1}{2}\int_{t_0}^{t_1} r^2(\theta(t))\theta'(t)\,dt = \frac{1}{2}(t_1 - t_0)h \tag{5}$$

Thus, $\mathbf{r}(t)$ sweeps out equal areas in equal times, which implies that as the planet nears the sun, its angular velocity $\theta'$ must increase.

The next result follows from the ODEs in (2) and is also known as **Kepler's First Law**.

**Elliptical Orbits:** *The orbit of a planet is an ellipse with the sun at one focus.*

A solution of the system (2) is a pair of functions, $r = r(t)$, $\theta = \theta(t)$, satisfying the ODEs (2) for all $t$ in some time interval $I$. It is not easy to find $r$ and $\theta$ as explicit functions of time, and so we resort to a trick we have used elsewhere to treat autonomous systems. If $r = r(t)$, $\theta = \theta(t)$ is a solution of (2), then $r$ can be considered to be a function of $\theta$ by solving $\theta = \theta(t)$ for $t$ in terms of $\theta$, $t = t(\theta)$, and then replacing $t$ in $r(t)$ by $t(\theta)$ to obtain $r(t(\theta))$. (A tacit assumption in this procedure is that $d\theta/dt$ never vanishes.) So thinking of $r = r(\theta)$, we have the identity $r = r(\theta(t))$. Thus, from the Chain Rule, $dr/dt = (dr/d\theta)(d\theta/dt)$, or, recalling from (4) that $r^2(d\theta/dt) = h$, a constant of the motion, we have (after multiplication by $r^2$)

$$r^2\frac{dr}{dt} = h\frac{dr}{d\theta}$$

Differentiating this result with respect to $t$ and using the Chain Rule we have

$$2r\left(\frac{dr}{dt}\right)^2 + r^2\frac{d^2r}{dt^2} = h\frac{d}{d\theta}\left(\frac{dr}{d\theta}\right)\frac{d\theta}{dt}$$

or, after replacing $d^2r/dt^2$ from (2) and applying the Chain Rule to $dr/dt$

$$2r\left(\frac{dr}{d\theta}\right)^2\left(\frac{d\theta}{dt}\right)^2 + r^2\left(r\left(\frac{d\theta}{dt}\right)^2 - \frac{G(M+m)}{r^2}\right) = h\frac{d^2r}{d\theta^2}\frac{d\theta}{dt}$$

Multiplying the last equation through by $r^2$, dividing by $h^2$, and again using equation (4), we obtain

$$\frac{d^2r}{d\theta^2} - \frac{2}{r}\left(\frac{dr}{d\theta}\right)^2 = r - \frac{G(M+m)}{h^2}r^2 \tag{6}$$

Equation (6) is a second order, nonlinear differential equation for $r(\theta)$. Fortunately, the change of variable $r = 1/u$ reduces (6) to the constant coefficient linear equation

$$\frac{d^2u}{d\theta^2} + u = \frac{G(M+m)}{h^2}$$

The general solution of this equation is

$$u = \frac{G(M+m)}{h^2} + K\cos(\theta - \delta)$$

for arbitrary constants $K > 0$ and $\delta$ (determined later by initial data). Thus, we finally have solutions for $r(\theta)$ in the form

$$r = \frac{1}{G(M+m)/h^2 + K\cos(\theta - \delta)}$$

which is the equation of a conic section in polar form with the origin at a focus. Let $e = Kh^2/G(M+m)$, $p = h^2/G(M+m)e$, and rewrite $r(\theta)$ in the more familiar form

$$r = \frac{ep}{1 + e\cos(\theta - \delta)} \tag{7}$$

where $e$ is the **eccentricity** and $p$ is the distance from a focus to its directrix. See Figure 5.11.2.

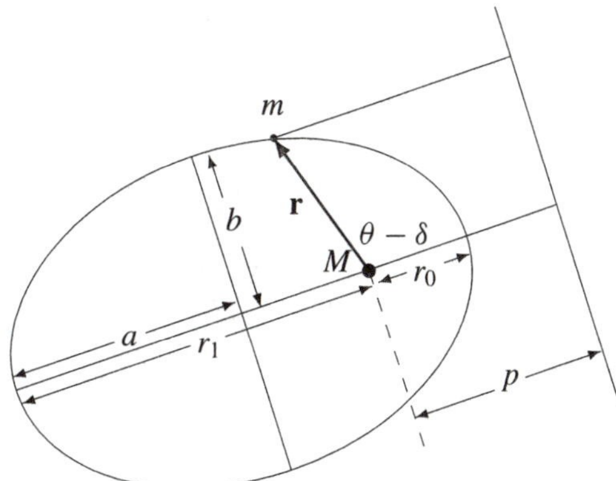


*Figure 5.11.2 Geometry of an elliptical orbit.*

Equation (7) is the polar form of an equation of an ellipse if (and only if) $e < 1$, a parabola if $e = 1$, and a hyperbola if $e > 1$. There is no way to decide on purely mathematical grounds which of these alternatives holds for a given body. On observational grounds, however, it is obvious that planets, asteroids, and returning comets must have elliptical orbits. The orbits of some comets, however, appear to be parabolic or hyberbolic. These comets approach the sun only once and then wander out into interstellar space.

Note that we have not solved the ODEs (2) for the location $\mathbf{r} = \mathbf{r}(t)$ of the planet as a function of time, nor have we said anything at all about the motion of the sun.

If the motion of the planet takes place in an elliptical orbit, then it is not hard to show that the motion is periodic. Remarkably, the period can be expressed rather simply in terms of the semimajor axis of the elliptical orbit (**Kepler's Third Law**).

**Period and Semimajor Axis:** *The period $T$ and semimajor axis $a$ of an elliptical orbit satisfy*

$$T^2 = \frac{4\pi^2 a^3}{G(M+m)} \tag{8}$$

The points on the orbit where $r$ has the least and greatest values are called the **perihelion** and the **aphelion** , respectively. Referring to Figure 5.11.2 and (7) note that the length of the major axis is

$$2a = r_0 + r_1 = \frac{ep}{1+e} + \frac{ep}{1-e} = \frac{2ep}{1-e^2} \tag{9}$$

and since the length of the minor axis is given by $2b = 2a(1-e^2)^{1/2}$, it follows from the expression (9) that

$$2b = 2a(1-e^2)^{1/2} = 2(epa)^{1/2} \tag{10}$$

From (5), (10), the definitions of $e$ and $p$, and the fact that the area of an ellipse is $\pi ab$, we finally achieve

$$\tfrac{1}{2}Th = \pi ab = \pi a(epa)^{1/2} = \frac{\pi h a^{3/2}}{[G(M+m)]^{1/2}}$$

The mass $m$ of the satellite, whether natural or artificial, is usually so small in comparison to $M$ that $m$ is omitted from (8). The resulting quotient, $4\pi^2/GM = T^2/a^3$, is characteristic for all satellite orbits about the massive body. The quotient for the sun is approximately $2.96 \times 10^{-19}\mathrm{s}^2/\mathrm{m}^3$, and measurements of $T^2/a^3$ for each of the nine planets approximate this value to within 0.5%.

***Observation*** **(Accuracy of the model)** Even in Newton's day it was known that the elliptical orbits of the two-body problem could not be the exact orbits of the actual planets. The other planets and their moons exert gravitational forces upon one another that must be taken into account in the differential equations of motion. The true state of affairs is that of a multibody problem, not of a two-body problem. The attempt to correct for the effects of the other heavenly bodies led to much new mathematics in the 18th and 19th centuries. Today, computers allow extremely accurate approximate solutions of the full equations of the motion of the planets, the moons, and the other satellites of the solar system. In this sense we may say that the problem has been solved even though no explicit general solution of the multibody problem (or even of the three-body problem) is yet known.

***Observation*** **(Frictional effects)** Finally, we have ignored entirely the natural degradation of any real system due to friction. This can hardly be of much significance for the planets, but it is of considerable importance for low altitude satellites of the earth. The atmosphere at an altitude of one hundred miles may be tenuous in the extreme, but it is enough to slow down a large satellite to the point where it "falls out" of orbit and burns-up or crashes on the surface of the earth. To model all this, a frictional term may be added to the equations of motion, which can then be solved approximately by computer. The time and place of burn up or crash may consequently be predicted.

***Observation*** **(Relativisitic correction)** The planet $m$ moves in a periodic elliptical orbit described by (7) with the sun $M$ at one focus. This orbit remains fixed in space as seen from a frame fixed to the sun. Einstein's Theory of General Relativity predicts an orbital equation that is a slight modification of (6):

$$\frac{d^2r}{d\theta^2} - \frac{2}{r}\left(\frac{dr}{d\theta}\right)^2 = r - \frac{G(M+m)}{h^2}r^2 - \frac{3G(M+m)}{c^2} \tag{11}$$

The ODEs (6) and(11) differ only in the last term, which is very small since $c$, the speed of light, is about $3 \times 10^{10}$ cm/s. The orbits generated by (11) are very nearly ellipses, which precess very slowly in time within the plane of motion. For large gravitational fields, this precession is observable. Indeed, the motion of the planet Mercury, the closest planet to the sun, has been observed to precess at 43 seconds of arc per century. This observation essentially agrees with the amount predicted by the ODE (11). This fact provided one of the earliest corroborations of General Relativity Theory.

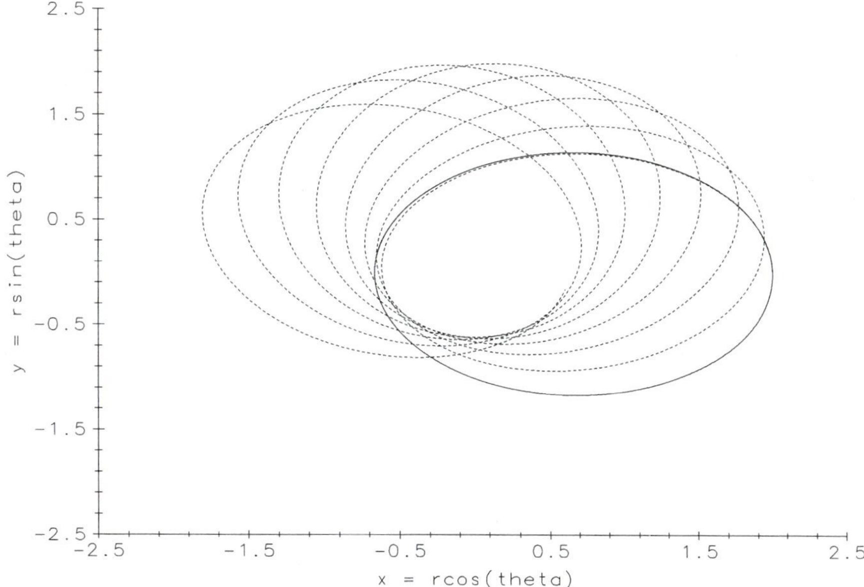


*Figure 5.11.3 An elliptical orbit and a precessing orbit.*

***Example 1*** **(Precession)** Figure 5.11.3 displays a schematic plot (greatly exaggerated) of a precessing orbit ($a_1 = 0.05$) compared to a fixed elliptical orbit ($a_1 = 0$). To see this, make the change of dependent variable $u = 1/r$ in the ODE (11) to obtain the transformed ODE

$$\frac{d^2u}{d\theta^2} = -u + \frac{3G(M+m)}{c^2}u^2 + \frac{G(M+m)}{h^2} \tag{12}$$

Using the scaling $u = G(M+m)v/h^2$, ODE (12) becomes

$$\frac{d^2v}{d\theta^2} = 1 - v + \frac{3G^2(M+m)^2}{h^2c^2}v^2 \tag{13}$$

Setting $a_1 = 3G^2(M+m)^2/h^2c^2$ yields the system in Figure 5.11.3. The axes in the plot ignore the scaling between $u$ and $v$ (i.e., $u$ and $v$ are identical). If $v$ is replaced by its scaled values $h^2u/G(M+m)$, the plot looks the same except that the $x$ and $y$-axes are differently scaled.

The planar system displayed in Figure 5.11.3 is autonomous, but the graph displayed is a curve that intersects itself. Nevertheless, this does not violate the non-intersection property for orbits of such systems. The reason for this is that the state variables $x_1$ and $x_2$ are not displayed on the axes. In fact, $x_1$ denotes $v$ and $x_2$ denotes $dv/d\theta$ in ODE (13). Thus $x = \cos\theta/x_1$ and $y = -\sin\theta/x_1$ are displayed on the axes in Figure 5.11.3.

***References*** Consult the material below for additional background and details.

1. Marion, J.B., Thornton, S.T. *Classical Dynamics of Particles and Systems*, 3rd ed. Harcourt Brace Jovanovich, New York, 1988.

_______________/_______________
hardware software

_______________/_______________
name date

Answer questions in the space provided,
or on attached sheets or carefully labeled graphs.

_______________/_______________
course section

## 5.11 *Elements of Orbital Mechanics*

***Abstract*** Some implications of Newton's Laws of Planetary Motion are examined, along with a relativistic extension of these laws.

**1.** The initial values $\mathbf{r}_0(0) = \mathbf{r}_0$, $\mathbf{r}'(0) = \mathbf{r}'_0$ cause a planet's orbit to lie in a plane orthogonal to $\mathbf{r}_0 \times \mathbf{r}'_0$, which contains the sun. Introducing polar coordinates $(r, \theta)$ in the orbital plane with the sun at the origin, the equations of motion of the planet relative to the sun are given by the system (2). The orbit of the planet $r = r(\theta)$ is given by (7), where the constants $h$, $K$, and $\delta$ are determined by the initial conditions.

**(a)** Show that for any solution of the system (2) there are constants such that

$$v^2 - \frac{2G(M+m)}{r} \equiv \text{constant}, \qquad r^2\theta' \equiv \text{constant}$$

where $r = \|\mathbf{r}\|$, $v = \|\mathbf{r}'\|$. (Hint: Multiply the first ODE in (2) through by $r'$ and integrate. Write the left-hand side of the second equation in (2) as a derivative.)

**(b)** Taking convenient values for the constant $G(M+m)$ and the initial data $\mathbf{r}(0)$, $\mathbf{r}'(0)$, solve the system (2) and verify directly the assertions in (a). (Hint: Use (3) and (4) to obtain initial values for $\theta$ and $\theta'$.)

**2.** Use the same definition of terms as in Problem 1.

**(a)** If at some point $P_0$ on the orbit of a planet the values $r_0$ and $v_0$ are known, then show that

$$1 - e^2 = \left[\frac{2G(M+m)}{r_0} - v_0^2\right] \frac{h^2}{G^2(M+m)^2}$$

where $e$ is the eccentricity of the orbit.

**(b)** The formula in (a) determines the eccentricity of an orbit from initial data. Setting $v_c^2 = 2G(M+m)/r_0$, show that the orbit is an ellipse, a parabola, or a hyperbola if $v_0 < v_c$, $v_0 = v_c$, or $v_0 > v_c$, respectively. Using convenient values for $G(M+m)$ and the initial data, solve numerically for the orbit and verify this property. Why is $v_c$ considered to be an escape velocity for the planet? See Experiment 1.10.

**3.** Use the relativistic correction to the planetary ODE (11) and perform the tasks below.

**(a)** Reproduce Figure 5.11.3. Then generate component graphs and calculate the precession of the relativistic orbit.

**(b)** Make a table of the values of parameter $a_1$ versus the orbit precession. Any conclusions?

* **(c)** Use realistic data to obtain the parameter $a_1$ and try to find the precession by solving ODE (13) numerically. Any conclusions?

# Stability and Lyapunov Functions

***Purpose*** To see how the stability type and orbital portrait of a planar autonomous linear system are affected by the addition of nonlinear terms.

***Keywords*** Eigenvalues, stability, nonlinear perturbations, total derivative, Lyapunov functions

***See also*** Experiments 5.2–5.3, 5.7–5.8

***Background*** The eigenvalues of the system matrix of an autonomous linear system determine the stability type and the main features of the orbital portrait (see Experiments 5.2–5.3). Now we consider the problem of determining the stability type and the general features of the orbits near an equilibrium point of an autonomous *nonlinear* system. Suppose that $(x^*, y^*)$ is an equilibrium point of the system

$$\begin{aligned} x' &= f(x, y) \\ y' &= g(x, y) \end{aligned} \tag{1}$$

where $f$ and $g$ are continuously differentiable functions near $(x^*, y^*)$. To address the problem posed above, a detour through some definitions and theorems is needed.

• **Stability.** Suppose that $(x^*, y^*)$ is an equilibrium point for the planar autonomous system (1). Then the system (1) is said to be **stable** at $(x^*, y^*)$ if for every number $E > 0$ there is a number $D > 0$ such that for any $(x_0, y_0)$ with $((x_0 - x^*)^2 + (y_0 - y^*)^2))^{1/2} < D$ the solution $x(t)$, $y(t)$ of the IVP

$$\begin{aligned} x' &= f(x, y), \qquad x(0) = x_0 \\ y' &= g(x, y), \qquad y(0) = y_0 \end{aligned} \tag{2}$$

is such that $((x(t) - x^*)^2 + (y(t) - y^*)^2)^{1/2} < E$, for all $t \geq 0$. Alternatively, we say that the point$(x^*, y^*)$ is stable. The point $(x^*, y^*)$ is **asymptotically stable** if it is stable and if there is a region $R$ about $(x^*, y^*)$ such that any orbit of (1) originating in $R$ tends to $(x^*, y^*)$ as $t \to +\infty$. A point $(x^*, y^*)$ for which such a region $R$ exists is called an **attractor**. The point $(x^*, y^*)$ is **globally asymptotically stable** if the region $R$ can be chosen to be the entire $xy$-plane (see Example 1); otherwise $(x^*, y^*)$ is **locally asymptotically stable** (see Example 3). The largest region $R$ that is "attracted" to an asymptotically stable equilibrium point $(x^*, y^*)$ is called its **basin of attraction**. An equilibrium point $(x^*, y^*)$ of (1) is **neutrally stable** if it is stable, but not asymptotically stable; thus, $(x^*, y^*)$ is stable, but not an attractor. For example, the systems

$$\text{(a)}\ \begin{aligned} x' &= y \\ y' &= -x \end{aligned} \qquad \text{(b)}\ \begin{aligned} x' &= y^3 \\ y' &= -x^3 \end{aligned} \tag{3}$$

are neutrally stable at the origin. See Example 4. System (1) is **unstable** at the equilibrium point $(x^*, y^*)$ if it is not stable. For example, the systems

$$\text{(a)}\ \begin{aligned} x' &= x \\ y' &= -y \end{aligned} \qquad \text{(b)}\ \begin{aligned} x' &= x^3 \\ y' &= -y^3 \end{aligned} \tag{4}$$

are unstable at the origin. See Example 5. From this point on it is assumed that the equilibrium point is at the origin.

• **Strong Lyapunov functions: asymptotic stability.** For brevity, the state vector $[x\ y]^T$ will be denoted by $u$. Suppose that $V(u)$ is a continuously differentiable, real-valued scalar function of $u$ defined on a neighborhood $R$ of the equilibrium point $u_0 = 0$. The **total derivative** $V'$ of $V(u)$ with respect to an orbit $u(t)$ of system (1) is defined to be $d/dt(V(u(t))$. Using (1) and the Chain Rule for a multivariable function, we have that

$$V'(u(t)) = V'(x(t), y(t)) = \frac{\partial V}{\partial x}x'(t) + \frac{\partial V}{\partial y}y'(t) = \frac{\partial V}{\partial x}f + \frac{\partial V}{\partial y}g \tag{5}$$

Thus, the total derivative can be computed at a point $u$ without knowing the orbit through the point. A scalar function $V$ of $u$ is **positive definite** on $R$ if it has only positive values on $R$ except at the origin, where $V(0) = 0$; **negative definite** is defined similarly. A continuously differentiable scalar function $V$ is a **strong Lyapunov function**[1] for system (1) on $R$ if $V$ is positive definite and $V'$ is negative definite on $R$. We can now state a stability theorem.

**Lyapunov's First Theorem:** *Suppose that there is a strong Lyapunov function $V(u)$ for system (1) in a region $R$ around the origin. Then system (1) is asymptotically stable at the origin.*

The conclusion of this theorem is plausible, since the values of $V(u(t))$ of the strong Lyapunov function must continually diminish as $t$ increases (since $V'$ is negative definite). This means that the orbit $u = u(t)$ must move through points with ever smaller values of $V$. In fact, $\lim_{t\to\infty} V(u(t)) = 0$, which implies that $u(t) \to 0$ as $t \to \infty$, since $u(t)$ and $V(u)$ are continuous, while $V = 0$ only at the origin.

The next theorem shows how to find a strong Lyapunov function if the rate functions have a special form. But first, another definition is required. A vector function $G(u)$ has **order of magnitude** $k$ at the origin if there is a positive constant $M$ such that for all $u$ in a neighborhood $R$ of the origin,

$$\|G(u)\| \leq M \|u\|^k \tag{6}$$

where $k$ is taken to be the least positive number for which (6) holds. A similar definition holds at any $u_0$ if u on the right-hand side of (6) is replaced by $u - u_0$.

**Lyapunov's Second Theorem:** *Let $A$ be a real matrix whose eigenvalues have negative real parts, and let $G(u)$ be a continuously differentiable vector function whose order of magnitude at the origin is at least 2. Then the system*

$$u' = Au + G(u) \tag{7}$$

*is asymptotically stable at the origin.*

Lyapunov's Second Theorem is proved in the following way. First, it is shown that under the stated conditions the matrix equation $A^T B + BA = -I$, where $I$ is the identity matrix, has a unique matrix solution $B$, and that $B$ is symmetric. Then it is shown that $V(u) = u^T Bu$ is a strong Lyapunov function for system (7) in some neighborhood of the origin $u_0 = 0$.

*Example 1* **(Asymptotic stability, linear system)** Figure 5.12.1 displays the orbits of an asymptotically stable linear system. The tilted ellipses are level sets of the strong Lyapunov function that is constructed in Example 2. The basin of attraction of the origin is the entire plane (the origin is a global attractor) since the system is linear and the eigenvalues of $A$ have negative real parts.

*Example 2* **(Constructing a strong Lyapunov function)** Consider system (8) with system matrix $A$

$$\begin{aligned} x' &= y \\ y' &= -x - y \end{aligned} \qquad A = \begin{bmatrix} 0 & 1 \\ -1 & -1 \end{bmatrix} \tag{8}$$

The same system with an additional perturbation with order of magnitude 3 is

$$\begin{aligned} x' &= y \\ y' &= -x - y + x^3 + y^3 \end{aligned} \tag{9}$$

Typical orbits near the origin of these two systems appear in Figures 5.12.1 and 5.12.2. The eigenvalues of $A$ are the complex conjugates $-1/2 \pm i/2$. Thus, from the fact that the eigenvalues of $A$ have negative real parts, system (8) is globally asymptotically stable at the

1. The Russian A.M. Lyapunov, the Frenchman H. Poincaré, and the American G.D. Birkhoff laid the foundations for most contemporary work on stability and dynamical systems.

origin. The symmetric matrix $B$ used to construct a strong Lyapunov function $V = u^T Bu$ is formed by first solving

$$\begin{bmatrix} 0 & -1 \\ 1 & -1 \end{bmatrix}\begin{bmatrix} b_{11} & b_{12} \\ b_{12} & b_{22} \end{bmatrix} + \begin{bmatrix} b_{11} & b_{12} \\ b_{12} & b_{22} \end{bmatrix}\begin{bmatrix} 0 & 1 \\ -1 & -1 \end{bmatrix} = \begin{bmatrix} -1 & 0 \\ 0 & -1 \end{bmatrix} \tag{10}$$

for $b_{11}$, $b_{12}$, and $b_{22}$. After doing the matrix multiplications and addition on the left of (10) and equating corresponding matrix elements on the left and right sides, we have

$$-2b_{12} = -1, \quad b_{11} - b_{12} - b_{22} = 0, \quad 2b_{12} - 2b_{22} = -1$$

whose solution is $b_{11} = \frac{3}{2}$, $b_{12} = \frac{1}{2}$, $b_{22} = 1$. Hence the quadratic form

$$V = [\ x \quad y\ ]\begin{bmatrix} \frac{3}{2} & \frac{1}{2} \\ \frac{1}{2} & 1 \end{bmatrix}\begin{bmatrix} x \\ y \end{bmatrix} = 3x^2/2 + xy + y^2 \tag{11}$$

is positive definite on the entire $xy$-plane, which may be verified by recalling that the quadratic form $Ax^2+Bxy+Cy^2$ is positive definite if and only if $A > 0$ and its **discriminant** $B^2 - 4AC$ is negative. In this case, $A = \frac{3}{2}$ and the discriminant is $-6$. The total derivative of $V$ with respect to system (8) is

$$V' = (3x + y)x' + (x + 2y)y' = (3x + y)y + (x + 2y)(-x - y) = -x^2 - y^2$$

which is negative definite on the plane. Thus, $V(x, y)$ in (11) is indeed a strong Lyapunov function for the system (8). Note that each nonconstant orbit cuts across the level sets $V(x, y) = K$, moving (as $t$ increases) through points of diminishing positive values of $K$ toward the origin where $K = 0$. Figure 5.12.1 displays several level sets.

*Example 3* **(Local asymptotic stability)** The function $V$ defined in (11) is also a strong Lyapunov function for the perturbed system (9), but only near the origin. The total derivative of $V$ with respect to (9) is

$$V' = (3x+y)+(x+2y)(-x-y+x^3+y^3) = -x^2-y^2+(x^4+2x^3y+xy^3+2y^4) \tag{12}$$

Let $x = r\cos\theta$, $y = r\sin\theta$ in (12) to obtain

$$V' = -r^2 + r^4(\cos^4\theta + 2\cos^3\theta\sin\theta + \cos\theta\sin^3\theta + 2\sin^4\theta) \tag{13}$$

Since $|\cos\theta| \leq 1$ and $|\sin\theta| \leq 1$, the magnitude of the trigonometric expression in (13) is at most 6. Hence,

$$V' \leq -r^2 + 6r^4 = r^2(-1 + 6r^2)$$

Thus, $V'$ is negative definite in a region of the $xy$-plane that includes the interior of the circle $r = 1/\sqrt{6}$. Thus, system (9) is (locally) asymptotically stable at the origin. Observe from Figure 5.12.2 that the orbits of (9) move across the inner tilted ellipses (the level sets of $V$) with increasing $t$ and toward the origin. Near the origin, the orbits of (8)and (9) are much alike since the cubic terms in (9) are very small in magnitude and have little effect on orbital behavior. However, out on the ellipse $V = 8$ the cubic terms have a strong effect, even turning some orbits completely away from the origin. A comparison of Figures 5.12.2 and 5.12.1 is revealing in this regard. The basin of attraction for the origin for system (9) consists of a region near the origin together with two spikelike regions to the upper left and lower right (see Figure 5.12.2).

• **Weak Lyapunov function: neutral stability**. Suppose that the system (1) has an equilibrium point at the origin. A continuously differentiable function $V(u)$ is a **weak Lyapunov function** for (1) if $V$ is positive definite on a region $R$ about the origin and if the total derivative $V'$ is negative semidefinite on $R$ (i.e., $V'(u) \leq 0$ for all $u$ in $R$ and $V'(0) = 0$). The following theorem holds

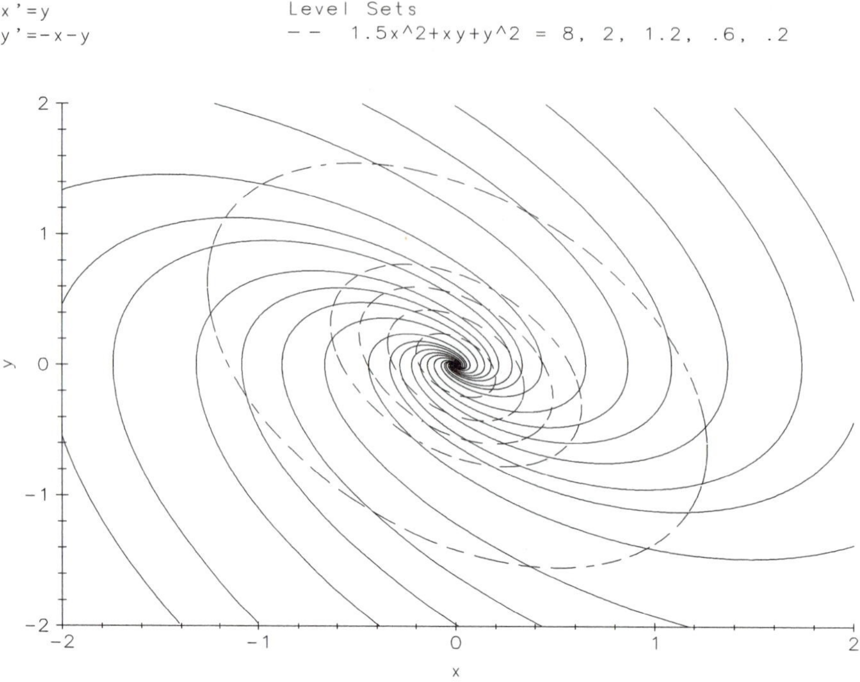


*Figure 5.12.1 Global asymptotic stability: level sets of a strong Lyapunov function (Examples 1 and 2).*

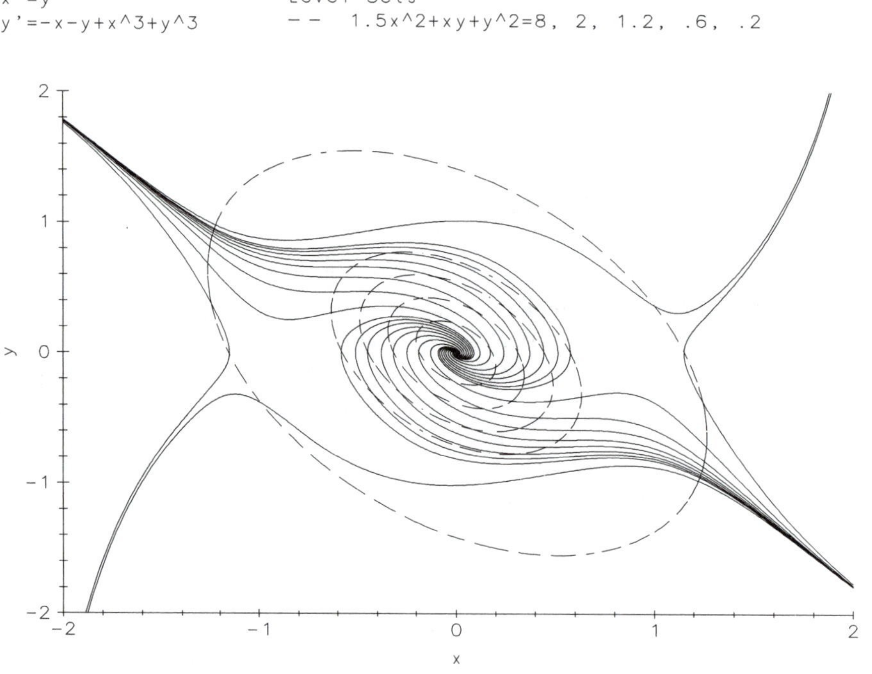


*Figure 5.12.2 Local asymptotic stability: level sets of a strong Lyapunov function (Examples 2 and 3).*

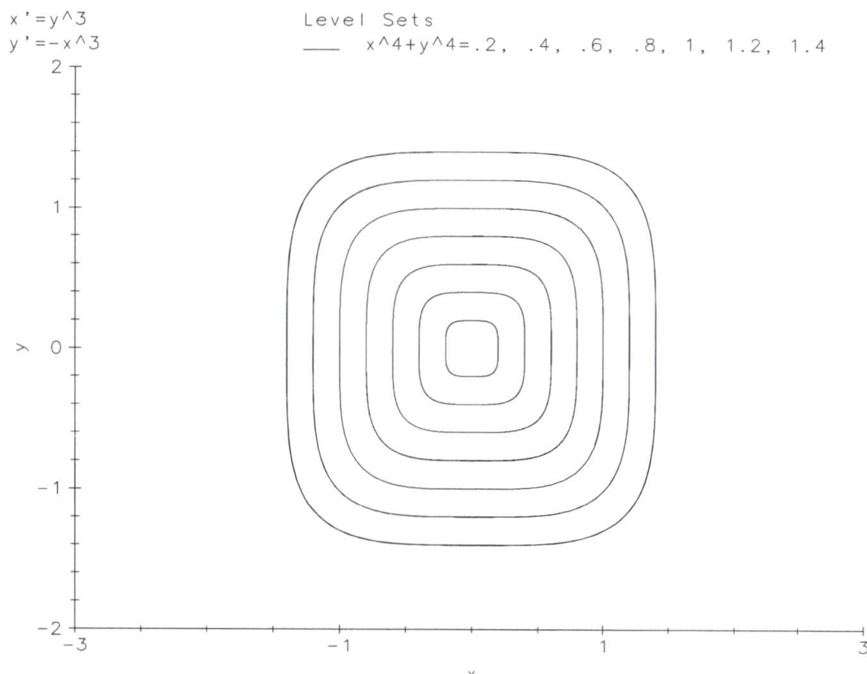


*Figure 5.12.3 Neutral stability: level sets of Lyapunov function $x^4 + y^4 = C$ are also orbits (refer to Example 4).*

**Lyapunov's Third Theorem:** *Suppose that there is a weak Lyapunov function for (1) on a region R about the origin. Then (1) is stable at the origin (but the stability may be neutral rather than asymptotic).*

***Example 4*** **(Neutral stability)** $V = x^2 + y^2$ is a weak Lyapunov function for system (3a) since it is positive definite on the plane, and its total derivative is $V' = 2xx' + 2yy' = 2x(y) + 2y(-x) \equiv 0$. Observe that the level sets of $V$ are the circles $x^2 + y^2 =$ constant, and these are precisely the orbits of (3a). For (3b) $V = x^4 + y^4$ is a weak Lyapunov function since $V$ is positive definite on the plane and $V' = 4x^3x' + 4y^3y' = 4x^3(y^3) + 4x^3(-x^3) \equiv 0$. The level sets of $V = x^4 + y^4$ once more coincide with the orbits. See Figure 5.12.3.

***Observation*** If $V$ is positive definite near an equilibrium point and if its total derivative vanishes identically near that point, $V$ is a weak Lyapunov function and the point is neutrally stable.

• We define a **Lyapunov instability function** $V$ for (1) in a neighborhood $R$ of the origin to be an indefinite function (i.e., $V$ takes on positive and negative values in every neighborhood of the origin and $V(0) = 0$) such that $V'$ is positive definite on $R$.

**Lyapunov's Fourth Theorem:** *Suppose that there is a Lyapunov instability function V for (1) in a neighborhood R of the origin. Then (1) is unstable at the origin.*

***Example 5*** **(Instability)** $V = x^2 - y^2$ is an indefinite function since $V(x, 0) > 0$ and $V(0, y) < 0$ for $x \neq 0$, $y \neq 0$. For (4a), $V' = 2xx' - 2yy' = 2x^2 + 2y^2$, which is positive definite. For (4b), $V' = 2xx' - 2yy' = 2x^4 + 2y^4$, which is also positive definite. Both systems are unstable at the origin. Figure 5.12.4 displays orbits of (4b) and level sets of $V$ (the dashed curves).

***Observation*** **(Constructing Lyapunov functions)** It may not be easy to decide whether a system is asymptotically stable, neutrally stable, or unstable. Perhaps the easiest approach is to first use computer graphics to get an idea of the stability properties of the system at an equilibrium

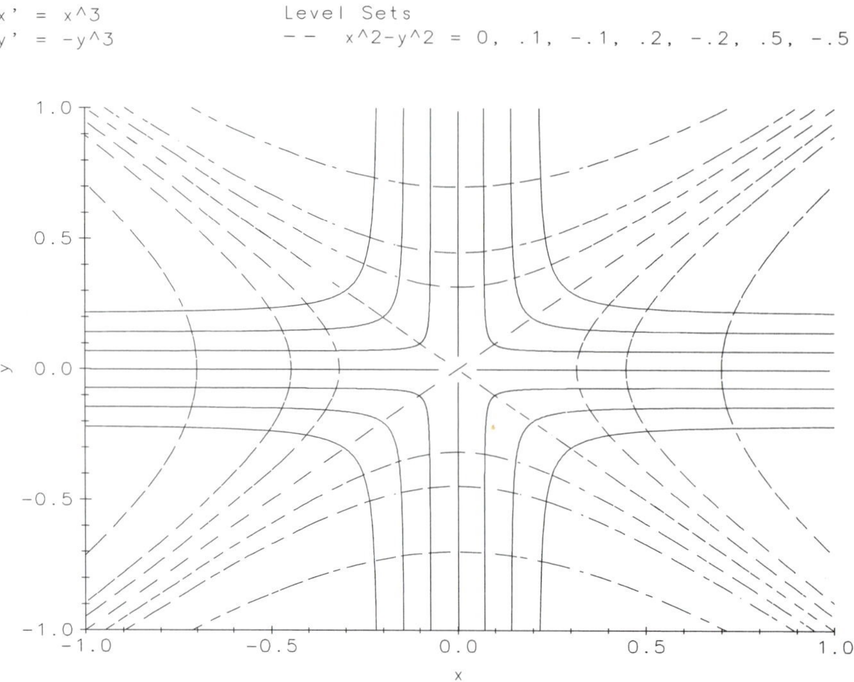


*Figure 5.12.4 An unstable system and level sets of a Lyapunov instability function.*

point. Then one might try to find a $V$ function of the form $Ax^2+Bxy+Cy^2$. This **quadratic form** is respectively (a) positive definite, (b) positive semidefinite, (c) negative definite, (d) negative semidefinite, or (e) indefinite if and only if (a) $B^2 < 4AC$ and $A > 0$, (b) $B^2 \le 4AC$ and $A$ and $C$ nonnegative, (c) $B^2 < 4AC$ and $A < 0$, (d) $B^2 \le 4AC$ and $A$ and $C$ non-positive, or (e) $A$ and $C$ have opposite signs. Finally, use the rate functions to compute the total derivative of $V$ and then adjust the coefficients $A$, $B$, $C$ until the hypothesis of one of Lyapunov's theorems is satisfied. Of course, there are no guarantees with this approach either. Over the years about fifty different methods for finding Lyapunov functions have been proposed, mostly by engineers concerned about the stability of a mechanical or electrical system. The total energy of a physical system is sometimes a good candidate for a Lyapunov function, and this approach is considered in the pendulum experiments (Experiments 5.7–5.8).

***Observation*** **(Effect of a perturbation)** It should be noted that the perturbation term in (7) has little effect on orbital behavior near the origin, a node remaining a node, a focus staying a focus. However, a higher order perturbation may destabilize a neutrally stable equilibrium point or, alternatively, make it asymptotically stable.

_______________/_______________
hardware software

_______________/_______________
name date

_______________/_______________
course section

Answer questions in the space provided,
or on attached sheets or carefully labeled graphs.

## 5.12 *The Effect of a Perturbation*

***Abstract*** The effect of adding higher order pertubations $G(u)$ to a stable linear system $u' = Au$ is explored.

**1.** Select one of the systems below and perform the tasks that follow, taking the origin to be the equilibrium point in each case.

☐ **A.** $x' = -4x + y$, $y' = -3y + (xy)$
☐ **B.** $x' = -x + (y^2)$, $y' = -y + (x^2)$
☐ **C.** $x' = -x + 10y + (x^2)$, $y' = -10x - y - (x^2 \sin x)$
☐ **D.** $x' = -2x + y + (xy)$, $y' = -2y + (x \sin y)$

**(a)** Draw an orbital portrait of the linear system obtained by deleting the perturbations (i.e., the terms in parentheses). Find the eigenvalues of the system matrix and write them on the graph. Is the equilibrium point a node or a focus? Is it globally asymptotically stable?

**(b)** Draw an orbital portrait of the perturbed system. Do you think the asymptotic stability is still global? Shade the part of the portrait that lies within the basin of attraction.

* **(c)** What is the order of the perturbation? Explain. (Hint: In C and D replace $\sin x$ by a Taylor series.) Construct a strong Lyapunov function $V$ as suggested in Lyapunov's Second Theorem. Draw level sets of $V$ on the orbital portraits of (a) and (b). Conclusions?

**2.** Choose a system from the list below and perform the tasks that follow.

☐ **A.** $x' = y$, $y' = -6x - y - 3x^2$
☐ **B.** $x' = -y - x(x^2 + y^2)$, $y' = x - y(x^2 + y^2)$
☐ **C.** $x' = -x + y^2$, $y' = x + y$
☐ **D.** $x' = -x + y^2$, $y' = -y + x^2$

**(a)** Find all the equilibrium points of the system. Produce an orbital portrait of the system in a region containing the equilibrium points. Make a conjecture about the stability type of the system at each equilibrium point. Explain. Write on the graphs.

* **(b)** At each equilibrum point $u_0$ write the system in the form $u' = A(u - u_0) + G(u)$, where $A$ is a constant matrix and the perturbation $G$ is of order at least 2 at $u_0$ (Hint: Expand the rate functions in a Taylor series about $u_0$). Identify the stability type of each equilibrium point. Compare with your conjecture in (a).

**3.** Use the computer to determine the stability type of the following systems.

(a) $x' = y + x(x^2 + y^2)$, $y' = -x + y(x^2 + y^2)$ (b) $x' = y - x(x^2 + y^2)$, $y' = -x - y(x^2 + y^2)$ (c) $x' = y$, $y' = -x - x^3$

**4.** Duplicate Atlas Plate `Planar System B`. Find the three equilibrium points, and determine their stability types.

______________ / ______________
hardware software

______________ / ______________
name date

Answer questions in the space provided,
or on attached sheets or carefully labeled graphs.

______________ / ______________
course section

## 5.13 *Stability and Lyapunov Functions*

***Abstract*** The stability types of planar autonomous systems are determined by constructing Lyapunov functions. Orbits and level sets of the Lyapunov functions are plotted.

***Special instructions*** Sketch the level sets of the Lyapunov functions on the orbital portraits (by hand if your solver does not have this capability).

**1.** Duplicate Figures 5.12.1, 5.12.2, and 5.12.3, insert arrowheads to show the direction of increasing time, and mark all equilibrium points and cycles. Identify each equilibrium point as asymptotically stable, neutrally stable, or unstable. Shade the basin of attraction.

**2.** Select one of the systems below and do the problems that follow.

☐ **A.** $x' = y,\ y' = -9x$

☐ **B.** $x' = y,\ y' = -\sin x$

☐ **C.** $x' = -\sin y,\ y' = \sin x$

**(a)** Draw an orbital portrait on the screen, $|x| \leq y$, $|y| \leq 4$. Mark all equilibrium points. What kind of stability (or instability) does each equilibrium point appear to have?

* **(b)** Divide the rate equations to obtain an ODE of the form $dy/dx = A(x)B(y)$. Separate variables and solve. Use your answer to construct a weak Lyapunov function $V$ on a neighborhood of the origin. Verify that your $V$ is positive definite and $V'$ is negative semidefinite. Why is the system neutrally stable at the origin?

**3.** Select one of the systems below. Find a $V$ function of the form $V = Ax^{2N} + By^{2N}$ to determine the stability properties at the origin. Verify that your $V$ satisfies the conditions of one of Lyapunov's theorems. Draw an orbital portrait.

☐ **A.** $x' = -2y^3,\ y' = 2x - y^3$

☐ **B.** $x' = -x^3 + y^3,\ y' = -x^3 - y^3$

* **4.** Select one of the Atlas Plates listed below. Duplicate the orbital portrait, inserting arrowheads to show the direction of increasing time, and marking the equilibrium points and cycles. Identify the stability type of the system at each equilibrium point. Give as good a mathematical justification for your answer as you can, using Lyapunov functions if possible. Shade the basins of attraction of asymptotically stable equilibrium points.

☐ **A.** `Pendulum A, B`

☐ **B.** `Planar System D`

☐ **C.** `Planar System E`

# Cycles and Limit Cycles

***Purpose*** To search for cycles and limit cycles, to create a system with limit cycles, and to model an electric circuit with limit cycle behavior.

***Keywords*** Cycles, limit cycles, electric circuit, van der Pol system, planar autonomous systems, equilibrium states

***See also*** Experiments 3.9, 5.5, 5.8, 5.10; Atlas Plates `First Order I`, `Limit Cycle A-C`, `Limit Set A`; Appendix B.4, B.5

***Background*** **Cycles** are the orbits of (nonconstant) periodic solutions of autonomous systems. A cycle represents the system in an active equilibrium state, while an equilibrium point depicts a static state. A cycle may belong to a continuous band of cycles (Atlas Plate `Pendulum A`), or it may be isolated from other cycles (Figure 5.14.1). A cycle of a linear autonomous system must belong to a band of cycles of a common period because multiples of a solution of such a system are also solutions (Atlas Plate `Spring A`). Isolated cycles can only appear in nonlinear autonomous systems. If the system is planar, then an isolated cycle is said to be a **limit cycle** because all nearby orbits spiral toward the cycle as $t \to \infty$ or spiral away as $t$ increases from $-\infty$. A limit cycle is an **attractor** (or **asymptotically stable**) if the nearby orbits on both sides spiral toward the cycle as time increases, a **repeller** (or **asymptotically unstable**) if the nearby orbits spiral away, or **semistable** if nearby orbits spiral toward on one side and away on the other.

• **Cycles and polar coordinates**. Sometimes the existence of a cycle or a limit cycle is revealed by the form of the rate equations, especially if the $xy$-equations are transformed into $r\theta$-coordinates. For example, suppose that $a$ and $b$ are constants, $b \neq 0$, and $h(r)$ is a

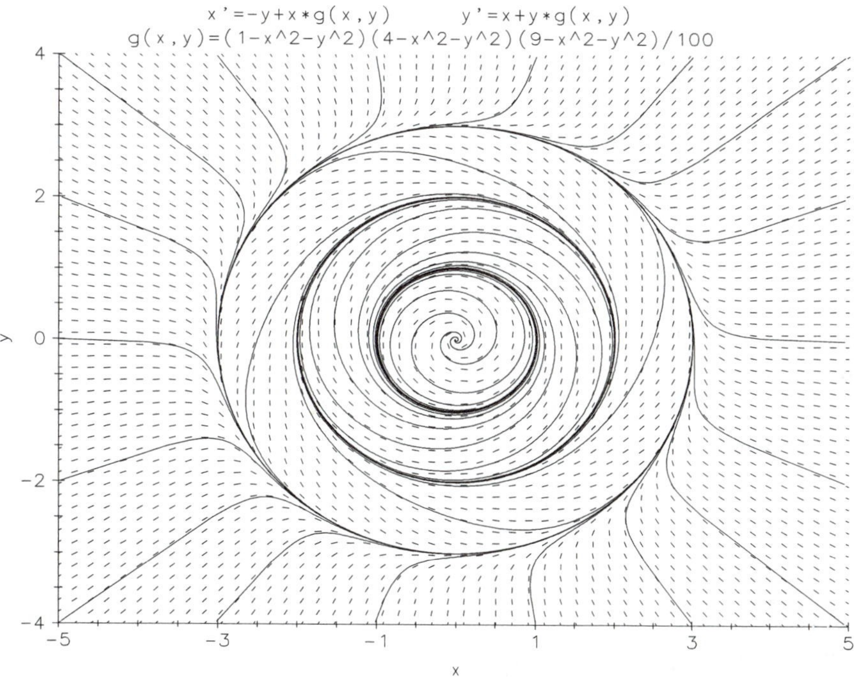


*Figure 5.14.1 A repelling limit cycle and two attracting limit cycles (Example 1).*

differentiable function. Rewrite the $xy$-system

$$\begin{aligned} x' &= ax - by + xh(r) \\ y' &= bx + ay + yh(r) \end{aligned} \tag{1}$$

in polar coordinates, obtaining

$$\begin{aligned} r' &= r(a + h(r)) \\ \theta' &= b \end{aligned} \tag{2}$$

From (2) we see that there is a circular cycle, $x^2 + y^2 = r_0^2$, corresponding to each positive root $r_0$ of the function $a + h(r)$. If $r_0$ is an isolated root, the cycle is an attracting or a repelling cycle, according to whether $h'(r_0)$ is negative or positive. If $h(r)$ does not change sign as $r$ increases through $r_0$, then $r = r_0$ is a semistable cycle. The cycle turns clockwise about the origin if $b < 0$ and counterclockwise if $b > 0$.

***Example 1*** **(Three limit cycles)** The system of Figure 5.14.1

$$\begin{aligned} x' &= -y + x(1 - x^2 - y^2)(4 - x^2 - y^2)(9 - x^2 - y^2)/100 \\ y' &= x + y(1 - x^2 - y^2)(4 - x^2 - y^2)(9 - x^2 - y^2)/100 \end{aligned}$$

is a special case of (1) with $a = 0$, $b = 1$, and $h(r) = (1 - r^2)(4 - r^2)(9 - r^2)/100$. The system is simpler in polar coordinates:

$$\begin{aligned} r' &= r(1 - r^2)(4 - r^2)(9 - r^2)/100 \\ \theta' &= 1 \end{aligned}$$

There are three limit cycles: a repelling cycle ($r = 2$) enclosed by a pair of attracting cycles ($r = 1$, $r = 3$). The nearby orbits are attracted to these cycles so strongly that after a time they appear to merge with them.

***Observation*** **(Cycles, forced oscillations)** The cycles and limit cycles of a given planar autonomous system are sometimes hard to detect. Locating equilibrium points and drawing a direction field may help since a cycle must enclose at least one equilibrium point, while "eddies" in the flow lines of the direction field indicate some kind of rotational motion. Warning: On a graphics screen a tightly coiled spiral or a thin band of cycles may appear to be a single limit cycle. A periodically driven system such as $x' = y$, $y' = -x - y + \cos t$ may have a periodic solution ($x = \sin t$, $y = \cos t$ in this case), but the corresponding closed-curve orbit in state space is not customarily called a cycle or a limit cycle. Indeed, there is no generally accepted term for such an orbit, although the periodic solution itself is called a **forced oscillation**. In these experiments, the term "cycle" is used only in connection with an autonomous system, and "limit cycle" is reserved for the orbit of an isolated cycle of a planar autonomous system.

***Example 2*** **(The van der Pol system)** Limit cycle behavior appears in a host of physical phenomena, but it is nonlinear electric circuits that are of interest here. In the 1920s Balthazar van der Pol detected limit cycle behavior in the current and voltages of early radio sets. These circuits are $RLC$ loops, but with the passive resistor of Ohm's Law replaced by an active element: an array of vacuum tubes then, a semiconductor device now. Unlike a passive resistor, which dissipates energy at all current levels, a semiconductor operates as if it were pumping energy into the circuit at low current levels, but absorbing energy at high levels. The interplay between energy injection and energy absorption results in a periodic oscillation in voltages and currents. Figure 5.14.2 shows a typical van der Pol circuit.

Suppose that a power supply is attached to the circuit in Figure 5.14.2 and the circuit is energized. How do the voltages and current change after the power source is removed? Suppose that the voltage drop across the semiconductor is given by the nonlinear function

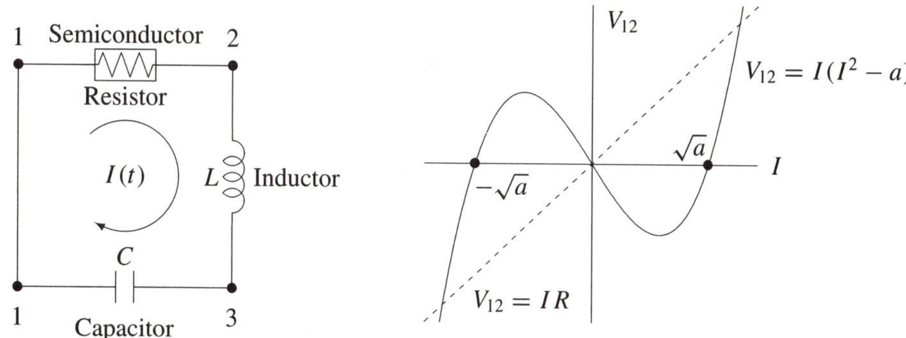


*Figure 5.14.2 A van der Pol circuit with nonlinear current/voltage characteristic $V_{12} = I(I^2 - a)$ across the semiconductor.*

$V_{12} = I(I^2 - a)$ of the current $I$ and a positive parameter $a$. In comparison with the linear voltage drop of $V_{12} = IR$ given by Ohm's Law, the nonlinear function models a region of negative resistance at current levels $|I| < \sqrt{a}$. The circuit laws outlined in Appendix B.4 relate all the voltage drops, the current, and the circuit constants:

| | |
|---|---|
| $V_{12} = I(I^2 - a)$ | (current/voltage characteristic across semiconductor) |
| $V_{23} = LI'$ | (Law of Induction) |
| $V'_{31} = I/C$ | (differential form of Coulomb's Law) |
| $V_{12} + V_{23} + V_{31} = 0$ | (Kirchhoff's Voltage Law) |

With $I$ and $V = V_{31}$ as state variables, the **van der Pol system** is

$$\begin{aligned} I' &= \frac{1}{L}V_{23} = \frac{1}{L}(-V_{12} - V_{31}) = \frac{a}{L}I - \frac{1}{L}V - \frac{1}{L}I^3 \\ V' &= \frac{1}{C}I \end{aligned} \tag{3}$$

For fixed, positive values of $a$, $L$, and $C$, an attracting limit cycle surrounding a repelling equilibrium point is clearly visible in the $IV$-plane. Atlas Plate `First Order I` shows integral curves of an equivalent form of (3). The limit cycle resembles a tilted parallelogram; the resemblance is even more marked for large values of $a$ (Atlas Plate `Limit Cycle A`). The corresponding periodic solution is sometimes called a **relaxation oscillation** because it depicts a current or voltage that stays at a nearly constant level and then suddenly relaxes and changes sign, after which the process repeats.

The proof that (3) has a unique limit cycle that attracts all other nonconstant orbits is not given here. It is based on two facts: (a) the eigenvalues of the matrix of coefficients of the linear terms in (3) have positive real parts (hence, orbits near the origin move away), and (b) for large values of $|I|$ the nonlinear term, $-I^3/L$ in (3), turns orbits inward. The limit cycle lies in a region between the outward and the inward moving orbits.

***Observation*** The **van der Pol equation** is the second order ODE

$$\frac{d^2x}{ds^2} + (3x^2 - \mu)\frac{dx}{ds} + x = 0$$

It is obtained by differentiating the first ODE of (3) with respect to $t$, replacing $V'$ by $I/C$, and setting $x = (C/L)^{1/4}I$, $s = (LC)^{-1/2}t$, $\mu = a(C/L)^{1/2}$. See Appendix B.5.

_______________/_______________ _______________/__________
hardware software name date

Answer questions in the space provided, _______________/__________
or on attached sheets or carefully labeled graphs. course section

## 5.14 *The van der Pol System*

***Abstract*** Van der Pol systems are studied, orbital portraits are drawn, and properties of the limit cycles are determined.

**1.** Duplicate Atlas Plates `Limit Cycle A` and `B`, inserting arrowheads of time, marking limit cycles and equilibrium points, and identifying the type of the latter (i.e., unstable focus, unstable node, etc.) by finding the eigenvalues of the matrix of coefficients of the linear terms of the corresponding system. Duplicate Atlas Plate `First Order I`, write out a corresponding van der Pol system, and identify the values of $a$, $L$, and $C$.

**2.** Rescaling reduces the number of parameters in the van der Pol system (3) from three to one and makes a sensitivity study of the system straightforward (see Appendix B.5).

**(a)** Show that the system (3) transforms to $dx/ds = \mu x - y - x^3, dy/ds = x$, if $x = I(C/L)^{1/4}$, $y = V(C/L)^{3/4}$, $s = t(LC)^{-1/2}$, and $\mu = a(C/L)^{1/2}$. Observe that the $xys$-system is system (3) with $L = 1$ and $C = 1$. (Also see Atlas Plate `First Order I`.)

**(b)** Use the computer to graph the limit cycle (and its $x$ and $y$-component graphs) of the $xys$-system of (a) for enough large and small values of the positive parameter $\mu$ that you can answer the following questions. Does the period increase, decrease, or neither as $\mu$ increases? What happens to the amplitude of the cycle (i.e., the maximum diameter of the cycle) as $\mu$ increases? What are the limiting values of the amplitude and period as $\mu \to 0$? As $\mu \to \infty$? What happens to the shape of the cycle as $\mu$ increases? Write explanations on the graphs. (Hint: See Atlas Plate `Limit Cycle B`.)

**3.** Replace $V_{12} = I(I^2 - a)$ by one of the functions below and perform the tasks that follow for the altered van der Pol system (3) with $L = 1$, $C = 1$, and $a$ a nonnegative constant.

☐ **A.** $V_{12} = I(I^4 - a)$

☐ **B.** $V_{12} = \begin{cases} I + 2a & \text{if } I \le -a \\ -I & \text{if } |I| < a \\ I - 2a & \text{if } |I| \ge a \end{cases}$

☐ **C.** $V_{12} = \begin{cases} I + \pi a & \text{if } I \le -\pi a \\ -\sin(I/a) & \text{if } |I| < \pi a \\ I - \pi a & \text{if } |I| \ge \pi a \end{cases}$

* **(a)** Use the computer to sketch an orbital portrait if $a = 1$ ; mark the limit cycle and equilibrium points; draw component graphs for the limit cycle and sketch the graphs.

* **(b)** Use the computer to make a study of the orbits as the parameter $a$ varies from 0 to large positive values. Is there always an attracting limit cycle? How do the period, amplitude, and shape of the limit cycle change as the parameter changes? How does the equilibrium point at the origin change as $a$ changes? Explain your answers; write your explanations on the graphs.

_______________/_______________
hardware software

_______________________/__________
name date

Answer questions in the space provided,
or on attached sheets or carefully labeled graphs.

_______________________/__________
course section

## *5.15 Systems with Cycles and Limit Cycles*

***Abstract*** Planar autonomous systems are examined for the presence of cycles and limit cycles. Periods are estimated. Systems with prescribed limit cycles are constructed.

**1.** Select one of the diagrams below and do the problems that follow.

- ☐ **A.** Figure 5.14.1
- ☐ **B.** Atlas Plate `Limit Cycle C`
- ☐ **C.** Atlas Plate `Two-Cycled System C`

**(a)** Duplicate the orbital portrait, insert arrowheads of time, and mark equilibrium points and limit cycles. Identify the type of (nonlinear) equilibrium points (focus, node, center; stable, unstable); identify the limit cycles as attractors, repellers, or semistable.

**(b)** Graph component curves of the limit cycles and estimate the periods.

**2.** Select one of the systems below and do the problems that follow.

- ☐ **A.** Atlas Plate `Pendulum A`
- ☐ **B.** Atlas Plate `Planar System D`
- ☐ **C.** Atlas Plate `Spring B`
- ☐ **D.** Atlas Plate `Spring C`

**(a)** Duplicate the orbital portrait, insert arrowheads of time, and mark equilibrium points and cycles.

**(b)** Graph component curves of the cycles and estimate the periods. How do the periods change with increasing cycle amplitude?

* **(c)** Explain why the system has bands of cycles, but no limit cycles. (Suggestion: Divide one rate equation by the other to obtain a first order separable equation of the form $dy/dx = A(x)B(y)$. Separate variables, integrate, and find the equation of the orbits.) Then explain why the form of the equation precludes isolated cycles. See also Problem 5.

**3.** Select one of the options below and construct a planar autonomous system with the indicated array of cycles. Draw an orbital portrait. Write a justification of your conclusions on the portrait. (Suggestion: Construct systems of the form of (1).)

- ☐ **A.** A system with one semistable, one attracting, and one repelling limit cycle.
- ☐ **B.** A system with limit cycles located at $r = 1, 2, \ldots$, the cycles alternately repelling and attracting.
- ☐ **C.** A system with a band of cycles and one attracting limit cycle. (Suggestion: In (1) take $h(r) = 0$ for $0 \le r \le 1$, $h(r) = (1 - r^2)^2(4 - r^2)$ for $r > 1$.)

**4.** Select one of the options below and do the problems that follow.

☐ **A.** $x' = -y - xr^2 \sin(1/r)$, $y' = x - yr^2 \sin(1/r)$

☐ **B.** $x' = -y - x(1 - r^2)\sin(1 - r^2)^{-1}$, $y' = x - y(1 - r^2)\sin(1 - r^2)^{-1}$

**(a)** Write the system in $r\theta$ coordinates; find all cycles, limit cycles, and equilibrium points. Determine the stability properties of each.

**(b)** The system has infinitely many cycles in the region $x^2 + y^2 \leq 2$. Where do they accumulate, and what are the limiting periods?

**5.** Suppose that $f(x, y)$ and $g(x, y)$ are continuously differentiable functions. The autonomous system $x' = f$, $y' = g$ may be written in differential form as $-g\,dx + f\,dy = 0$. If the latter equation is exact for all $x$, $y$, then there is a continuously differentiable, nonconstant function $k(x, y)$ called an **integral** that has a constant value on each orbit (Experiment 3.7).

* **(a)** Suppose that $k(x, y)$ is a continuous function that is nonconstant on every open set. Show that the system has no isolated cycles. (Suggestion: Show that if it did, $k(x, y)$ would have the same value on the isolated cycle and on all the nearby spiraling orbits because $k(x, y)$ is continuous. Then show that this property contradicts the nonconstancy property of $k$.)

* **(b)** Select one of the following, duplicate the orbital portrait, and explain why the visible cycles must belong to bands and cannot be isolated. Write your argument on the graphs. (Suggestion: Use (a) with $k(x, y)$ given by an appropriate integral.)

☐ **A.** $(e^x \sin y - 2y \sin x)dx + (y^2 + e^x \cos y + 2\cos x)dy = 0$ (Atlas Plate `First Order G`)

☐ **B.** $(\cos x + 2x \cos x^2 \cos 2y)dx + (\sin y - 2\sin x^2 \sin 2y)dy = 0$ (Atlas Plate `First Order H`)

☐ **C.** $x' = xy$, $y' = \cos(x^2 + y^2)$ (Atlas Plate `Planar System E`)

# The Poincaré-Bendixson Alternatives

***Purpose*** To study the limit behavior of orbits of planar autonomous systems.

***Keywords*** Positively bounded orbit, limit set, equilibrium point, limit cycle, cycle-graph

***See also*** Experiment 5.1; Atlas Plates `Limit Set A-C`, `Planar Systems A-E`

***Background*** Planar autonomous systems with continuously differentiable rate functions $f$ and $g$

$$x' = f(x, y), \qquad y' = g(x, y) \tag{1}$$

have a striking variety of orbital portraits. However, as $t \to \infty$ the limit behavior of a single positively bounded orbit takes one of only three forms. First, some notation and a definition. The orbit of (1) that passes through the point $p$ at time 0 is denoted by $\phi_t(p)$, and so $\phi_0(p) = p$. The orbit is **positively bounded** if, for all $t \geq 0$, $\phi_t(p)$ stays inside a closed and bounded region $R$ in the $xy$-plane. Suppose that the region $R$ contains finitely many equilibrium points and that the orbit $\phi_t(p)$ is positively bounded. Then, $\phi_t(p)$ must do one of the following as $t \to \infty$:

- tend to an equilibrium point (or be an equilibrium point)
- spiral toward a limit cycle (or be a cycle)
- spiral toward a cycle-graph.

These **Poincaré-Bendixson alternatives** exhaust the possibilities. The alternatives also apply to a negatively bounded orbit as $t \to -\infty$. Cycles and limit cycles are treated in Experiment 5.15; cycle-graphs are defined in Experiment 5.1. See Figure 5.16.1 for an example of orbits tending to a cycle-graph.

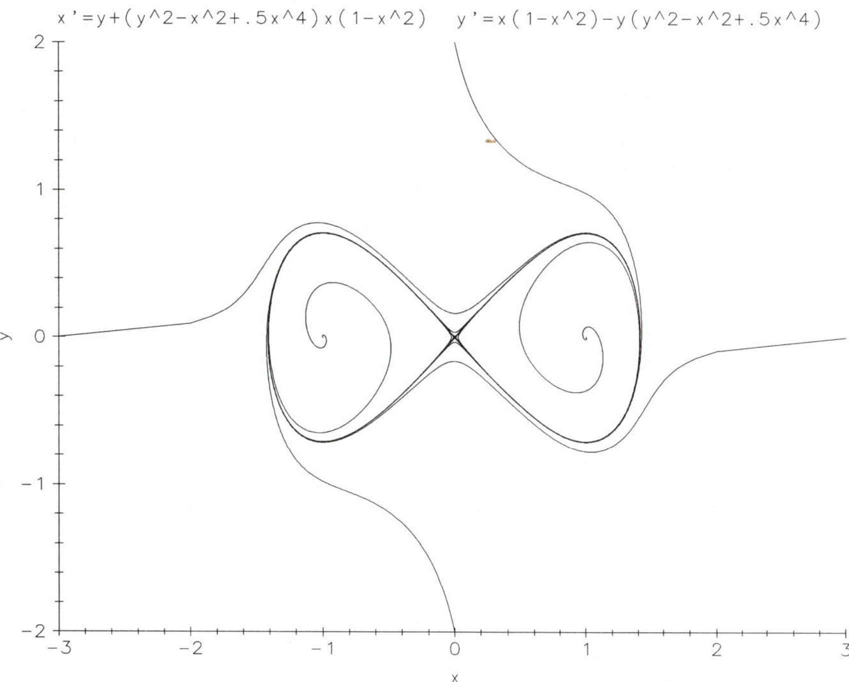


*Figure 5.16.1 The lazy-eight cycle-graph attracts outside orbits. Each nonconstant interior orbit spirals out from an equilibrium point and toward a cycle-graph lobe, which is a subgraph of the lazy eight.*

__________ / __________
hardware software

__________ / __________
name date

Answer questions in the space provided,
or on attached sheets or carefully labeled graphs.

__________ / __________
course section

## 5.16 *The Poincaré-Bendixson Alternatives*

***Abstract*** The limit behavior of orbits of system (1) is explored.

***Special instructions*** Include a discussion of the points outlined below in a team report.

- The **positive** (or **omega**) **limit set** $\omega(p)$ of a positively bounded orbit $\phi_t(p)$ is the set of points $q$ such that for some sequence $\{t_n\}$, $\lim_{n\to\infty} t_n = \infty$ and $\lim_{n\to\infty} \phi_{t_n}(p) = q$. Define the **negative** (or **alpha**) **limit set** $\alpha(p)$ of $\phi_t(p)$. Search Atlas Plates `Limit Set A-C` and `Planar Systems A-E` for positively or negatively bounded orbits; label the limit sets and explain the limit behavior.
- Duplicate Figure 5.16.1. Find all equilibrium points. Explain why $y^2 = x^2(1 - 0.5x^2)$ defines a cycle-graph that attracts outside orbits. There are two homoclinic orbits (see Experiment 5.17 for the definition) in the figure—find them. Explain why all nonconstant orbits inside a lobe have $\alpha$ limit set an equilibrium point and $\omega$ limit set a cycle-graph consisting of another equilibrium point and a homoclinic orbit.
- Sketch all possible combinations of $\alpha$ and $\omega$ limit sets for a bounded orbit (for example, $\alpha(p)$ and $\omega(p)$ each a distinct equilibrium point, $\alpha(p)$ a limit cycle and $\omega(p)$ a cycle-graph, $\alpha(p) \equiv \omega(p)$ a cycle, and so on). Now find a specific planar system for each combination. (Hint: Look through the graphs in the experiments and the Atlas.) Justify your conclusions.
- Find systems that have an unbounded orbit such that (a) the $\alpha$ and $\omega$ limit sets are empty; (b) the $\alpha$ limit set is empty and the $\omega$ limit set consists of a line segment of limit points on the $x$-axis; (c) the $\alpha$ limit set is the origin and the $\omega$ limit set consists of two vertical lines.
- Give a plausible definition for a chaotic set of orbits. Test your definition by examining some of the orbits of a driven Duffing Equation (Experiment 5.10). Explain why a bounded set of orbits of a planar autonomous system with finitely many equilibrium points cannot be chaotic. (Suggestion: Read the Chapter 6 Notes.)
- Use conditionally-defined rate functions to construct a planar autonomous system with an orbit whose positive limit set is the silhouette of the head of a mouse (see Experiment 3.8).

***References*** Consult the material below for additional background and details.

1. Hirsch, M.W., Smale, S. *Differential Equations, Dynamical Systems and Linear Algebra.* Academic Press, New York, 1974.
2. Perko, L. *Differential Equations and Dynamical Systems*, Springer-Verlag, New York, 1991.

# The Hopf Bifurcation

***Purpose*** To see how a stable nonlinear focus loses its stability and spawns an attracting limit cycle as a parameter is changed, and to study a model of satiable predation.

***Keywords*** Hopf bifurcation, equilibrium point, limit cycle, stability, sensitivity, homoclinic orbit

***See also*** Experiments 5.14, 6.7; Atlas Plates `Bifurcation A-C`

***Background*** Cycles and equilibrium points are distinctive features of the portrait of the orbits of autonomous systems. If the field vector of the system is changed slightly, one expects the new portrait to resemble the old. A cycle might contract a little, an equilibrium point shift somewhat, a spiral tighten or loosen, but the dominant features of the portrait would remain—or so one might expect. This seems reasonable, but it is not necessarily true. Small changes in a coefficient may mean the disappearance of one feature and the appearance of another that is altogether different. That is, a system might be highly sensitive to a parameter change. The particular phenomenon studied in this experiment is the bifurcation of an asymptotically stable equilibrium point into an attracting limit cycle enclosing a destabilized equilibrium point. The transfer of stability and the birth of a limit cycle occur as a parameter passes through a critical value. The phenomenon is usually called a **Hopf bifurcation** and is now widely used in nonlinear autonomous systems to model the sudden appearance of a limit cycle out of an equilibrium point.

***Example 1*** **(Birth of a limit cycle)** The system

$$\begin{aligned} x' &= ax + y - x(x^2 + y^2) \\ y' &= -x + ay - y(x^2 + y^2) \end{aligned} \tag{1}$$

of Atlas Plates `Bifurcation A-C` displays a Hopf bifurcation as the parameter $a$ traverses the critical value $a = 0$. To see this, rewrite (1) in polar coordinates (see Experiment 3.9):

$$\begin{aligned} r' &= r(a - r^2) \\ \theta' &= -1 \end{aligned} \tag{2}$$

Since $\theta = -t + C$, as $t$ increases orbits of (1) turn clockwise around the origin, which is an isolated equilibrium point of (1). By sign analysis, we see that:

- If $a < 0$, orbits of (1) spiral toward the origin as $t \to \infty$ since $r' < 0$ whenever $r > 0$.
- If $a = 0$, orbits still spiral toward the origin, but not as rapidly.
- If $a > 0$, then $r' > 0$ for $0 < r < \sqrt{a}$, and $r' < 0$ for $r > \sqrt{a}$. As $t \to \infty$ all nonconstant orbits spiral toward the attracting limit cycle $r = \sqrt{a}$.
- The origin is an asymptotically stable nonlinear focus if $a \le 0$; it is an unstable nonlinear focus surrounded by an attracting limit cycle if $a > 0$.

System (1) is an example of a system of linear plus higher order nonlinear terms

$$u' = A(a)u + Q(u), \qquad u = \begin{bmatrix} x \\ y \end{bmatrix} \tag{3}$$

where $A(a)$ is the system matrix and $Q(u)$ contains the higher order terms. In Example 1 we have

$$A(a) = \begin{bmatrix} a & 1 \\ -1 & a \end{bmatrix}, \qquad Q(u) = \begin{bmatrix} -x(x^2 + y^2) \\ -y(x^2 + y^2) \end{bmatrix}$$

In this case the eigenvalues of the system matrix are the complex conjugates $a \pm i$; as $a$ increases through 0, the eigenvalues leave the left half of the complex plane, cut across the imaginary axis, and move into the right half-plane.

• **The Hopf conditions.** With examples like this in mind, Eberhard Hopf formulated a bifurcation theorem for the general system (3). Suppose first that the component functions of $Q$ have Taylor series in powers of $x$ and $y$ that converge in a region containing the origin (polynomials in $x$ and $y$ have this property) and that these series have no terms of degree less than 2 (the terms of $Q$ in Example 1 are all of degree 3). Next suppose that eigenvalues of the system matrix are the complex conjugates $\alpha(a) \pm i\beta(a)$ for all $a$ in some open interval $I$ containing $a = 0$. Suppose that $\alpha(a)$ and $\beta(a)$ are continuously differentiable functions of $a$ and that $\alpha(0) = 0$, $\alpha'(0) > 0$, and $\beta(0) \neq 0$ (the eigenvalues of the system matrix of Example 1 have these properties). Suppose that for each $a \in I$ the origin is an isolated equilibrium point of (3) and, finally, that (3) is asymptotically stable at the origin if $a = 0$. The nonlinear terms of (3) are essential for meeting the last condition; without them the system has a neutrally stable linear center at the origin for $a = 0$. Then we have the

**Hopf Bifurcation Theorem:** *Under the above conditions, as the parameter $a$ increases through the critical value $a = 0$, the orbits of (3) behave as follows:*

- *For $a \leq 0$ and $|a|$ small, (3) has a stable nonlinear focus at the origin.*
- *For $a > 0$ and $|a|$ small, (3) has an unstable nonlinear focus at the origin, surrounded by an attracting limit cycle whose amplitude grows like $\sqrt{a}$.*

***Observation*** **(Other bifurcations)** The Hopf bifurcation described above is **supercritical.** If the system is asymptotically unstable for $a = 0$ (instead of stable), then a **subcritical bifurcation** takes place in which the origin absorbs a repelling limit cycle as $a$ increases through 0. More generally, the system may have the form $u' = G(u, a)$ with a moving equilibrium point $u_0(a) = (x_0(a), y_0(a))$, which is isolated from other equilibrium points for each $a$ in some open interval $I$. The critical value of the parameter need not be $a = 0$, but some other value $a_0$ in $I$. However, the eigenvalues of the system matrix (i.e., of the matrix of coefficients of the linear terms of the Taylor series expansions of the components of $G$ in powers of $x - x_0(a)$ and $y - y_0(a)$) must have the properties given earlier but with critical value $a_0$. If the number of state variables exceeds two, it is assumed that two of the eigenvalues have the prescribed properties and all others have negative real parts. Finally, we note the existence of conditions on the coefficients of the above-mentioned Taylor series that imply the desired stability property of the system for the critical value of the parameter. These conditions are lengthy, however, and are omitted. From a graphical standpont, one can collect visual evidence of a bifurcation by plotting orbits for several values of a system parameter without explicitly checking each of the Hopf conditions.

***Example 2*** **(Hopf, homoclinic bifurcations)** The system

$$\begin{aligned} x' &= ax - y + x^2 \\ y' &= x + ay + x^2 \end{aligned} \tag{4}$$

has a Hopf bifurcation at $a = 0$. All of the Hopf conditions, except for asymptotic stability when $a = 0$, are easily verified. Figure 5.17.1 displays the visual evidence that a Hopf bifurcation occurs at $a = 0$. Observe that for $a = 0.10$ the plotted orbit spirals outward toward a pointed, oval-shaped limit cycle, but as $a$ is increased to 0.11, the limit cycle vanishes. System (4) has a second equilibrium point, which for some value $a = a^*$, $0.10 < a^* < 0.11$, becomes a nonlinear saddle point with a **homoclinic orbit** (an orbit that leaves the point as $t$ increases from $-\infty$ and reenters as $t \to +\infty$). As $a$ increases through $a^*$, the homoclinic orbit absorbs the limit cycle in a so-called **homoclinic bifurcation**. Hopf and homoclinic bifurcations are but two of several ways a limit cycle can be created or destroyed by changing a parameter. Figure 5.17.2 displays the limit cycle with $a = 0.05$.

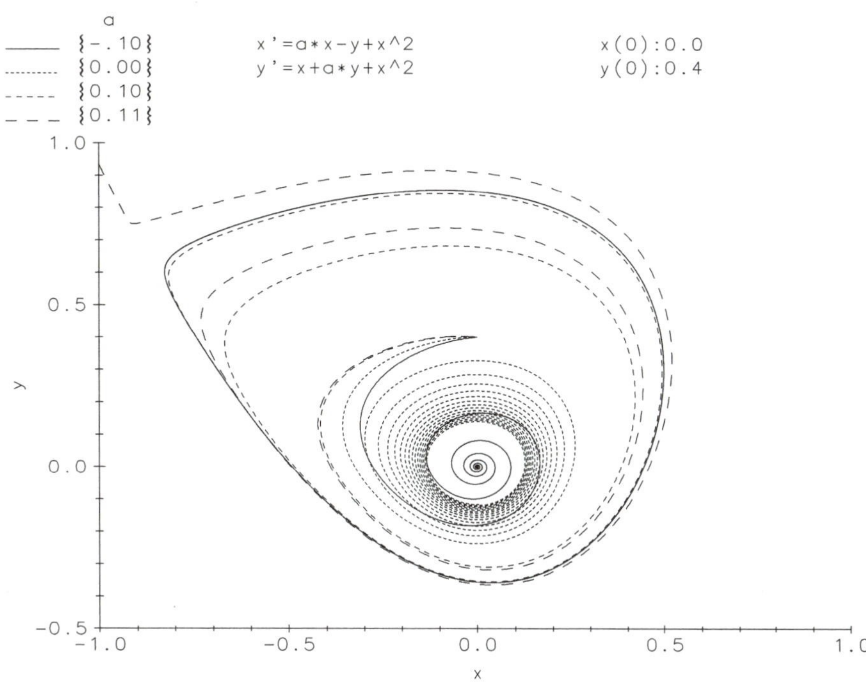


*Figure 5.17.1 Birth and disappearance of a limit cycle (Example 2).*

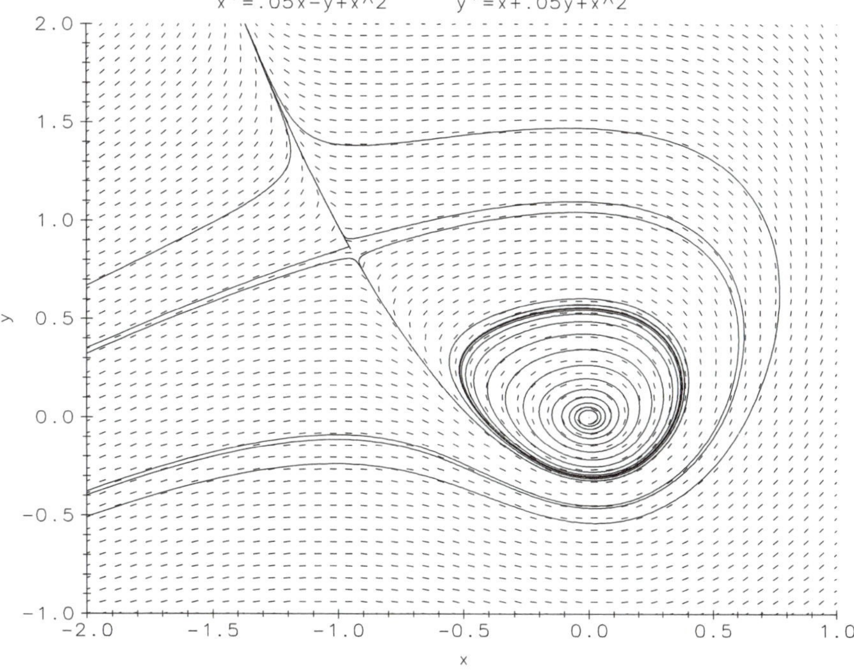


*Figure 5.17.2 The attracting limit cycle after the Hopf bifurcation (Example 2).*

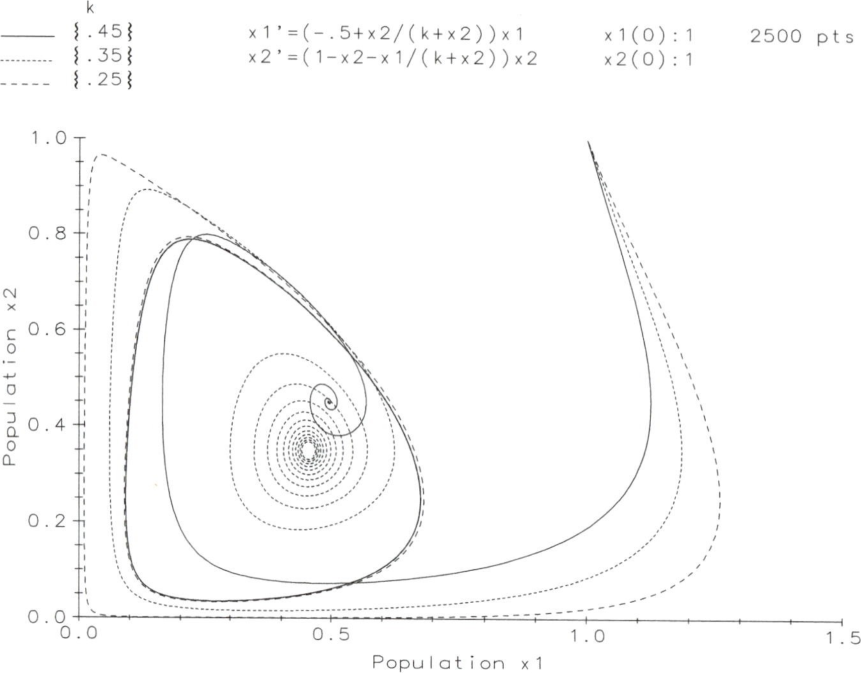


*Figure 5.17.3 Change in a satiation parameter creates a population limit cycle.*

***Example 3*** **(Satiable predation)** A model of satiable predation is given in Experiment 5.5–5.6. The ODEs are

$$\begin{aligned} x' &= (-a + by/(k + y))x \\ y' &= (c - \gamma y - dx/(k + y))y \end{aligned} \tag{5}$$

where $x$ and $y$ are the respective predator and prey populations, and $a$, $b$, $c$, $d$, $k$, and $\gamma$ are positive parameters. Atlas Plate `Predator-Prey B` displays the orbits in the population quadrant for one set of values of the parameters. Observe that there is an attracting limit cycle that encloses an unstable equilibrium point. The limit cycle is spawned by a Hopf bifurcation from the equilibrium point. Figure 5.17.3 shows one orbit for each of three values of the parameter $k$. The Hopf bifurcation occurs at some value of $k$ between 0.35 and 0.25.

_______________/_______________ hardware / software

_______________/_______ name / date

Answer questions in the space provided,
or on attached sheets or carefully labeled graphs.

_______________/_______ course / section

## *5.17 The Hopf Bifurcation*

***Abstract*** Orbits are computed and graphed for evidence of a Hopf bifurcation.

***Special instructions*** For each graph in this experiment, insert arrowheads to show the direction of advancing time. In addition, mark all cycles and equilibrium points. Show all written work directly on the graphs.

**1.** Duplicate Atlas Plates `Bifurcation A-C`.

**2.** Choose a system from the list below. Rewrite the system in polar coordinates. Find the eigenvalues of the system matrix of the linear terms. What happens to the eigenvalues as $a$ increases through 0? Verify that the general conditions for a Hopf bifurcation are satisfied. Plot several orbits for $a = -1, 0, 0.5, 1$.

- ☐ **A.** $x' = ax + 2y - x(x^2 + y^2),\ y' = -2x + ay - y(x^2 + y^2)$
- ☐ **B.** $x' = y - x^3,\ y' = -x + ay - y^3$
- ☐ **C.** $x' = ax + y - x^3,\ y' = -16x + ay - y^3$

**3.** Rewrite the following system using polar coordinates centered at the equilibrium point $(a, 0)$: $x = a + r\cos\theta,\ y = r\sin\theta$.

$$x' = a(x - a) + y - (x - a)[(x - a)^2 + y^2]$$
$$y' = -(x - a) + ay - y[(x - a)^2 + y^2]$$

Verify explicitly that all Hopf bifurcation conditions are met. Graph $xy$-orbits for the values $a = -1, 0, 0.5, 1$. Find equations for the limit cycles, and describe their behavior as $a \to \infty$.

**4.** The system of Example 2 is explored in this problem.

**(a)** Duplicate Figures 5.17.1 and 5.17.2, writing out the verification of as many Hopf conditions as possible on the graph.

**(b)** For each of the four values of $a$ in Figure 5.17.1 draw a full portrait of orbits on a screen large enough to include both equilibrium points. Describe the behavior of the orbits near each of these points.

* **5.** Construct orbital portraits of the scaled van der Pol equation in Experiment 5.14: $x' = \mu x - y - y^3,\ y' = x$ for $\mu = -1, -0.5, 0, 0.5, 1$. Show that all the conditions of the Hopf bifurcation are satisfied, and write your argument on the graph. (Suggestion: At $\mu = 0$ rewrite the system in $r\theta$-coordinates, refer to the Poincaré-Bendixson alternatives of Experiment 5.16, and show that the system is asymptotically stable at the origin.)

_______________/_______________
hardware software

_______________/_______________
name date

Answer questions in the space provided,
or on attached sheets or carefully labeled graphs.

_______________/_______________
course section

## 5.18 *Satiable Predation: Bifurcation to a Cycle*

***Abstract*** If the predator's appetite is satiated then under certain circumstances the predator-prey orbits do not tend toward an equilibrium point but instead spiral toward a limit cycle. This is an example of a Hopf bifurcation.

**1.** Duplicate Atlas Plate `Predator-Prey B`. Insert arrowheads on the orbits to show the direction of increasing time. Plot component curves for $0 \le t \le 20$.

**2.** Consider the model of satiation given by system (5).

**(a)** Explain each of the rate functions of (5) in terms of population behavior.

* **(b)** Beginning with the values of the rate constants of system (5) used in Atlas Plate `Predator-Prey B` (i.e., $a = 0.5$, $b = c = \gamma = d = 1$, $k = 0.25$), vary $k$ up and down and observe what happens. Does the amplitude of the limit cycle shrink? Grow? Find a value of $k$ at which the cycle disappears. What happens once $k$ has changed beyond that value? Duplicate Figure 5.17.3, insert arrowheads on orbits to show increasing time, and mark the equilibrium points for each of the three systems. Explain why the limit cycle appears as $k$ is lowered. What are the visual clues that a Hopf bifurcation has occurred?

## *Eigenelements and Linear Systems*

The eigenvalues and eigenvectors of a matrix $A$ can be used to construct solutions of the linear system $u' = Au$, $u = [x\ y]^T$. Examples 1 through 3 illustrate the process.

***Example 1*** **(Real, distinct eigenvalues)** The eigenvalues of the system matrix

$$A = \begin{bmatrix} 5 & 3 \\ -1 & 1 \end{bmatrix}$$

are the roots 2 and 4 of the characteristic polynomial $p(\lambda) = \det[A - \lambda I] = \lambda^2 - 6\lambda + 8$. A corresponding collection of independent eigenvectors is $[1\ -1]^T$ and $[3\ -1]^T$. The **general (real-valued) solution** $u$ of $u' = Au$ is

$$u = C_1 e^{2t} \begin{bmatrix} 1 \\ -1 \end{bmatrix} + C_2 e^{4t} \begin{bmatrix} 3 \\ -1 \end{bmatrix}$$

where $C_1$ and $C_2$ are arbitrary real numbers.

***Example 2*** **(Complex eigenvalues)** The eigenelements of the system matrix

$$A = \begin{bmatrix} -1 & 2 \\ -2 & -1 \end{bmatrix}$$

are $\lambda_1 = -1 + 2i$ with $v^1 = [-i\ 1]^T$ and $\lambda_2 = \bar{\lambda}_1 = -1 - 2i$ with $v^2 = \bar{v}^1 = [i\ 1]^T$. The **general complex-valued solution** of $u' = Au$ is

$$u = C_1 e^{(-1+2i)t} \begin{bmatrix} -i \\ 1 \end{bmatrix} + C_2 e^{(-1-2i)t} \begin{bmatrix} i \\ 1 \end{bmatrix}$$

where $C_1$ and $C_2$ are arbitrary complex constants. Since $A$ is real, the general real-valued solution is ususally preferred. Using Euler's formula, $e^{a+bi} = e^a(\cos b + i \sin b)$ and the fact that linear combinations of solutions of $u' = Au$ are also solutions, we obtain the **general real-valued solution** of $u' = Au$ in the form

$$u = C_1 \text{Re}(e^{(-1+2i)t} v^1) + C_2 \text{Im}(e^{(-1+2i)t} v^2)$$

or

$$u = C_1 e^{-t} \begin{bmatrix} \sin 2t \\ \cos 2t \end{bmatrix} + C_2 e^{-t} \begin{bmatrix} -\cos 2t \\ \sin 2t \end{bmatrix}$$

where $C_1$ and $C_2$ are arbitrary real constants. This method works because the real and imaginary parts of any solution of a homogeneous linear ODE with real coefficients are independent, *real*-valued solutions of that ODE.

***Example 3*** **(Deficient eigenspace)** The characteristic polynomial of

$$\begin{bmatrix} 1 & -1 \\ 1 & 3 \end{bmatrix}$$

has a double root $\lambda = 2$ but the corresponding eigenspace $V_2$ is only one dimensional—spanned by $v = [1\ -1]^T$. Although we have one solution $u = e^{2t}[1\ -1]^T$ of $u' = Au$, we cannot hope to find a second independent solution in this way. It can be shown that if the **generalized eigenvector** $w$ defined by $(A-2I)w = v$ is computed, then a second and independent solution of $u' = Au$ is given by $te^{2t}v + e^{2t}w$. In this case one possible $w$ (there are many) is $w = [1\ -2]^T$. Hence, the general solution of $u' = Au$ in this deficiency case is

$$= C_1 e^{2t}\begin{bmatrix} 1 \\ -1 \end{bmatrix} + C_2 \left\{ te^{2t}\begin{bmatrix} 1 \\ -1 \end{bmatrix} + e^{2t}\begin{bmatrix} 1 \\ -2 \end{bmatrix}\right\}$$

where $C_1$ and $C_2$ are arbitrary real constants.

***Observation*** These three examples illustrate the fact that solving the system $u' = Au$ is largely an algebraic problem of finding the eigenelements of $A$ and using them to build the solution space. The solution space is in any case a two-dimensional space of real-valued column vector functions of time.

## *Rescaling Time*

A practical problem is often encountered when using a computer to solve ODEs: The rates are so high that the orbit quickly leaves the screen. Even when the orbit is on the screen, the high speed may place successive plotted points so far apart that the orbit resembles a jagged polygonal path rather than a smooth curve. One way to remedy this difficulty for an autonomous system is to rescale time. This process was introduced in Experiment 2.1. Here it is amplified in the setting of a planar autonomous system.

Consider the system

$$\begin{aligned} dx/dt &= f(x, y) \\ dy/dt &= g(x, y) \end{aligned} \tag{1}$$

Suppose that there is a curve $(X(s), Y(s))$ parameterized by $s$ with the property that for some positive scalar function $r(X(s), Y(s))$ we have

$$\begin{aligned} dX/ds &= r(X(s), Y(s))f(X(s), Y(s)) = r^*(s)f(X(s), Y(s)) \\ dY/ds &= r(X(s), Y(s))g(X(s), Y(s)) = r^*(s)g(X(s), Y(s)) \end{aligned}$$

where $r^*(s) = r(X(s), Y(s))$. Thus, the tangent to the curve "fits" the direction field of (1) everywhere. Now we shall show that we can find a change of variable $s = s(t)$ such that $x(t) = X(s(t))$, $y(t) = Y(s(t))$ defines an orbit of the planar system (1). Observe from the Chain Rule that

$$\begin{aligned} \frac{dx}{dt} &= \frac{dX}{ds}\frac{ds}{dt} = r^*(s)\frac{ds}{dt}f(X(s), Y(s)) \\ \frac{dy}{dt} &= \frac{dY}{ds}\frac{ds}{dt} = r^*(s)\frac{ds}{dt}g(X(s), Y(s)) \end{aligned}$$

Thus, if $s = s(t)$ solves the separable first order ODE

$$r^*(s)\frac{ds}{dt} = 1$$

then $x(t) = X(s(t))$, $y(t) = Y(s(t))$ is an orbit of system (1). The same curve is traced by $(x(t), y(t))$ and by $(X(s), Y(s))$, thus finishing the proof.

Rescaling time by a positive scalar function of position has no effect on orbits. Component graphs however, are another matter and Atlas Plates `Limit Set A,B` show the marked changes that can occur when time is rescaled. On a practical level if one is interested only in the orbits of an equation, a judiciously chosen rescaling factor can eliminate some of the computational difficulties mentioned above. For example, if orbits of (1) are moving too fast through some region $R$ of the $xy$-plane, a positive scalar function whose values in $R$ are less than 0.1 can be used to slow down the orbits by a factor of at least 10. To avoid abrupt changes in the apparent speed at which an orbit is generated, the scaling factor is usually chosen to be continuous..

CHAPTER 6

# Higher Dimensional Systems

The dynamical phenomena of nature can be so intricate and complex that the mathematical models describing them must have many state variables. The scope of the earlier chapters is broadened to include the multidimensional state spaces needed to model systems that use at least three state variables such as the vibrations of coupled mass-spring oscillators, the concentrations of several chemical species in a complex reaction, the amounts of lead in the compartments of the body, and the voltages and current of a multiloop circuit.

The extra dimensions of higher dimensional state spaces allow orbits of autonomous systems to move in surprising ways that never occur if there are only one or two state variables. The new orbital phenomena are seen mostly in nonlinear systems, since linear systems and their orbits possess a (relatively) narrow range of possible behaviors.

## *Systems of ODEs*

The IVPs of this chapter have the form

$$\begin{aligned} x' &= f(t, x, a) \\ x(t_0) &= x^0 \end{aligned} \tag{1}$$

where the **state**, **rate**, **parameter**, and **initial vectors**, $x$, $f$, $a$, and $x^0$ are given by:

$$x(t) = \begin{bmatrix} x_1(t) \\ \vdots \\ x_n(t) \end{bmatrix}, \quad f(t, x, a) = \begin{bmatrix} f_1(t,\ x_1, \ldots, x_n,\ a_1, \ldots, a_k) \\ \vdots \\ f_n(t,\ x_1, \ldots, x_n,\ a_1, \ldots, a_k) \end{bmatrix}$$

$$a = \begin{bmatrix} a_1 \\ \vdots \\ a_k \end{bmatrix}, \quad x^0 = \begin{bmatrix} x_1^0 \\ \vdots \\ x_n^0 \end{bmatrix}$$

The **state dimension** is $n$ and the **parameter dimension** is $k$ ($k \geq 0$). The parameters often model coefficients that can be changed, (e.g., the resistance in a circuit, the frequency of a sinusoidal driving force, the rate coefficient of a chemical reaction). The **sensitivity** of a system is a measure of how much an orbit changes if a parameter is changed. The approach to sensitivity in this chapter is qualitative (rather than quantitative) and is based on visual evidence of changes in the computed graphs.

IVP (1) has a solution $x = x(t)$ that is defined on some time interval $I$ containing $t_0$. This solution is unique for each initial state vector $x^0$ and parameter vector $a$ chosen from respective open regions in state and parameter space in which the rate vector $f$, and its first order partial derivatives $\partial f_i / \partial x_j$ are continuous. In this chapter $t_0$ is usually 0; $x^0$ is occasionally denoted by $p$ and the corresponding solution, by $x = \varphi_t(p)$ instead of $x(t)$. As $t$ traverses $I$, the point $\varphi_t(p)$ generates an **orbit** through $p$ in state space. Except in the constant coefficient linear case, it is difficult to find formulas for this solution, but the approach in these experiments is mainly graphical and does not rely on solution formulas.

### Linear Systems

A **driven linear IVP** with **system matrix** $A$ and **driving force** or **input** $F$ has the form

$$\begin{aligned} x' &= Ax + F(t) \\ x(t_0) &= x^0 \end{aligned} \tag{2}$$

where

$$x(t) = \begin{bmatrix} x_1(t) \\ \vdots \\ x_n(t) \end{bmatrix}, \qquad A = \begin{bmatrix} a_{11} & \cdots & a_{1n} \\ \vdots & & \vdots \\ a_{n1} & \cdots & a_{nn} \end{bmatrix}, \qquad F(t) = \begin{bmatrix} F_1(t) \\ \vdots \\ F_n(t) \end{bmatrix}$$

Although the elements $a_{ij}$ of the system matrix of a linear system may be functions of $t$, it is assumed in this chapter that they are (real) constants. IVP (2) has a unique solution that is defined on the largest time interval containing $t_0 = 0$ on which the components of $F(t)$ are continuous. The Chapter 6 Notes explore solution techniques for (2).

### Nonlinear Systems

In a truly nonlinear system (1) (i.e., where $f(t, x, a)$ does not have the form $Ax + F(t)$), there is no linear structure to use to represent a solution. Computer graphics, however, may be employed to track the course of the orbit or the components of the solution curve. Of particular interest is the asymptotic behavior of an orbit as time increases, the behavior in the "long run" after initial transients have died out. Certain distinguished orbits play a prominent role in this study. These are most easily studied in the autonomous case. For the present, reference to any parameter vector $a$ in the rate function is also suppressed. Thus, we are interested in solutions of

$$x' = f(x) \tag{3}$$

A point $p$ in $n$-dimensional state space is an **equilibrium point** (also **rest**, **stationary**, or **critical point**) of (3) if $f(p) = 0$, that is, if $x(t) \equiv p$ is a constant solution. The equilibrium points for the linear system $x' = Ax$ are given by the vectors $p$ for which $Ap = 0$. If $A$ is a nonsingular matrix, then $p = 0$ is the only equilibrium point. If $A$ is singular, then there is an equilibrium point $p \neq 0$ and in fact a line of equilibrium points $x = rp$, $-\infty < r < \infty$. For a general system an equilibrium point is **stable** if orbits that pass nearby remain nearby as time increases. A stable equilibrium point that attracts all nearby orbits is **asymptotically stable**. An equilibrium point is **unstable** if it is not stable. The precise stability definitions are direct extensions of those for planar systems (see Experiments 5.12–5.13).

Another kind of limit orbit of (3) of considerable significance is the nonconstant **periodic orbit** corresponding to a solution $x(t)$ for which $x(t+T) = x(t)$ for some positive constant $T$ and for all $t$. The smallest such number $T$ is the **period** of $x(t)$. It often happens that all nearby orbits are pulled toward a periodic orbit as time increases. The periodic orbit is called an **orbitally stable limit cycle** in this case. A limit cycle is **orbitally unstable** if there are orbits passing arbitrarily close to the cycle but which are not attracted to it as time increases.

In recent years certain sets of limit orbits called strange attractors (defined in the Chapter 6 Notes) have received much attention. These orbit sets usually contain equilibrium points and periodic orbits, but they also contain orbits that appear to wander "chaotically." Several experiments take up the computer study of system that appear to have strange attractors.

# Undriven Linear Systems

***Purpose*** To visualize the orbits and component curves of first order, undriven, constant coefficient linear systems in three or more state variables and to study the asymptotic behavior of solutions and the connection with the eigenelements of the coefficient matrix.

***Keywords*** Linear systems with constant coefficients, orbital portraits, component curves, system matrix, eigenvalues, eigenvectors, stability

***See also*** Experiments 5.1–5.3; Chapter 6 Notes

***Background*** The solutions of the undriven, constant coefficient linear system

$$x' = Ax \tag{1}$$

where $A = [a_{ij}]$ is an $n \times n$ system matrix of real constants $a_{ij}$, are defined in terms of the eigenelements of $A$. (See Examples 1–3 of the Chapter 6 Notes.) Representative systems can be characterized according to the structure of the set of eigenvalues of $A$ (i.e., the **spectrum** of $A$). For example, systems whose spectrum consists of distinct negative real numbers belong to the category entitled **asymptotically stable generalized nodes**. A system whose spectrum contains only pairs of distinct pure imaginaries is a **(neutrally) stable generalized center**. If the state dimension $n$ is large, the number of distinct spectral types is very large, and it would be a formidable task to list them all and to portray the orbits of systems whose system matrix has a given spectral type. Consequently, only some representative samples are explored.

The eigenelements of $A$ dominate the orbital behavior of (1), and this is particularly evident over the long term (i.e., as $t \to \infty$). Experiment 6.2 takes up this question in connection with the stability properties of (1). The definition of stability given in Experiments 5.2–5.3 applies here as well. Recall that there are three general categories: **asymptotic stability** in which (1) is stable and all orbits tend to the origin as $t \to \infty$ (real parts of all eigenvalues of $A$ are negative); **neutral stability** in which (1) is stable but not asymptotically stable (real parts of eigenvalues are nonpositive, at least one eigenvalue with zero real part, but no such eigenvalue with eigenspace dimension less then the multiplicity of the eigenvalue); and **instability**, which simply means "not stable." An unstable system (1) has at least one orbit, which becomes unbounded as $t \to \infty$. System (1) is unstable if at least one eigenvalue has a positive real part.

***Example 1*** **(Tornado)** A linear system and its system matrix $A$ are given by

$$\begin{aligned} x' &= -x/2 \\ y' &= -y + 20z \\ z' &= -20y - z \end{aligned} \qquad A = \begin{bmatrix} -\frac{1}{2} & 0 & 0 \\ 0 & -1 & 20 \\ 0 & -20 & -1 \end{bmatrix} \tag{2}$$

The eigenvalues of $A$ are $-1/2$, $-1 + 20i$, $-1 - 20i$. All orbits approach the origin of the three-dimensional state space. Figure 6.1.1 displays one of the orbits.

***Example 2*** **(Orbit on a cylinder)** The system matrix $A$ of the system

$$\begin{aligned} x' &= -x/2 \\ y' &= 20z \\ z' &= -20y \end{aligned} \qquad A = \begin{bmatrix} -\frac{1}{2} & 0 & 0 \\ 0 & 0 & 20 \\ 0 & -20 & 0 \end{bmatrix} \tag{3}$$

has eigenvalues $-1/2$, $20i$, $-20i$. The system is neutrally stable. Each nonconstant orbit lies on a cylinder, $y^2 + z^2 =$ constant. Figure 6.1.2 shows one of these orbits.

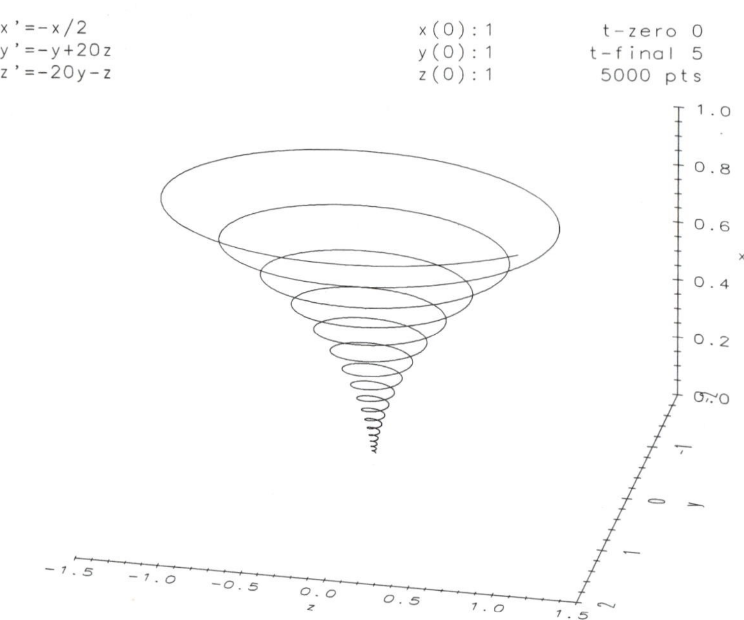


*Figure 6.1.1 An orbit of (2) spirals toward the asymptotically stable point at the origin.*

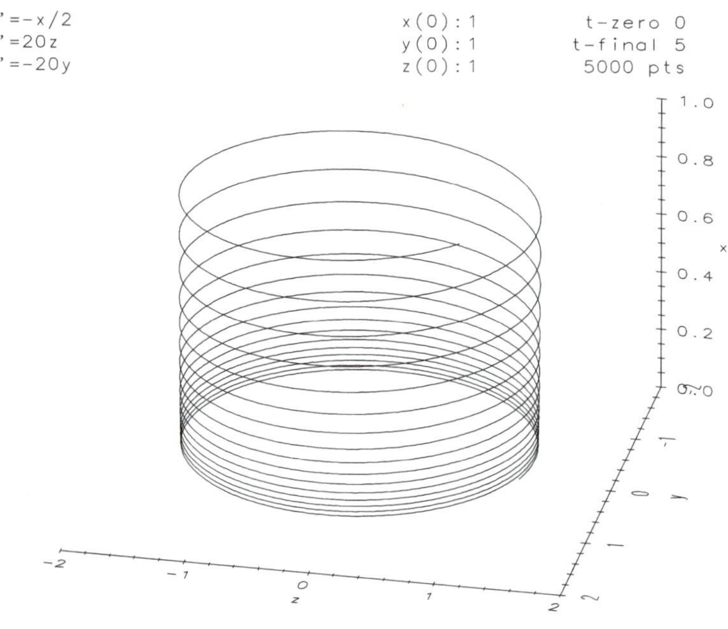


*Figure 6.1.2 An orbit of (3) that spirals on a cylinder.*

_______________/_______________ hardware / software

_______________/_______________ name / date

Answer questions in the space provided, or on attached sheets or carefully labeled graphs.

_______________/_______________ course / section

## 6.1 *Portraits of Undriven Linear Systems*

***Abstract*** Orbital portraits and component graphs are constructed for undriven linear systems $x' = Ax$, where $x$ is a state vector with at least three components.

***Special instructions*** Orbital portraits and projections should be drawn in 3D perspective if the local software/hardware supports that capability. Otherwise, use projections onto various planes. For example, if state space is three-dimensional ($xyz$-space), orbits may be projected onto the $xy$, $xz$, $yz$ or other planes. If the state dimension exceeds three, the projections should be carefully chosen to display the characteristic behavior of the orbits.

**1.** For each of Examples 1–3 in the Chapter 6 Notes select suitable initial points and sketch orbits and component curves. Write on the graphs the stability types and descriptions of how orbits behave near the origin. What happens to the orbits as $t \to \infty$? As $t \to -\infty$? Repeat for the systems of Examples 1 and 2 on the cover sheet.

**2.** Construct a gallery of orbital portraits and component graphs for the 3D case of $x' = Ax$, where $A$ is a $3 \times 3$ real matrix. Write the stability type on each graph. Include in your collection at least the following seven spectral cases:

- $\lambda_1 < \lambda_2 < \lambda_3 < 0$
- $\lambda_1 < \lambda_2 = \lambda_3 < 0$
- $\lambda_1 < \lambda_2 < 0 < \lambda_3$
- $0 < \lambda_1 < \lambda_2 < \lambda_3$
- $\lambda_1 = \alpha + i\beta = \bar{\lambda}_2 \quad (\alpha < 0, \beta \neq 0), \quad \lambda_3 < 0$
- $\lambda_1 = \alpha + i\beta = \bar{\lambda}_2 \quad (\alpha = 0, \beta \neq 0), \quad \lambda_3 < 0$
- $\lambda_1 = \alpha + i\beta = \bar{\lambda}_2 \quad (\alpha > 0, \beta \neq 0), \quad \lambda_3 < 0$

**3.** Consider the system in 4D

$$x_1' = px_2 \qquad x_3' = qx_4$$
$$x_2' = -px_1 \qquad x_4' = -qx_3$$

where $p$ and $q$ are positive real numbers.

**(a)** Set $p = q = 1$ and plot orbital projections and component curves if the initial value of every state variable is 1. Is the corresponding solution periodic? If so, what is the period?

**(b)** Repeat (a) but with $p = 2$, $q = 3$. Duplicate Figure 6.1.3.

**(c)** Repeat (a) with $p = 1$, $q = \pi$.

* **(d)** Find conditions on $p$ and $q$ that imply that all solutions are periodic. (Suggestion: Suppose $p/q$ is a rational number.)

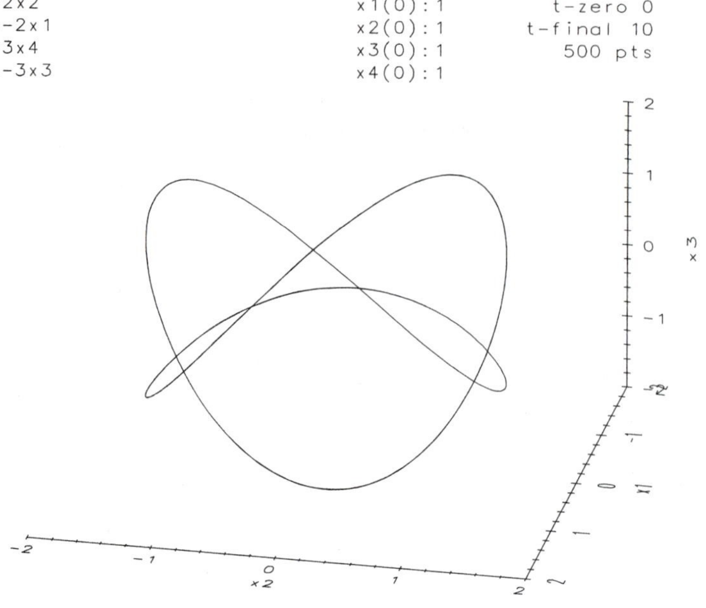


*Figure 6.1.3 Projection of an orbit in 4-space into 3-space.*

**4.** Plot enough orbits or component curves of the system $x' = Ax$ to decide whether the system is asymptotically stable, neutrally stable, or unstable. Use component graphs to estimate the imaginary parts of all complex eigenvalues of $A$, if there are any. (Suggestion: Oscillatory behavior of a component graph indicates complex eigenvalues. The time between successive local maximum values of an oscillating component can be used to determine $\beta$ where $\lambda = \alpha + i\beta$ is an eigenvalue of $A$, since $\sin \beta t$ and $\cos \beta t$ are periodic terms in the components of some solutions.)

**(a)**

$$A = \begin{bmatrix} 2 & 1 & 1 \\ 2 & 3 & 2 \\ 1 & 1 & 2 \end{bmatrix}$$

**(b)**

$$A = \begin{bmatrix} -3 & -1 & 0 \\ 6 & -7 & 1 \\ 8 & 1 & -10 \end{bmatrix}$$

**(c)**

$$A = \begin{bmatrix} -1 & -2 & 2 \\ 2 & -2 & -1 \\ 2 & 1 & 2 \end{bmatrix}$$

**(d)**

$$A = \begin{bmatrix} 0 & 1 & 0 & 0 \\ 0 & 0 & 1 & 0 \\ 0 & 0 & 0 & 1 \\ 1 & 1 & 1 & 1 \end{bmatrix}$$

## *6.2 Asymptotic Behavior and Eigenelements*

***Abstract*** The signs of the real parts of the eigenvalues of the system matrix $A$ are the primary factors in the stability properties of $x' = Ax$. This experiment takes up this question.

**1.** For each of the system matrices $A$ given below, graph several orbits (or their projections) and component curves of $x' = Ax$. Write the eigenvalues of $A$ and their multiplicities on the graph. Identify the stability properties of the system.

**(a)**

$$A = \begin{bmatrix} -3 & 5 & 0 \\ -5 & -3 & 0 \\ 0 & 0 & -2 \end{bmatrix}$$

**(b)**

$$A = \begin{bmatrix} -1 & 2 & 3 \\ 0 & -1 & 2 \\ 0 & 0 & -1 \end{bmatrix}$$

**(c)**

$$A = \begin{bmatrix} -3 & 15 & 0 & 0 \\ -15 & -3 & 0 & 0 \\ 0 & 0 & -3 & 15 \\ 0 & 0 & -15 & -3 \end{bmatrix}$$

**(d)**

$$A = \begin{bmatrix} -3 & 15 & 1 & 0 \\ -15 & -3 & 0 & 1 \\ 0 & 0 & -3 & 15 \\ 0 & 0 & -15 & -3 \end{bmatrix}$$

**2.** Duplicate Figures 6.2.1 and 6.2.2. Draw component graphs. Are the eigenvalues of the system matrix real or complex? Explain. Is the system asymptotically stable, neutrally stable, or unstable? Explain.

**3.** Construct a linear system $x' = Ax$ of state dimension 6 that has one negative and one positive eigenvalue, a conjugate pair of pure imaginary eigenvalues, and a conjugate pair of complex eigenvalues with negative real parts. Write out your system below and list the eigenvalues. Sketch orbital components and projections. Suggestion: Construct a tridiagonal matrix $A$ with the real eigenvalues on the diagonal and each pair of complex eigenvalues $\alpha \pm i\beta$ represented by the block

$$\begin{bmatrix} \alpha & \beta \\ -\beta & \alpha \end{bmatrix}$$

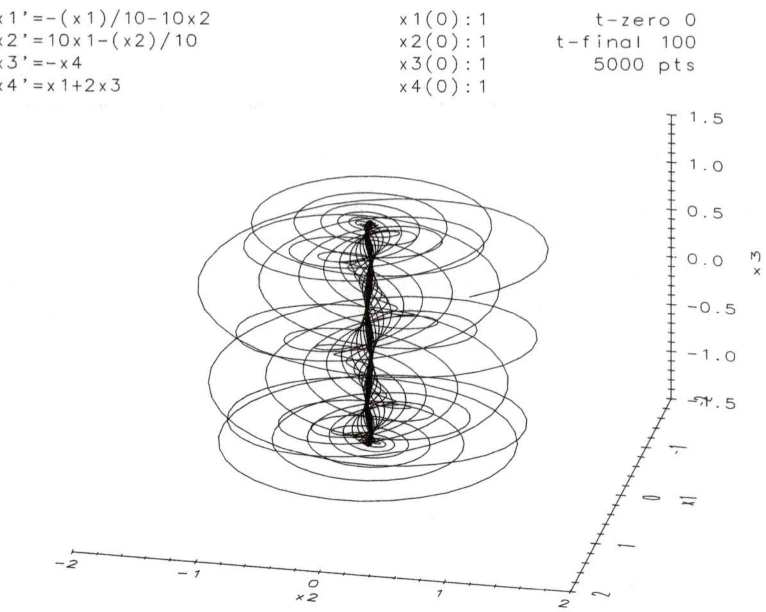


*Figure 6.2.1 Three dimensional projection of an orbit in 4-space.*

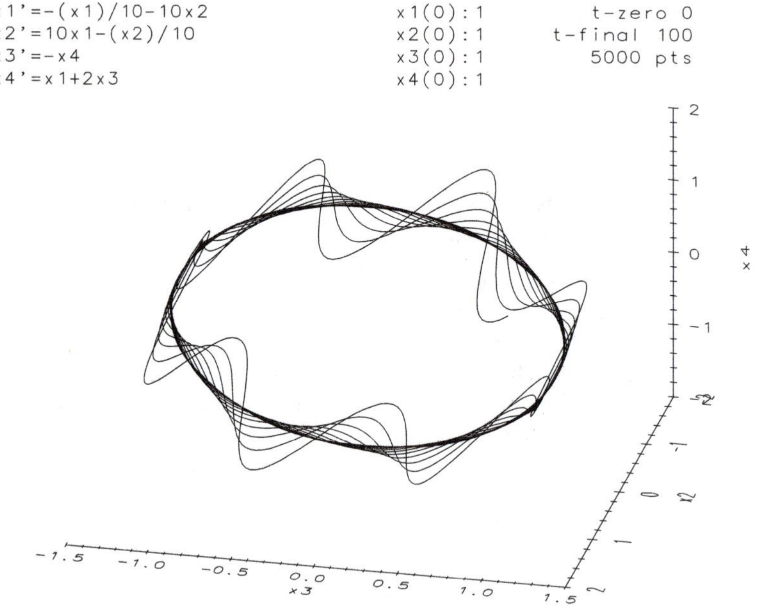


*Figure 6.2.2 The orbit of Figure 6.2.1 projected into another 3-space.*

# Driven Linear Systems

***Purpose*** To study the effects of a periodic driving force on the orbits of a constant coefficient linear system, and to see how coupled and undamped oscillators respond when driven.

***Keywords*** Linear systems, driving force, steady state, transients, resonance, coupled oscillators

***See also*** Experiments 4.3, 4.6, 6.1–6.2; Chapter 6 Notes

***Background*** Most IVPs for driven linear systems have the form

$$\begin{aligned} x' &= Ax + F(t) \\ x(0) &= x^0 \end{aligned} \tag{1}$$

where the state vector $x$ has $n$ components, $A$ is the $n \times n$ real system matrix $[a_{ij}]$, and the components of the driving force vector $F(t)$ are continuous (or piecewise continuous) on a time interval $I$. As stated in the Chapter 6 Notes, the solution of (1) is given by the formula

$$x(t) = e^{tA}x^0 + e^{tA}\int_0^t e^{-sA}F(s)ds \tag{2}$$

The total response $x(t)$ is the sum of the response to the data and the response to the input $F$. However, effective use of the formula requires calculation of the matrix exponential, integration of the components of $e^{-sA}F(s)$, and matrix-vector multiplication. The response $e^{tA}x^0$ to the input is **transient** and tends to 0 as $t \to \infty$ if the eigenvalues of $A$ have negative real parts. The term with the integral (i.e., the response to the input) consists of another transient (corresponding to the lower limit $t = 0$ in the integration) and the **steady state** (corresponding to the upper limit on the integral). After the transients have decayed to the point that they are no longer detectable, the steady state term dominates and can easily be seen on the orbital or component graphs if the time span is long enough (Figure 6.3.1). If the input is bounded for $t \geq 0$ (i.e., if $|F(t)| \leq M$ for some constant $M$ and all $t \geq 0$), then the total response, or output, $x(t)$ is also bounded.

If the system matrix $A$ has pure imaginary eigenvalues ($\pm i\beta$), then the undriven system has periodic solutions of (angular) frequency $\beta$ and period $T = 2\pi/\beta$. If the driving force also has period $T$, resonance may occur and (1) would then have oscillatory solutions whose amplitudes become unbounded as $t \to \infty$ (Figure 6.3.2). Even if there is some damping in the system, there may be a high gain if the driving frequency is close to a natural frequency. The graphical evidence of such a phenomenon would be a high amplitude response to a low amplitude driving force. Physically, this phenomenon occurs with masses coupled by springs when a periodic external force is applied to one of the masses. The same phenomenon occurs with coupled or double pendulums.

***Observation*** Transients, steady states, resonance, and oscillations are types of linear phenomena that occur in planar linear systems and in single scalar second order linear ODEs. The new features that appear in the higher dimensional case are, first of all, that these systems model a much wider variety of physical phenomena and, second, that it is possible for a driving frequency to match a natural frequency without causing resonance. The latter point is pursued in Experiment 6.3.

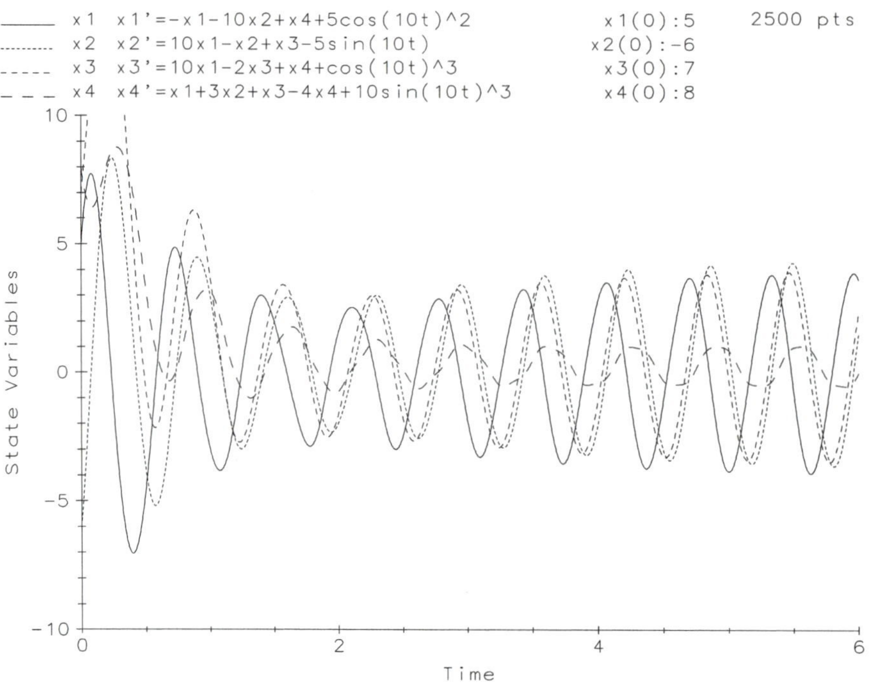


*Figure 6.3.1 Transients decay and the steady state persists.*

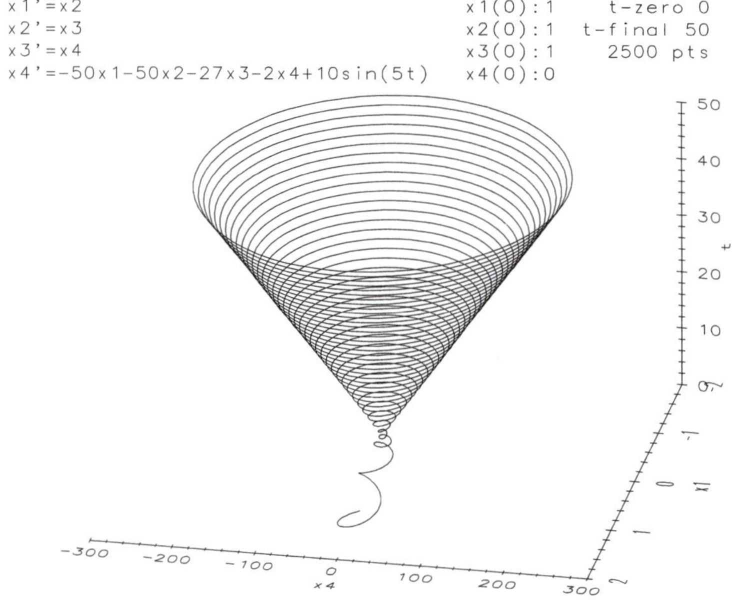


*Figure 6.3.2 Resonance.*

_______________/_______________ hardware software

_______________/_______________ name date

Answer questions in the space provided,
or on attached sheets or carefully labeled graphs.

_______________/_______________ course section

## 6.3 *Transients, Steady States, Resonance*

***Abstract*** Transients, steady states, and resonance are illustrated in this experiment for a variety of system matrices and periodic driving forces.

***Special instructions*** Use orbital projections and component graphs to determine the behavior of the solutions if the state dimension exceeds 3 or if the local hardware/software is inadequate for 3D graphs. Write your descriptions of orbital behavior and periods on the computer graphs.

**1.** Consider the driven system

$$x' = -4x + 2y + 2z + F_1(t)$$
$$y' = 9x - 17y - 6z + F_2(t)$$
$$z' = -4x + 6y - 5z + F_3(t)$$

where $F_1$, $F_2$, and $F_3$ have the values given below.

**(a)** Graph enough orbits or component curves to convince yourself that all solutions of the undriven system (i.e., with $F_1 = F_2 = F_3 = 0$) are transients.

**(b)** Suppose that $F_1(t) = 2\cos t$, $F_2(t) = \cos 2t$, $F_3(t) = \sin 3t$. Show that the steady state response tends to a periodic solution as $t$ increases by graphing component curves over a time span long enough that the transients have almost vanished. Use initial data $x(0) = y(0) = z(0) = 0$. What is the period?

**(c)** If the input is a constant vector $F^0$, then the steady state response for (2) is the constant vector $-A^{-1}F^0$. Use the computer to graph the steady state response in the case where $F^0 = [1\ 2\ 3]^T$.

**(d)** Drive the system with an input vector whose components have incommensurate periods, $F(t) = [\cos t, \sin \pi t, 1]^T$. Describe the steady state in this case. Is it periodic?

**2.** Consider the system

$$x' = 28x - 48y - 58z + \cos(5t)$$
$$y' = 100x - 170y - 200z + \sin(5t)$$
$$z' = -71x + 121y + 141z + 10\cos(10t)$$

with a periodic input.

**(a)** Describe the behavior of the orbits of the corresponding undriven system. Are there periodic solutions? If so, estimate the periods.

**(b)** Describe the nature of the orbits of the driven system. Do you see resonance, (i.e., oscillations of growing amplitude)?

**(c)** Replace the term $\sin(5t)$ by $\sin(\pi t)$. What is the long-term behavior of orbits now?

**3.** Consider the two undriven systems

$$\begin{aligned} x_1' &= x_2 \\ x_2' &= -x_1 \\ x_3' &= x_4 \\ x_4' &= -x_3 \end{aligned} \qquad\qquad \begin{aligned} x_1' &= x_2 + x_3 \\ x_2' &= -x_1 + x_4 \\ x_3' &= x_4 \\ x_4' &= -x_3 \end{aligned}$$

* **(a)** Graph component curves for each. Describe their behavior. Find the eigenvalues of the two system matrices. Although there are no periodic input terms, the orbits in the $x_1x_2$ plane of the second system seem to exhibit resonance in that their amplitude grows without bound. Explain.

* **(b)** Replace the last rate equation in the first system above by $x_4' = -2x_3 + \sin t$. Plot component curves for the new system. Show that the natural angular frequencies are 1 and $\sqrt{2}$. Explain why there is no resonance even though the angular frequency of the driving force matches a natural frequency.

_______________/_______________ hardware software

_______________/_______________ name date

Answer questions in the space provided,
or on attached sheets or carefully labeled graphs.

_______________/_______________ course section

## *6.4 Coupled Oscillators*

***Abstract*** Masses that are coupled by springs to each other and to walls slide freely on a horizontal plane. A horizontal force acts on one of the masses. The motion of the masses is studied.

**1.** Consider the system of masses shown schematically below.

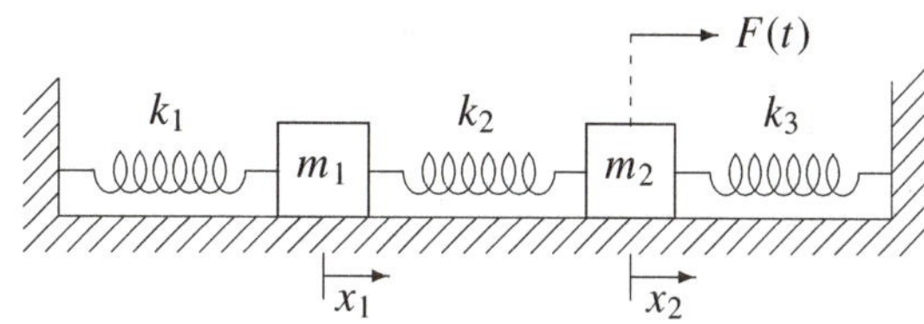


**(a)** Formulate the coupled system of second order ODEs in the displacements $x_1$ and $x_2$ from equilibrium of the two masses. The external force $F(t)$ acts horizontally on the second mass. The constants $k_i$, $m_j$ are the Hooke's Law restoring force coefficients and the masses, respectively. Assume frictional forces that act opposite to the direction of motion and are of magnitude proportional to speed. (Suggestion: The first ODE is $m_1 x_1'' = -k_1 x_1 + k_2(x_2 - x_1) - c_1 x_1'$.) Introduce velocities as state variables and write the equivalent system of four first order ODEs.

**(b)** Set $m_2 = 2m_1 = 2$, $k_1 = k_2 = k_3 = 8$, assume no friction, and assume that $F = 0$. For nonzero initial data, graph $x_1(t)$ and $x_2(t)$ over a time sufficiently long to permit you to describe the steady state behavior of the oscillating masses.

**(c)** Now let $F(t) = \cos(\pi t)$. How do the masses move? Repeat but with $F(t) = \cos(2t)$. What happens now?

**(d)** Choose a periodic driving force $F(t)$ that creates resonance and destroys the system.

**(e)** Repeat (b) and (c) but with light friction, $c_1 = c_2 = 0.01$. For each of several periodic driving forces $F(t)$, graph the steady state response. Is it periodic? If so, what is it period? (Suggestion: Solve over a long enough time interval that the transients decay.)

**2.** Suppose that the linear springs of Problem 1 are replaced by nonlinear soft springs whose restoring force is given by $F(x_i) = -k_i x_i + \alpha_i x_i^3$, where $k_i$ and $\alpha_i$ are positive constants and $x_i$ is the displacement of the $i^{\text{th}}$ mass from its equilibrium position.

* **(a)** Write a coupled system of second order ODEs in $x_1$ and $x_2$. Write an equivalent system of four first order ODEs in the state variables $x_1, x_2, x_3 = x_1', x_4 = x_2'$.

* **(b)** Let $F(t) = A\cos\omega t$. Choose the constants $k_i$, $\alpha_i$, $m_i$, $c_i$, $A$, and $\omega$ such that the masses move in an apparently chaotic way. (Suggestion: See Experiment 5.10 and Atlas Plates `Duffing A-C`.) Graph orbits and component curves. Let $\Delta t = 2\pi/\omega$ and plot Poincaré time sections for several initial conditions. Explain what you see.

# A Compartment Model: Lead in the Body

***Purpose*** To examine a model for absorption of lead by the human body.

***Keywords*** Compartment model, linear system of ODEs, Balance Law, steady state

***See also*** Experiment 6.3; Atlas Plates `Compartment Model A, B`

***Background*** Lead enters the body in food, air, and water contaminated by automobile and industrial emissions. It accumulates in the blood, in tissues, and, especially, in the bones. Some lead is excreted through the urinary system and by hair, nails, and sweat, but enough may remain in the body to impair mental and motor capacity. A model of this system is obtained by tracking the concentrations of lead in each of the three **compartments**, blood, tissue and bones, over time. Numbering these compartments as 1, 2 and 3, respectively, the schematic diagram of Figure 6.5.1 aids the modeling process:

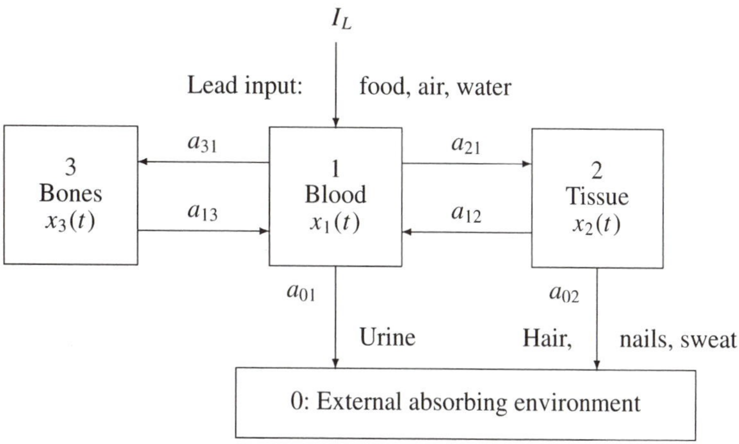


*Figure 6.5.1 The ingestion, distribution, and excretion of lead.*

The amount of lead in compartment $i$ at time $t$ is denoted by $x_i(t)$. The rate of transfer of lead from compartment $j$ to compartment $i$ at time $t$ follows a first order rate law, hence is proportional to $x_j(t)$; the constant of proportionality is denoted by $a_{ij}$. Note that $a_{ij} \geq 0$ and that $a_{ij} = 0$ if no transfer takes place. A reverse transfer from compartment $i$ to compartment $j$ is possible, but $a_{ji}$ need not be equal to $a_{ij}$. A mathematical model for the flow of lead through the body is based on the basic

**Balance Law:** *Net rate = rate in − rate out*

applied to the rate of change of the amount of lead in each compartment. This yields the linear system

$$\begin{aligned} &\text{(blood)} \quad x_1' = -(a_{01} + a_{21} + a_{31})x_1 + a_{12}x_2 + a_{13}x_3 + I_L \\ &\text{(tissue)} \quad x_2' = a_{21}x_1 - (a_{02} + a_{12})x_2 \\ &\text{(bones)} \quad x_3' = a_{31}x_1 - a_{13}x_3 \end{aligned} \tag{1}$$

where $I_L$ is the ingestion rate of lead into the bloodstream. The external absorbing compartment 0 is usually called the **environmental compartment**. Its rate equation is omitted since it does not send the substance being tracked back into the system.

***Example 1*** **(Smog)** Rabinowitz, Wetherill, and Kopple studied the lead intake and excretion of a healthy volunteer living in an area of heavy smog (Los Angeles). Their work was reported in *Science* (**182**, 725–727 (1973)), and extended by Batschelet, Brand, and Steiner *J. Math. Biol.*, (**8**, 15–23 (1979)). The data from this study were used to estimate the rate constants for the compartment model (1). Lead is measured in micrograms and time in days. For example, the rate 49.3 in (2) below is the ingestion rate $I_L$ of lead in micrograms per day, while the coefficient 0.0361 is the sum of the three compartment transfer coefficients, $a_{01}$, $a_{21}$, and $a_{31}$, of lead from the blood into, respectively, the excretory system, tissue, and bones. The full system is given by

$$\begin{aligned} x_1' &= -0.0361x_1 + 0.0124x_2 + 0.000035x_3 + 49.3 \\ x_2' &= 0.0111x_1 - 0.0286x_2 \\ x_3' &= 0.0039x_1 - 0.000035x_3 \end{aligned} \tag{2}$$

With (2) we can study the effect of doubling (or halving) the input rate of lead into the bloodstream or the effect of drugs that increase the backward diffusion rate of lead out of bone. See Atlas Plates `Compartment Model A, B` for component graphs of a modified model.

***Observation*** **(Compartment systems)** System (1) is a special case of the linear system

$$x' = Ax + F(t) \tag{3}$$

where the state $x(t)$ is a column vector with $n$ component functions, the driving term $F(t)$ is also a column vector with $n$ component functions, and $A = (a_{ij})$ is a $n \times n$ constant matrix. System (3) is a **compartment system** if the matrix $A$ satisfies the following conditions: $a_{ij} \geq 0$ if $i \neq j$, and $0 \geq \sum_{i=1}^{n} a_{ij} = -a_{0j}$, for each $j = 1, 2, \ldots, n$ (such matrices are called **compartment matrices**). The compartments of the system are numbered $i = 1, 2, \ldots, n$, and $i = 0$ denotes the absorbing environment. A compartment $j$ is **open to the environment** if there is a sequence of compartments $i_1, i_2, \ldots, i_r$, for which $a_{i_1 j}, a_{i_2 i_1}, \ldots, a_{i_r i_{r-1}}$, and $a_{0 i_r}$ are all positive. It is not hard to show that in system (1) all components are open to the environment. In general, if all compartments of (3) are open to the environment, it can be shown that all the eigenvalues of $A$ have negative real parts. If, in addition, $F$ is a constant vector, the general solution of (3) is

$$x(t) = e^{tA}c - A^{-1}F \tag{4}$$

where $c$ is an arbitrary constant vector. Since the eigenvalues of $A$ all have negative real parts, it follows that, as $t \to \infty$, $x(t)$ tends to the steady state solution vector $-A^{-1}F$, whose components are the equilibrium levels in the individual compartments.

A compartment system is **closed** if the entries in each column of the compartment matrix sum to zero. It may be shown that any compartment system can be closed by attaching an absorbing (environmental) compartment in an appropriate way. For example, lead system (2) can be closed by including the compartment labeled "0" as shown in Figure 6.5.1. If $x_4(t)$ is the amount of lead in the new compartment, the augmented model consists of (2) and the ODE

$$x_4' = 0.0221x_1 + 0.0162x_2 \tag{5}$$

Note that $x_1' + x_2' + x_3' + x_4' = 49.3$, for all $t$. As might be expected, in a closed system the substance being tracked through the compartments accumulates in the absorbing compartment.

__________/__________ hardware software

__________/__________ name date

Answer questions in the space provided, or on attached sheets or carefully labeled graphs.

__________/__________ course section

## 6.5 A Compartment Model: Lead in the Body

***Abstract*** Using the model system, the effects of changing the data are analyzed; in particular the effect on the body of moving from polluted to lead-free surroundings is studied. Some drugs speed up the rate of lead removal from the bones, and this effect is modeled.

**1.** The following tasks concern model system (2).

**(a)** Find the values of all the transfer coefficients $a_{ij}$. (Hint: Find $a_{12}, a_{13}, a_{21}$, and $a_{31}$ first.)

**(b)** Find the equilibrium levels of lead (in micrograms) in each of the three body compartments.

**(c)** Reproduce Atlas Plate `Compartment Model A` but only for 400 days. Assuming no lead initially in each compartment, how far from their equilibrium values are the amounts of lead in the compartments at the end of a 400-day period?

**2.** Suppose that the compartments contain no lead initially, and that after 400 days of lead exposure, all lead is removed from the environment.

**(a)** How long will it take for the levels in the compartments to drop back to no more than half their values at 400 days? Reproduce Atlas Plate `Compartment Model A`.

**(b)** Assuming that all compartments contain no lead initially, find the minimum time required for all compartments to attain 90% of their equilibrium levels.

**(c)** Certain drugs alleviate the effects of lead poisoning by increasing the rate of removal of lead from the bones back into the bloodstream. The rate constant for that removal is given as 0.000035 in the third rate equation in system (2). What should that rate constant be if the goal is to cut in half within a year the amount of lead in the bones?

* **3. (Washout Theorem)** It can be shown that if compartment $k$ is open to the environment and if $F = 0$, then for any initial data, $x_k(t) \to 0$ as $t \to \infty$ (i.e., the substance being tracked eventually "washes out" of compartment $k$). What happens when $F$ is a constant vector? (Hint: Use (4).) Use this result to explain Atlas Plate `Compartment Model A`. Increase the removal rate constant by an order of magnitude after 400 days and reproduce Atlas Plate `Compartment Model B`. Invent some other open systems, simulate their behavior, and show that they are consistent with theory.

**4. (Closed compartment systems)** If (3) is closed, and if $F \equiv 0$, then it can be shown that for any initial data $x(0) \geq 0$ there exists an equilibrium point $b \geq 0$ of the system such that $x(t) \to b$ as $t \to \infty$.

**(a)** Show that $\sum x_i(t) \equiv \sum x_i(0) = \sum b_i$, for all $t \geq 0$. Invent a closed compartment system for $n = 3$ and illustrate the result quoted above. Estimate the values of the coordinates of the equilibrium point. Graph the components of the solutions of your system for a long enough time that the components have essentially reached equilibrium.

**(b)** Show that the lead system modeled by the ODEs (2) and (5) is closed. Graph the four components for $0 \leq t \leq 800$ days. Repeat, but with $I_L = 0$ for $t > 400$, and use your graphs to estimate the equilibrium levels of lead in each compartment.

# Lorenz System: Sensitivity

***Purpose*** To show that changes in a parameter in a rate function of the nonlinear Lorenz system induce instability, chaotic wandering, period doubling, and bifurcations.

***Keywords*** Lorenz system, strange attractor, chaotic wandering, period doubling, Hopf bifurcation

***See also*** Chapter 6 Introduction; Experiments 2.3, 5.17; Chapter 6 Notes; Atlas Plates `Lorenz A-C`

***Background*** In 1963 the meteorologist and mathematician E. N. Lorenz published numerical studies of the solutions of a simplified model for atmospheric turbulence in a vertical planar air cell beneath a thunderhead. The model equations of the Lorenz system are

$$\begin{aligned} x' &= -ax + ay \\ y' &= rx - y - xz \\ z' &= -bz + xy \end{aligned} \tag{1}$$

where $a$, $r$, and $b$ are positive constants denoting certain dimensionless physical parameters, $x$ is the amplitude of the convective air currents in the cell, $y$ is the temperature difference between the rising and falling currents, and $z$ is the deviation from normal of the temperature in the cell. See Figure 6.6.1 for a typical plot of solutions $x(t)$, $y(t)$, $z(t)$ of the system (1) versus $t$. Lorenz observed strange, almost random behavior in the computed orbits, small changes in the initial conditions leading to marked and (apparently) unpredictable changes in the orbits (see Atlas Plate `Lorenz C`). Mathematical theories and experimental evidence to explain and support the theoretical results for (1) and similar systems have developed rapidly, but much still remains a mystery. Computer graphics is at the heart of many of the studies, providing visual demonstrations of the unexpected twists and turns of the orbits.

The Lorenz system has the properties summarized in the following observations (recall that $a$, $b$, $r$ are positive constants).

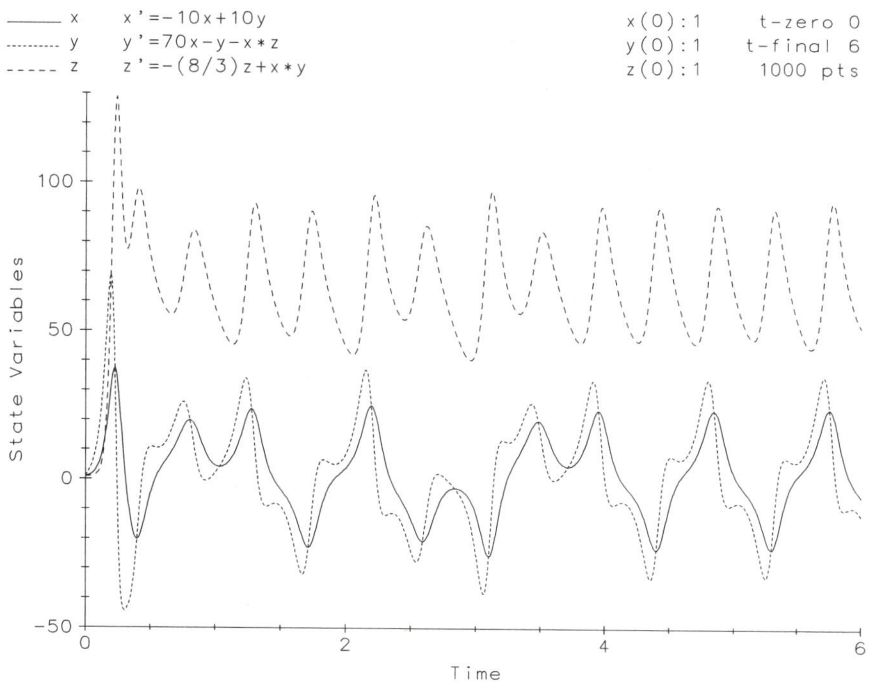


*Figure 6.6.1 Component plots of a Lorenz orbit.*

***Observation*** The assertions below are mathematical theorems about the orbits of the Lorenz system.

- **Orbital symmetry.** If $(x(t), y(t), z(t))$ is an orbit, so is $(-x(t), -y(t), z(t))$. See Figure 6.6.2.
- **Bifurcation of equilibrium points.** For $0 < r \leq 1$, the origin $O$ is the equilibrium point. As $r$ increases through 1, the equilibrium point at the origin bifurcates into three equilibrium points, $O$, $P_1(c, c, r-1)$ and $P_2(-c, -c, r-1)$, where $c = (br - b)^{1/2}$.
- **Bounded orbits.** Every orbit remains bounded as $t$ increases.
- **Dissipative property.** Let $V(R)$ be the volume of a bounded region $R$ of $xyz$-space. Let $R_t$ be the region obtained by following each point of $R$ forward $t$ units of time along the orbit through the point. Then $\lim_{t\to\infty} V(R_t) = 0$.
- **Stability.** The origin $O$ is asymptotically stable if $0 < r < 1$ but is unstable if $r > 1$. $P_1$ and $P_2$ are asymptotically stable for $1 < r < r_{cr} \equiv a(a+b+3)/(a-b-1)$ and unstable for $r > r_{cr}$; $r_{cr}$ is the **critical value**. In Figures 6.6.2–6.6.3, $a = 10$ and $b = 8/3$; hence $r_{cr} = 470/19 \approx 24.74$.
- **Hopf bifurcation.** As $r$ increases through $r_{cr}$, a subcritical Hopf bifurcation at $P_1$ occurs for fixed positive $a$ and $b$, with $P_1$ "swallowing" an unstable periodic orbit and destabilizing in the process. $P_2$ has the same property.

***Observation*** For the properties below, there is strong computational evidence, but as yet, no proof.

- **Sensitivity.** For most values of $r > r_{cr}$, small changes in the initial data cause large changes in the corresponding orbits. See Atlas Plate `Lorenz C`.
- **Strange attractor.** For many values of $r$, the Lorenz system (1) appears to possess a strange attractor $A$ (Figure 6.6.3), which is a bounded collection of orbits containing periodic orbits of arbitrarily large period, uncountably many nonperiodic orbits, and an orbit that is dense in $A$ (i.e., an orbit that passes arbitrarily close to every point of $A$). Every nearby orbit tends to $A$ as $t \to \infty$.
- **Period doubling.** System (1) has periodic orbits that slightly change their position in state space and double their periods as $r$ decreases through a sequence of positive values $r_1 > r_2 > \ldots$, a **period doubling cascade**. See Figures 6.7.1–6.7.3. It is thought that the Lorenz system possesses a strange attractor at $r = r^*$, where $r^* = \lim_{n\to\infty} r_n$.
- **Chaotic wandering.** For $r > r_{cr}$ most orbits seem to exhibit chaotic wandering from a neighborhood about $P_1$ to another about $P_2$ and back (see Atlas Plates `Lorenz A, B`).

***Observation*** Lorenz system (1) has other remarkable properties such as **homoclinic bifurcations** (homoclinic orbits turning into periodic orbits as $r$ changes), but these are not treated here. The two experiments focus mainly on computational aspects.

***References*** Consult the material below for additional background and details.

1. Perko, L. *Differential Equations and Dynamical Systems.* Springer-Verlag, New York, 1991.
2. Sparrow, C. *The Lorenz Equation: Bifurcations, Chaos, and Strange Attractors*, Vol. 41 of *Applied Mathematical Sciences*. Springer-Verlag, New York, 1982.

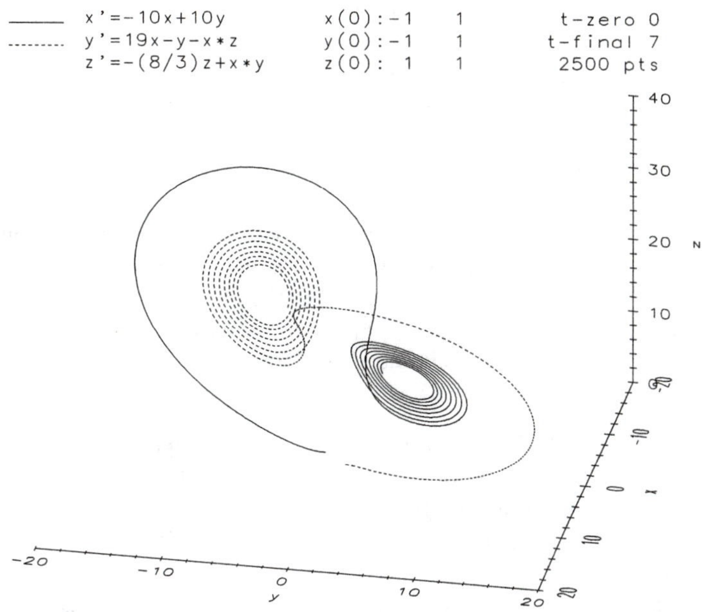


*Figure 6.6.2 Symmetric orbits.*

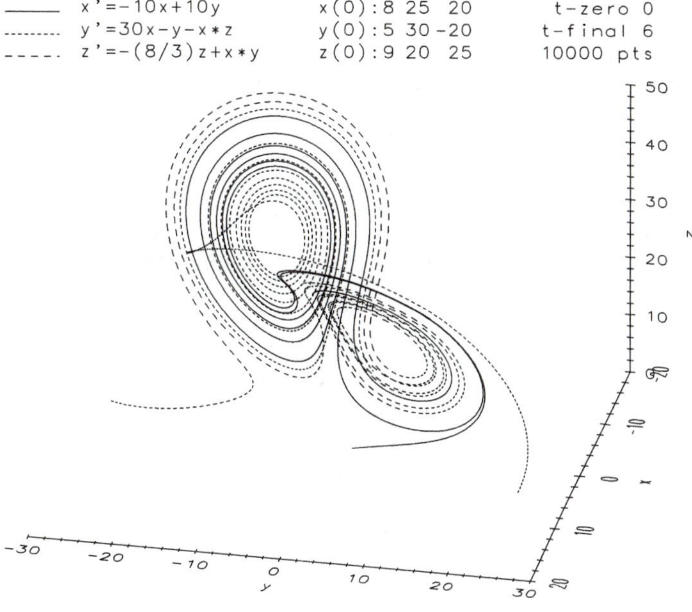


*Figure 6.6.3 Orbits tend to a folded orbit set, believed to be a strange attractor.*

_______________/_______________ hardware / software

Answer questions in the space provided, or on attached sheets or carefully labeled graphs.

_______________/_______________ name / date

_______________/_______________ course / section

## 6.6 *Inducing Chaos*

***Abstract*** As the parameter $r$ in the Lorenz system of ODEs increases through a sequence of positive values, the stable equilibrium point $O$ at the origin destabilizes and emits a pair of stable equilibrium points $P_1$ and $P_2$ in a pitchfork bifurcation (Experiment 2.3). $P_1$ and $P_2$ then destabilize and absorb unstable periodic orbits, and chaotic orbits appear and wander back and forth between neighborhoods of $P_1$ and $P_2$. These orbits display marked sensitivity to small changes in initial data.

***Special instructions*** Unless noted otherwise, $a = 10$ and $b = 8/3$ in the Lorenz system (1). Computer graphs may be 3D $xyz$-orbit plots, planar orbital projections, or component plots, whichever are clearest on the user's computer system. $J(P)$ denotes the Jacobian matrix of the rate functions of (1) evaluated at $P$ (see Chapter 6 Notes). Problem 5 must be completed before Problem 6.

**1.** Duplicate (as far as possible) Figures 6.6.1–6.6.3 and Atlas Plates `Lorenz A-C`. Which figures illustrate sensitivity to changes in initial data and which illustrate chaotic wandering? Explain.

**2.** $(0 < r < 1)$ For each of two values of $r$ in this range, plot several orbits over a time span long enough that they can be seen to approach $O$ as $t$ increases.

**3.** $(1 < r < 1.34)$ For a value of $r$ in this range, plot several orbits over a time span sufficiently long that $O$ is seen to be unstable and $P_1$ and $P_2$ asymptotically stable. Why do the graphs suggest that the eigenvalues of $J(P_1)$ and $J(P_2)$ are negative real numbers?

**4.** $(2 < r < 24.74)$ For each of three values of $r$ in this range, plot several orbits that suggest that orbits near $P_i$, $i = 1, 2$, spiral toward $P_i$ as time increases. What does this imply about the eigenvalues of $J(P_i)$? Try to find wandering orbits that move back and forth between neighborhoods of $P_1$ and $P_2$.

**5.** **(Chaotic wandering,** $r > 24.74$) For each of several values of $r$, plot orbits that suggest the instability of $O$, $P_1$, and $P_2$ and show chaotic wandering back and forth between neighborhoods of $P_1$ and $P_2$. Choose initial points that are close together.

**6.** **(Symbol strings)** Label each orbit found in Problem 5 with a symbol string of 1's and 2's according to whether the orbit loops "once around" $P_1$ or $P_2$. For example, 111221 represents the behavior of an orbital segment that loops three times around $P_1$, then loops twice around $P_2$, and returns for one loop around $P_1$.

## 6.7 *Search for Cycles*

***Abstract*** The Lorenz system appears to have several period doubling cascades, but they are hard to find. Some of the stable cycles in two of these cascades are computed and displayed. The values for $r$ in the cascades of Problems 1 and 2 were found by Sparrow (See References).

***Special instructions*** Read Problem 6 in Experiment 6.6 before starting this experiment. In all the problems $a = 10$ and $b = 8/3$ in the Lorenz system (1).

**1.** There is a period doubling cascade whose first three values are $r_1 = 260$, $r_2 = 222$, $r_3 = 216.2$. Duplicate Figures 6.7.1–6.7.3, which show the $xz$-projections of attracting periodic orbits for these three values of $r$. Make $tx$-component graphs, determine the periods, and verify graphically that period doubling occurs. Write the symbol string over the period of each cycle (e.g., the symbol string for the first periodic orbit is 1212...).

* 

**2.** Graph the first three cycles in the period doubling cascade $r_1 = 160$, $r_2 = 148.5$, $r_3 = 147.5$. What are the periods? Describe what happens at $r = 130$.

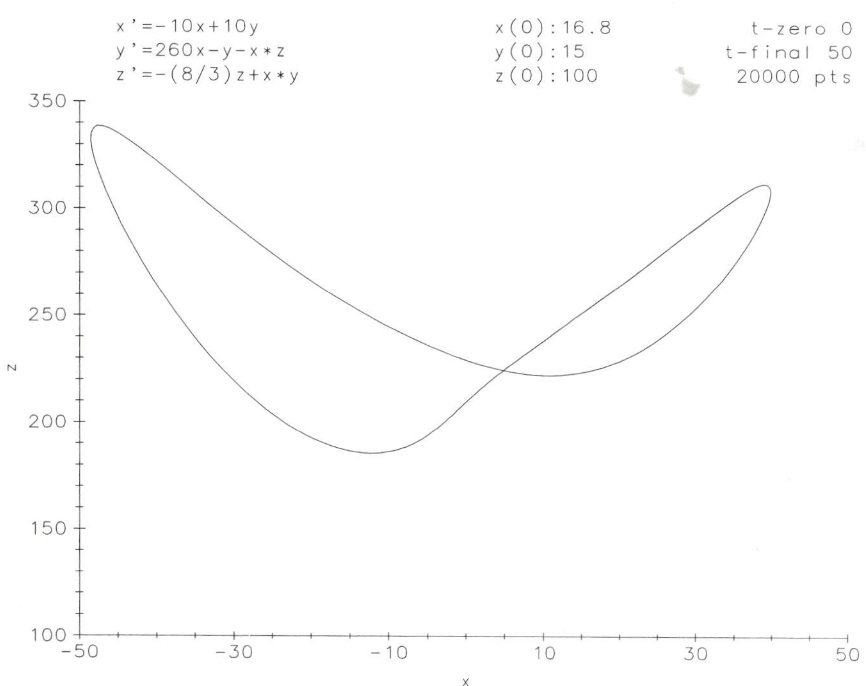


*Figure 6.7.1 Projection of an attracting periodic orbit:* $r_1 = 260$.

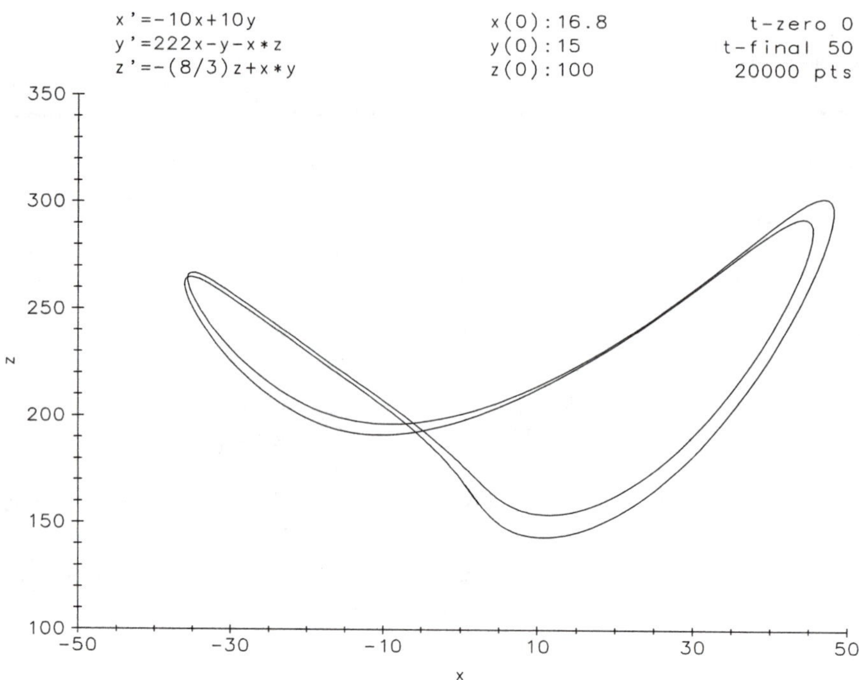


*Figure 6.7.2 Period has doubled:* $r_2 = 222$.

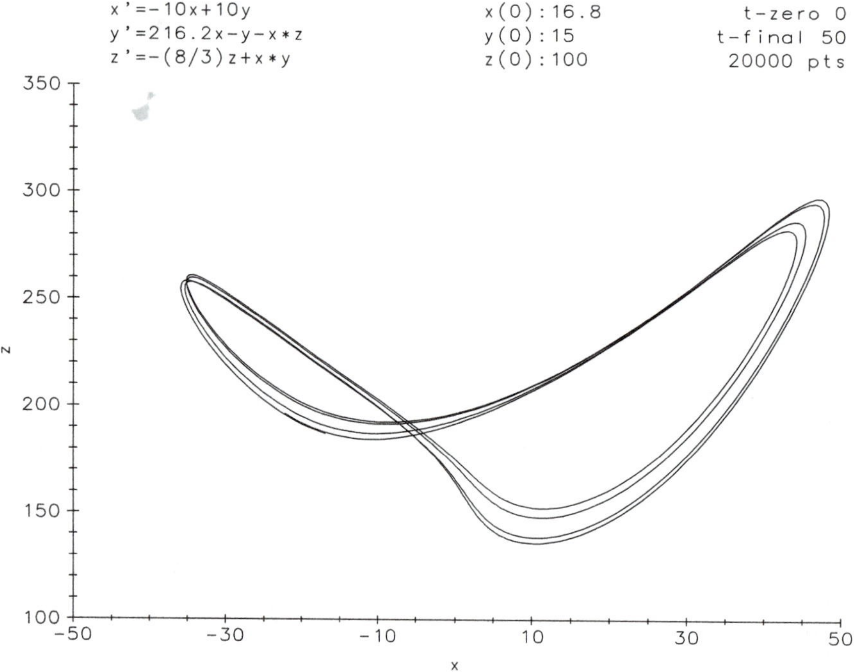


*Figure 6.7.3 Period has doubled again:* $r_3 = 216.2$.

# Rössler System: Period-doubling

***Purpose*** To explore the Rössler system as a parameter changes, to study the folded band type of chaotic wandering, to observe period doubling and approach to a strange attractor.

***Keywords*** Folded band, period-doubling, sensitivity, chaotic wandering

***See also*** Experiments 6.2, 6.6–6.7; Chapter 6 Notes; Atlas Plates `Rössler A-C`

***Background*** The Rössler system is

$$\begin{aligned} x' &= -y - z \\ y' &= x + ay \\ z' &= b - z(c - x) \end{aligned} \tag{1}$$

where $a$, $b$, and $c$ are positive parameters. O.E. Rössler wanted to build the "simplest possible" system that would model the phenomena displayed by the somewhat complex Lorenz system. System (1), now named after the inventor, is the result. In the parameter region of interest, this "model of a model" has two equilibrium points and one nonlinear term rather than the three equilibrium points and two nonlinear terms of the Lorenz system. As in the Lorenz system, there appears to be a "folded" strange attractor, spiraling, chaotic wandering, and period doubling. Figure 6.8.1 shows the $xz$-projections of the first three periodic orbits of a period doubling sequence for $a \equiv 0.300, 0.350, 0.375$. There are other such sequences, some apparently terminating with the creation of a strange attractor.

***References*** Consult the material below for additional background and details.

1. Thompson, J.M.T., Stewart, H.B. *Nonlinear Dynamics and Chaos*. John Wiley & Sons, New York, 1986.

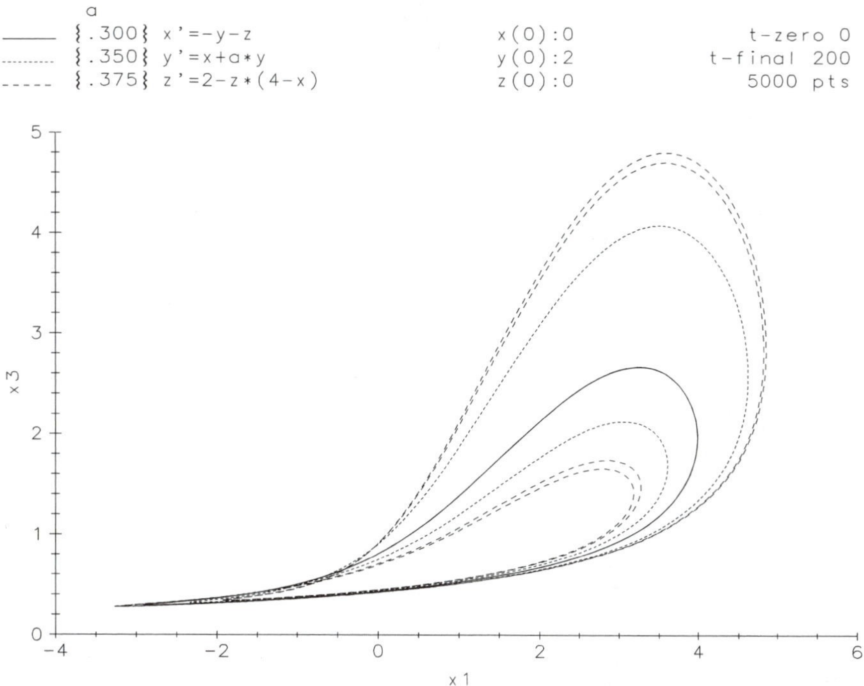


*Figure 6.8.1 A period doubling sequence for the Rössler system (plotted for* $175 \le t \le 200$*).*

_______________/_______________
hardware software

_______________/_______________
name date

Answer questions in the space provided,
or on attached sheets or carefully labeled graphs.

_______________/_______________
course section

## 6.8 *Rössler System: Sensitivity*

***Abstract*** Period doubling and chaotic wandering toward an apparent strange attractor are studied for the orbits of the Rössler system.

***Special instructions*** This experiment is designed as a team project. Consult Appendix A before writing the team report.

---

The orbits and component curves of the Rössler system are to be studied as one of the three parameters is varied. For most of the study, fix the parameters $b$ and $c$ at respective values 2 and 4, and vary $a$. Look for period doubling sequences, perhaps interspersed with values of $a$ for which there are orbits of other periods. Check for chaotic spiraling, and especially for the folded strange attractor. If possible, plot orbits in $xyz$-space from different points of view. Plot orbital projections on various component planes and use component plots to "see" periodic orbits. If possible, plot only over an end segment of the time interval for which the system is actually solved, (e.g., if the system is solved for $0 \leq t \leq 1000$, plot for $900 \leq t \leq 1000$). This way any transient behavior is likely to have disappeared and the orbit will be essentially in the strange attractor. Some other suggestions:

- Duplicate the graphs of Figure 6.8.1 and Atlas Plates `Rössler A-C`.
- Plot orbits and component curves for the sequence of values of $a$= 0.3, 0.35, 0.375, 0.386, 0.3909, 0.398, 0.4, 0.411. Use $x_0 = 0$, $y_0 = 2$, $z_0 = 0$ and identify periodic orbits and their periods, period doubling, chaotic spiraling, and any other significant phenomena. What happens if different initial data are used?
- For arbitrary positive values of $a$, $b$, $c$, find the equilibrium points of the system. For $b = 2$, $c = 4$, and the values of $a$ given above, study orbital behavior near each equilibrium point.
- Explain why, if $|z|$ is very small, the subsystem $x' = -y$, $y' = x + ay$ has an unstable focus at the origin if $0 < a < 2$. Explain why this leads to a spreading apart of nearby orbits (a central part of chaotic wandering). This spreading is not unbounded, however. Explain from the third equation of the Rössler system how if $x < c$, then the coefficient of $z$ is negative and the $z$-system stabilizes near $b/(c - x)$. If $x > c$, the $z$-system "diverges" (assuming $b > 0$). Now look at the $x$-system and explain why orbits alternately lie in a plane parallel to the $xy$-plane, then are thrown upward, and in turn are folded back and reinserted closer to the origin.
- Vary $b$ and $c$ as well as $a$ and analyze the effect on the orbits. Any period doubling, folding, chaotic wandering? Explain.

# The Rotational Stability of a Tennis Racket

***Purpose*** To examine the stability properties of steady rotations of a tennis racket about each of its principal axes.

***Keywords*** Lyapunov stability, Principal Axes Theorem, coordinate reference frames, orthogonal matrix, angular velocity, integrals of motion, time derivatives in moving and fixed frames, inertia tensor, Newton's Laws of Motion

***See also*** Experiments 5.12–5.13; Appendix B.2

***Background*** The axes of symmetry of a tennis racket are indicated in Figure 6.9.1 by the starred frame $\{\hat{\mathbf{i}}^*, \hat{\mathbf{j}}^*, \hat{\mathbf{k}}^*\}$ with the origin at its center of mass, $C$. Every tennis player has had the following experience when trying to spin the racket in the air about one of its axes of symmetry: The racket appears to spin about the $\hat{\mathbf{i}}^*$ and $\hat{\mathbf{k}}^*$ axes without wobbling, but try as one might, wobbling always occurs when the racket is spun about the $\hat{\mathbf{j}}^*$ axis. Try it and see. The same effect can be observed when spinning a book in the air (after the judicious use of a rubber band). This experiment will explain why bodies behave this way when rotating.

The equations of motion of a rigid body are derived in Appendix B.2, and the reader is urged to read that derivation before proceeding. Those equations are

$$\begin{aligned} M\frac{d^2\mathbf{R}}{dt^2} &= \mathbf{F} \\ \frac{d\mathbf{L}_C}{dt} &= \mathbf{N}_C \end{aligned} \tag{1}$$

where $M$ is the mass of the body, $\mathbf{R}(t)$ the position vector (in a fixed frame) of $C$, the body's center of mass, $\mathbf{F}$ the resultant external force acting on the body, $\mathbf{L}_C = \mathcal{I}\boldsymbol{\omega}$ the angular momentum of the body about $C$, $\mathcal{I}$ the inertia tensor of the body, $\boldsymbol{\omega}$ the angular velocity vector of a frame fixed in the body with respect to an inertial frame, and $\mathbf{N}_C$ the external torque with respect to $C$. The vector $\mathcal{I}\boldsymbol{\omega}$ has a coordinate column vector in a frame $\{\hat{\mathbf{i}}, \hat{\mathbf{j}}, \hat{\mathbf{k}}\}$ given by the matrix product $I\omega$, where $\omega$ is the coordinate vector of $\boldsymbol{\omega}$ in the frame and $I$ is the $3\times3$ matrix

$$I = \int_B (x^T x E - x x^T)\rho\, dV \tag{2}$$

where $\rho$ is the density function for the body $B$, $[x_1, x_2, x_3]^T$ is the coordinate column vector in the given frame, and $E$ is the $3 \times 3$ identity matrix.

To be specific, we consider the motion of a tennis racket when thrown into the air, assuming that the gravitational field is constant (hence there is no torque due to gravity).

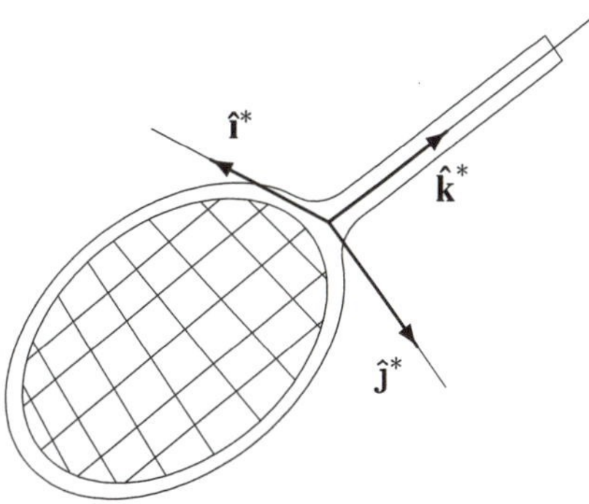


*Figure 6.9.1 Principal axes of a tennis racket.*

Since $\mathbf{N}_C$ vanishes, it is clear that the equations of motion of the center of mass and of the racket orientation decouple. Since the motion of the center of mass does not interest us, we shall consider only the rotational motion.

First, we find a suitable coordinate system to express the rotational motion. Since $\mathbf{L}_C = \mathcal{I}\omega$, it is natural to ask whether there exists a coordinate system in which the angular momentum vector $\mathbf{L}_C$ is somehow simplest. Now the inertia tensor $\mathcal{I}$ is easily seen to have the property that its representation $I$ is such that $I^T = I$. Tensors with this property are caled **symmetric tensors**. But $\mathcal{I}$ also has the property that the representation $I$ is **positive definite**, that is, $y^T I y > 0$ for all vectors $y \neq 0$. Now Appendix B.2 shows that if $I$ and $I^*$ are the representations of the inertia tensor $\mathcal{I}$ given by (2) in the frames $\{\hat{\mathbf{i}}, \hat{\mathbf{j}}, \hat{\mathbf{k}}\}$ and $\{\hat{\mathbf{i}}^*, \hat{\mathbf{j}}^*, \hat{\mathbf{k}}^*\}$, respectively, then

$$I^* = UIU^T \quad \text{where} \begin{bmatrix} \hat{\mathbf{i}}^* \\ \hat{\mathbf{j}}^* \\ \hat{\mathbf{k}}^* \end{bmatrix} = U \begin{bmatrix} \hat{\mathbf{i}} \\ \hat{\mathbf{j}} \\ \hat{\mathbf{k}} \end{bmatrix} \tag{3}$$

Thus the representation $I$ of the inertia tensor $\mathcal{I}$ is known in any frame if it is known in even one frame. The orthogonal matrix $U$ in (3) relating the starred and unstarred frames is the key to this claim. A basic result in linear algebra is the

**Principal Axes Theorem:** *Let $A$ be a $3 \times 3$ symmetric matrix. Then the eigenvalues $\lambda_1$, $\lambda_2$, $\lambda_3$ of $A$ are real and there are eigenvectors of $A$ that form an orthogonal coordinate frame. Moreover, there exists an orthogonal matrix $U$ such that*

$$UAU^T = \begin{bmatrix} \lambda_1 & 0 & 0 \\ 0 & \lambda_2 & 0 \\ 0 & 0 & \lambda_3 \end{bmatrix} \tag{4}$$

*If $A$ is positive definite, then $\lambda_1 > 0$, $\lambda_2 > 0$, $\lambda_3 > 0$. (The rows of $U$ are the eigenvectors of $A$. The orthogonal matrix $U$ is not unique, but the numbers $\lambda_1$, $\lambda_2$, $\lambda_3$ always appear on the diagonal in some order.)*

Now observe that the representation $I$ of the inertia tensor $\mathcal{I}$ in the inertial frame $\{\hat{\mathbf{i}}, \hat{\mathbf{j}}, \hat{\mathbf{k}}\}$ is positive definite. Thus by (3), (4) the representation $I^*$ of the inertia tensor $\mathcal{I}$ is given by

$$I^* = \begin{bmatrix} I_1 & 0 & 0 \\ 0 & I_2 & 0 \\ 0 & 0 & I_3 \end{bmatrix}$$

in the frame

$$\begin{bmatrix} \hat{\mathbf{i}}^* \\ \hat{\mathbf{j}}^* \\ \hat{\mathbf{k}}^* \end{bmatrix} = U \begin{bmatrix} \hat{\mathbf{i}} \\ \hat{\mathbf{j}} \\ \hat{\mathbf{k}} \end{bmatrix}$$

where $I_1 > 0$, $I_2 > 0$, $I_3 > 0$. The frame $\{\hat{\mathbf{i}}^*, \hat{\mathbf{j}}^*, \hat{\mathbf{k}}^*\}$ attached to the center of mass determines the **principal axes** of the body $B$, and $I_1$, $I_2$, $I_3$ are the **principal inertias** about the principal axes determined by $\hat{\mathbf{i}}^*$, $\hat{\mathbf{j}}^*$, and $\hat{\mathbf{k}}^*$, respectively. Without loss of generality, we may assume that the frame vectors $\{\hat{\mathbf{i}}^*, \hat{\mathbf{j}}^*, \hat{\mathbf{k}}^*\}$ have been so ordered that the starred frame is right-handed and $I_1 \geq I_2 \geq I_3$.

Returning to the tennis racket now, it can be shown that the principal axes are just the axes of symmetry, and the labeling of the starred frame in Figure 6.9.1 has anticipated this fact. Notice also that this "principal axes" frame is right-handed, and that $I_1 > I_2 > I_3$. Recall that no external torque is acting on the spun tennis racket, so $\mathbf{N}_C = 0$; hence from (1) $d\mathbf{L}_C/dt = 0$. Replacing $\mathbf{R}$ by $\mathbf{L}_C$ in (21) of B.2 we obtain

$$\frac{d\mathbf{L}_C}{dt} + \omega \times \mathbf{L}_C = 0 \tag{5}$$

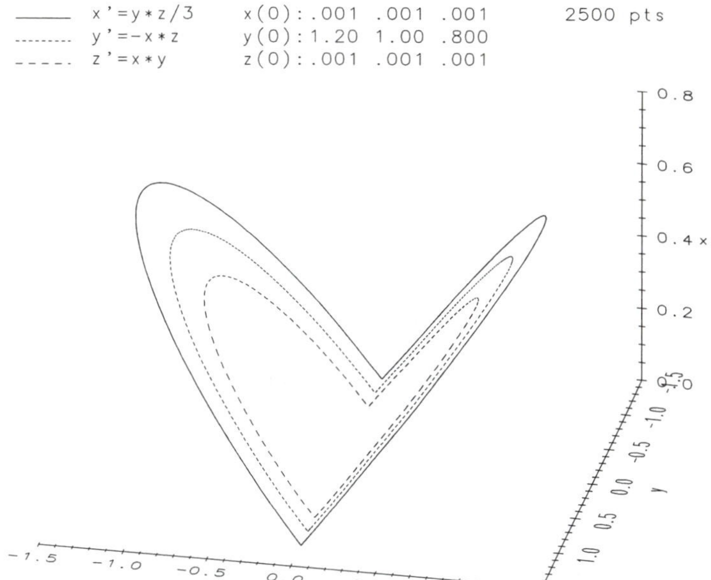


*Figure 6.9.2 Orbits of system (6) due to changing initial data near the equilibrium point* (0, 1, 0).

or expessing $\mathbf{L}_C$ and $\omega$ in the principal axes frame we have $\mathbf{L}_C = I_1\omega_1^*\hat{\mathbf{i}}^* + I_2\omega_2^*\hat{\mathbf{j}}^* + I_3\omega_3^*\hat{\mathbf{k}}^*$, and so (5) becomes

$$\begin{aligned} d\omega_1^*/dt &= I_1^{23}\omega_2^*\omega_3^* \\ d\omega_2^*/dt &= I_2^{31}\omega_1^*\omega_3^* \\ d\omega_3^*/dt &= I_3^{12}\omega_1^*\omega_2^* \end{aligned} \tag{6}$$

where we have used the notation $I_\alpha^{\beta\gamma} = (I_\beta - I_\gamma)/I_\alpha$ for $\alpha, \beta, \gamma = 1, 2, 3$. Observe that $I_1^{23} > 0$, $I_2^{31} < 0$, $I_3^{12} > 0$. The equilibrium points of (6) assume one of the following forms:

$$\omega^* = \begin{bmatrix} A \\ 0 \\ 0 \end{bmatrix} \quad \text{or} \quad \begin{bmatrix} 0 \\ B \\ 0 \end{bmatrix} \quad \text{or} \quad \begin{bmatrix} 0 \\ 0 \\ C \end{bmatrix}$$

where $A$, $B$, $C$ are arbitrary real numbers. Physically, the equilibrium points of (6) can be characterized as the totality of steady rotations about a principal axis. Figures 6.9.2–6.9.4 show motions of the system (6) due to perturbations about the (unstable) equilibrium point (0, 1, 0). Notice that the orbits are periodic.

To determine which steady rotations are stable, we restrict ourselves to $\omega^*$-space, regard these motions as equilibrium points in this space, and make use of Lyapunov's Stability Theorems. The ODEs are autonomous, and there exist two integrals:

$$\phi_1(\omega^*) \equiv (\omega_2^*)^2 - \frac{I_2^{31}}{I_3^{12}}(\omega_3^*)^2 \qquad \phi_2(\omega^*) \equiv (\omega_1^*)^2 - \frac{I_1^{23}}{I_2^{31}}(\omega_2^*)^2 \tag{7}$$

From these integrals we can construct other integrals that are Lyapunov functions for some of the equilibrium points. Consider, for example, the integral

$$F(\omega^*) \equiv \phi_1(\omega^*) + (\phi_2(\omega^*) - A^2)^2 \tag{8}$$

where $A$ is some real number. Further, let $\mathcal{N}$ be any neighborhood of the point $(A, 0, 0)$ that does not contain the point $(-A, 0, 0)$. Then we claim that $F$ is a weak Lyapunov function for the equilibrium point $(A, 0, 0)$ in the neighborhood $\mathcal{N}$. Hence all rotations about the axis of greatest inertia (i.e., the $\hat{\mathbf{i}}^*$-axis) are (neutrally) stable.

Similarly, consider the integral

$$G(\omega^*) \equiv \phi_2(\omega^*) + \left(\phi_1(\omega^*) + \frac{I_2^{31}}{I_3^{12}} C^2\right)^2 \tag{9}$$

where $C$ is some real number. Further, let $\mathcal{N}$ be any neighborhood of the point $(0, 0, C)$ that does not contain $(0, 0, -C)$. Then, $G$ can be shown to be a weak Lyapunov function for the equilibrium point $(0, 0, C)$ in the neighborhood $\mathcal{N}$. Hence, steady rotations about the axis of least inertia (i.e., the $\hat{\mathbf{k}}^*$-axis) are (neutrally) stable.

Now we show that all steady rotations about the remaining axis (the $\hat{\mathbf{j}}^*$-axis) are *unstable.* Consider the function

$$W(\omega^*) = \omega_1^* \omega_3^* \tag{10}$$

We claim that there exists a neighborhood, $\mathcal{N}$, about $(0, B, 0)$ with $B \neq 0$ such that $W$ satisfies the conditions of the Lyapunov Instability Theorem. Hence we conclude that these rotations are unstable as was asserted.

***Observation*** **(Integrals and orbits)** Note that if $\omega^*(t)$ is any solution of system (6) then $\phi_1(\omega^*(t)) \equiv$ constant and $\phi_2(\omega^*(t)) \equiv$ constant, where $\phi_1$ and $\phi_2$ are the integrals defined in (7). Note that the level surfaces $\phi_1(\omega^*) =$ constant, and $\phi_2(\omega^*) =$ constant, are elliptical cylinders whose axes are the $\omega_1^*$-axis and the $\omega_3^*$-axis respectively. Notice that the orbit traced out by any solution $\omega^*(t)$ must lie simultaneously on the two level surfaces $\phi_1(\omega^*) =$ constant and $\phi_2(\omega^*) =$ constant in $\omega^*$-space, hence on the intersection of the two cylinders. Alternatively, the curves of intersection of level surfaces define an orbit (or orbits) of system (6). Figure 6.9.2 shows some orbits that lie on the intersections of these cylinders.

***Observation*** **(Fixed axis rotations)** If $\{\hat{\mathbf{i}}, \hat{\mathbf{j}}, \hat{\mathbf{k}}\}$ is a fixed frame and $\{\hat{\mathbf{i}}^*, \hat{\mathbf{j}}^*, \hat{\mathbf{k}}^*\}$ is a moving frame (think of the frames as "attached" to the same point), then the formula (see (21) in Appendix B.2)

$$\frac{d\mathbf{R}}{dt} = \frac{d^*\mathbf{R}}{dt} + \boldsymbol{\omega} \times \mathbf{R} \tag{11}$$

relates the time derivatives of the position vector $\mathbf{R}$ in these two frames. Equation (11) can be used to show that if a body is rotating about the fixed axis $L$ through the center of mass, then $\boldsymbol{\omega}$ is directed along $L$ and $\|\boldsymbol{\omega}\| = \theta'(t)$, where the angle $\theta$ is measured from a fixed reference line in a plane orthogonal to $L$.

***References*** Consult the material below for additional background and details.

1. Marion, J.B., Thornton, S.T. *Classical Dynamics of Particles and Systems*, 3rd ed. Harcourt Brace Jovanovich, New York, 1988.

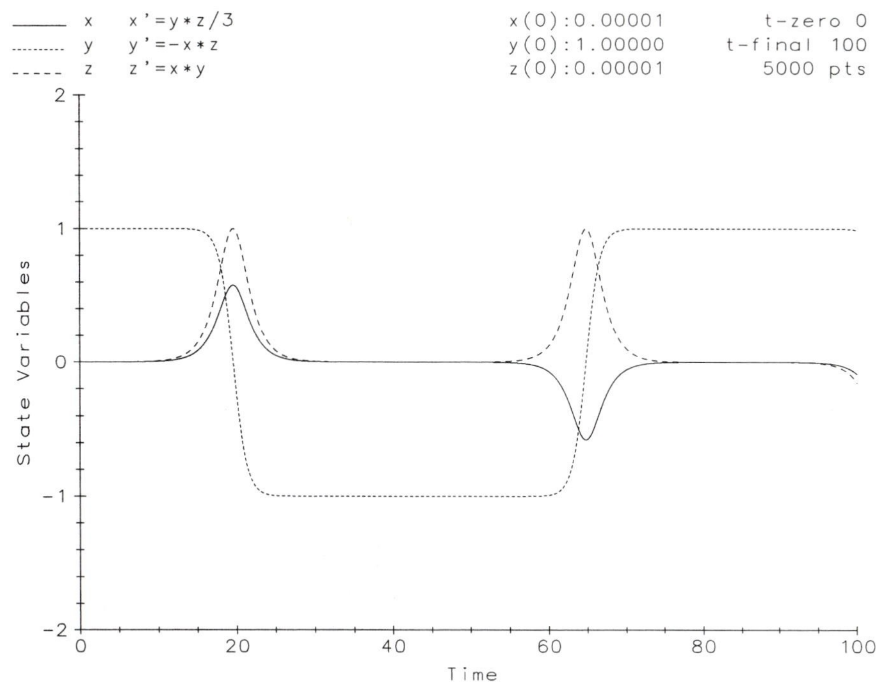


*Figure 6.9.3 Component graphs of an orbit of system (6) with initial data near* (0, 1, 0).

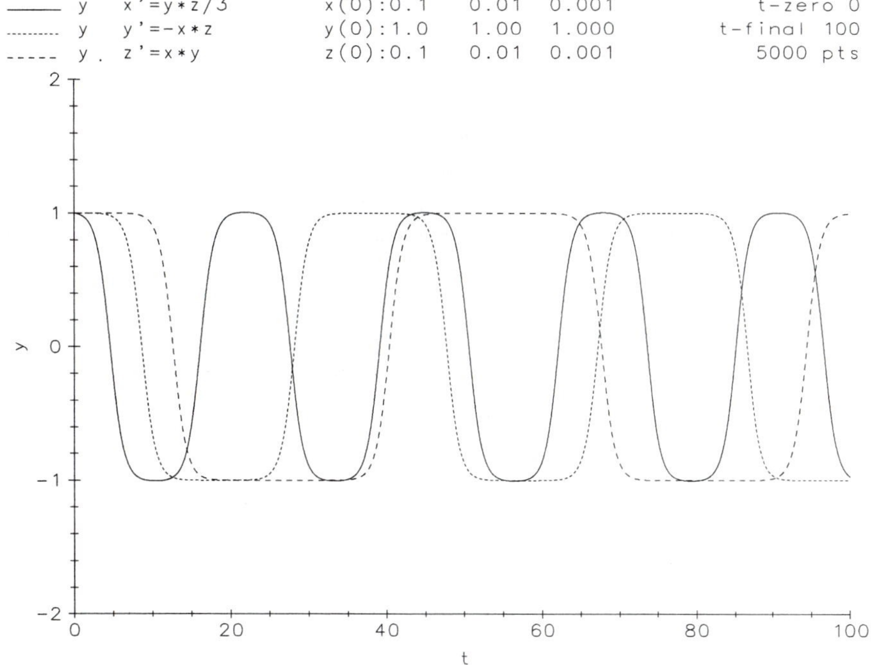


*Figure 6.9.4 The y-components of orbits of system (6) with initial data near* (0, 1, 0).

_______________ / _______________
hardware software

_______________ / _______________
name date

Answer questions in the space provided,
or on attached sheets or carefully labeled graphs.

_______________ / _______________
course section

## 6.9 *The Rotational Stability of a Tennis Racket*

***Abstract*** The stability of steady rotations of a tennis racket about a principal axis are examined both computationally and theoretically. The goal is to show that steady rotations about the principal axes of greatest and least inertia are stable, while rotations about the remaining principal axis are unstable.

***Special instructions*** Read Appendix B.2 before embarking on this experiment, especially the section on the motion of rigid bodies. Follow the instructions in Appendix A for writing the laboratory report.

Supply proofs or plausibility arguments (whichever is appropriate) for assertions in the cover sheet that lack supporting arguments. Use computer simulation to illustrate the stability properties of steady rotations. Some suggestions:

- Prove the assertion in the last observation of the cover sheet concerning the angular velocity vector $\boldsymbol{\omega}$ for rotations about a fixed axis.
- Prove that the inertia tensor $\mathcal{I}$ is symmetric and positive definite.
- Use a plausibility argument to establish that the principal inertias $I_1$, $I_2$, and $I_3$ about the principal axes $\hat{\mathbf{i}}^*, \hat{\mathbf{j}}^*, \hat{\mathbf{k}}^*$ in Figure 6.9.1 are related by $I_1 > I_2 > I_3$.
- Show that $\phi_1(\omega^*)$ and $\phi_2(\omega^*)$ in (7) are integrals of the system (6).
- Show that $\phi_3(\omega^*) \equiv I_1(\omega_1^*)^2 + I_2(\omega_2^*)^2 + I_3(\omega_3^*)^2$ is an integral for the system (6). What are the level surfaces of this integral?

Show (or demonstrate) the stability/instability properties of steady rotations about a principal axis in some (or all) of the ways indicated below.

- **Geometrical approach**. The level surfaces $\phi_1$ =const. and $\phi_2$ =constant are cylinders in $\omega$-space. Show that the intersections of these cylinders define orbits of system (6). Use this fact to show that steady rotations about the $\omega_2^*$-axis are unstable. (Hint: The intersections should resemble Figure 6.9.2.)
- **Geometrical approach**. Use the integral $\phi_3(\omega^*)$ in turn with the integrals $\phi_1(\omega^*)$ and $\phi_2(\omega^*)$.
- **Simulation**. Use "slight" perturbations of the equilibrium solutions of (6) as initial data, and solve the system (6) for $\boldsymbol{\omega}^*(t)$. Use the component map plots of $\omega_n(t)$ versus $t$ to verify the assertions made about the stability of steady rotations about a principal axis.
- **Linearization, eigenvalues**. Linearize system (6) about an equilibrium point and compute the eigenvalues of the coefficient matrix.
- **Lyapunov approach**. Prove that $F(\omega^*)$ and $G(\omega^*)$ in (8) and (9) are weak Lyapunov functions. Show that $W(\omega^*)$ in (10) is a Lyapunov instability function.
- **Experimental approach**. Try spinning other objects in the air, for example, a book (after judicious use of a rubber band), and describe your observations about the stability of steady rotations. Draw diagrams, if helpful.

# Nonlinear Systems and Chemical Reactions

***Purpose*** To construct the nonlinear systems of ODEs that model multispecies chemical reactions and to observe how the approach to steady state depends on the rate constants.

***Keywords*** Chemical reactions, steady state, nonlinear systems, Law of Mass Action, rate constants

***See also*** Experiments 3.16–3.17; Appendix B.3

***Background*** In a chemical reaction, molecules of various species (the reactants) interact and generate other species (the products). If the reaction takes place in a constant volume reactor, amounts are specified by concentrations, where $[X(t)]$ denotes the concentration of species $X$ at time $t$. If the reactions are elementary, a chemical diagram such as

$$A + B \to C + D$$

means that one molecule of species $A$ reacts with one of $B$ to create one molecule of product $C$ and one of $D$, $[A]$ and $[B]$ each diminishing at the same rate that $[C]$ and $[D]$ increase (a kind of conservation law). By the **Law of Mass Action** (valid for elementary reactions), the production rate of species $C$ is proportional to the product of the concentrations of species $A$ and $B$,

$$\frac{d[C(t)]}{dt} = k_1[A][B]$$

where $k_1$ is a positive rate constant.

In a more complex reaction, several elementary reactions may occur simultaneously. An example is the triple reaction,

$$D + O \underset{k_{-1}}{\overset{k_1}{\rightleftharpoons}} C \overset{k_2}{\rightarrow} N + R \tag{1}$$

Species $C$ is the product of an elementary bimolecular reaction between $D$ and $O$. $C$ also breaks down into $D$ and $O$ in a reverse reaction and decomposes into species $N$ and $R$ in a third reaction. The rates of decay of $C$ in the second and third reaction are assumed to be elementary. For example, $[C]'$ is proportional to $[C]$ in each reaction, and $k_{-1}$ and $k_2$ are the respective rate constants of proportionality. These considerations and the basic

**Balance Law:** *Net rate = rate of creation − rate of destruction*

may be used to write out the nonlinear system of first order ODEs that models the triple reaction (1), one ODE for each concentration:

$$\begin{aligned}
[D]' &= -k_1[D][O] + k_{-1}[C] \\
[O]' &= -k_1[D][O] + k_{-1}[C] \\
[C]' &= k_1[D][O] - k_{-1}[C] - k_2[C] \\
[N]' &= k_2[C] \\
[R]' &= k_2[C]
\end{aligned} \tag{2}$$

Once the three rate constants and the initial concentrations of the five species have been given, computer solvers and graphics may be used to display the five solution curves that model the concentrations. From (1) one might expect $[C(t)]$ and $[D(t)]$ (or $[O(t)]$) to tend to zero as time increases. At these zero steady states the five rates in (2) vanish. The steady state values for the concentrations of the remaining species, $N$, $R$ and $O$ (or $D$), depend upon the initial concentrations of all five species.

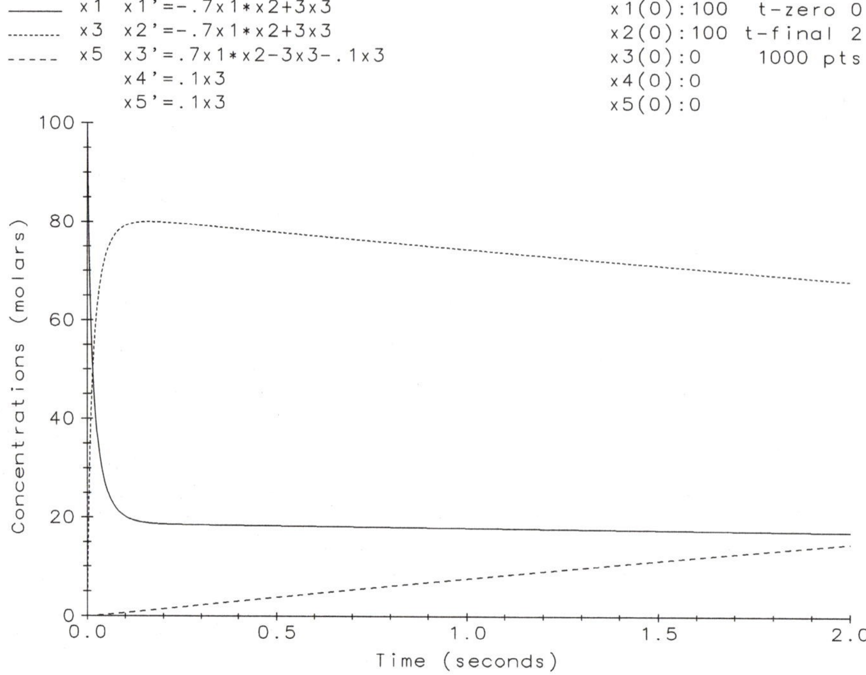


*Figure 6.10.1 Changing concentrations (Example 1).*

***Example 1*** **(Data sets)** A set of data that can be used in solving system (2) is $k_1 = 0.7\ (\mathrm{M}\cdot\mathrm{s})^{-1}$, $k_{-1} = 3\ (\mathrm{s})^{-1}$, $k_2 = 0.1\ (\mathrm{s})^{-1}$, $[D] = [O] = 100$ M and $[C] = [N] = [R] = 0$ at $t = 0$, where M denotes molarity. The solution interval is $0 \le t \le 2$ seconds. Since the rates and initial values for species $D$ and $O$ coincide (and, likewise for species $N$ and $R$) we have that $[O(t)] \equiv [D(t)]$ and $[N(t)] \equiv [R(t)]$. In Figure 6.10.1, the concentrations of $D$, $O$, $C$, $N$ and $R$ are respectively denoted by $x_1$, $x_2$, $x_3$, $x_4$, and $x_5$, but only $x_1$, $x_3$, and $x_5$ are plotted, since $x_2 = x_1$ and $x_4 = x_5$.

***Observation*** The linear dependence of the rate functions in (2) may be used to reduce the number of ODEs to be solved. For example, since

$$[O]' \equiv [D]', \quad [N]' \equiv [R]', \quad [D]' + [C]' + [R]' \equiv 0$$

an integration gives

$$\begin{aligned} [O(t)] - [O(0)] &\equiv [D(t)] - [D(0)] \\ [N(t)] - [N(0)] &\equiv [R(t)] - [R(0)] \\ [D(t)] + [C(t)] + [R(t)] &\equiv [D(0)] + [C(0)] + [R(0)] \end{aligned} \tag{3}$$

The first identity of (3) may be used to write the rate equations for $[D]$ and $[C]$ in (2) solely in terms of $[D]$ and $[C]$ and the initial values. Once the concentrations $[D(t)]$ and $[C(t)]$ have been computed, the concentrations of the other three species can be determined by using (3).

Finally observe that in the five-dimensional state space of concentrations, system (2) has infinitely many equilibrium points. These equilibria are generated by taking $[C] = 0$, $[D] = 0$ (or $[O] = 0$) with $[N]$ and $[R]$ arbitrary. For given initial values, the concentrations tend to the coordinates of one of these points as time increases. These limiting concentrations are the **steady states**.

_____/_____ hardware software

_____/_____ name date

Answer questions in the space provided, or on attached sheets or carefully labeled graphs.

_____/_____ course section

## 6.10 Approach to Equilibrium: Five Species

***Abstract*** The five-species chemical reaction given by (1) is studied and extended. The effects on the steady states of changing a rate constant are explored.

**1.** Use the data set from Example 1.

**(a)** What are the limiting values of the concentrations as $t$ increases? What is the largest value of $[C(t)]$? Explain why these values are independent of the constants $k_1$, $k_{-1}$, and $k_2$.

**(b)** Use computer solvers and graphics to solve (2), first for $0 \leq t \leq 1$ and then for $0 \leq t \leq 50$. Identify on the latter graphs the limiting values of the concentrations.

**(c)** Change $k_1$ to $0.07(\mathrm{M} \cdot \mathrm{s})^{-1}$ and the time interval to $0 \leq t \leq 100$. Solve (2) and identify on the graphs the limiting values of the concentrations.

**2.** Consider the reaction $D + O \underset{k_{-1}}{\overset{k_1}{\rightleftharpoons}} C \underset{k_{-2}}{\overset{k_2}{\rightleftharpoons}} N + R$.

**(a)** Write the five ODEs modeling the reaction. Do the steady state concentrations change if the rate constants are changed?

**(b)** Let $k_1 = 0.2(\mathrm{M}\cdot\mathrm{s})^{-1}$, $k_{-1} = 1(\mathrm{s})^{-1}$, $k_2 = 0.3(\mathrm{s})^{-1}$; $[D] = [O] = 100\mathrm{M}$ and $[C] = [N] = [R] = 0$ at $t = 0$. Plot solutions of the system for values of $k_{-2}$ = 0.09, 0.045, 0.9, 1.35, 1.9, and 5 $(\mathrm{M} \cdot \mathrm{s})^{-1}$. Solve over a sufficiently long time span that the values of the steady states can be determined—write these values on the graphs.

_______________ / _______________
hardware software

_______________ / _______________
name date

Answer questions in the space provided,
or on attached sheets or carefully labeled graphs.

_______________ / _______________
course section

## *6.11 Approach to Equilibrium: Four Species*

***Abstract*** A complex chemical reaction involving four species is modeled and the corresponding concentrations are computed and graphed.

***Special instructions*** For the problems below, use the chemical reaction given by:

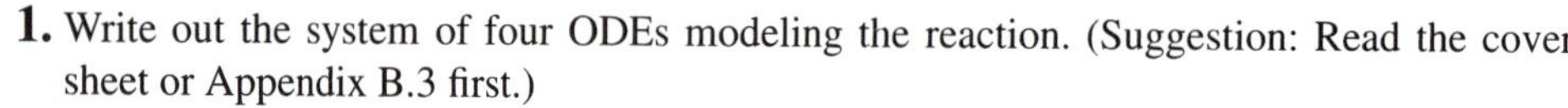

$$A + B \underset{k_{-1}}{\overset{k_1}{\rightleftharpoons}} C + D \qquad C + B \overset{k_2}{\rightarrow} D$$

Problem 1 must be done before Problems 2 and 3. For Problems 2 and 3 set $k_1 = 0.7(\text{M}\cdot\text{s})^{-1}$, $k_2 = 0.55(\text{M}\cdot\text{s})^{-1}$; $[A] = [B] = 10\text{M}$, $[C] = 8\text{M}$, $[D] = 0\text{M}$ at $t = 0$.

**1.** Write out the system of four ODEs modeling the reaction. (Suggestion: Read the cover sheet or Appendix B.3 first.)

**2.** Set $k_{-1} = 0$. Plot the four concentrations for $0 \le t \le 1$ seconds, $0 \le t \le 4$ seconds, and finally for $0 \le t \le 50$ seconds. What is the limiting value of $[C]$ as $t$ increases?

**3.** Let $k_{-1}$ be a positive constant. What are the effects on the limiting concentrations of the four species if $k_{-1}$ is varied through the values 0, 0.3, 0.6, 0.9, 1.2, 1.5 $(\text{M}\cdot\text{s})^{-1}$? Explain what you see. Write directly on the graphs.

# Oscillating Chemical Reactions

***Purpose*** To see how oscillations are created or annihilated by changing a parameter in the nonlinear system modeling an autocatalytic chemical reaction.

***Keywords*** Oscillation, Hopf bifurcation, nonlinear system, chemical reaction, autocatalysis, Belousov-Zhabotinskii reaction, dimensionless variables, Balance Law

***See also*** Experiments 1.11, 3.16–3.17, 5.17, 6.10–6.11; Appendices B.3, B.5; Atlas Plates `Autocatalator A-C`

***Background*** The concentrations of the species in most chemical reactions tend to characteristic equilibrium levels (steady states). After an initial swing or two, the approach to equilibrium is monotonic, each concentration rising (or falling) to its own equilibrium. However, some reactions pass through a striking intermediate phase during which concentrations suddenly exhibit large-amplitude oscillations. After a while these oscillations decay and the monotonic approach to equilibrium resumes. In Belousov-Zhabotinskii reactions (named for the discoverers), the oscillatory phase lasts so long that it is taken to be the (nonconstant) steady state of the reaction.

Various kinetic mechanisms for these unusual reactions have been proposed and tested. The two mechanisms taken up in these experiments are comparatively simple and meet the requirements of chemical consistency. The modeling systems of ODEs are nonlinear in the state variables (i.e., in the concentrations) and autonomous. Depending on the values of the rate constants and the initial data, the vector of state variables tends asymptotically to an equilibrium point or to a limit cycle, the latter corresponding to a persistent oscillation (the oregonator). Even if the state vector approaches equilibrium, it may pass through an oscillatory phase (the autocatalator), again depending on the values of the parameters and the data. Mathematically, the underlying mechanism is a Hopf bifurcation in which an equilibrium point in state space loses its stability and spawns a stable limit cycle in the process—all of this occurring if a parameter is changed. The mathematical model mimics the observed behavior of the chemical reactions so closely that there is considerable confidence in the proposed kinetic mechanisms and the model ODEs. Figures 6.12.1 and 6.12.2 show the characteristic behavior of the oscillations of, respectively, the autocatalator and the oregonator.

***References*** Consult the material below for additional background and details.

1. Gray, P. Review Lecture: Instabilities and Oscillations in Chemical Reactions in Closed and Open Systems. *Proc. R. Soc. London* (**A415**, 1–34 (1988)).
2. Gray, P. and Scott, S. K. *Chemical Oscillations and Instabilities*. Clarendon Press, Oxford, 1990.
3. Murray, J. D. *Lectures on Nonlinear-Differential-Equation Models in Biology*. Clarendon Press, Oxford, 1977.
4. Murray, J. D. *Mathematical Biology*. Springer-Verlag, Berlin, 1989.

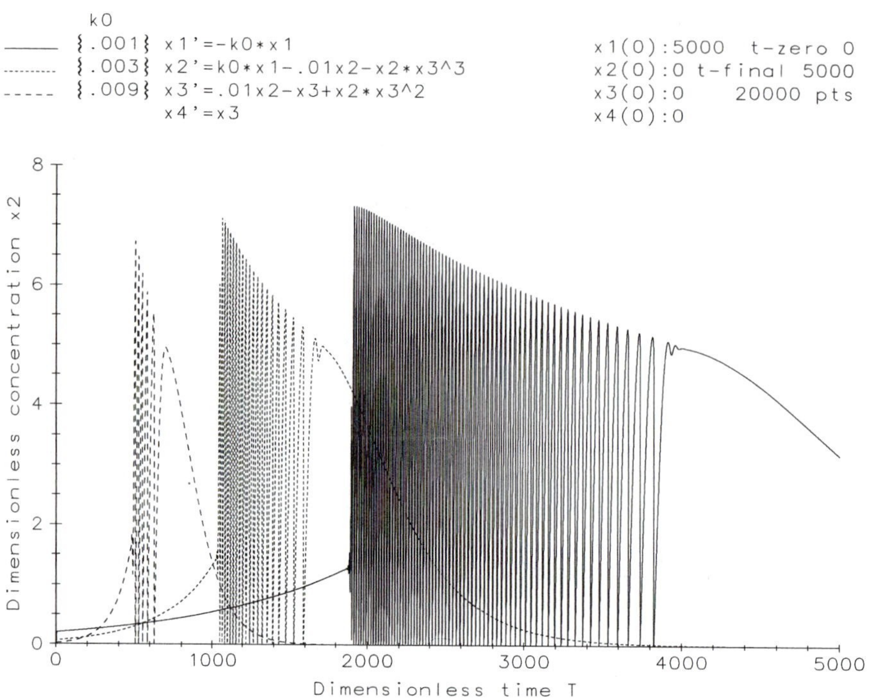


*Figure 6.12.1 Times of onset and shutdown of autocatalator oscillations change with the constant* $k_0$.

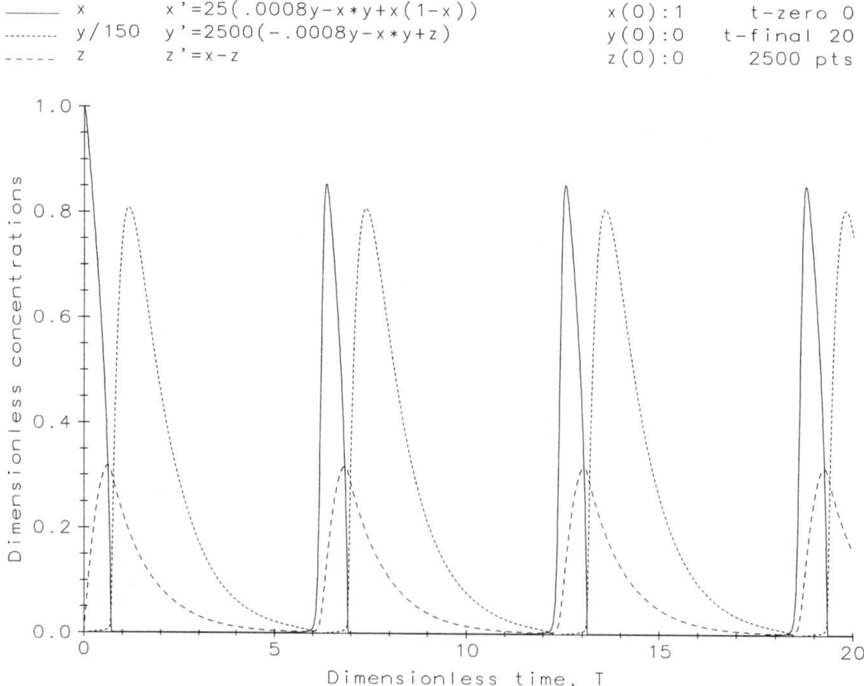


*Figure 6.12.2 Periodic bursts of oregonator concentrations.*

hardware / software

name / date

Answer questions in the space provided, or on attached sheets or carefully labeled graphs.

course / section

## *6.12 On/Off Oscillations: Autocatalator*

***Abstract*** The autocatalator models a closed chemical reaction in which the concentrations of intermediate species exhibit pronounced oscillations on the path to equilibrium.

***Special instructions*** The nature of this experiment makes it a candidate for a team approach. Consult Appendix A before writing the laboratory report. Concentrations are written in square brackets $[\cdots]$.

A precursor chemical $R$ in a closed constant volume reactor (i.e., no inflow and no outflow) triggers the autocatalytic reaction shown in the table.

| Nature of Step | Symbols | Rate Law |
|---|---|---|
| Precursor | $R \rightarrow A$ | $k_0[R]$ |
| Uncatalyzed step | $A \rightarrow B$ | $k_1[A]$ |
| Autocatalysis | $A + 2B \rightarrow 3B$ | $k_2[A][B]^2$ |
| Decay to product | $B \rightarrow C$ | $k_3[B]$ |

Model the autocatalator system with ODEs, describe the way oscillations appear and then vanish, and explain the relationship to a Hopf bifurcation. Some suggestions:

- **Rate equations**. Write out the rate equations for $[R]$, $[A]$, $[B]$, and $[C]$. Explain the terms. Explain (mathematically) why the sum of the four concentrations is constant over time, and (chemically) why this must be so in a closed reaction. (Hint: One of the rate equations is $[A]' = k_0[R] - k_1[A] - k_2[A][B]^2$).
- **Nondimensionalizing**. Following the suggestions in Appendix B.5, rewrite the rate equations in dimensionless terms using the variables and parameters $x_1 = \sqrt{k_2/k_3}\,[R]$, $x_2 = \sqrt{k_2/k_3}\,[A]$, $x_3 = \sqrt{k_2/k_3}\,[B]$, $x_4 = \sqrt{k_2/k_3}\,[C]$, $T = k_3 t$, $a_1 = k_0/k_3$, $a_2 = k_1/k_3$. Explain why the new quantities are dimensionless. (Hint: $dx_2/dT = a_1x_1 - a_2x_2 - x_2x_3^2$ is the second dimensionless rate equation.)
- **Parameter studies**. Duplicate Figure 6.12.1 and Atlas Plates `Autocatalator A-C`. Explain the graphs and the relationship of the ODEs of the Atlas Plates to the ODEs of Figure 6.12.1. Now vary the parameters $a_1, a_2, x_1(0)$, the span of dimensionless time $T$, and the number of solution points, to determine how and when the oscillations of $x_2$ (and $x_3$) are turned on and off. Identify the corresponding parameter ranges. Replace the term $a_1x_1$ in the rate equation for $x_2$ by a positive constant $A$. Find the $x_2x_3$-equilibrium points in the positive quadrant in terms of $a_1$, $a_2$, and $x_1(0)$. Explain the birth and death of the limit cycles as $A$ is changed in terms of a Hopf bifurcation about an equilibrium point.
- **Computational sensitivity**. Look at the systems of Figures 1.11.9 and 1.11.10. Explain the connection with the autocatalator. It appears that a slight change in the number of plotted points causes a dramatic change in the solution components. Does this happen on your hardware/software platform? Any explanation of the change? If your solver uses adaptive code, try changing the maximum internal step size. What happens? Explain.

_______________/_______________
hardware software

_______________/_______________
name date

Answer questions in the space provided,
or on attached sheets or carefully labeled graphs.

_______________/_______________
course section

## *6.13 Persistent Oscillations: The Oregonator*

***Abstract*** The concentrations of the chemicals in the open reactor of the oregonator oscillate for a very long time. The oscillations correspond to those of a model system of ODEs that has undergone a Hopf bifurcation.

***Special instructions*** This open-ended experiment is best done as a team project. Consult Appendix A before writing the lab report.

The chemical species in a simplified model of the **Belousov-Zhabotinskii reaction** are

$$X = \mathrm{HBrO_2}, \quad Y = \mathrm{Br^-}, \quad Z = \mathrm{Ce^{4+}}$$
$$A = \mathrm{BrO_3^-}, \quad P = \mathrm{HOBr}, \quad B = \text{organic species}$$

where H, Br, O, and Ce respectively denote hydrogen, bromine, oxygen, and cerium. The five reactions of the kinetic mechanism are

$$A + Y \xrightarrow{k_3} X + P, \quad X + Y \xrightarrow{k_2} 2P,$$
$$A + X \xrightarrow{k_5} 2X + 2Z, \quad 2X \xrightarrow{k_4} A + P, \quad B + Z \xrightarrow{k_0} \tfrac{1}{2} fY$$

where $f$ is a chemical constant whose value is approximately 1. Chemists at the University of Oregon simplified the Belousov-Zhabotinskii reaction by assuming that reactants $A$ and $B$ are in such large supply that their concentrations are constant. Some suggestions for studying the oregonator:

- **Rate equations**. Write rate equations for the concentrations $[X]$, $[Y]$, and $[Z]$, assuming that $[A]$ and $[B]$ are positive constants. (Hint: The first rate equation is $d[X]/dt = k_3[A][Y] - k_2[X][Y] + k_5[A][X] - 2k_4[X]^2$.)
- **Nondimensionalizing**. Introduce the dimensionless variables and parameters $x = 2k_4[X]/k_5[A]$, $y = k_2[Y]/k_5[A]$, $z = k_4k_0[B][X]/k_5^2[A]^2$, $T = k_0[B]t$, $a_1 = k_5[A]/k_0[B]$, $a_2 = k_2k_5[A]/2k_0k_4[B]$, $a_3 = 2k_3k_4/k_2k_5$, and derive the dimensionless rate equations

$$dx/dT = a_1(a_3 y - xy + x(1 - x))$$
$$dy/dT = a_2(-a_3 y - xy + fz)$$
$$dz/dT = x - z$$

  Explain why the variables and parameters are dimensionless.
- **Parameter studies**. Duplicate the graphs of Figure 6.12.2. Vary some of the parameters $a_1$, $a_2$, $a_3$, $f$, and $x(0)$ away from the values of the figures and see what happens to the $xyz$-limit cycle whose component graphs appear in the figure. Can you find sets of parameter values that turn the limit cycles on or off? Explain in terms of a Hopf bifurcation about an equilibrium point in the positive $xyz$-orthant. In this 3D setting, a Hopf bifurcation may occur for a set of parameter values if the Jacobian of the system of ODEs at the equilibrium point has one negative eigenvalue and a pair of conjugate pure imaginary eigenvalues.

# Bifurcations and Chaos in a Nonlinear Circuit

***Purpose*** To see how various bifurcations and chaotic wanderings occur as circuit parameters are varied in a nonlinear electric circuit.

***Keywords*** Electric circuits, bifurcations, chaos, torus, solid torus, strange attractor

***See also*** Experiment 5.14; Chapter 2 Notes; Atlas Plates `Scroll A-C`; Appendices B.4, B.5

***Background*** The diagram in Figure 6.14.1 shows a circuit with two capacitors (one with a negative capacitance $-C_1$), a nonlinear resistor, and an inductor. Because of the shape of the projected orbit shown in Atlas Plate `Scroll B`, it is now called the **scroll circuit**.

The circuit in Figure 6.14.1 is modeled by the system of dimensionless ODEs:

$$\begin{aligned} x' &= -cf(y-x) \\ y' &= -f(y-x) - z \\ z' &= ky \end{aligned} \tag{1}$$

where $x = V_1/E_1$, $y = V_2/E_1$, $z = I/(C_2E_1)$, $c = C_2/C_1$, $k = 1/(LC_2)$, $f(v) = -m_0v/C_2 + .5(m_0 + m_1)(|v+1| - |v-1|)/C_2$, and $-m_0$, and $m_1$ are the respective slopes of the falling and the rising segments of the characteristic in Figure 6.14.2. With the parameters fixed except for the dimensionless parameter $c$, the circuit may be "tuned" by changing $c$. The resulting graphs display period doubling, chaotic motion, and a stable periodic orbit that bifurcates first to a torus, and then to a solid torus that appears to contain a strange attractor. Figures 6.14.2–6.14.3 show a toroidal orbit in three dimensions for $c = 6$ and the projection of an orbit on a strange attractor for $c = 15$. These orbits are solved over $0 \leq t \leq 1000$, but plotted over $900 \leq t \leq 1000$ and $500 \leq t \leq 1000$, respectively.

***References*** Consult the material below for additional background and details.


1. Matsumoto, T., Chua, L.D., Tokunaga, R. Chaos via torus breakdown. *IEEE Trans. Circuits Syst.*, (**CAS–34(3)**: 240–53 (March 1987)).
2. Parker, T.S., Chua, L.O. *Practical Numerical Algorithms for Chaotic Systems*. Springer-Verlag, New York, 1989.


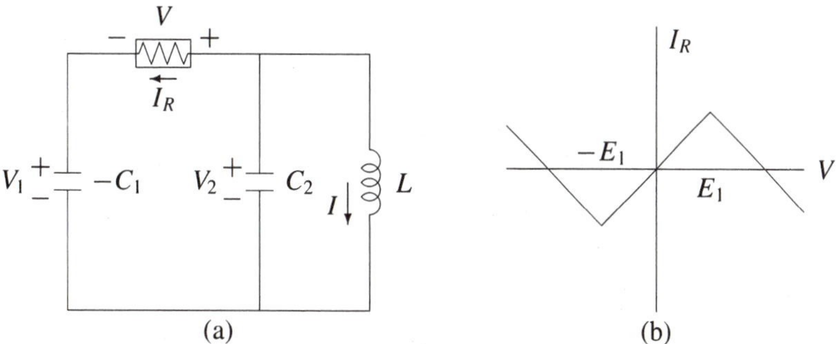


*Figure 6.14.1 (a) The scroll circuit. (b) Voltage/current characteristics for the nonlinear resistor.*

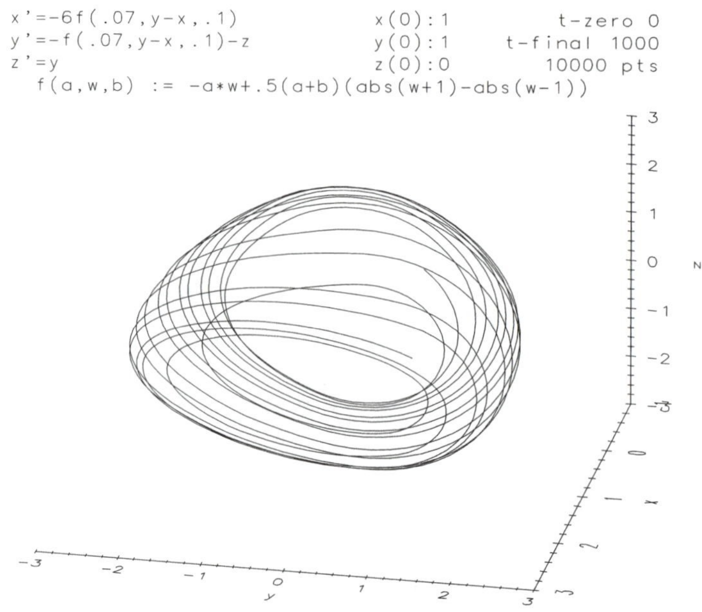


*Figure 6.14.2 A toroidal orbit. All nearby orbits tend to the torus.*

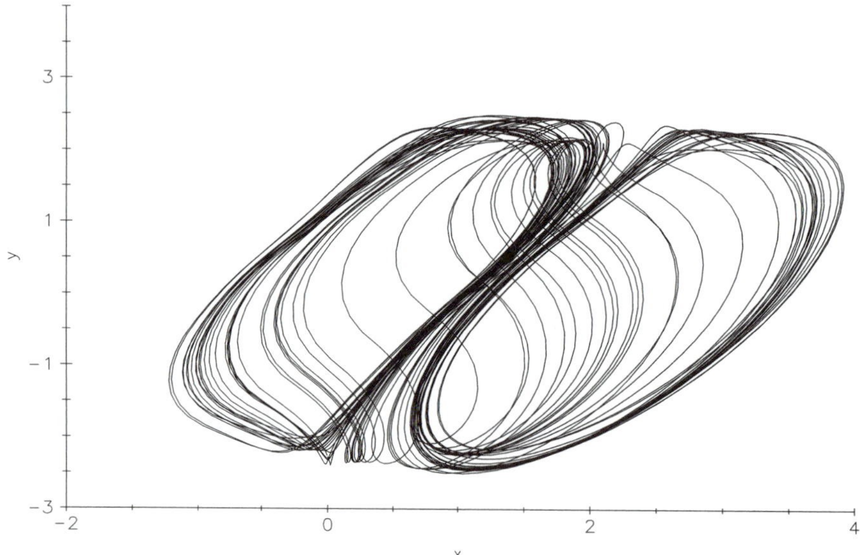


*Figure 6.14.3 Projection of an orbit on a folded strange attractor.*

_______________ / _______________
hardware software

Answer questions in the space provided,
or on attached sheets or carefully labeled graphs.

_______________ / _______________
name date

_______________ / _______________
course section

## *6.14 Bifurcations and Chaos in a Nonlinear Circuit*

***Abstract*** The circuit equations of the nonlinear circuit shown on the cover sheet are derived from circuit laws, nondimensionalized to computable form, and then solved for various values of a dimensionless circuit parameter. Bifurcations from a periodic orbit, to a torus, then to a strange attractor inside a torus are seen.

***Special instructions*** Since the limiting behavior of orbits is of primary interest, solve the ODEs over a long time interval (e.g., $0 \le t \le 1000$). If possible, plot the graphs only over an end segment of that interval.

**1.** Duplicate Figures 6.14.2 and 6.14.3 and Atlas Plates `Scroll A-C`. What are the corresponding values of the parameter $c$? Draw similar graphs for $c = 1, 6, 20, 50$. Use these graphs to estimate the value of $c$ at which a bifurcation to a strange attractor occurs.

**2.** Let $a = 0.07$, $b = 0.1$, $k = 1$. Solve (1) over the time interval [0,1000] but plot over the time interval [500,1000] to eliminate transient initial behavior. Use $x_0 = 1$, $y_0 = 1$, $z_0 = 0$, and the following values for the parameter $c$: 0.1, 0.5, 2, 8.8, 9.6, 10.8, 12, 13, 13.4, 13.45, 13.52, 15, 33. Graph the projections of the orbits on the $xz$-plane and in $xyz$-space (if possible). Then graph the components against $t$ to pick out periodic behavior. Identify periods and any period doubling you detect, as well as the approach to a strange attractor.

**3.** With $a = 0.07$, $b = 0.1$, $c = 0.1$, $k = 1$, $x_0 = 1$, $y_0 = 1$, $z_0 = 0$, *decrease* time from $t = 0$ and see if you pick up a repelling toroidal limit set.

**4.** A power source (not shown) is connected to the circuit in Figure 6.14.1. After it has energized the circuit, the source is removed. Use the basic circuit laws given in Appendix B.4 to derive the system of ODEs for the scroll circuit:

$$\begin{aligned} C_1 V_1' &= -g(V_2 - V_1) \\ C_2 V_2' &= -g(V_2 - V_1) - I \\ LI' &= V_2 \end{aligned}$$

where $g(v) = -m_0 v + .5(m_0 + m_1)(|v + E_1| - |v - E_1|)$.

**5.** Derive system (1) from the system in Problem 4 by using the dimensionless quantities defined on the cover sheet.

* **6.** Do a parameter study of the scroll system by varying the parameters $a$, $b$, $c$, and $k$ independently ($c$ is varied in Problem 1). Begin with $a = 0.07$, $b = 0.1$, $c = 2$, $k = 1$.

The paragraphs below outline some of the basic concepts of linear and nonlinear systems mentioned in the Chapter 6 Introduction.

### *Undriven Linear Systems*

The **undriven linear system**

$$x' = Ax \tag{1}$$

has the property that linear combinations of solution vectors are solution vectors. If $x^1(t)$ and $x^2(t)$ are solutions of (1) and $C_1$ and $C_2$ are constants, then $x = C_1x^1 + C_2x^2$ is also a solution of (1). For, we have that

$$x' = (C_1x^1 + C_2x^2)' = C_1(x^1)' + C_2(x^2)' = C_1Ax^1 + C_2Ax^2 = A(C_1x^1 + C_2x^2) = Ax$$

where properties of matrix-vector and matrix-scalar multiplication have been used. This linear property suggests that the set $S$ of all solutions of (1) may be constructed by taking linear combinations of a few. In fact, any $n$ linearly independent solutions of (1) generate all of $S$ in this way. If $A$ is a matrix of constants, then one way to construct such a collection of independent solutions is through **eigenelements** (i.e., **eigenvalues** and **eigenvectors**) of $A$. Let $\lambda$ be an **eigenvalue** and $v$ a corresponding **eigenvector** of $A$ (i.e., $Av = \lambda v$, $v \neq 0$). Then $x(t) = e^{\lambda t}v$ is a solution of (1). For, we have

$$x' = (e^{\lambda t}v)' = \lambda e^{\lambda t}v = e^{\lambda t}(\lambda v) = e^{\lambda t}Av = A(e^{\lambda t}v) = Ax$$

If $\lambda_1, \ldots, \lambda_n$ are eigenvalues and $\{v^1, \ldots, v^n\}$ a corresponding set of independent eigenvectors of $A$, then the solutions of (1), $e^{\lambda_1 t}v^1, \ldots, e^{\lambda_n t}v^n$, generate all solutions by taking linear combinations. The **general solution** of (1) is in this case

$$x = C_1e^{\lambda_1 t}v^1 + \cdots + C_ne^{\lambda_n t}v^n \tag{2}$$

where $C_1, \ldots, C_n$ are arbitrary constants. Every vector function $x(t)$ of the form given in (2) is a solution of (1), and every solution of (1) has the form of (2) for some values of the constants.

The eigenvalues $\lambda_1, \ldots, \lambda_r$ of $A$ are the (distinct) roots of the $n^{\text{th}}$ degree **characteristic polynomial** of $A$, $p(\lambda) = \det[A - \lambda I]$, where det denotes **determinant** and $I$ is the $n \times n$ diagonal matrix with 1's on the diagonal. Once the eigenvalues and their multiplicities as roots of $p(\lambda)$ have been determined, eigenvectors can be found by solving the corresponding linear algebraic systems $Av = \lambda_j v$ for $v$. For each $\lambda_j$, a maximal independent set of eigenvectors may be found. These eigenvectors span the **eigenspace** $V_j$ corresponding to the eigenvalue $\lambda_j$. If $\lambda_j$ has multiplicity $m_j$ as a root of $p(\lambda)$, then the eigenspace $V_j$ is a linear space of dimension $k_j$, where $1 \leq k_j \leq m_j$. If $m_j = 1$ (i.e., $\lambda_j$ is a simple root), then only a single (nonzero) eigenvector need be found for $\lambda_j$.

***Example 1*** **(Distinct real eigenvalues)** The eigenvalues of the system matrix

$$A = \begin{bmatrix} 5 & 3 & 0 \\ -1 & 1 & 0 \\ 0 & 0 & -1 \end{bmatrix}$$

are the roots $-1$, 2, 4 of the characteristic polynomial $p(\lambda) = \det[A - \lambda I] = -\lambda^3 + 5\lambda^2 - 2\lambda - 8$. A corresponding collection of independent eigenvectors is $v^1 = [0\ 0\ 1]^T$,

$v^2 = [1\ -1\ 0]^T$, and $v^3 = [3\ -1\ 0]^T$, where $T$ denotes the transpose of the (row) vector. The **general (real-valued) solution** (i.e., the set of all real-valued solutions) of $x' = Ax$ is

$$x = C_1 e^{-t} v^1 + C_2 e^{2t} v^2 + C_3 e^{4t} v^3$$

where $C_1$, $C_2$ and $C_3$ are arbitrary real numbers. If an initial condition $x(0) = x^0$ is imposed, these constants must be particularized accordingly.

*Example 2* **(Real and complex eigenvalues)** The eigenelements of the system matrix

$$A = \begin{bmatrix} -1 & 2 & 0 \\ -2 & -1 & 0 \\ 0 & 0 & -2 \end{bmatrix}$$

are $\lambda_1 = -1 + 2i$, with $v^1 = [-i\ 1\ 0]^T$; $\lambda_2 = \bar{\lambda}_1 = -1 - 2i$, with $v^2 = \bar{v}^1 = [i\ 1\ 0]^T$; and $\lambda_3 = -2$, with $v^3 = [0\ 0\ 1]^T$. The complex-valued solutions

$$x^1 = e^{(-1+2i)t} v^1, \quad x^2 = \bar{x}^1 = e^{(-1-2i)t} v^2, \quad x^3 = e^{-2t} v^3$$

generate the **general (complex-valued) solution** of $x' = Ax$ by taking all complex linear combinations of $x^1$, $x^2$, and $x^3$. Since $A$ is a real system matrix, an independent set of real solutions is often preferred. Using Euler's formula, $e^{a+bi} = e^a \cos b + ie^a \sin b$, and the fact that $\text{Re}[z(t)]$ and $\text{Im}[z(t)]$ are real-valued solutions for any solution $z(t)$, a collection of real-valued generating solutions may be found:

$$\text{Re}[x^1] = e^{-t} \begin{bmatrix} \sin 2t \\ \cos 2t \\ 0 \end{bmatrix}, \quad \text{Im}[x^1] = e^{-t} \begin{bmatrix} -\cos 2t \\ \sin 2t \\ 0 \end{bmatrix}, \quad x^3 = e^{-2t} \begin{bmatrix} 0 \\ 0 \\ 1 \end{bmatrix}$$

The **general (real-valued) solution** for $x' = Ax$ is thus given by

$$x = C_1 \,\text{Re}[x^1] + C_2 \,\text{Im}[x^1] + C_3 x^3$$

where $C_1$, $C_2$, and $C_3$ are any real numbers.

Example 2 illustrates how complex eigenvalues and the corresponding complex eigenvectors still produce real solutions of (1). The graphical evidence of complex eigenvalues is spiraling orbits and oscillating component curves (because of the sine and cosine terms).

The problem of multiple eigenvalues and **deficient eigenspaces** (where the dimension $k_j$ of the eigenspace $V_j$ corresponding to $\lambda_j$ is less than $m_j$, the multiplicity of $\lambda_j$) is harder to solve. An example illustrates how one must introduce **generalized eigenvectors** to generate enough independent solutions to form a general solution.

*Example 3* **(Deficient eigenspace)** The characteristic polynomial of the system matrix

$$\begin{bmatrix} 1 & -1 & 0 \\ 1 & 3 & 0 \\ 0 & 0 & -2 \end{bmatrix}$$

is $p(\lambda) = (1 - \lambda)(3 - \lambda)(-2 - \lambda) - (2 + \lambda) = -\lambda^3 + 2\lambda^2 + 4\lambda - 8 = -(\lambda - 2)^2(\lambda + 2)$. An eigenvector corresponding to $\lambda_1 = -2$ is $v^1 = [0\ 0\ 1]^T$. The eigenspace corresponding to the double eigenvalue $\lambda_2 = 2$ is only one-dimensional (i.e., $k_2 = 1$ and $m_2 = 2$) and is spanned by the eigenvector $v^2 = [1\ -1\ 0]^T$. Thus, we have two independent solutions of $x' = Ax$, $x^1 = e^{-2t} v^1$ and $x^2 = e^{2t} v^2$, but a third is needed to generate all solutions. A vector $w$ is a **generalized eigenvector** of the matrix $A$ with eigenvalue $\lambda$ and corresponding eigenvector $v$ if it satisfies

$$(A - \lambda I)w = v, \text{ that is, } Aw = \lambda w + v$$

It can be shown that if $w$ is a generalized eigenvector, then $x = te^{\lambda t}v + e^{\lambda t}w$ is a solution of $x' = Ax$ that is independent of $x = e^{\lambda t}v$. In this example, $Aw = \lambda w + v$ becomes

$$\begin{bmatrix} 1 & -1 & 0 \\ 1 & 3 & 0 \\ 0 & 0 & -2 \end{bmatrix} w = 2w + \begin{bmatrix} 1 \\ -1 \\ 0 \end{bmatrix}$$

and one solution is $w = [1\ -2\ 0]^T$. The general solution of $x' = Ax$ in this case is

$$x = C_1 e^{-2t} \begin{bmatrix} 0 \\ 0 \\ 1 \end{bmatrix} + C_2 e^{2t} \begin{bmatrix} 1 \\ -1 \\ 0 \end{bmatrix} + C_3 \left\{ te^{2t} \begin{bmatrix} 1 \\ -1 \\ 0 \end{bmatrix} + e^{2t} \begin{bmatrix} 1 \\ -2 \\ 0 \end{bmatrix} \right\}$$

The process becomes more involved for more severe deficiencies.

Thus the general solution of (1) is found by a linear algebraic technique involving the eigenelements of $A$. Sometimes it is useful to construct a **fundamental solution matrix** $M(t)$ from $n$ independent solution vectors $x^1(t), \dots, x^n(t)$, of (1). Construct the matrix $M(t)$ whose columns are these vectors. From this matrix a second fundamental solution matrix, denoted by the **matrix exponential** $e^{tA}$, can be constructed:

$$e^{tA} = M(t)M^{-1}(0)$$

The reason for the exponential notation for this particular fundamental matrix is that $(e^{tA})' = Ae^{tA}$ and $e^{0A} = I$, characteristic properties of exponential functions. That $e^{tA}$ is a solution matrix of $n$ independent column vector solutions of (1) is a consequence of the way $e^{tA}$ is defined as a matrix whose columns are independent linear combinations of the columns of a fundamental solution matrix, $M(t)$. Using $e^{tA}$, it is easy to find the solution of the IVP

$$\begin{aligned} x' &= Ax \\ x(0) &= x^0 \end{aligned} \tag{3}$$

That solution is

$$x(t) = e^{tA}x^0 \tag{4}$$

Once again, we see that algebraic calculations (the construction of the matrix product $M(t)M^{-1}(0)$) lead to the solution of a problem in differential equations.

### *Driven Linear Systems*

The solution of the driver linear IVP

$$x' = Ax + F(t), \qquad x(0) = x^0 \tag{5}$$

where $A$ is a matrix of real constants and $F(t)$ is a vector driving force (the **input**), may be written in terms of the matrix exponential:

$$x(t) = e^{tA}x^0 + e^{tA} \int_0^t e^{-sA} F(s)\, ds \tag{6}$$

The term under the integral is a vector whose components must be integrated from $s = 0$ to $s = t$. The first term on the right-hand side is the **response to the data** and is the solution of IVP (3). The second term, a particular solution of the driven system with vanishing initial data, is the **response to the input**. The **total response** or **output** $x(t)$ is the sum of the two.

### *Nonlinear Systems*

The equilibrium points and periodic orbits (if any) of an autonomous nonlinear system,

$$x' = f(x) \tag{7}$$

have been defined in the introductory text of this chapter. Here we give the definitions of chaotic sets and attractors, but first some preliminary definitions are needed. A subset $S$ of state space is **closed** if the limit point of any convergent sequence of points of $S$ also lies in $S$. $S$ is **bounded** if $S$ lies inside a ball in state space. $S$ is **invariant** if the entire orbit $\varphi_t(p)$ of (7) lies in $S$ whenever $p$ is in $S$. An orbit in $S$ is **dense in** $S$ if it passes arbitrarily close to every point of $S$ (i.e., if every point of $S$ is the limit of some sequence of points on the dense orbit). A closed and bounded invariant set displays **sensitive dependence on initial data** if there is a positive number $d$ such that if $p$ is any point of $S$, then every neighborhood of $p$ has at least one point $q$ such that the distance from $\varphi_t(q)$ to $\varphi_t(p)$ exceeds $d$ at some positive value of $t$. A closed and bounded invariant set is **chaotic** if it has more than one orbit, contains a dense orbit, and has sensitive dependence on initial data.

Chaotic sets in many systems of contemporary interest attract all nearby orbits. The word "attractor" is defined as follows. A closed and bounded invariant set $A$ is an **attracting set** if it has a neighborhood $\mathcal{N}$ with the property that for all $p$ in $\mathcal{N}$ $\varphi_t(p)$ lies in $\mathcal{N}$ for all $t \geq 0$ and $\varphi_t(p) \to A$ as $t \to \infty$. An **attractor** is an attracting set with a dense orbit. An attractor that is also chaotic is a **strange attractor**. It is very hard to prove that a given nonlinear system has a chaotic set or a strange attractor. Much of the evidence is computational, and that is as far as we can go in the experiments. Since the alternatives of Poincaré and Bendixson (Experiment 5.16) imply that autonomous systems with only one or two state variables have no strange attractors or chaotic sets, these phenomena can occur only in autonomous systems with at least three state variables, or in nonautonomous systems with at least two state variables and periodic dependence on time.

Some of the experiments in Chapter 6 refer to rate functions with parameters $a$,

$$x' = f(x, a) \tag{8}$$

As $a$ is changed, chaotic dynamics may suddenly appear, or a Hopf bifurcation or some other nonlinear phenomenon may occur. The Lorenz system, the Rössler system, the oscillating chemical reactions, and the scroll circuit (Experiments 6.6–6.8, 6.12–6.14) illustrate what can happen if a parameter changes.

### *Jacobian Matrix*

The **Jacobian matrix** of the rate function $f(x)$ is the matrix $J(x) = [\partial f_i/\partial x_j]$ of the first partial derivatives of the component functions $f_i$ of $f$. Observe that the Taylor expansion of $f(x)$ about a point $x^0$ has the form

$$f(x) = f(x^0) + J(x^0)(x - x^0) + \text{ higher order terms in } (x - x^0)$$

If $f(P) = 0$ (i.e., if $P$ is an equilibrium point of (7)), then the eigenelements of $J(P)$ play a central role in determining the behavior of orbits and the stability properties of system (7) near $P$. Observe that if $f(x) = Ax$, then $J(0)$ is just the system matrix $A$.

# APPENDIX A

# Team Laboratory Reports

Some projects in this workbook are designed for a team. The ideal size for a team is three to five members—enough to handle a job of broad scope, without having so many that people get in each other's way. Once laboratory projects have been assigned, the team may have to narrow the topic, specifying exactly which aspect of it the team will study. Teams often will not be able to do everything because of a lack of time or resources. The team must reach an agreement about who will work on what. Some overlap is fine, and it may be a good idea to have two or more members assigned to the same topic if involved analysis is required. Topics the team may wish to consider include:

- Behavior of numerical solutions as the number of solution points changes
- Location of equilibrium points of a first order system of autonomous ODEs
- Behavior of orbits of an autonomous first order system near equilibrium points
- Behavior of solutions of a system as $t \to \infty$
- How solution of a system changes as data change; sensitivity, parameter dependency
- Prediction of system behavior followed by data and analysis
- Examination of a system for periodic solutions, chaotic behavior, or approach to a strange attractor
- Discovery of unexpected or new properties, conjectures
- Comparison of output from several solvers on the same problem

The final report may be jointly written by the team, or students may be required to submit their own reports, but in any case the report(s) should contain everything important about the project. Reports must be readable, literate, and informative. Since the work of the team is jointly done, a single grade based on the common final report may be assigned to all team members. Alternatively, the grades assigned to students may be based on each individual's project report.

## *Suggested Outline for Team Report*

**Title page.** Authors, date, course number, and title.

**Abstract.** Brief summary of the project, method of attack, and results of the project.

**Introduction.** Background, details of the problem studied, and discussion of the project's emphasis (i.e., theoretical or experimental). Description of the procedure(s) used in treating the problem, such as analytical treatment or computer simulation.

**Body.** Contents of the body of the final report will depend on the type of project, but will probably include some or all of the following:

- If mathematical modeling is involved, the physical phenomenon should be described. Explain why the equations portray the physical system. Derive the model and discuss its accuracy, limitations, and possible improvements.
- The "point" of the problem.
- Mathematical principles used.

- A description of the method of solution employed.
- A brief description of the "experimental" procedure (leave the detailed procedure for an appendix).
- The results of the "experiment." Do the results match the expectations? If not, how can they be justified?

**Conclusion.** Summarize results; explain in mathematical terms and also give a physical interpretation if appropriate.

**References.** List references and sources alphabetically by author.

**Appendices.** Should contain all material necessary to support the body and conclusion of the project report, including:

- Carefully labeled figures, graphs, and tables.
- Detailed derivations, calculations, or procedural notes.
- Identification of all software and hardware used in the project.

# Mathematical Modeling

## Introduction

Practitioners in the sciences or engineering know what the modeling process is all about even if they don't feel the need to articulate it well. Modeling is an important part of their work-a-day activity. Briefly described, the modeling process goes something like this. The problem usually arrives posed in vague and nontechnical language, giving some background details and stating a goal to be achieved. A first step is to determine mechanisms affecting the behavior of the system and to isolate the features that are important in terms of the goal to be achieved, whittling the problem down to manageable size. Next, laws and constraints applicable to the specialized problem must be determined. The basic elements of a model are (1) a logical structure of some sort (calculus, differential equations, linear algebra, etc), (2) an interpretation of the variables in that structure, and (3) a characterization within the logical structure of all laws and constraints pertinent to the problem. Sometimes it is useful to modify the model to produce a simpler one that can more easily be treated (i.e., special cases, limiting cases, etc.). Once a model has been obtained, it can be analyzed through its own internal logical structure so that the behavior of its variables can be predicted (see Figure B.1).

The word "modeling" is a fairly recent addition to the mathematician's vocabulary but the modeling process itself is as old as rational thought.

> *Philosophy is written in this grand*
> *book of the universe,*
> *which stands continually open to our gaze . . .*
> *It is written in the language of mathematics.*
>
> —*Galileo*

A mathematical model of a physical system then is a portrait in the symbolic language of mathematics. Like all portraits, the model will emphasize some features of the original and ignore or distort others. Thus, the modeling process is as much an art as it is a logical process.

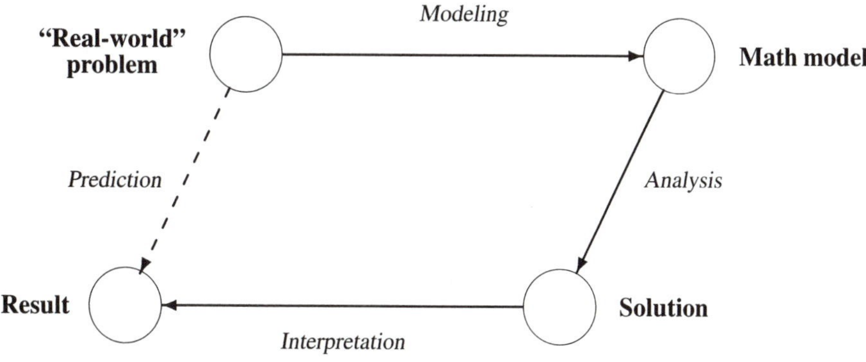


*Figure B.1 The modeling process.*

*Art is the lie that helps us to see the truth.*
*—Picasso*

Nevertheless, a skillfully constructed model often can provide more insight than direct observation of the system itself. Indeed, for many of the systems of modern science and technology, accurate direct analysis is impossible and our only perception of their realities is via mathematical or other models based on partial data, scientific common sense, experience, and intuition.

The power of abstraction is apparent in mathematical modeling. A mathematical model has a life of its own quite independent of any physical system being portrayed. To mathematicians, working almost entirely with the models, this is self-evident. To the physical scientist or the engineer, it may seem strange to analyze the abstract mathematical model with little reference to its nonmathematical subject. And yet a strong argument can be made for doing just that.

### *How Is a Mathematical Model Created?*

That is as easy to explain as to tell an artist how to paint. A study of completed models is helpful: the so-called **case study approach**, in which the modeling process is described in a series of examples that are more-or-less self-contained. The examples are designed to bring out the features of the modeling process as well as to provide information about basic models within a discipline.

A second approach differs from the case study approach in that the modeling process is learned via "hands-on" experience. This approach is "open-ended" because it is not clear at the outset what kind of model will be successful. An interesting sidelight of this approach is that the models selected depend not only on the modeler's breadth of knowledge, but also on time constraints and computer (and other) resources available. Because of its "hands-on" nature, this approach has recently come to be referred to as the **experiential approach** to learning about the modeling process. A great deal of evidence over the past decade shows that the experiential approach works and that students find it a rewarding experience, which also sharpens their communication skills. Several successful formats for experiential modeling projects have emerged—all seem to use the team approach. In this workbook we have opted for an approach midway between the case study and the experiential.

## *B.1 Population and Rate Models*

### *Introduction*

Let $P(t)$ denote the **population** at time $t$ of a species in a community. The values of $P(t)$ are integers, and they change by integer amounts as time goes on. However, for a large population an increase by one or two over a short time span is "infinitesimal" relative to the total, and we may think of the population as changing continuously instead of by discrete jumps. Once we assume that $P(t)$ is continuous, we might as well smooth off any corners on the graph of $P(t)$ and assume that the function is differentiable. If we had let $P(t)$ denote the population density (i.e., the number per unit area or volume of habitat), the continuity and differentiability of $P(t)$ would have seemed more natural. However, we shall ususally interpret $P(t)$ as the size of the population, rather than as the density. The underlying principle of population change is the simple

**Balance Law:** *Net rate of change = rate in − rate out.*

The "rate in" term is the sum of the birth and the immigration rates, while the "rate out" term is the sum of the death and the emigration rates. Let us regroup the rates into an "internal" birth minus death rate and "external" **migration** rate $M$ (immigration minus emigration). Averaged over all classes of age, sex, and fertility, a "typical" individual makes a net contribution $R$ to the internal rate of change. Thus the internal rate of change at time $t$ follows a simple rate law and is given by $RP(t)$ = (individual's contribution) ×

(number of individuals). The **intrinsic rate coefficient** $R$ will differ from species to species, but it always denotes the average individual contribution to the rate. Thus, the Balance Law becomes

$$P'(t) = RP(t) + M \tag{1}$$

and the study of population changes depends upon solving (1), given the rate coefficient $R$, the migration rate $M$, and the initial population. $R$ and $M$ may depend on $P$ and $t$, but in the simplest cases that dependence will either disappear or be linear in $P$.

## *Exponential Growth*

*It may safely be pronounced, therefore, that population, when unchecked, goes on doubling itself every twenty-five years.*

—Malthus[1]

The Malthusian principle of explosive growth of human population has become one of the classic "laws" of population change. The principle follows directly from (1) if we set $M = 0$ and $R$ to be a positive constant, say $r$. In this case, equation (1) is linear:

$$P' = rP \tag{2}$$

Solutions of (2) are the exponential functions

$$P(t) = P_0 e^{rt} \tag{3}$$

where $P_0$ is the population at time $t = 0$. Malthus speaks of a population "doubling itself." This **doubling time** $T$ is given by

$$T = \frac{1}{r}\ln 2 \tag{4}$$

To see this, observe from (3) and (4) that for any time $t$ we have that

$$P(t+T) = P_0 e^{r(t+T)} = (P_0 e^{rt})e^{rT} = (P(t))e^{rT} = (P(t))e^{\ln 2} = 2P(t)$$

Thus the population doubles every $T$ units of time.

Malthus claimed a doubling time of 25 years for the human population, which implies that the rate coefficient $r = (1/T)\ln 2 = \frac{1}{25}\ln 2 \cong 0.0277$, corresponding to a 2.8% annual increase in population. Although Malthus's figure for $r$ is too high for our late-20$^{\text{th}}$-century world, individual countries may have intrinsic rate coefficients as high as 0.033. The corresponding doubling time shrinks from 25 years to 21 years when $r = 0.033$.

## *Logistic Growth*

*The positive checks to population are extremely various and include... all unwholesome occupations, severe labor and exposure to the seasons, extreme poverty, bad nursing of children, great towns, excesses of all kinds, the whole train of common diseases and epidemics, wars, plague, and famine.*[2]

The unbridled growth of a population as predicted by the simple Malthusian law of exponential increase cannot continue forever. Malthus claimed that resources grow at most arithmetically (i.e., the net increase in resources each year does not exceed a fixed constant). An exponential increase in the size of a population must soon outstrip the resources available to support the population. The resulting hardships would increase the death rate and put a damper on growth.

---

1. Thomas Robert Malthus (1766–1834) was a professor of history and political economy in England. The quotation is from "An Essay on the Principle of Population as It Affects the Future Improvement of Society." Malthus's views have had a profound effect on Western thought. Darwin and Wallace each said that reading Malthus which led him to the theory of evolution.
2. Malthus, op. cit.

The simplest way to model restricted growth within the context of the rate equation (2) with no net migration is to assume that the rate coefficient $R$ has the form $r_0 - r_1 P$, where $r_0$ and $r_1$ are positive constants. In effect, we are now taking the first two terms of a Maclaurin series expansion of $R$ as a function of population, rather than just the first term as in the earlier model. The term $-r_1 P$ represents a restraint on the growth rate, whereas $+r_1 P$ would lead to accelerated growth. It is customary to write the rate coefficient as $r(1 - P/K)$, where $r$ and $K$ are positive constants, rather than as $r_0 - r_1 P$. We have the **logistic equation** with initial condition

$$P' = r(1 - \frac{P}{K})P, \quad P(0) = P_0 \tag{5}$$

The IVP (5) may be solved by setting $z = 1/P$ to obtain an IVP for a first order linear rate equation,

$$z' = -rz + \frac{r}{K}, \quad z(0) = \frac{1}{P_0} \tag{6}$$

Using the integrating factor $e^{rt}$ to solve (6), we find after some algebraic manipulation

$$P(t) = \frac{1}{z(t)} = \frac{K}{1 + Ce^{-rt}}, \quad C = \frac{K}{P_0} - 1$$

Unlike the exponential growth curves of the Malthusian model, the logistic population curves are bounded, tending toward $K$, the **saturation level** or **carrying capacity**.

### *Multispecies Communities*

Suppose several species interact within a community. Let $x_1(t), \ldots, x_n(t)$ denote the respective populations. For each population there is a rate equation with the form of (1),

$$x_i'(t) = R_i x_i(t) + M_i$$

where $R_i$ is the average individual contribution to the rate of change of the population $x_i$ and $M_i$ is the net migration rate of that population. For simplicity, suppose that there are just two species with populations $x(t)$ and $y(t)$, respectively, and that the migration rates are negligible. Thus, the rate equations for this two-species community are

$$\begin{aligned} x' &= R_1 x \\ y' &= R_2 y \end{aligned}$$

We saw in the single species logistic model that the rate coefficients $R_1$ and $R_2$ need not be constants. Consider, first, the coefficient $R_1$. Since we assume that the two species do interact, the coefficient must be a function of $y$. It may also depend on $x$. In the absence of any information about the form of the dependence of $R_1$ on $x$ and $y$, we recall the analyst's adage, "When in doubt, linearize!" Suppose that $R_1$ is the linear polynomial

$$R_1 = a + bx + cy$$

where $a$, $b$, and $c$ are constants. The rate equation for the $x$-species is

$$x' = (a + bx + cy)x \tag{7}$$

There is a similar derivation for the $y$-species.

The derivation of the rate equation has been theoretical, but the terms have biological interpretations. The constant $a$ in (7) is the **natural growth coefficient** of the $x$-species, while $bx$ measures the effect of the size of the population on an individual's contribution. If $b$ is negative, that effect is negative, and $b$ is the **overcrowding** or **self-limiting coefficient.**

If $b$ is positive, $b$ is the **mutualism coefficient**, and $bx$ indicates that an increase in the fertility occurs with population growth.

The coefficient $c$ is a measure of the effect of species $y$ on the growth rate of the $x$-population. If $c$ is negative, the $x$-species "loses" in any encounter between the two species and $x$ is the **prey** of the **predator** $y$. An alternative interpretation is that both species are in competition for the same resources and an increase in the numbers of either lowers the rate of growth (one would expect $b$ as well as $c$ to be negative in this case). If $c$ is positive, the predator-prey relationship is reversed, or else the $y$-species contributes in some other way to the "well-being" of the $x$-population. The terms in the rate equation for the $y$-species have similar interpretations. Whatever the signs of the coefficients may be, rate terms such as $bx^2$, $cxy$, or $dy^2$ are called **social** or **mass action** terms. From this point on all coefficients are taken to be nonnegative, and a minus sign is used when a negative rate is intended.

***Example 1*** **(Predator-Prey)** The simplest models of a predator and prey association include only natural growth and decay and the predator-prey interaction itself. All other forms of intra- or interspecific relationships are negligible and are omitted. We assume that the prey species would expand exponentially in the absence of predation, while the predator species would decline exponentially if there were no prey to consume. The predator-prey interaction is modeled by mass action terms proportional to the product of the two populations. The rate equations for such a simplified model are

$$\begin{aligned} x' &= (-a + by)x \\ y' &= (c - dx)y \end{aligned}$$

where $x$ is the predator, $y$ is the prey, and all coefficients are positive.

***Example 2*** **(Prey overcrowding)** Suppose that the rate equations for the populations of a pair of interacting species are

$$\begin{aligned} x' &= (a - bx - cy)x \\ y' &= (-d + ex)y \end{aligned}$$

where the coefficients are positive. What types of biological dynamics are modeled by these equations? The first species has natural exponential growth (corresponding to the rate term $ax$), but this is tempered by overcrowding (through the term $-bx^2$) and by predation by the $y$-species (modeled by the term $-cxy$). The $y$-species will die out exponentially if there is no food (corresponding to the term $-dy$), but $y$ converts individuals of the $x$-species into increased $y$ fertility ($+exy$), and the $y$-species may survive, and even thrive, if the conversion efficiency $e$ is high enough.

***Example 3*** **(Mutualism)** The rate equations

$$\begin{aligned} x' &= (a + by)x \\ y' &= (c + dx)y \end{aligned} \tag{8}$$

model a pair of interacting species, each with a natural, exponentially growing population (the consequence of the terms $ax$ and $cy$), whose mutual association is beneficial to both (modeled by the rate terms $+byx$ and $+dxy$). Equations such as (8) cannot model real populations for any great length of time, since the consequent explosive growth quickly exceeds reasonable bounds.

Solutions $x(t)$ and $y(t)$ of the rate equations for two interacting species are of interest only in the first quadrant (the **population quadrant**), $x \geq 0$ and $y \geq 0$. The curve in the population quadrant defined parametrically by $x = x(t)$ and $y = y(t)$ is a **population orbit**. The vanishing of $x(t)$ or of $y(t)$ for some $t$ means that the corresponding species vanishes. In reality, our models lose their significance if either $x(t)$ or $y(t)$ becomes small.

Then individual behavior determines the fate of the species, and the averaging used to justify the rate equations is meaningless.

***References*** Consult the material below for additional background and details.


1. Edelstein-Keshet, L. *Mathematical Models in Biology*. New York, Random House, 1988.
2. May, R. *Stability and Complexity in Model Ecosystems*. Princeton University Press, Princeton, NJ, 1973.
3. Murray, J. D. *Mathematical Biology*. Springer-Verlag, New York, 1989.
4. Roughgarden, J. *Theory of Population Genetics and Evolutionary Ecology: An Introduction*. Macmillan, New York, 1979.


## B.2 Mechanics

Mechanics is the study of the motion of material bodies that is due to the action of their surroundings. It is the oldest, and perhaps the most thoroughly developed, area in which mathematical models involving differential equations have been used. Mechanics and differential equations "grew up together," and their histories are intimately intertwined.

The motion of a falling body, a vibrating mass attached to a spring, an oscillating pendulum, the planets orbiting the sun, or a space probe in the cosmos can all be described by the principles of mechanics. In these systems there is no "balance of rates," no "growth or decay of populations," no "flow of electrical currents," but there are changes in position and velocity. Mechanics supplies some different principles, which relate these changes in the motion of a body to the circumstances of its environment. Building on the work of Galileo, Isaac Newton[3] saw to the heart of the matter and formulated three fundamental laws of motion that relate the acceleration of a material body to its **mass** and the **resultant force** acting on that body. To do this, Newton (in effect) introduced the **vector concept** as a modeling device in order to express his Laws of Dynamics with an elegant simplicity that transcends any particular reference frame or coordinates in these frames.

### The Vector Concept

Basic physical concepts such as **velocity** and **acceleration** are thought of as directed line segments (i.e., **geometric vectors**) in the familiar 3-space of Euclid. In this workbook the symbols for geometric vectors are boldface letters. Two vectors **v** and **w** are identical if and only if they can be made to coincide by translations (which preserve length and direction of vectors). Thus there is no need to have a new name for any vector that can be identified with another vector that already has a name. The length of of a vector **v** is denoted by $\|\mathbf{v}\|$. The **zero vector**, denoted by **0**, is the vector of zero length (direction is irrelevant).

Velocity and acceleration vectors of a material body have been observed to have the following property: If a body is measured to have the velocities (or accelerations) **v** and **w** in two directions at the same time, then it has the resultant velocity (or acceleration), denoted by $\mathbf{v} + \mathbf{w}$, defined by the **parallelogram law** as follows. Find a vector equivalent to **v** whose "tail" coincides with **w**'s "head." Then $\mathbf{v} + \mathbf{w}$ is the vector whose "tail" is **w**'s and whose "head" is **v**'s ( a diagonal of the parallelogram formed by **v** and **w**). This prompts the definition:

**Basic Operations for Geometric Vectors:** *Let* **v** *and* **w** *be any two geometric vectors and r any real number. The* **sum** $\mathbf{v} + \mathbf{w}$ *is the vector produced by the parallelogram law. The* **product** $r\mathbf{v}$ *is the vector of length* $|r|\,\|\mathbf{v}\|$*, which points in the direction of* **v** *if* $r > 0$ *and in the direction opposite to* **v** *if* $r < 0$.

---

3. Isaac Newton was born Christmas Day 1642 (the year Galileo died), and died in 1727. In addition to setting forth the basic laws of motion, Newton formulated the Law of Universal Gravitation and invented the calculus.

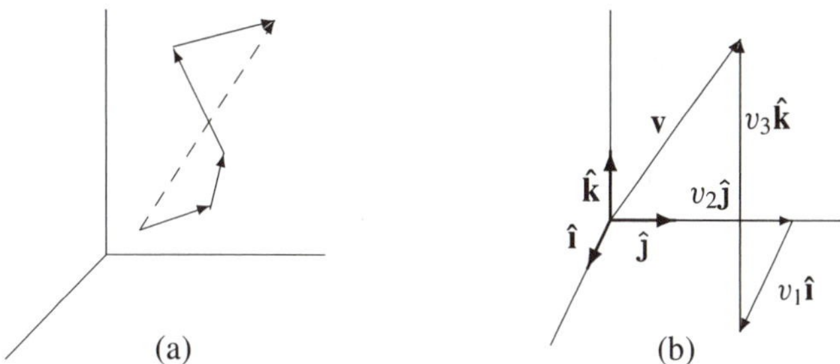


*Figure B.2.1 (a) Addition of vectors. (b) Resolution of a vector as a sum.*

Note that $r\mathbf{v} = \mathbf{0}$ if either $r = 0$ or $\mathbf{v} = \mathbf{0}$. Also note that $(+1)\mathbf{v} = \mathbf{v}$ and that $(-1)\mathbf{v}$ has the property that $\mathbf{v} + (-1)\mathbf{v} = 0$, for any $\mathbf{v}$. The vector $(-1)\mathbf{v}$ is commonly denoted by $-\mathbf{v}$. Note that $-\mathbf{v}$ has the same length as $\mathbf{v}$ but points in the opposite direction. When a number of vectors are to be added together, the parallelogram law is used in succession—the order in which this is done is immaterial. See Figure B.2.1. Other computations with vectors are defined geometrically as follows:

- **Angle**. Two nonzero, nonparallel vectors $\mathbf{u}$ and $\mathbf{v}$ form a plane. In this plane two vectors equivalent to $\mathbf{u}$ and $\mathbf{v}$ can be found which have coincident "tails." The angle less than or equal to 180° between these two vectors is said to be the **angle** between $\mathbf{u}$ and $\mathbf{v}$. The angle between nonzero parallel vectors is either 0° or 180°.
- **Dot product**. The **dot product** of two vectors $\mathbf{u}$ and $\mathbf{v}$ denoted by $\mathbf{u} \cdot \mathbf{v}$ is the real number $\|\mathbf{u}\| \cdot \|\mathbf{v}\| \cos\theta$ where $\theta$ is the angle between $\mathbf{u}$ and $\mathbf{v}$. Two vectors are **orthogonal** if and only if their dot product is zero.

When geometric notions have served their purpose in constructing a model, the idea of coordinates for vectors is introduced to facilitate computation. A (**coordinate**) **frame** is a triple of vectors $\{\hat{\mathbf{i}}, \hat{\mathbf{j}}, \hat{\mathbf{k}}\}$ that are mutually orthogonal and all of unit length. Some trigonometry reveals that every vector can be uniquely written as the sum of vectors parallel to $\hat{\mathbf{i}}$, $\hat{\mathbf{j}}$, and $\hat{\mathbf{k}}$. Thus for every vector $\mathbf{v}$ there is a unique set of real numbers $v_1$, $v_2$, and $v_3$ such that $\mathbf{v} = v_1\hat{\mathbf{i}} + v_2\hat{\mathbf{j}} + v_3\hat{\mathbf{k}}$. The elements of the ordered triple $(v_1, v_2, v_3)$ are called **coordinates** or **components** of $\mathbf{v}$ in the frame $\{\hat{\mathbf{i}}, \hat{\mathbf{j}}, \hat{\mathbf{k}}\}$. See Figure B.2.1. Coordinates are written in the same order of occurrence as vectors in the frame $\{\hat{\mathbf{i}}, \hat{\mathbf{j}}, \hat{\mathbf{k}}\}$. It can be shown that $\mathbf{v} \cdot \mathbf{w}$ can be calculated from the coordinates of $\mathbf{v}$ and $\mathbf{w}$ in any frame as

$$\mathbf{v} \cdot \mathbf{w} = v_1 w_1 + v_2 w_2 + v_3 w_3$$

The **derivative** of the vector $\mathbf{u} = \mathbf{u}(t)$ is denoted by $d\mathbf{u}/dt$ or $\mathbf{u}'(t)$ and is defined in the usual way as the limit of a difference quotient:

$$\mathbf{u}'(t) = \frac{d\mathbf{u}}{dt} = \lim_{h \to 0} \frac{\mathbf{u}(t+h) - \mathbf{u}(t)}{h} \tag{1}$$

where the limit exists if the difference between the vectors $d\mathbf{u}/dt$ and $[\mathbf{u}(t+h) - \mathbf{u}(t)]/h$ has length that tends to zero as $h \to 0$. Observe that

$$(\mathbf{u} \cdot \mathbf{v})' = \mathbf{u}' \cdot \mathbf{v} + \mathbf{u} \cdot \mathbf{v}'$$
$$(r(t)\mathbf{u}(t))' = r'\mathbf{u} + r\mathbf{u}'$$

which is very reminiscent of the usual product differentiation rule. If $\mathbf{u}$ is a constant vector, note that $\mathbf{u}' = 0$.

There are many reference frames in the 3-space of our experience. One can imagine frames that move in space or frames that are fixed. Suppose that $\{\hat{\mathbf{i}}, \hat{\mathbf{j}}, \hat{\mathbf{k}}\}$ is a fixed frame and that a particle moves in a manner described by the **position vector**

$$\mathbf{R} = \mathbf{R}(t) = x(t)\hat{\mathbf{i}} + y(t)\hat{\mathbf{j}} + z(t)\hat{\mathbf{k}}$$

If **R** is differentiable, it follows from (1) that

$$\mathbf{R}'(t) = x'(t)\hat{\mathbf{i}} + y'(t)\hat{\mathbf{j}} + z'(t)\hat{\mathbf{k}} \tag{2}$$

where primes indicate time derivatives. The derivative $\mathbf{R}'(t) = \mathbf{v}(t)$ is the **velocity vector** of the particle at time $t$, and $\mathbf{v}(t)$ is tangential to the path of the particle's motion at the point $\mathbf{R}(t)$. Furthermore, if $\mathbf{R}'(t)$ is differentiable,

$$\mathbf{R}''(t) = \mathbf{v}'(t) = x''(t)\hat{\mathbf{i}} + y''(t)\hat{\mathbf{j}} + z''(t)\hat{\mathbf{k}} \tag{3}$$

is the **acceleration vector** for the body's center of mass. (Computation of the velocity and acceleration vectors in given coordinate frames when the path of motion is assumed known is called **kinematics**.)

### *Forces, Newton's Laws*

Newton developed Galileo's central idea that the environment creates "forces," which act on bodies causing them to accelerate. Galileo's principle, now named for Newton, is stated as follows

**Newton's First Law:** *A body remains in a state of rest or of uniform motion in a straight line if there are no external forces acting on it.*

Forces can be measured without regard to reference frames. Scientists of Newton's time were able to show experimentally that forces behave like geometric vectors (i.e., that they satisfy the parallelogram law). The effect of Newton's First Law is to identify all frames of reference that are either fixed in space or undergoing a translation at a constant velocity with respect to a fixed frame. Such frames are called **inertial frames**, and they are extremely important in the modeling of moving bodies. Practically speaking, how do we know when we are dealing with an inertial frame? According to Newton's First Law, a frame is inertial if and only if a body is unaccelerated with respect to that frame whenever the resultant (sum) force acting on the body vanishes.

Next we have the basic and central principle in dynamics,

**Newton's Second Law:** *For a body* in any inertial frame *we have*

$$\mathbf{F} = m\mathbf{a} \tag{4}$$

*where $m$ denotes the mass (assumed to be constant),* **a** *is the acceleration of the body, and* **F** *is the net resultant of all external forces acting on the body, where all these quantities are measured in a consistent set of units*[4] *(see Figure B.2.2.).*

Applied to a particle, or to a rigid body, (4) states that the acceleration **a** is proportional to the net force **F** acting on it, and that the proportionality constant is the mass $m$ of the particle. Since (4) is a vector equation, this implies that **a** has the same direction as **F** and that the magnitudes $\|\mathbf{a}\|$ and $\|\mathbf{F}\|$ of the acceleration and force, respectively, are related by $\|\mathbf{F}\| = m\|\mathbf{a}\|$. Depending on the type of motion, reference frames may be introduced such that (4) may correspond to one, two, or three scalar equations. Examples are provided in the experiments.

The terms in Newton's laws must be quantified in some system of units. In the **mks system** (meter, kilogram, second) the unit of mass is 1 kilogram, and (4) is used to define a

4. This form (4) of Newton's Second Law is for so-called point masses. For some "distributed" masses it is permissible to apply (4) as if all of the mass of the body were concentrated at its center of mass

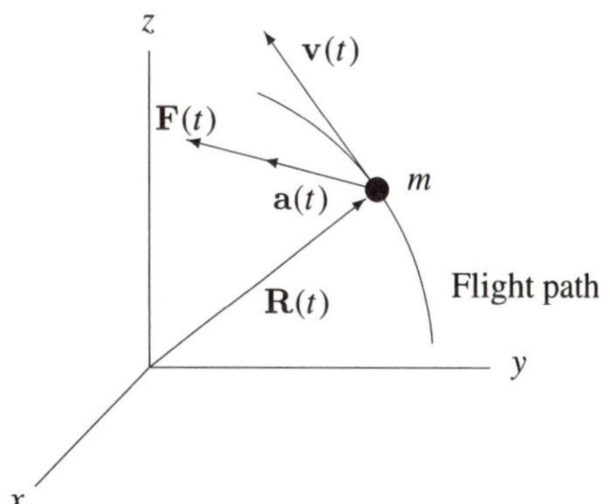


*Figure B.2.2 Geometry of Newton's Second Law.*

unit of force that will accelerate a 1 kg mass at 1 m/s$^2$. This unit of force is called a **newton**. In the **cgs system** (centimeter, gram, second) the unit of force is a **dyne** (1dyne $= 10^{-5}$ newton) and is that force which accelerates a 1 g mass at 1 cm/s$^2$. In these metric units, mass, length, and time are considered to be fundamental quantities, while force is secondary. We avoid the use of any particular system of units as much as possible. When units must be used, they are labeled explicitly.

The Third Law of Newton gives us some insight as to how we may treat multiple-body systems.

**Newton's Third Law:** *If a body A exerts a force* **F** *on body B, then body B exerts a force* $-$**F** *on body A.*

To give these simply stated principles some operational content, we would need to specify ways of calculating the acceleration and the "external force" acting on each body of a system in an inertial reference frame. These techniques form the central core of the field known as **dynamics**, and we cannot do more here than give the barest of introductions to this, the oldest branch of physics. As we shall see, a major consideration in the application of Newton's Second Law is the frame of reference and a coordinate system within that frame. The examples considered in the experiments give a general idea of how Newton's Second Law is applied.

## *Gravitation*

From (3) we see that Newton's Second Law is a differential equation that somehow must be formulated in a way that accommodates solutions for the velocity and position of the body as a function of time. Nothing can be done, of course, until the resultant force **F** has been specified. Gravitational force is discussed below; spring forces and frictional forces are discussed in the experiments.

Using observations of falling bodies and the extensive astronomical work and empirical laws of Tycho Brahe and Johannes Kepler concerning the orbits of the moon and of the planets, Newton focused attention on just one force: gravity. His Law of Universal Gravitation has to do with with the gravitational effect of one body on another.

**Newton's Law of Universal Gravitation:** *The force* **F** *between any two particles having masses* $m_1$ *and* $m_2$ *and separated by a distance* $r$ *is attractive, acts along the line joining the two particles (i.e., tends to pull the particles together), and has magnitude*

$$\|\mathbf{F}\| = G\frac{m_1 m_2}{r^2} \tag{5}$$

*where $G$ is a universal constant independent of the nature and masses of the particles. In the mks system, $G = 6.67 \times 10^{-11}\,\mathrm{N} \cdot \mathrm{m/kg}^2$.*

Newton also showed that massive bodies affect one another as if the mass of each body were concentrated at its "center of mass," if the masses of those bodies are distributed in a spherically symmetric way (it is not true in general). In this case, $r$ is the distance between the centers of mass.

A small object (e.g., a ball of lead) is dropped and falls toward the surface of the earth. We suppose that the only significant force acting on the body is the gravitational attraction of the earth (we ignore the resistive force, if any, of the medium through which the body falls). We shall fix a $z$-coordinate line on the surface of the earth directly below the body and pointing upward. A positive value of $z$ indicates a location above the surface of the earth. A negative value of $v = dz/dt$ means that the body is moving downward (in the direction of the negative $z$-axis). A negative value of force acting on the body means that the force acts in the direction of the negative $z$-axis, and so on. We shall suppose that at time $t = 0$ the body is at height $z(0) = h$ and has velocity $v_0$. If the body is falling, then $v < 0$. In calculating the gravitational attraction of the earth on this body, we imagine that the mass of the earth is concentrated at its center. This means that the earth attracts the body with a force whose magnitude is

$$\|\mathbf{F}\| = \frac{GMm}{(R+z)^2}$$

where $M$ is the mass of the earth, $m$ is the mass of the falling body, $R$ the radius of the earth, and $z$ the distance from the earth's surface to the center of mass of the falling body. For motion close to the surface of the earth the change in the magnitude of the force as the body falls is negligible. We can therefore define a constant $g = GM/R^2$ and express the gravitational force on the body as $-mg$. Note that $g \approx 9.8\,\mathrm{m/s}^2$ in the mks system. The equation of motion of a falling body is then simply written $mz'' = -mg$ with the initial conditions $z(0) = h$, $z'(0) = v_0$.

### *The Two-Body Problem*

The two giants of 17th century science, Galileo and Newton, tried to write the book of nature in mathematical deductions. An early success of this approach was the mathematical derivation of the approximate elements of the orbits of the earth, its moon, and the planets. Current calculations of the orbits of artificial satellites and space probes have their roots in that early work. Below we derive the equations of motion for the simplest case of "planetary motion," that of two bodies in a central force field created by their mutual attraction.

To be specific, say that the force field arises from the mutual gravitational attraction of a sun of mass $M$ and a planet of mass $m$. In an inertial frame of reference, let $\mathbf{X}$ denote the position vector of the sun, $\mathbf{x}$ the position vector of the planet, and $\mathbf{r}$ the radius vector from the center of the sun to the center of the planet. See Figure B.2.3. The gravitational force $\mathbf{F}$ of the sun on the planet acts along the line through the centers of mass and has the magnitude given by the Law of Universal Gravitation (5). Thus, Newton's Second Law as applied to the planet gives

$$m\mathbf{x}'' = \mathbf{F} = -\frac{GMm}{r^3}\mathbf{r} \tag{6}$$

where $r = \|\mathbf{r}\|$ and $\mathbf{r}/r$ is a unit vector pointing from the sun to the planet. According to Newton's Third Law, the gravitational force on the sun is $-\mathbf{F}$. Thus

$$M\mathbf{X}'' = -\mathbf{F} = \frac{GMm}{r^3}\mathbf{r} \tag{7}$$

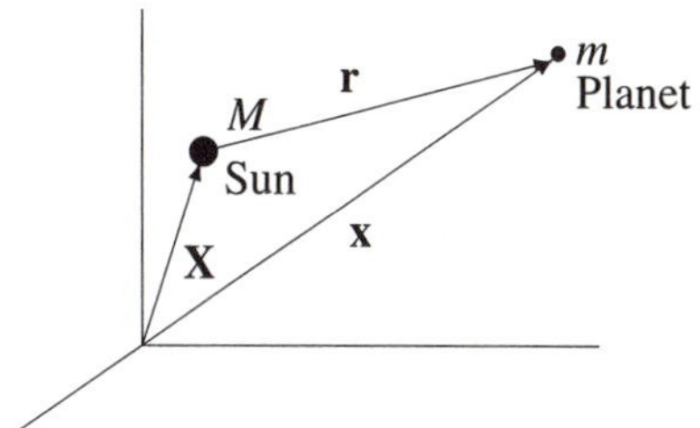


*Figure B.2.3 Inertial frames for a sun-planet system.*

Dividing (6) by $m$ and (7) by $M$ and then subtracting the two equations, we have

$$\mu \mathbf{r}'' = -\frac{GMm}{r^3}\mathbf{r} \tag{8}$$

where the **reduced mass** $\mu = mM/(M+m)$. Thus, we may imagine the sun to be fixed at the origin, while a planet of reduced mass $\mu$ moves around it under the influence of the **central force field** $(-GMm/r^3)\mathbf{r}$ which acts along the line through the centers of the sun and planet. The changing location of the planet relative to the sun is a solution of (8). The parametrically defined curve $\mathbf{r} = \mathbf{r}(t)$ is said to be the **orbit** of the planet (this is the traditional meaning of the word, although it differs from the word used elsewhere in the workbook).

By Newton's time it was clear that a planet moves in a plane containing the sun. Each planet travels in its own plane, but all are very close to the **plane of the ecliptic**, which contains the orbit of the earth.[5] The challenge was to prove the planarity of the orbit directly from the equation of motion (8).

First recall that the **cross product** $\mathbf{v} \times \mathbf{w}$ of two geometric vectors is a vector whose magnitude is $\|\mathbf{v}\| \cdot \|\mathbf{w}\| \sin\theta$, where $\theta$ is the angle between the vectors, and whose direction is given by the right-hand rule. It can be shown that $\mathbf{v} \times \mathbf{w}$ can be calculated from the components of $\mathbf{v}$ and $\mathbf{w}$ in any right-handed frame (i.e., a frame $\{\hat{\mathbf{i}}, \hat{\mathbf{j}}, \hat{\mathbf{k}}\}$ where $\hat{\mathbf{i}} \times \hat{\mathbf{j}} = \hat{\mathbf{k}}$) by evaluating a determinant as follows:

$$\mathbf{v} \times \mathbf{w} = \begin{vmatrix} \hat{\mathbf{i}} & \hat{\mathbf{j}} & \hat{\mathbf{k}} \\ v_1 & v_2 & v_3 \\ w_1 & w_2 & w_3 \end{vmatrix}$$

Thus, if $\mathbf{v}(t)$ and $\mathbf{w}(t)$ are differentiable vector functions, then so is $\mathbf{v}(t) \times \mathbf{w}(t)$, and its derivative follows the usual product rule

$$(\mathbf{v}(t) \times \mathbf{w}(t))' = \mathbf{v}'(t) \times \mathbf{w}(t) + \mathbf{v}(t) \times \mathbf{w}'(t)$$

Now we show that the "planet" moves in a plane that includes the "sun." Observe from (8) that $\mathbf{r} \times \mu\mathbf{r}'' = \mathbf{0}$, the zero vector. Thus,

$$\frac{d}{dt}(\mathbf{r} \times \mu\mathbf{r}') = \mathbf{r}' \times \mu\mathbf{r}' + \mathbf{r} \times \mu\mathbf{r}'' = \mathbf{r} \times \mu\mathbf{r}'' = \mathbf{0}$$

---

5. Pluto's orbital plane has the greatest inclination to the ecliptic: 17°.

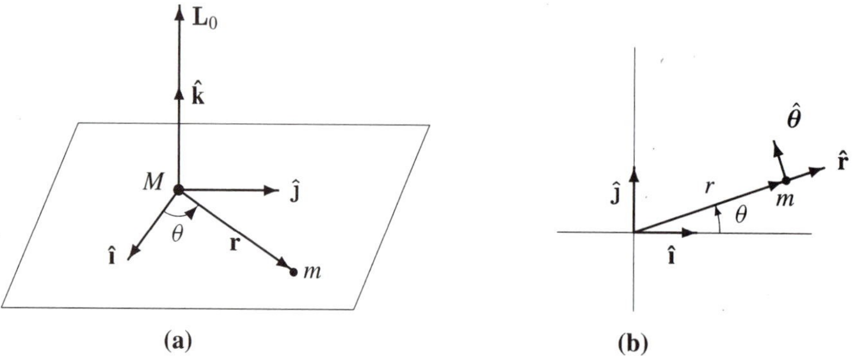


*Figure B.2.4 (a) Coordinate frame for the two-body problem. (b) Rotating frame in the plane.*

for all $t$, and the **angular momentum vector** $\mathbf{r} \times \mu\mathbf{r}'$ is a constant vector, say, $\mathbf{L}_0$. Hence, $\mathbf{r}$ and $\mathbf{r}'$ are always in a plane orthogonal to the constant vector $\mathbf{L}_0$.

Finally, we derive the equations of motion of the mass $m$ relative to the mass $M$. From the center of mass $M$ "draw" the constant angular momentum vector $\mathbf{L}_0$ and a plane through the center of $M$ orthogonal to $\mathbf{L}_0$. Find an inertial frame with $\hat{\mathbf{i}}$ and $\hat{\mathbf{j}}$ in this plane and $\hat{\mathbf{k}}$ parallel to $\mathbf{L}_0$, and introduce polar coordinates $(r, \theta)$ in the plane as shown in Figure B.2.4. Next, set up a moving reference frame with unit vectors $\hat{\mathbf{r}}$, $\hat{\boldsymbol{\theta}}$ in the plane of the planet's orbit as shown in Figure B.2.4. Namely, $\hat{\mathbf{r}}$ points in the direction of $\mathbf{r}$ and $\hat{\boldsymbol{\theta}}$ arises by a 90° counterclockwise rotation of $\hat{\mathbf{r}}$. Hence, $\hat{\mathbf{r}} = \cos\theta\hat{\mathbf{i}} + \sin\theta\hat{\mathbf{j}}$, and $\hat{\boldsymbol{\theta}} = -\sin\theta\hat{\mathbf{i}} + \cos\theta\hat{\mathbf{j}}$, and so we have the following differentiation formulas:

$$\hat{\mathbf{r}}' = \theta'\hat{\boldsymbol{\theta}} \qquad \text{and} \qquad \hat{\boldsymbol{\theta}}' = -\theta'\hat{\mathbf{r}} \tag{9}$$

Thus, using (9) we have

$$\mathbf{r}' = (r\hat{\mathbf{r}})' = r'\hat{\mathbf{r}} + r\hat{\mathbf{r}}' = r'\hat{\mathbf{r}} + r\theta'\hat{\boldsymbol{\theta}} \tag{10}$$

$$\mathbf{r}'' = (\mathbf{r}')' = (r'' - (\theta')^2 r)\hat{\mathbf{r}} + (\theta'' r + 2\theta' r')\hat{\boldsymbol{\theta}} \tag{11}$$

Now comparing (8) and (11) and matching coefficients of $\hat{\mathbf{r}}$ and $\hat{\boldsymbol{\theta}}$ we finally obtain

$$\begin{aligned} r'' - (\theta')^2 r &= -\frac{G(M+m)}{r^2} \\ \theta'' r + 2\theta' r' &= 0 \end{aligned} \tag{12}$$

which are **Newton's equations of planetary motion.** Observe that the equations are coupled, nonlinear, and autonomous, and that the corresponding state space is four-dimensional (the state variables are $r$, $r'$, $\theta$, and $\theta'$). Experiment 5.11 has further details.

## *Rotational Motion of Rigid Bodies*

The motion of a rigid body relative to an inertial frame in 3-space can be described by (1) the position vector of its center of mass and (2) the orientation of a fixed coordinate frame on the body attached to the center of mass. Newton's Laws of Dynamics can be used to derive the equations of motion of a rigid body, a task we initiate here and complete in Experiment 6.9.

## *Derivatives in Fixed and Moving Frames*

Let $\{\hat{\mathbf{i}}, \hat{\mathbf{j}}, \hat{\mathbf{k}}\}$ and $\{\hat{\mathbf{i}}^*, \hat{\mathbf{j}}^*, \hat{\mathbf{k}}^*\}$ be two frames and let $(x_1, x_2, x_3)$ and $(x_1^*, x_2^*, x_3^*)$ be the coordinates of the vector $\mathbf{x}$ in each of these frames. Using the formalism of matrix algebra, if coordinates are always written as column vectors, we can simply write

$$\mathbf{x} = \begin{bmatrix} \hat{\mathbf{i}} & \hat{\mathbf{j}} & \hat{\mathbf{k}} \end{bmatrix} \begin{bmatrix} x_1 \\ x_2 \\ x_3 \end{bmatrix} = \begin{bmatrix} \hat{\mathbf{i}}^* & \hat{\mathbf{j}}^* & \hat{\mathbf{k}}^* \end{bmatrix} \begin{bmatrix} x_1^* \\ x_2^* \\ x_3^* \end{bmatrix} \tag{13}$$

Denoting the coordinate column vectors by $x$ and $x^*$, we derive a relation between them. There is a matrix $U$ such that

$$\begin{bmatrix} \hat{\mathbf{i}}^* \\ \hat{\mathbf{j}}^* \\ \hat{\mathbf{k}}^* \end{bmatrix} = U \begin{bmatrix} \hat{\mathbf{i}} \\ \hat{\mathbf{j}} \\ \hat{\mathbf{k}} \end{bmatrix} = \begin{bmatrix} \alpha_1 & \alpha_2 & \alpha_3 \\ \beta_1 & \beta_2 & \beta_3 \\ \gamma_1 & \gamma_2 & \gamma_3 \end{bmatrix} \begin{bmatrix} \hat{\mathbf{i}} \\ \hat{\mathbf{j}} \\ \hat{\mathbf{k}} \end{bmatrix} \tag{14}$$

using the usual matrix multiplication scheme. From the fact that $\{\hat{\mathbf{i}}, \hat{\mathbf{j}}, \hat{\mathbf{k}}\}$ and $\{\hat{\mathbf{i}}^*, \hat{\mathbf{j}}^*, \hat{\mathbf{k}}^*\}$ are frames, we see that (denoting the rows of $U$ by the row vectors $\alpha$, $\beta$, and $\gamma$)

$$\begin{aligned} \alpha\alpha^T &= \beta\beta^T = \gamma\gamma^T = 1 \\ \alpha\beta^T &= \alpha\gamma^T = \beta\gamma^T = 0 \end{aligned} \tag{15}$$

hence $U$ is an **orthogonal matrix**. Thus, $UU^T = U^TU = E$, where $E$ is the $3 \times 3$ identity matrix and $U^T = U^{-1}$. Now (13) and (14) show that for all vectors $\mathbf{x}$

$$x^* = Ux$$

It is useful to note that in any frame, $\mathbf{x} \cdot \mathbf{y}$ and $\mathbf{x} \times \mathbf{y}$ can be expressed as

$$\mathbf{x} \cdot \mathbf{y} = x^T y \tag{16a}$$

and

$$\mathbf{x} \times \mathbf{y} = x^T \begin{bmatrix} 0 & -y_3 & y_2 \\ y_3 & 0 & -y_1 \\ -y_2 & y_1 & 0 \end{bmatrix} \begin{bmatrix} \hat{\mathbf{i}} \\ \hat{\mathbf{j}} \\ \hat{\mathbf{k}} \end{bmatrix} \tag{16b}$$

Now we shall begin our introduction to rigid body mechanics with little more than the formalism given above.

Let the frame $\{\hat{\mathbf{i}}, \hat{\mathbf{j}}, \hat{\mathbf{k}}\}$ be *fixed* and the frame $\{\hat{\mathbf{i}}^*, \hat{\mathbf{j}}^*, \hat{\mathbf{k}}^*\}$ *move* smoothly with time. Then the elements of $U$ are differentiable functions of $t$; that is, the coordinates of $\hat{\mathbf{i}}^*$, $\hat{\mathbf{j}}^*$, and $\hat{\mathbf{k}}^*$ in the fixed frame are differentiable functions of time. Now for vectors we shall introduce two time derivatives: time derivatives in the fixed frame shall be denoted by $d/dt$, whereas time derivatives in the moving frame shall be denoted by $d^*/dt$. These time derivatives are obviously related, and it is important in the applications to display this relation explicitly. First note that for any vector function of time, $\mathbf{R}(t)$, we compute $d\mathbf{R}/dt$ and $d^*\mathbf{R}/dt$ as follows: if

$$\mathbf{R} = R_1\hat{\mathbf{i}} + R_2\hat{\mathbf{j}} + R_3\hat{\mathbf{k}} = R_1^*\hat{\mathbf{i}}^* + R_2^*\hat{\mathbf{j}}^* + R_3^*\hat{\mathbf{k}}^*$$

then by definition of the derivatives $d/dt$ and $d^*/dt$, we have

$$\frac{d\mathbf{R}}{dt} = \frac{dR_1}{dt}\hat{\mathbf{i}} + \frac{dR_2}{dt}\hat{\mathbf{j}} + \frac{dR_3}{dt}\hat{\mathbf{k}} \qquad \frac{d^*\mathbf{R}}{dt} = \frac{dR_1^*}{dt}\hat{\mathbf{i}}^* + \frac{dR_2^*}{dt}\hat{\mathbf{j}}^* + \frac{dR_3^*}{dt}\hat{\mathbf{k}}^* \tag{17}$$

Now by applying $d/dt$ to both sides of (14) we observe that

$$\begin{bmatrix} d\hat{\mathbf{i}}^*/dt \\ d\hat{\mathbf{j}}^*/dt \\ d\hat{\mathbf{k}}^*/dt \end{bmatrix} = \frac{dU}{dt} \begin{bmatrix} \hat{\mathbf{i}} \\ \hat{\mathbf{j}} \\ \hat{\mathbf{k}} \end{bmatrix} = \frac{dU}{dt} U^T \begin{bmatrix} \hat{\mathbf{i}}^* \\ \hat{\mathbf{j}}^* \\ \hat{\mathbf{k}}^* \end{bmatrix} \tag{18}$$

Using the orthogonality of the rows of $U$, from (15), we see that

$$\frac{dU}{dt}U^T = \begin{bmatrix} \frac{d\alpha}{dt}\alpha^T & \frac{d\alpha}{dt}\beta^T & \frac{d\alpha}{dt}\gamma^T \\ \frac{d\beta}{dt}\alpha^T & \frac{d\beta}{dt}\beta^T & \frac{d\beta}{dt}\gamma^T \\ \frac{d\gamma}{dt}\alpha^T & \frac{d\gamma}{dt}\beta^T & \frac{d\gamma}{dt}\gamma^T \end{bmatrix} = \begin{bmatrix} 0 & \frac{d\alpha}{dt}\beta^T & -\frac{d\gamma}{dt}\alpha^T \\ -\frac{d\alpha}{dt}\beta^T & 0 & \frac{d\beta}{dt}\gamma^T \\ \frac{d\gamma}{dt}\alpha^T & -\frac{d\beta}{dt}\gamma^T & 0 \end{bmatrix}$$

Now let $\omega$ be a vector whose components in the starred frame are

$$\omega_1^* = -\frac{d\beta}{dt}\cdot\gamma, \qquad \omega_2^* = -\frac{d\gamma}{dt}\cdot\alpha, \qquad \omega_3^* = -\frac{d\alpha}{dt}\cdot\beta \tag{19}$$

The equations are known as the **equations of Poisson**. Then using (16), the identity (18) can be summarized as follows:

$$\frac{d\hat{\mathbf{i}}^*}{dt} = \omega\times\hat{\mathbf{i}}^*, \qquad \frac{d\hat{\mathbf{j}}^*}{dt} = \omega\times\hat{\mathbf{j}}^*, \qquad \frac{d\hat{\mathbf{k}}^*}{dt} = \omega\times\hat{\mathbf{k}}^* \tag{20}$$

Now applying $d/dt$ to $\mathbf{R} = R_1^*(t)\hat{\mathbf{i}}^* + R_2^*(t)\hat{\mathbf{j}}^* + R_3^*(t)\hat{\mathbf{k}}^*$ and using (20), we have

$$\begin{aligned} \frac{d\mathbf{R}}{dt} &= \frac{dR_1^*}{dt}\hat{\mathbf{i}}^* + \frac{dR_2^*}{dt}\hat{\mathbf{j}}^* + \frac{dR_3^*}{dt}\hat{\mathbf{k}}^* + R_1^*\frac{d\hat{\mathbf{i}}^*}{dt} + R_2^*\frac{d\hat{\mathbf{j}}^*}{dt} + R_3^*\frac{d\hat{\mathbf{k}}^*}{dt} \\ &= \frac{d^*\mathbf{R}}{dt} + R_1^*\omega\times\hat{\mathbf{i}}^* + R_2^*\omega\times\hat{\mathbf{j}}^* + R_3^*\omega\times\hat{\mathbf{k}}^* \end{aligned}$$

$$\frac{d\mathbf{R}}{dt} = \frac{d^*\mathbf{R}}{dt} + \omega\times\mathbf{R} \tag{21}$$

which gives the connection between the two derivatives $d/dt$ and $d^*/dt$. The vector $\omega(t)$ is called the **angular velocity** of the moving frame with respect to a fixed frame at time $t$.

## *Dynamics of Mass Systems*

Now consider a system of $N$ particles of masses $m_1, m_2, \ldots, m_N$ which obey Newton's laws. Choose an origin in an inertial frame and measure all position vectors from this point. See Figure B.2.5. Let $\mathbf{R}$ be the position vector for the center of mass $C$. Then

$$M\mathbf{R} = \sum_{k=1}^{N} m_k\mathbf{r}_k \tag{22}$$

where $\mathbf{r}_k$ is the position vector for the $k^{\text{th}}$ particle, and $M = \sum m_k$. Let $\mathbf{F}_k^e$ and $\mathbf{F}_k^i$ denote the forces on $m_k$ due to external and internal sources, respectively. Summing Newton's Equations of Motion for each particle, we obtain

$$\sum_{k=1}^{N} m_k\frac{d^2\mathbf{r}_k}{dt^2} = \sum_{k=1}^{N}\mathbf{F}_k^e + \sum_{k=1}^{N}\mathbf{F}_k^i \tag{23}$$

Now using (22) and the fact that $\sum\mathbf{F}_k^i = 0$ (from Newton's Third Law), we see that (23) becomes

$$M\frac{d^2\mathbf{R}}{dt^2} = \sum_{k=1}^{N}\mathbf{F}_k^e = \mathbf{F}$$

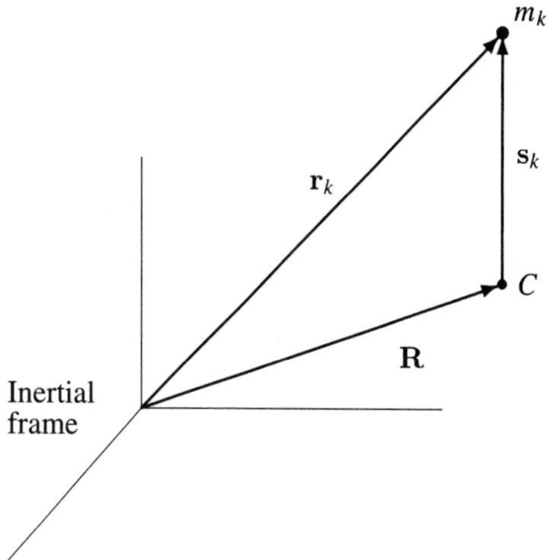


*Figure B.2.5 Location of $m_k$ from center of mass C of the system.*

The total angular momentum of the system about the center of mass $C$ is given by

$$\mathbf{L}_C = \sum_{k=1}^{N} m_k \mathbf{s}_k \times \frac{d\mathbf{s}_k}{dt}$$

where $\mathbf{s}_k = \mathbf{r}_k - \mathbf{R}$ locates the $k^{\text{th}}$ particle from the center of mass $C$. See Figure B.2.5. Observe that

$$\frac{d\mathbf{L}_C}{dt} = \sum_{k=1}^{N} m_k \mathbf{s}_k \times \frac{d^2\mathbf{s}_k}{dt^2} \tag{24}$$

On the other hand, Newton's equations for the $k^{\text{th}}$ particle imply that

$$\mathbf{s}_k \times m_k \left( \frac{d^2\mathbf{R}}{dt^2} + \frac{d^2\mathbf{s}_k}{dt^2} \right) = \mathbf{s}_k \times (\mathbf{F}_k^e + \mathbf{F}_k^i) \tag{25}$$

and summing over all the particles, we obtain

$$\left( \sum_{k=1}^{N} m_k \mathbf{s}_k \right) \times \frac{d^2\mathbf{R}}{dt^2} + \sum_{k=1}^{N} m_k \mathbf{s}_k \times \frac{d^2\mathbf{s}_k}{dt^2} = \sum_{k=1}^{N} \mathbf{s}_k \times \mathbf{F}_k^e + \sum_{k=1}^{N} \mathbf{s}_k \times \mathbf{F}_k^i$$

but the first term vanishes, since $\sum m_k \mathbf{s}_k = 0$ (why?) and the last term vanishes also (from Newton's Third Law again). By definition, the **external torque** acting on the system with respect to $C$ is

$$\mathbf{N}_C = \sum_{k=1}^{N} \mathbf{s}_k \times \mathbf{F}_k^e$$

and so (25) becomes

$$\sum_{k=1}^{N} m_k \mathbf{s}_k \times \frac{d^2\mathbf{s}_k}{dt^2} = \sum_{k=1}^{N} \mathbf{s}_k \times \mathbf{F}_k^e = \mathbf{N}_C \tag{26}$$

Using (26) we see that (24) becomes

$$\frac{d\mathbf{L}_C}{dt} = \mathbf{N}_C$$

## Rigid Bodies

Everything above is valid for any system of particles. If the particles compose a rigid body, then we can find a somewhat different expression for $\mathbf{L}_C$. Now fix a frame in the rigid body, having its origin at the center of mass, and let $\boldsymbol{\omega}$ be the angular velocity of the moving frame relative to the fixed one. Then we evaluate $d\mathbf{s}_k/dt$ as follows:

$$\frac{d\mathbf{s}_k}{dt} = \frac{d^*\mathbf{s}_k}{dt} + \boldsymbol{\omega} \times \mathbf{s}_k$$

Recalling that $\mathbf{s}_k$ is fixed in the body (hence $d^*\mathbf{s}_k/dt = 0$) it follows that

$$\mathbf{L}_C = \sum_{k=1}^{N} m_k \mathbf{s}_k \times \frac{d\mathbf{s}_k}{dt} = \sum_{k=1}^{N} m_k \mathbf{s}_k \times (\boldsymbol{\omega} \times \mathbf{s}_k) = \sum_{k=1}^{N} m_k \{(\mathbf{s}_k \cdot \mathbf{s}_k)\boldsymbol{\omega} - (\mathbf{s}_k \cdot \boldsymbol{\omega})\mathbf{s}_k\}$$

Rewriting the last term in the frame $\{\hat{\mathbf{i}}, \hat{\mathbf{j}}, \hat{\mathbf{k}}\}$, we obtain

$$(\mathbf{s}_k \cdot \boldsymbol{\omega})\mathbf{s}_k = \mathbf{s}_k(\mathbf{s}_k \cdot \boldsymbol{\omega}) = [\ \hat{\mathbf{i}} \quad \hat{\mathbf{j}} \quad \hat{\mathbf{k}}\ ]\, s_k s_k^T \omega$$

Therefore in the frame $\{\hat{\mathbf{i}}, \hat{\mathbf{j}}, \hat{\mathbf{k}}\}$, $\mathbf{L}_C$ has the form

$$\mathbf{L}_C = [\ \hat{\mathbf{i}} \quad \hat{\mathbf{j}} \quad \hat{\mathbf{k}}\ ] \left( \sum_{k=1}^{N} m_k (s_k^T s_k E - s_k s_k^T) \right) \omega \tag{27}$$

where $s_k$ and $\omega$ are column vectors of coordinates of $\mathbf{s}_k$ and $\boldsymbol{\omega}$ in the $\{\hat{\mathbf{i}}, \hat{\mathbf{j}}, \hat{\mathbf{k}}\}$ frame, and $E$ is the $3 \times 3$ identity matrix. Observe that in the frame $\{\hat{\mathbf{i}}^*, \hat{\mathbf{j}}^*, \hat{\mathbf{k}}^*\}$ we would have

$$\mathbf{L}_C = [\ \hat{\mathbf{i}}^* \quad \hat{\mathbf{j}}^* \quad \hat{\mathbf{k}}^*\ ] \left( \sum_{k=1}^{N} m_k (s_k^{*T} s_k^* E - s_k^* s_k^{*T}) \right) \omega^*$$

where $s_k^*$ and $\omega^*$ are column vectors of the coordinates of $\mathbf{s}_k$ and $\boldsymbol{\omega}$ in the starred frame $\{\hat{\mathbf{i}}^*, \hat{\mathbf{j}}^*, \hat{\mathbf{k}}^*\}$. Now let us define an "object" $\mathcal{I}$, which acts on a vector $\boldsymbol{\omega}$ to produce another vector denoted by $\mathcal{I}\boldsymbol{\omega}$, whose coordinate column vector in any frame $\{\hat{\mathbf{i}}, \hat{\mathbf{j}}, \hat{\mathbf{k}}\}$ is given by the matrix product $I\omega$, where $\omega$ is the column vector of $\boldsymbol{\omega}$ in the $\{\hat{\mathbf{i}}, \hat{\mathbf{j}}, \hat{\mathbf{k}}\}$ frame, and the $3 \times 3$ matrix $I$ is given by

$$I = \sum_{k=1}^{N} m_k (s_k^T s_k E - s_k s_k^T) \tag{28}$$

Thus, for rigid bodies we have

$$\mathbf{L}_C = \mathcal{I}\boldsymbol{\omega} \tag{29}$$

and the equations of motion are

$$\begin{aligned} M\frac{d^2\mathbf{R}}{dt^2} &= \mathbf{F} \\ \frac{d\mathbf{L}_C}{dt} &= \mathbf{N}_C \end{aligned} \tag{30}$$

where $\mathbf{F}$ is the resultant external force acting on the body and $\mathbf{N}_C$ is the external torque (defined earlier) acting on the body about its center of mass $C$.

To see that the equations (30) completely describe the motion of the rigid body, we must first locate the body frame with reference to the fixed frame by any variety of parameters (for example, Euler angles, $\theta_1$, $\theta_2$, and $\theta_3$). Then the elements of the orientation matrix $U$ are known functions of these parameters. Thus, if the angular velocity vector $\boldsymbol{\omega}(t)$ is known by solving the system (30), the equations of Poisson (19) will determine the elements of the orientation matrix $U$ as functions of time. Thus, if the vector functions $\mathbf{R}(t)$ and $\boldsymbol{\omega}(t)$

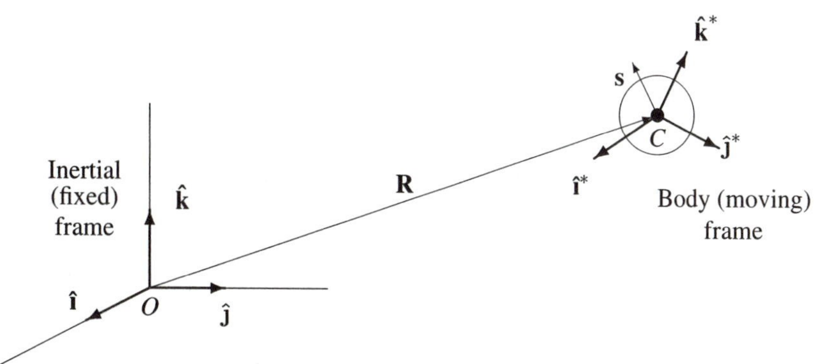


*Figure B.2.6 Fixed and moving frames for a rigid body.*

solve the system (30), the center of mass $C$ of the rigid body is known, and from $\omega(t)$ the orientation of the body relative to the fixed frame is known. See Figure B.2.6.

Several special cases are worth noting. If torques due to external forces vanish, then angular momentum is conserved; that is, $\mathbf{L}_C$ is an integral of the second equation in (30). As another special case, assume that the external force on the body is independent of the angular velocity and that the torques about the center of mass due to external forces are independent of the position of the center of mass. In this case the two equations in (30) decouple and thus may be solved separately.

Two final remarks. Consider the $3 \times 3$ matrix representations $I$ and $I^*$ of the "object" $\mathcal{I}$ in the frames $\{\hat{\mathbf{i}}, \hat{\mathbf{j}}, \hat{\mathbf{k}}\}$ and $\{\hat{\mathbf{i}}^*, \hat{\mathbf{j}}^*, \hat{\mathbf{k}}^*\}$, respectively, as described in (28). Suppose these frames are related by the orthogonal matrix $U$ as in (14). Then the coordinate column vectors $s_k$ and $s_k^*$ of the vector $\mathbf{s}_k$ in the two frames are related by $s_k^* = Us_k$. Thus,

$$\begin{aligned} I^* &= \sum_{k=1}^{N} m_k(s_k^{*T} s_k^* E - s_k^* s_k^{*T}) = \sum_{k=1}^{N} m_k(s_k^T U^T U s_k E - U s_k s_k^T U^T) \\ &= U\left\{\sum_{k=1}^{N} m_k(s_k^T s_k E - s_k s_k^T)\right\} U^T = U I U^T \end{aligned} \tag{31}$$

since $s_k^T s_k$ is a scalar and $U^T U = U U^T = E$. Any "object" like $\mathcal{I}$ whose representation $I$ in a frame transforms from frame to frame as above is called a **Cartesian tensor**. The "object" $\mathcal{I}$ with its representation $I$ defined by (28) in a given frame is called the **inertia tensor** of the body.

If the mass of the body $B$ is distributed continuously with density $\rho$, then the representation $I$ of the inertia tensor $\mathcal{I}$ in a frame (with origin at $C$, the center of mass) is given by

$$I = \int_B (x^T x E - x x^T) \rho \, dV$$

where $[x_1 \; x_2 \; x_3]^T$ is the coordinate column vector in a given frame.

## B.3 Chemical Reactions

Iron blackens when heated in air. Aspirin reduces inflammation. Enzymes turn starches into simple sugars. These are examples of chemical reactions. In this part of the Appendix the basic ideas of chemical reactions are presented and the connection with ODEs explained.

## Net Reactions

In any chemical reaction some species, the **reactants**, interact and form other species, the **products**. For example, iron (Fe) and molecular oxygen ($O_2$) react when heated to form black iron oxide ($Fe_3O_4$). This reaction is denoted by

$$3\text{Fe} + 2\text{O}_2 \rightarrow \text{Fe}_3\text{O}_4 \tag{1}$$

The arrow shows the direction of the reaction. The reactants are at the foot of the arrow and the products at the head. The coefficients show the relative amounts: three units of iron and two units of oxygen are used to produce one unit of iron oxide. More generally, a **net reaction** views only the initial reactants and the final products and ignores the intermediate chemical species produced in the process. A net reaction is denoted by

$$r_1 R_1 + r_2 R_2 + \cdots + r_n R_n \rightarrow p_1 P_1 + \cdots + p_k P_k \tag{2}$$

where the (stoichiometric) coefficients $r_i$ and $p_j$ (integers or fractions) are the relative amounts of the reactants $R_i$ and the products $P_j$.

In reality most chemical reactions involve a multitude of net reactions (called **steps**), each of which has the form of (2). In many reactions the products themselves simultaneously re-form into the original reactants, thus constituting a reverse reaction. The chemical process shown below involves three steps, all of which occur simultaneously.

$$E + S \rightleftharpoons C \rightarrow P + E$$

Observe that the first reaction is reversible and that $E$ is both a reactant and a product. More complex reactions may involve hundreds of species and steps.

## Reaction Velocity

The **reaction velocity** (or **rate**) of a step is the rate of formation of any one of the products in that step divided by the corresponding coefficient in the net reaction (2), or, equivalently, the negative of the rate of disappearance of any one reactant divided by its corresponding coefficient. For example, the velocity of reaction (1) is given by any of three quantities:

$$\text{Velocity} = \frac{d[\text{Fe}_3\text{O}_4]}{dt} = -\frac{1}{3}\frac{d[\text{Fe}]}{dt} = -\frac{1}{2}\frac{d[\text{O}_2]}{dt}$$

where $[\ldots]$ denotes concentration and $t$ is time. The coefficients $\frac{1}{3}$ and $\frac{1}{2}$ reflect the fact that iron and oxygen are used up three times and twice as fast, respectively, as iron oxide is produced. In general, the reciprocal of the coefficient of a species appearing in a net reaction is used for finding the velocity. For example, the velocity in reaction (2) is

$$\text{Velocity} = -\frac{1}{r_1}\frac{d[R_1]}{dt} = \cdots = \frac{1}{p_k}\frac{d[P_k]}{dt}$$

Concentrations are measured in **molarities**, one molarity being a mole of species per liter of medium. Time is usually measured in seconds, although a fast reaction may be over in a microsecond and a slow reaction may last a millennium. The reaction is assumed to occur in a constant volume environment.

## Rate Laws

The velocity of a reaction step depends upon the concentrations of the reactants and, perhaps, of the products. The **rate law** of the reaction is the explicit formulation of this dependence. For reactions of certain kinds, called **elementary reactions** the rate law can be found directly from the net reaction. There is a basic principle behind elementary reactions:

**Law of Mass Action:** *The velocity of an elementary reaction is proportional to the product of the concentrations of the reactants.*

The Law of Mass Action is based on the assumption that elementary reactions occur when molecules of all the different reactants are in simultaneous physical (or chemical) contact. Hence, the higher the concentrations, the higher the velocity. Three elementary reactions for distinct chemical species are listed below:

1. **Unimolecular**: $A \xrightarrow{k} P$

   Velocity $= \dfrac{d[P]}{dt} = -\dfrac{d[A]}{dt}$ Rate law: $\dfrac{d[P]}{dt} = k[A]$

2. **Bimolecular**: $A + B \xrightarrow{k} P$

   Velocity $= \dfrac{d[P]}{dt} = -\dfrac{d[A]}{dt} = -\dfrac{d[B]}{dt}$ Rate law: $\dfrac{d[P]}{dt} = k[A][B]$

3. **Termolecular**: $A + B + C \xrightarrow{k} P$

   Velocity $= \dfrac{d[P]}{dt} = -\dfrac{d[A]}{dt} = -\dfrac{d[B]}{dt} = -\dfrac{d[C]}{dt}$ Rate law: $\dfrac{d[P]}{dt} = k[A][B][C]$

The coefficient $k$ is the **rate constant** and is always taken to be positive. The units of the rate constants depend upon the nature of the reaction. For example, the respective units for uni-, bi-, and termolecular reactions are $(\text{time})^{-1}$, $(\text{molar})^{-1} \cdot (\text{time})^{-1}$, and $(\text{molar})^{-2} \cdot (\text{time})^{-1}$. The values of the rate constants must be determined by measurements of the rates and the concentrations involved in the specific reactions.

Unimolecular and bimolecular reactions are common. Termolecular reactions are rare, since the simultaneous collision of three molecules is unlikely.

### *ODEs for Elementary Reactions*

Since rates are associated with reactions, and rates are derivatives with respect to time, it is not surprising that the kinetics of a chemical reaction are modeled by ordinary differential equations. The ordinary differential equations that model the three elementary reactions described previously are listed below. A rate equation may be written out for each reactant and product. For brevity, lowercase letters are used for concentrations (e.g., $[A]$ is replaced by $a$). It is assumed that each reaction begins at time 0 with initial concentrations $a_0$, $p_0$, $b_0$, and so on.

1. **Unimolecular**: $A \xrightarrow{k} P$

$$\text{Rate equations:} \begin{cases} a' = -ka \\ p' = ka = k(a_0 + p_0 - p) \end{cases}$$

2. **Bimolecular**: $A + B \xrightarrow{k} P$

$$\text{Rate equations:} \begin{cases} a' = -kab = -ka(a - a_0 + b_0) \\ b' = -kab = -kb(b - b_0 + a_0) \\ p' = kab = k(a_0 + p_0 - p)(b_0 + p_0 - p) \end{cases}$$

3. **Termolecular**: $A + B + C \xrightarrow{k} P$

$$\text{Rate equations:} \begin{cases} a' = -kabc = -ka(a - a_0 + b_0)(a - a_0 + c_0) \\ b' = -kabc = -kb(b - b_0 + a_0)(b - b_0 + c_0) \\ c' = -kabc = -kc(c - c_0 + a_0)(c - c_0 + b_0) \\ p' = kabc = k(a_0 + p_0 - p)(b_0 + p_0 - p)(c_0 + p_0 - p) \end{cases}$$

Observe that each rate equation has been written in a form that involves a single state variable. For example, the first rate equation for the unimolecular reaction involves the state variable $a$ alone, while the second equation is only in terms of the state variable $p$, and the constants $p_0$ and $a_0$. This simplification is possible because for each of these reactions there is an underlying law:

**Law of Conservation:** *The sum of the concentrations of the product species and of any one of the reactants remains constant throughout the constant volume reaction.*

For example, in the bimolecular case, we have

$$b' + p' = -kab + kab = 0$$

from which it follows that

$$b(t) + p(t) = \text{constant} = b_0 + p_0$$

Hence, $b(t) = b_0 + p_0 - p(t)$. The other rate equations can be simplified in much the same way. These simplifications reduce the number of ODEs to be solved. For example, in the bimolecular reaction, once the rate equation

$$p' = k(a_0 + p_0 - p)(b_0 + p_0 - p), \quad p(0) = p_0$$

has been solved for $p(t)$, the concentrations of the reactants $A$ and $B$ can be found from the formulas

$$a(t) = a_0 + p_0 - p(t), \quad b(t) = b_0 + p_0 - p(t)$$

The elementary reactions for distinct species are represented quite simply by ODEs, but for nonelementary reactions the representation problem and corresponding Law of Conservation are more complicated. Generally speaking, additional chemical experiment and observation are needed to suggest ways to depict the overall reaction as a set of elementary steps whose rate equations can be written out.

## *Examples of Elementary Reactions*

***Example 1*** **(Radioactive decay)** A well known example of a unimolecular elementary rate law is radioactive decay. For example, an isotope of carbon $^{14}C$ decays to ordinary $^{14}N$, emitting an electron in the process:

$$^{14}\text{C} \xrightarrow{k} {}^{14}\text{N}$$

$$\text{Rate law:} \quad \frac{d[^{14}\text{C}]}{dt} = -k[^{14}\text{C}]$$

where $k = 1.245 \times 10^{-4}(\text{year})^{-1}$. This decay process is used to date samples of once-living tissue. See Experiment 1.6.

***Example 2*** **(Uranium series)** The slow decay of uranium $^{238}U$ to lead $^{206}Pb$ is an example of a long chain of first order elementary steps of radioactive decay. Only the initial and final steps are shown below:

$$^{238}\text{U} \xrightarrow{k_1} {}^{234}\text{Th} \xrightarrow{k_2} \cdots \xrightarrow{k_{14}} {}^{210}\text{Bi} \begin{array}{c} \overset{k_{15}}{\nearrow} {}^{210}\text{Po} \overset{k_{17}}{\searrow} \\ \underset{k_{16}}{\searrow} {}^{206}\text{Tl} \underset{k_{18}}{\nearrow} \end{array} {}^{206}\text{Pb}$$

The elements shown are uranium, thorium (Th), bismuth, polonium, thallium (Tl), and lead (Pb). The rate constants shown range from $k_1 = 1.540 \times 10^{-10}(\text{year})^{-1}$ to $k_{18} = 8.673 \times 10^4(\text{year})^{-1}$. A sample of uranium ore should contain large amounts of uranium and lead, the former because $k_1$ is so small (corresponding to a half-life of 4.5 billion years), the latter because lead $^{206}$Pb is stable and doesn't decay. There will be only a trace of thallium because $k_{18}$ is so large. The rate equations for the segments shown are

$$\frac{d[^{238}\text{U}]}{dt} = -k_1[^{238}\text{U}]$$

$$\frac{d[^{234}\text{Th}]}{dt} = k_1[^{238}\text{U}] - k_2[^{234}\text{Th}]$$

$$\vdots$$

$$\frac{d[^{210}\text{Bi}]}{dt} = k_{14}[\cdot] - (k_{15} + k_{16})[^{210}\text{Bi}]$$

$$\frac{d[^{210}\text{Po}]}{dt} = k_{15}[^{210}\text{Bi}] - k_{17}[^{210}\text{Po}]$$

$$\frac{d[^{206}\text{Tl}]}{dt} = k_{16}[^{210}\text{Bi}] - k_{18}[^{206}\text{Tl}]$$

$$\frac{d[^{206}\text{Pb}]}{dt} = k_{17}[^{210}\text{Po}] + k_{18}[^{206}\text{Tl}]$$

Care must be taken in solving the system on a computer, since the rate constants have such a wide range of magnitudes.

*Example 3* **(An elementary bimolecular reaction)** The hydrogen molecule $H_2$ reacts with the iodine molecule $I_2$ to form the halogen acid HI. The net reaction is described by

$$H_2 + I_2 \to 2HI$$

$$\text{Velocity} = -\frac{d[H_2]}{dt} = -\frac{d[I_2]}{dt} = \frac{1}{2}\frac{d[HI]}{dt}$$

Experiment has shown that this reaction is an elementary bimolecular reaction. Using the Law of Mass Action, the rate equation takes the form

$$\text{Velocity} = k[H_2] \cdot [I_2]$$

Using the various forms for the velocity, the ODEs for the three chemical species are

$$x' = -kxy$$
$$y' = -kxy$$
$$z' = 2kxy$$

where $x = [H_2]$, $y = [I_2]$, $z = [HI]$. Observe that $x(t) + y(t) + z(t) = x(0) + y(0) + z(0)$, since $x' + y' + z' = 0$. The coefficient 2 in the third ODE corresponds to the coefficient 2 in the bimolecular reaction. In addition, $x(t) - y(t) = x(0) - y(0)$, since $x' - y' = 0$. These two conservation laws may be used to reduce the system of three ODEs to one (e.g.), $x' = -kx(x - x(0) + y(0))$.

### *Order of a Reaction*

Each of the ordinary differential equations modeling a reaction step is first order, since only first derivatives appear. However, that does not mean that the corresponding chemical reaction is first order. Indeed, a bimolecular reaction is said to be second order (chemically), since the rate law invokes the product of the concentrations of two reactants. Correspondingly, a termolecular reaction is third order. Among the elementary reactions, only in the unimolecular case does the order of the reaction coincide with the order of the ODE.

There is a more general notion of chemical order that includes the cases of the three elementary reactions. For example, suppose that the rate law of the nonelementary net reaction $A + B \to P$ is

$$\frac{d[P]}{dt} = k[A]^m[B]^n[P]^q \tag{3}$$

which is a common type of net rate law. Then the sum $m + n + q$ of the exponents is the **order of the reaction**, while the respective **orders of the components** $A$, $B$, and $P$ are $m$, $n$, and $q$. Usually, the exponents are integers $0, \pm 1, \pm 2, \pm 3$ or simple fractions $\pm\frac{1}{2}$ or $\pm\frac{3}{2}$. The values of $m$, $n$, and $q$ can be determined only by chemical experiment and observation. In any case, several intermediate reactions are likely to be present, and (3) only approximates the net overall reaction.

### *Complex Reactions*

Most reactions have a more complex structure than any considered so far. That is, the net reaction is not elementary, but represents the overall effect of a number of intermediate reactions, each of which is elementary. In addition, not only do many reactions move from reactants to products, but the products dissociate back into reactants as well. Example 4 takes up an apparent termolecular reaction that is actually a composite of three elementary bimolecular intermediate reactions. It should be emphasized again that experimental observations are critical in determining the nature of the reactions and values of the chemical constants.

***Example 4*** **(A three-step reaction)** The net reaction that describes the interaction between nitric oxide NO and bromine $Br_2$ to form NOBr is

$$2NO + Br_2 \to 2NOBr$$

where the velocity of the reaction is given by any one of the three quantities

$$\frac{1}{2}\frac{d[NOBr]}{dt} = -\frac{1}{2}\frac{d[NO]}{dt} = -\frac{d[Br_2]}{dt}$$

Chemical experiments show that the overall order of this reaction is 3, while the order of NO is 2 and that of $Br_2$ is 1. Since the implied termolecular elementary reaction is unlikely, intermediate reactions have been proposed to account for these orders. Additional chemical experiments suggest that the most likely intermediate elementary reactions are:

$$\begin{aligned} NO + Br_2 &\underset{k_{-1}}{\overset{k_1}{\rightleftharpoons}} NOBr_2 \\ NOBr_2 + NO &\overset{k_2}{\to} 2NOBr \end{aligned} \tag{4}$$

For brevity, let $x$, $y$, $z$, and $w$ denote the concentrations [NO], $[Br_2]$, $[NOBr_2]$, and [NOBr], respectively. In these variables (4) becomes

$$\begin{aligned} x + y &\underset{k_{-1}}{\overset{k_1}{\rightleftharpoons}} z \\ z + x &\overset{k_2}{\to} 2w \end{aligned} \tag{5}$$

The Law of Mass Action applies to each of the elementary reactions shown in (5). Keeping track of the locations of each of the four species in the three reactions and taking into account the rate constants, we have

$$\begin{aligned} x' &= -k_1xy + k_{-1}z - k_2zx \\ y' &= -k_1xy + k_{-1}z \\ z' &= k_1xy - k_{-1}z - k_2zx \\ w' &= 2k_2zx \end{aligned} \tag{6}$$

where the coefficient 2 in the fourth equation of (6) comes from the second reaction of (5).

System (6) is a set of four coupled ODEs in the state variables $x$, $y$, $z$, and $w$. We now explain how we can deduce from system (6) and some additional experimental results that the reaction is third order overall, second order in $x$ [NO] and first order in $y$ [$Br_2$].

Experiment shows that the intermediate species $z$ [$NOBr_2$] is produced very rapidly but disappears almost as fast as it is formed. Hence, $z$ remains small in magnitude and approximately constant. Thus, $z' \approx 0$. Equating the right-hand side of the third equation in (6) to 0 and solving for this **steady state** value of $z$, we have

$$z_{ss} = \frac{k_1xy}{k_{-1} + k_2x} \tag{7}$$

In addition, measurement of the rates shows that $k_2x$ remains much smaller than $k_{-1}$. Hence, we have

$$z_{ss} \approx \frac{k_1}{k_{-1}}xy \tag{8}$$

Finally, using $z$ as given by (8) in the rate law (6) for the production of $w$, we have

$$\frac{d[\mathrm{NOBr}]}{dt} \equiv w' = 2k_2 \cdot \frac{k_1}{k_{-1}}xy \cdot x = 2\frac{k_1k_2}{k_{-1}}x^2y = 2\frac{k_1k_2}{k_{-1}}[\mathrm{NO}]^2 \cdot [\mathrm{Br}_2]$$

and the overall order and the individual species orders are just as asserted.

From the viewpoint of solving ODEs, we may use a computer to solve the four rate equations of system (6), assuming that the three rate constants $k_1$, $k_{-1}$, and $k_2$ are known. Alternatively, one could use the steady state approximation for $z$ given by (7) (or by (8)) wherever $z$ appears in the rate equations. With this approach, system (6) reduces to three rate equations, one for each of $x$, $y$, and $w$. In addition, "conservation laws" may be used to reduce the number of equations. For example, since we have from (6)

$$2x' - 2y' + w' \equiv 0$$

we see that

$$2x(t) - 2y(t) + w(t) = 2x(0) - 2y(0) + w(0)$$

Hence, $w(t)$ can be found in terms of $x(t)$, $y(t)$ and initial data. However, the capabilities of computers and software mean that, rather than reducing the number of equations before solving, it is just as easy to solve the full set of equations.

*Example 5* **(A five-step reaction)** Molecules of bromine $Br_2$ and of hydrogen $H_2$ react to form molecules of the halogen acid HBr:

$$\mathrm{H}_2 + \mathrm{Br}_2 \to 2\mathrm{HBr}$$

$$\text{Velocity} = \frac{1}{2}\frac{d[\mathrm{HBr}]}{dt} = -\frac{d[\mathrm{H}_2]}{dt} = -\frac{d[\mathrm{Br}_2]}{dt}$$

Experimental observation has shown that the rate law is far from elementary:

$$\frac{1}{2}\frac{d[\mathrm{HBr}]}{dt} = \frac{k_a[\mathrm{H}_2] \cdot [\mathrm{Br}_2]^{1/2}}{k_b + [\mathrm{HBr}] \cdot [\mathrm{Br}_2]^{-1}} \tag{9}$$

Although the reaction is first order in hydrogen $H_2$, there is no easy way to define the order in the acid HBr or in the bromine molecule $Br_2$, nor is it clear that the reaction has an overall order. To account for such a complex rate law, the net reaction of hydrogen and bromine must be composed of several elementary mechanisms. It is currently believed that there are five elementary steps involving the species $Br_2$, Br, $H_2$, H, and HBr. These species and their respective concentrations are denoted by $x$, $y$, $z$, $w$, and $p$. The five steps are

$$\begin{aligned} x &\underset{k_{-1}}{\overset{k_1}{\rightleftharpoons}} 2y \\ y+z &\underset{k_{-2}}{\overset{k_2}{\rightleftharpoons}} w+p \\ x+w &\overset{k_3}{\rightarrow} y+p \end{aligned}$$

The rate equations for the five species are

$$\begin{aligned} x' &= -k_1 x + k_{-1} y^2 - k_3 x w \\ y' &= 2k_1 x - 2k_{-1} y^2 + k_3 x w - k_2 y z + k_{-2} w p \\ z' &= -k_2 y z + k_{-2} w p \\ w' &= -k_3 x w + k_2 y z - k_{-2} w p \\ p' &= k_3 x w + k_2 y z - k_{-2} w p \end{aligned} \tag{10}$$

where the 2's in the second rate equation come from the 2 in the $k_1 k_{-1}$ reactions. The complex rate law (9) may be derived from equation (10) if concentrations $y$ and $w$ are assumed to be in the steady state. Equate the rates in the second and fourth lines of (10) to zero and solve for $y_{ss}$ and $w_{ss}$ in terms of $x$, $z$ and $p$. Rate law (9) is obtained by replacing $y$ and $w$ in the last line of (10) by $y_{ss}$ and $w_{ss}$.

### Conclusion

Chemical reactions extend over time, with the concentrations of the species involved rising or falling. Given enough experimental measurements, rate laws can be found and elementary intermediate reactions tested. Then first order differential equations in the concentrations can be written out and ODE solvers and graphers brought in. Experiments 3.16–3.17 and 6.10–6.13 show some results that can be found this way, both expected (chemically and mathematically) and unexpected. Recent chemical research has shown that concentrations in a reaction may experience rapid oscillations before approaching equilibrium values. Other reactions may even appear to progress chaotically.

***References*** Consult the material below for additional background and details.

1. Campbell, J.A. *Chemical Systems.* Freeman, San Francisco, 1970.
2. Gray, P. and Scott, S.K. *Chemical Oscillations and Instabilities.* Clarendon Press, Oxford 1990.
3. Laidler, K.J. *Chemical Kinetics*, 3rd ed. McGraw-Hill, New York, 1987.

## B.4 Circuits

Among the many physical processes that can be modeled by ODEs one of the most important is the flow of electrical energy in a circuit.

### Electrical Units

A **circuit** consists of a closed loop (or loops) of **resistors**, **capacitors**, and **inductors** (described below) connected by wires. Figure B.4.1 is a schematic diagram of a simple circuit.

The **current** $I$ in a circuit is proportional to the number of free electrons (each with a constant negative charge) moving through any given point in the circuit per second. By convention, current flow is described in terms of positive charge carriers whose movement is opposite to that of electrons; that is, if the current in Figure B.4.1 is flowing clockwise, the actual movement of the electrons is counterclockwise. Current is measured in **amperes** (amps), one of the basic electrical units in the SI (*Système Internationale*) system. A current of one amp corresponds to $6.2420 \times 10^{18}$ positive charge carriers moving past a given point in one second.

The SI unit of charge is the **coulomb**, defined to be the amount of charge flowing through the cross section of a wire in one second when a current of one amp is flowing. Thus, one amp is one coulomb per second. If the current $I$ is not constant in time, an instantaneous current $I(t)$ flowing past a point in a circuit is defined in the same way as other instantaneous physical quantities. Thus if $I(t)$ is a continuous function, the amount of charge flowing past a point in a time interval $[t_0, t_1]$ is given by $\int_{t_0}^{t_1} I(t)\,dt$.

As current moves through a circuit, the charge carriers either impart energy to a circuit element or receive energy from a circuit element. The energy per coulomb of charge that has been imparted (or received) by the charge carriers between points $a$ and $b$ (denoted by $V_{ab}$) is computed as $V_{ab} = V_a - V_b$, where $V_a$ and $V_b$ are the values of the **potential function** $V$ at points $a$ and $b$ of the circuit. The difference $V_{ab}$ is called the **voltage drop** or **potential difference** and is measured in joules per coulomb or **volts**.

Certain devices have the peculiar property that they can maintain a prescribed potential difference (or voltage) denoted by $\mathcal{E}$, between two terminals. Such devices, which include batteries and electric generators, behave like an external force when connected to a circuit and are in fact known as sources of **electromotive force** (EMF). Electromotive force $\mathcal{E}$ is also measured in volts. For batteries, the higher potential terminal is labeled with a plus sign and the lower potential terminal with a minus sign. The internal chemical energy supplied by the battery imparts a constant amount of energy per coulomb as positive charge carriers move through it, thus raising the potential function $V$ by the voltage rating of the battery.

## *Circuit Elements*

Each circuit is composed of one or more of three basic circuit elements: the resistor, the capacitor, and the inductor.

### *Resistors*

As a current flows through a conducting substance, the charge carriers lose energy (which then appears in other forms such as heat and light). Hence the potential at which the current emerges is lower than the potential at which the current enters the conductor. Although every conductor has this property, those that lose comparatively large amounts of energy are called **resistors** and are symbolized schematically as follows:

$R$

$a$ o—/\/\/\/—o $b$

Common examples of resistors are heating elements, toaster coils, and lightbulb filaments. The instantaneous voltage drop between the two terminals $a$ and $b$ of a resistor is often found experimentally to be proportional to the current flowing through the resistor at that instant. The constant of proportionality, denoted by $R$, is called the **resistance** of the resistor and in SI units is measured in **ohms**. Thus we have

**Ohm's Law:** $V_{ab} = IR$.

If the current is directed from $a$ to $b$ then the voltage at $b$ is lower then the voltage at $a$, hence $V_{ab} = IR > 0$. If the current is directed from $b$ to $a$, then $V_{ab} = -IR < 0$. The voltage drop across some semiconductor resistors is a nonlinear function of the current. In such cases, $V_{ab} = f(I)$ for some nonlinear function $f$. The van der Pol and scroll circuits are based on such devices (see Experiments 5.14 and 6.14; Atlas Plates `First Order I, Limit Cycle A, and Scroll A-C`).

*Inductors*

A variable electric current $I(t)$ creates a changing magnetic field near the conductor. This magnetic field then induces a voltage drop (or backward EMF) between the ends of the conductor that opposes the change in current. Conductors designed to produce such an opposition to changes in current are called **inductors** and are represented by the symbol:

L

a○——⌇⌇⌇——○b

Coils of wire are examples of inductors. It has been found experimentally that the voltage drop $V_{ab}$ across an inductor is proportional to the instantaneous rate of change of the current through the inductor. This constant of proportionality is called the **inductance** of the inductor and in SI units is measured in **henries**. Thus we have

**Law of Induction:** $V_{ab} = LI'(t)$.

Note that the voltage through the inductor drops in the direction of the current flow.

*Capacitors*

A **capacitor** consists of two conductors separated by an insulator (such as air). Capacitors are often visualized as a pair of conducting plates separated by a gap, represented schematically as follows:

C

a○——| |——○b

If the terminals $a$ and $b$ of a capacitor are connected across a voltage source, a negative charge will build up on the plate connected to the negative terminal and a positive charge on the plate connected to the positive terminal. The magnitude of the charge on either plate of the capacitor is denoted by $q(t)$. Observe that if $q_0(t)$ is the initial charge, then

$$q(t) = q(t_0) + \int_{t_0}^{t} I(s)\,ds$$

It has been observed experimentally that the instantaneous voltage drop across a capacitor is proportional to the charge on the capacitor. The constant of proportionality is $1/C$, where $C$ is the **capacitance** of the capacitor and is measured in the SI unit of **farads**. This leads to

**Coulomb's Law:** $V_{ab}(t) = \dfrac{1}{C}q(t) = \dfrac{1}{C}\left[q(t_0) + \displaystyle\int_{t_0}^{t} I(s)\,ds\right].$

Note again that the voltage drops in the direction of the current flow. The capacitance $C$ depends on the area of the plates, their separation, and the nature of the insulator between. As might be expected, it takes a high voltage to build up charge on the capacitor, and thus for a given charge the capacitance is usually very small (on the order of $10^{-5}$ or $10^{-6}$ farad).

## Kirchhoff's Laws

To find the current flowing in a circuit, it is necessary to have a relation between the voltage drops across the various components of a circuit. The following "energy conservation" law has been observed to hold.

**Voltage Law:** *Let the points $a_1, a_2, \ldots, a_n$ be an oriented closed loop in a circuit. Then $V_{a_1a_2} + V_{a_2a_3} + \cdots + V_{a_na_1} = 0$, where $V_{a_ia_{i+1}} \equiv V_{a_i} - V_{a_{i+1}}$ is the voltage drop between points $a_i$ and $a_{i+1}$.*

The second law has to do with "current conservation."

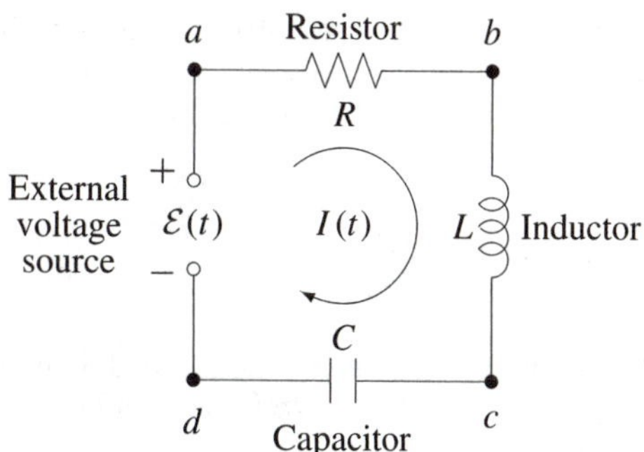


*Figure B.4.1 A simple circuit.*

**Current Law:** *At each point of a circuit, the sum of the currents entering that point is equal to the sum of the currents leaving.*

When applying Kirchhoff's Laws, it is important to keep in mind that it cannot be assumed in general that the current through every circuit element of a given loop will be the same. Thus the first thing to do is *assume* a direction for the current through every circuit element; the current through that element will be positive if it flows in the direction specified, and negative otherwise. If the assumed direction of current flow is incorrect for a given circuit element, the circuit equations will ultimately reveal that the current has a negative value for that element, indicating a current flow in the opposite direction. Next introduce the reference points $a, b, c, \ldots$, and determine the polarity of the external source by indicating the positive and negative terminals by "+" and "−" signs (see Figure B.4.1).

## *ODEs and Circuits*

The way the circuit laws lead to model ODEs is best illustrated by examples.

***Example 1*** **(Simple $RLC$ circuit)** To analyze the circuit shown in Figure B.4.1, identify the polarity of the external voltage source, assume a direction for the current $I(t)$, and label the reference points $a, b, c, d$. Application of Ohm's Law, the Law of Induction, Coulomb's Law, and then Kirchhoff's Laws yields

$$V_{ab} = RI(t)$$
$$V_{bc} = LI'(t)$$
$$V_{cd} = \frac{1}{C}\left[q(t_0) + \int_{t_0}^{t} I(s)\,ds\right]$$
$$\mathcal{E} = RI(t) + LI'(t) + \frac{1}{C}\left[q(t_0) + \int_{t_0}^{t} I(s)\,ds\right] \tag{1}$$

Equation (1) is an integrodifferential equation for the unknown current $I(t)$ in terms of $R$, $L$, $C$, $q(t_0)$, and $\mathcal{E}$. Equation (1) may be converted to two equivalent ODEs. In terms of the charge $q(t)$ on the capacitor, we have

$$Lq'' + Rq' + \frac{1}{C}q = \mathcal{E}(t) \tag{2}$$

since $q = q(t_0) + \int_{t_0}^{t} I(s)\,ds$. It is often more convenient to use $I(t)$ rather than $q(t)$ as the unknown. If $\mathcal{E}(t)$ is differentiable, we can differentiate (1) with respect to $t$ and obtain

$$LI'' + RI' + \frac{1}{C}I = \mathcal{E}' \tag{3}$$

The relevant initial value problem for the circuit equation in the form (3) is derived as follows. If we know $q_0$, the initial charge, and $I_0$, the initial current, then from (1) we have $RI_0 + LI'(t_o) + (1/C)q_0 = \mathcal{E}(t_0)$, which immediately determines $I'(t_0)$. We are thus led to solve the IVP

$$\begin{aligned} &LI'' + RI' + I/C = \mathcal{E}'(t) \\ &I(t_0) = I_0, \quad I'(t_0) = [\mathcal{E}(t_0) - RI_0 - q_0/C]/L \end{aligned}$$

The IVP in this case contains a single second order linear ODE and appropriate initial data. Similarly, we can convert ODE (2) into an IVP for the charge $q(t)$ by inserting initial data, $q(t_0) = q(0), q'(t_0) = I_0$.

*Example 2* **(Two-loop circuit)** Consider the two-loop circuit pictured in Figure B.4.2, where $I_1$, $I_2$, $I_3$, and $I_4$ are the currents through the elements indicated. From the Current Law at node point

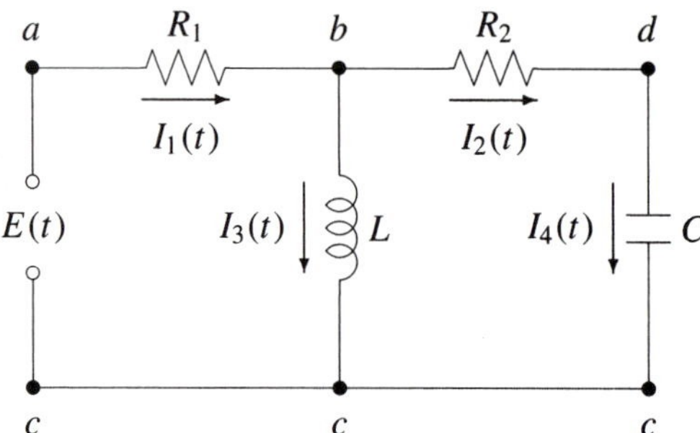


*Figure B.4.2 Two-loop circuit.*

$b$, the current through the inductor must be given by $I_3 = I_1 - I_2$. Applying the Voltage Law to the two loops, we have

$$\begin{aligned} R_1I_1 + L(I_1' - I_2') &= \mathcal{E}(t) \\ R_2I_2 + \tfrac{1}{C}\int_0^t I_2(s)\,ds + L(I_2' - I_1') &= 0 \end{aligned}$$

because $I_2 = I_4$ and because $q_0 = 0$ is assumed for the capacitor. Differentiating the second equation and rearranging, we have

$$\begin{aligned} L(I_1' - I_2') + R_1I_1 &= \mathcal{E}(t) \\ -L(I_1'' - I_2'') + R_2I_2' + \tfrac{1}{C}I_2 &= 0 \end{aligned}$$

Thus, the ODE model for the two-loop circuit is a linear system containing one first order and one second order linear ODE in the unknown currents $I_1$ and $I_2$. Initial data may be imposed before the system is solved for the currents.

### Conclusion

The circuit laws may be used to model complex multiloop circuits by systems of ODEs in the component currents, charges, or voltage drops. Even nonlinear resistors can be included if the Ohm's Law voltage drop $V = IR$ is replaced by the appropriate functional form $V = f(I)$.

## B.5 Scaling and Dimensionless Variables

The rate functions of a system of ODEs may involve many parameters. For example, the ODEs of population models, models of the motion of a vibrating spring or an oscillating pendulum, and models of currents and voltages in a circuit include as parameters damping coefficients, rate constants, spring force constants, and electrical resistances, capacitances, and inductances. In addition, solving an IVP involves the use of other parameters such as the initial data. Even an "abstract" ODE, one that does not necessarily arise as a model of a physical system, involves coefficients and other parameters. These coefficients, physical constants, numbers, and other parameters have an effect on the computed solutions. Change one, and the solutions may change, sometimes in drastic and unexpected ways—so called **sensitivity to the data**.

Are there ways to single out the significant parameters before computing? Can the number of parameters be reduced by rescaling the state variables or time? Can the variables and parameters be nondimensionalized before computing so that it is clear that orbital and solution behavior is inherent in the structure of the ODEs, not just a consequence of using a particular set of physical units, scales, and dimensions? Three examples address these questions.

***Example 1*** **(Radiocarbon dating)** According to Experiment 1.7, the model IVP for a radioactive carbon-14 dating process is

$$\frac{dx(t)}{dt} = -kx(t), \qquad x(0) = x_0, \qquad T \le t \le 0 \tag{1}$$

where $x(t)$ is the (dimensionless) fraction of $^{14}C$ at time $t$ (measured in years) out of the mass of all forms of carbon in a sample of once-living material, $k$ is a positive decay constant (measured in reciprocal years), $x_0$ is the current fraction of $^{14}C$ (at $t = 0$), $T$ is the time in the past when the material was last alive (the current time is $t = 0$), and $x_T$ is the fraction of $^{14}C$ in the sample at time $T$. The parameters here are $k$, $x_0$, $T$, and $x_T$. The fraction $x(t)$ is very small; to avoid working with small numbers, it is convenient to rescale $x(t)$ by dividing it by its current value $x_0$. Observe that both $x(t)$ and $x(t)/x_0$ are dimensionless. In dating problems the time span may be very large (thousands of years). Time can be rescaled (and nondimensionalized) by dividing by the half-life $t_{1/2} = 5568$ years of $^{14}C$. Denote the rescaled time by $s = t/t_{1/2}$. In terms of the rescaled state variable, $y = x/x_0$, and rescaled time, we have

$$y(s) = \frac{x(t)}{x_0} = \frac{x(t_{1/2}s)}{x_0} \tag{2}$$

From (1), (2), and the Chain Rule we have

$$\begin{aligned}\frac{dy(s)}{ds} &= \frac{1}{x_0}\frac{dx(t)}{dt}\frac{dt}{ds} = \frac{t_{1/2}}{x_0}\frac{dx(t)}{dt} \\ &= \frac{t_{1/2}}{x_0}(-kx(t)) = \frac{t_{1/2}}{x_0}(-kx_0y(s)) = -kt_{1/2}y(s) = -(\ln 2)y(s)\end{aligned}$$

where we have made use of the half-life relationship $kt_{1/2} = \ln 2$ from Experiment 1.7. After setting $S = T/t_{1/2}$, IVP (1) becomes

$$\frac{dy(s)}{ds} = -(\ln 2)y(s), \qquad y(0) = 1, \qquad S \le s \le 0 \tag{3}$$

Observe that by rescaling the state variable and time we have reduced the number of parameters to two, $S$ and $y_S (= x_T/x_0)$. The radiocarbon dating problem now reduces to finding the value $S < 0$ for which $y(S) = x_T/x_0$.

***Example 2*** **(Soft spring)** The IVP

$$m\frac{d^2x(t)}{dt^2} + c\frac{dx(t)}{dt} + kx(t) - bx^3(t) = A_0 \cos\omega t, \qquad x(0) = 0, \qquad x'(0) = 0 \tag{4}$$

models the changing displacement $x(t)$ from equilibrium of a mass $m$ suspended by a "soft" spring whose action on the mass is modeled by the spring force terms $kx(t) - bx^3(t)$ and which is subject to the external driving force $A_0 \cos \omega t$. The term $c\,dx(t)/dt$ represents a frictional damping force. The physical phenomenon and this model are discussed in Experiment 4.8. The variables $x$ and $t$ have respective units of length and time. The parameters $m$, $c$, $k$, $b$, $A_0$, and $\omega$ also have physical units. By rescaling $x$ and $t$ we can simultaneously introduce dimensionless variables and parameters and reduce the number of parameters from six to three. Let $u = x/A$ and $s = t/B$, where the scale factors $A$ and $B$ are to be determined. By the Chain Rule we have

$$\frac{dx(t)}{dt} = A\frac{du}{ds}\frac{ds}{dt} = \frac{A}{B}\frac{du}{ds}$$

$$\frac{d^2x(t)}{dt^2} = \frac{d}{dt}\left(\frac{dx(t)}{dt}\right) = \frac{d}{ds}\left(\frac{A}{B}\frac{du}{ds}\right)\frac{ds}{dt} = \frac{A}{B^2}\frac{d^2u}{ds^2}$$

Hence, equation (4) transforms to

$$m\frac{A}{B^2}\frac{d^2u}{ds^2} + c\frac{A}{B}\frac{du}{ds} + kAu - bA^3u^3 = A_0\cos(\omega Bs)$$

which is normalized by dividing by $mA/B^2$ to yield

$$\frac{d^2u}{ds^2} + \frac{c}{m}B\frac{du}{ds} + \frac{k}{m}B^2u - \frac{b}{m}(AB)^2u^3 = \frac{A_0}{m}\frac{B^2}{A}\cos(\omega Bs) \qquad (5)$$

Recalling that the scale factors $A$ and $B$ are yet to be determined, we decide which effects we want to focus on in a computer study of the solution of IVP (4). For example, we may want to see how the friction force, the soft spring force, and the driving force frequency affect the solution. In that case the coefficients of the terms involving $du/ds$ (friction), $-u^3$ (soft spring), and the frequency will be the new parameters, while $A$ and $B$ will be chosen to set the remaining coefficients at some convenient value, say 1. Thus, choose $A$ and $B$ so that

$$\frac{k}{m}B^2 = 1, \qquad \frac{A_0}{m}\frac{B^2}{A} = 1$$

that is,

$$B = \sqrt{\frac{m}{k}}, \qquad A = \frac{A_0}{k}$$

Renaming the other three coefficients such that $c_1 = cB/m$, $b_1 = b(AB)^2/m$, $\omega_1 = \omega B$, we have the IVP

$$\frac{d^2u}{ds^2} + c_1\frac{du}{ds} + u - b_1u^3 = \cos(\omega_1 s), \qquad u(0) = 0, \qquad u'(0) = 0 \qquad (6)$$

with three parameters $c_1$, $b_1$, and $\omega_1$. In fact, $u$, $s$, $c$, $b$, and $\omega$ are all dimensionless quantities. To show this, we begin by identifying the physical dimensions of the parameters and variables in (4), using $M$ for mass, $L$ for length, and $T$ for time dimensions. All the summands in (4) must have the force dimensions $ML/T^2$ of the first term on the left. This establishes the dimensions of the parameters and thus of $A$ and $B$. The appropriate dimensions for the parameters and variables are tabulated as follows:

| Parameter | Dimension | Parameter | Dimension |
|---|---|---|---|
| $x$ | $L$ | $t$ | $T$ |
| $dx/dt$ | $L/T$ | $d^2x/dt^2$ | $L/T^2$ |
| $m$ | $M$ | $c$ | $M/T$ |
| $k$ | $M/T^2$ | $b$ | $M/(LT)^2$ |
| $A_0$ | $ML/T^2$ | $\omega$ | $1/T$ |
| $A = A_0/k$ | $L$ | $B = \sqrt{m/k}$ | $T$ |

From this table, it is straightforward to show the nondimensionality of the variables and parameters of (6). For example, $u = x/A$ has the dimension of $x$ divided by that of $A$; in other words $L/L$, which is dimensionless, and $c_1 = cB/m$ has the dimensions of $c$ times the dimension of $B$ divided by that of $m$ (i.e., $(M/T)(T/M)$, which is again dimensionless).

Thus a computer study of the effect of the damping force, the soft spring force, and the driving frequency on solution behavior should use equation (6), varying the dimensionless parameters $c_1$, $b_1$, and $\omega_1$ from some standard values and computing the dimensionless displacement $u$ and velocity $du/ds$ over some dimensionless time interval $0 \leq s \leq S$. The results of such a study then apply to the behavior of the solutions of any damped, driven soft spring equation in any units.

Systems of ODEs can be rescaled and nondimensionalized in much the same way. Example 3 refers to the van der Pol system of Experiment 5.14, although the potassium-argon model of Experiment 1.8 or the chemical kinetics models of several experiments could equally well be used to illustrate scaling in systems.

***Example 3*** **(Van der Pol System)** The van der Pol system of Experiment 5.14 is

$$\begin{aligned} \frac{dI(t)}{dt} &= \frac{a}{L}I(t) - \frac{1}{L}V(t) - \frac{1}{L}I^3(t) \\ \frac{dV(t)}{dt} &= \frac{1}{C}I(t) \end{aligned} \tag{7}$$

where $V(t)$ and $I(t)$ are, respectively, the voltage (in volts) across a capacitor and the current (in amps) at time $t$ (in seconds) in a simple circuit with a nonlinear resistor. The circuit parameters $a$, $L$, and $C$ also have appropriate units. Suppose that we rescale current, voltage, and time by setting $x = I/\alpha$, $y = V/\beta$, $s = t/\gamma$ and then choose $\alpha$, $\beta$, and $\gamma$ to reduce the number of parameters. From (7) and the Chain Rule we have

$$\begin{aligned} \frac{\alpha}{\gamma}\frac{dx}{ds} &= \frac{a}{L}\alpha x - \frac{1}{L}\beta y - \frac{\alpha^3}{L}x^3 \\ \frac{\beta}{\gamma}\frac{dy}{ds} &= \frac{\alpha}{C}x \end{aligned}$$

Normalizing yields

$$\begin{aligned} \frac{dx}{ds} &= \frac{a}{L}\gamma x - \frac{\beta\gamma}{L\alpha}y - \frac{\alpha^2\gamma}{L}x^3 \\ \frac{dy}{ds} &= \frac{\alpha\gamma}{C\beta}x \end{aligned} \tag{8}$$

We can choose $\alpha$, $\beta$, and $\gamma$ so that three of the coefficient groups are, say, unity. For example, we may set

$$\frac{\beta\gamma}{L\alpha} = 1, \qquad \frac{\alpha^2\gamma}{L} = 1, \qquad \frac{\alpha\gamma}{C\beta} = 1$$

After some algebraic manipulation we see that

$$\alpha = \left(\frac{L}{C}\right)^{1/4}, \qquad \beta = \left(\frac{L}{C}\right)^{3/4}, \qquad \gamma = (CL)^{1/2}$$

Then, setting $a\gamma/L = \mu$ we have the scaled system with a single parameter $\mu$,

$$\begin{aligned} \frac{dx}{ds} &= \mu x - y - x^3 \\ \frac{dy}{ds} &= x \end{aligned} \tag{9}$$

Although we shall not verify the fact, $x$, $y$, $s$, and $\mu$ are dimensionless. A computer study of the dependence of orbits and solutions on $\mu$ can be carried out from (9) rather than trying to deal with the three parameters $a$, $L$, and $C$ of (7).

***Observation*** There are several ways to scale and nondimensionalize the variables and parameters of an ODE system. It is usually best to begin by rescaling each state variable and time by a scaling factor to be chosen later in the rescaling process. After the ODE system has been written in terms of the scaled variables and normalized, decide which of the new parameters and data should remain in parameter form and which can be scaled to convenient fixed values. If the original ODE system is dimensionally correct, this process will produce a new system in dimensionless variables and parameters.

***Observation*** **(Rescaling graphs)** There is another aspect of scaling that often arises when attempting to overlay one graph on another. The range of one state variable may be so different from that of another that a translation and a rescaling of one of the state variables is essential if both are to be plotted on the same graph. Atlas Plates `Autocatalator A` and `Compartment Model C` illustrate this kind of rescaling.

***Observation*** **(Log scales)** Finally, nonlinear scaling of axes is common in the applications. For example, log scales are used in Experiment 4.6.

APPENDIX C

# The Atlas

The Atlas plates show the variety and the complexity of orbits and solution components of differential systems. Experiments often reference plates in the Atlas. Other plates are included to suggest additional directions for study. Small changes in the data, the coefficients, the time span, or the number of solution points may lead to large changes in the graphs. Even with the same data, different software, solvers, hardware, and graphics can also produce quite different pictures. It is a challenge to decide whether the discrepancies are real or just artifacts of the particular devices being used. In spite of all this, the authors are (reasonably) confident that the portraits in the Atlas accurately portray the corresponding systems.

The plates appear in Appendix C in the alphabetic order below.

### *List of Atlas Plates*

| | |
|---|---|
| `Autocatalator A-C` | On/off oscillations, sensitivity to rate constant |
| `Bifurcation A-C` | Hopf bifurcation to a limit cycle |
| `Compartment Model A-C` | Buildup of lead; potassium-argon decay |
| `Duffing A-C` | Driven nonlinear oscillation; chaotic wandering |
| `First Order A-C` | Solution curves of nonlinear $1^{st}$ order ODEs |
| `First Order D-F` | Solution curves of linear ODEs; singularities |
| `First Order G-I` | Integral curves |
| `Limit Cycle A-C` | Planar limit cycles |
| `Limit Set A-C` | Cycle-graphs; rescaling time on orbits |
| `Lorenz A-C` | Chaotic wandering and the Lorenz attractor |
| `Pendulum A-C` | Damped, undamped pendulum; basin of attraction |
| `Pendulum (Upended) A-C` | Driven upended pendulum; chaotic wandering |
| `Planar Systems A-C` | Linear systems; perturbation |
| `Planar Systems D-F` | Quadratic rates; menagerie of orbits |
| `Planar Systems G-I` | Painlevé equation and system |
| `Poincaré A-C` | Time-stepping along an orbit; chaos |
| `Predator-Prey A-C` | Predation; satiation; competition |
| `Rössler A-C` | Period multiplying; banded chaos and folding |
| `Scroll A-C` | Voltages and current in a nonlinear circuit |
| `Second Order A-C` | Input/output graphs; resonance for linear ODEs |
| `Spring A-C` | Oscillations of linear, hard, and soft springs |
| `Two-cycled System A` | Quadratic system with two limit cycles |

## Autocatalator A — On/off oscillations

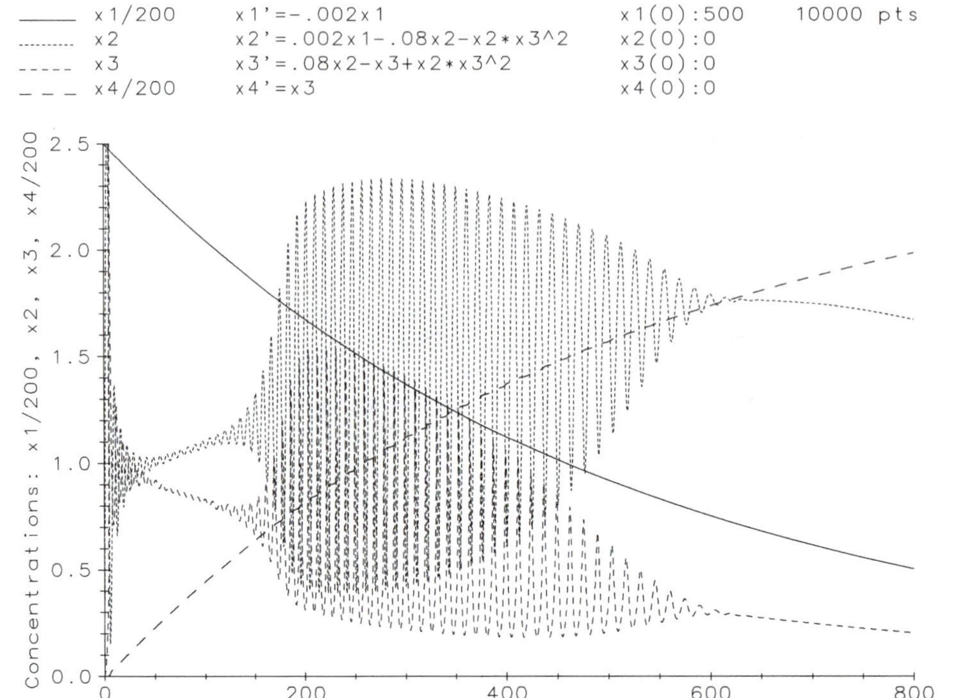


## Autocatalator B — An orbit of concentrations

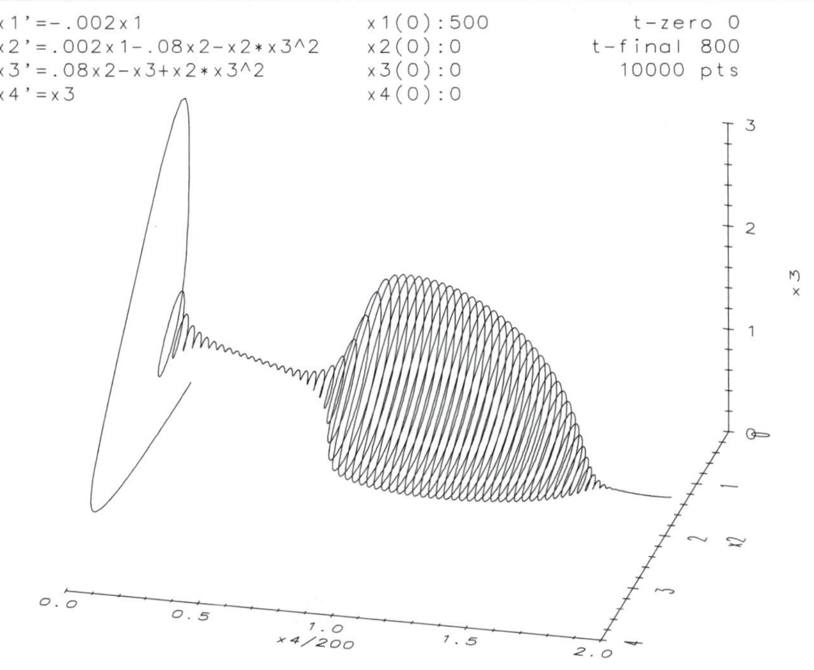

## Sensitivity to parameter changes — Autocatalator C

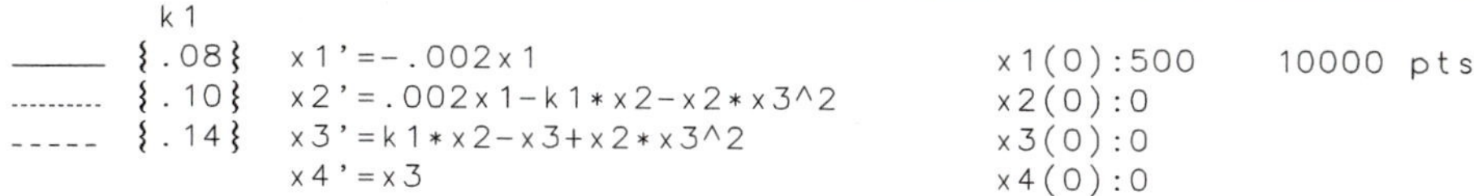


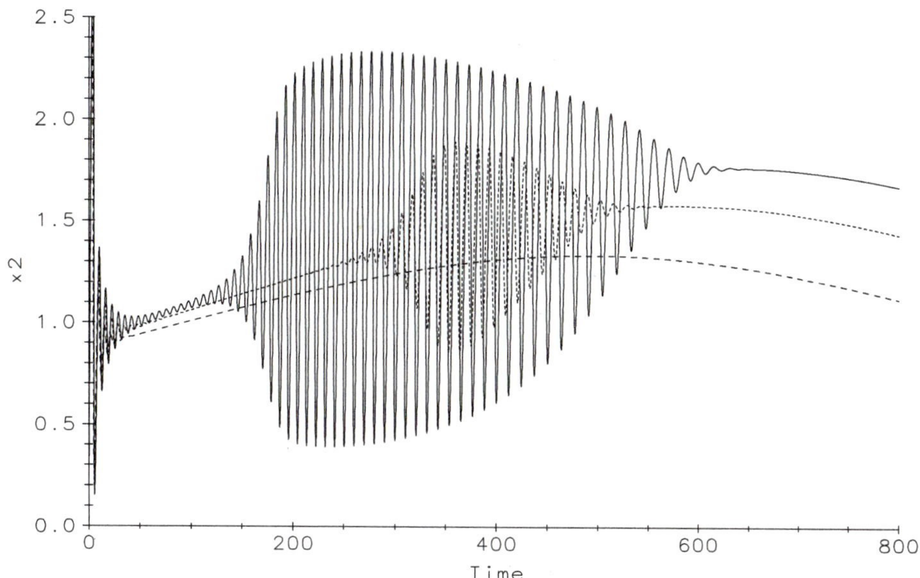


The autocatalator models a closed chemical reaction in which the concentration of a precursor $R$ decays exponentially, generating a new species $A$ in the process. Species $A$ decays to $B$ and at the same time reacts autocatalytically with $B$, the latter reaction creating more $B$ than is consumed. $B$ decays in turn to $C$. The net rate equations for the process are shown on the plates. They are based on the scheme:

$$R \xrightarrow{k_0} A, \quad A \xrightarrow{k_1} B, \quad A + 2B \xrightarrow{k_2} 3B, \quad B \xrightarrow{k_3} C$$

State variables $x_1$, $x_2$, $x_3$, and $x_4$ denote the dimensionless concentrations of the species $R$, $A$, $B$, and $C$. The values of the dimensionless rate constants are $k_0 = 0.002$, $k_1 = 0.08$, $k_2 = 1$, and $k_3 = 1$ for Plates A and B.

Plate A shows that as soon as the concentration of the precursor falls below a certain level, concentrations $x_2$ and $x_3$ begin to oscillate. The oscillations stop when the precursor concentration $x_1$ gets too low. Precursor and product concentrations change monotonically as expected (both are scaled by 200 in the graphs so that the component curves stay on the screen). Plate B shows the projected orbit of three concentrations. Plate C shows that the oscillations are shut off altogether if the rate constant $k_1$ is increased from 0.08 to 0.14.

***Reference*** Experiments 1.11, 6.12

## Bifurcation A — Before the Hopf bifurcation occurs

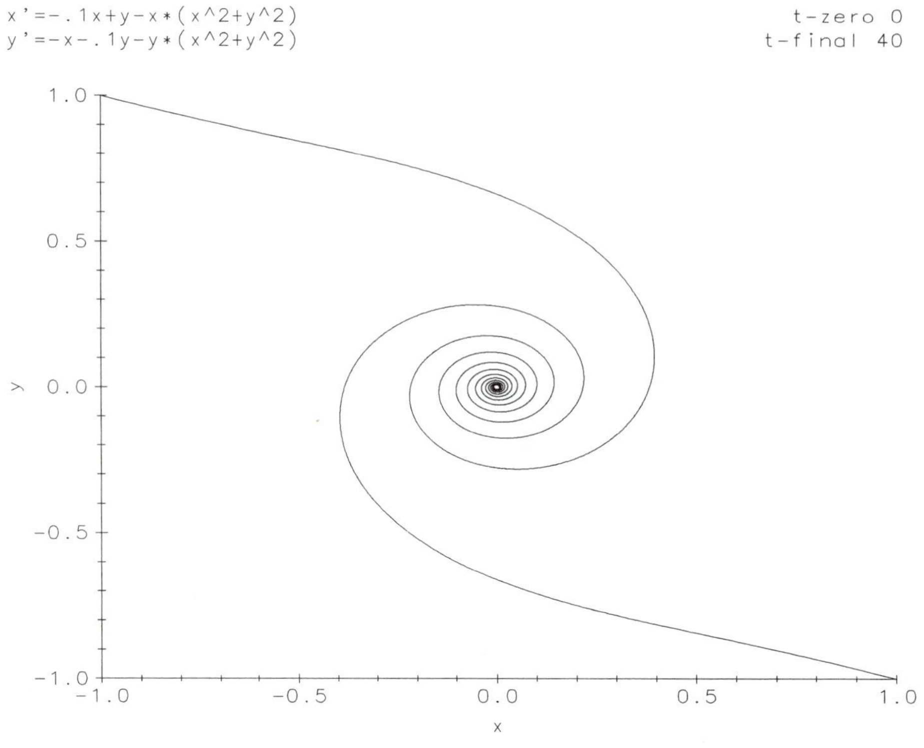


## Bifurcation B — At the Hopf bifurcation

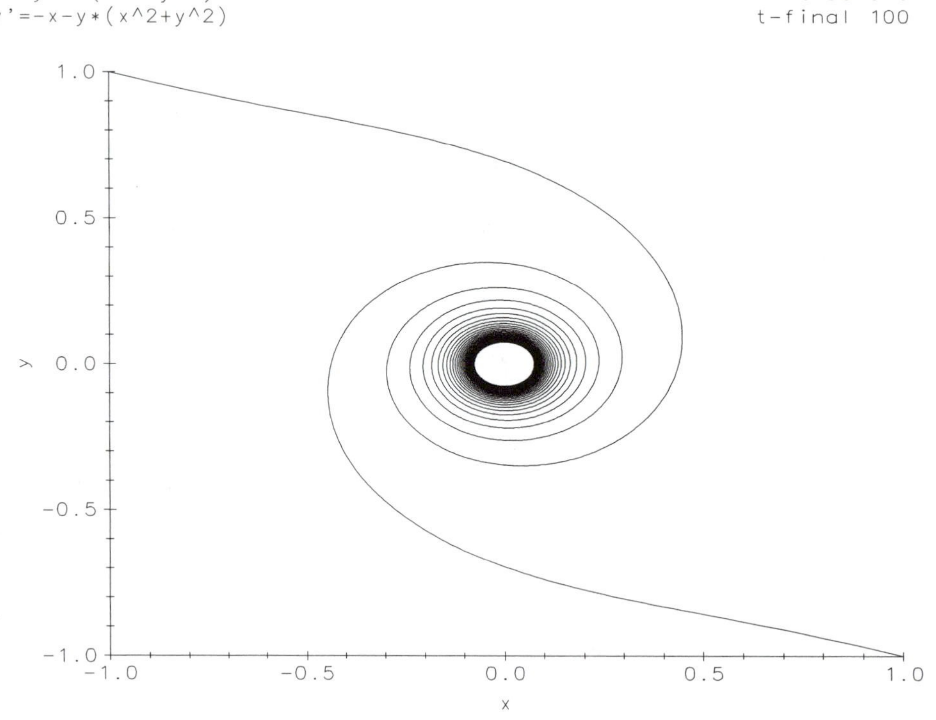

# After the bifurcation: limit cycle

## Bifurcation C

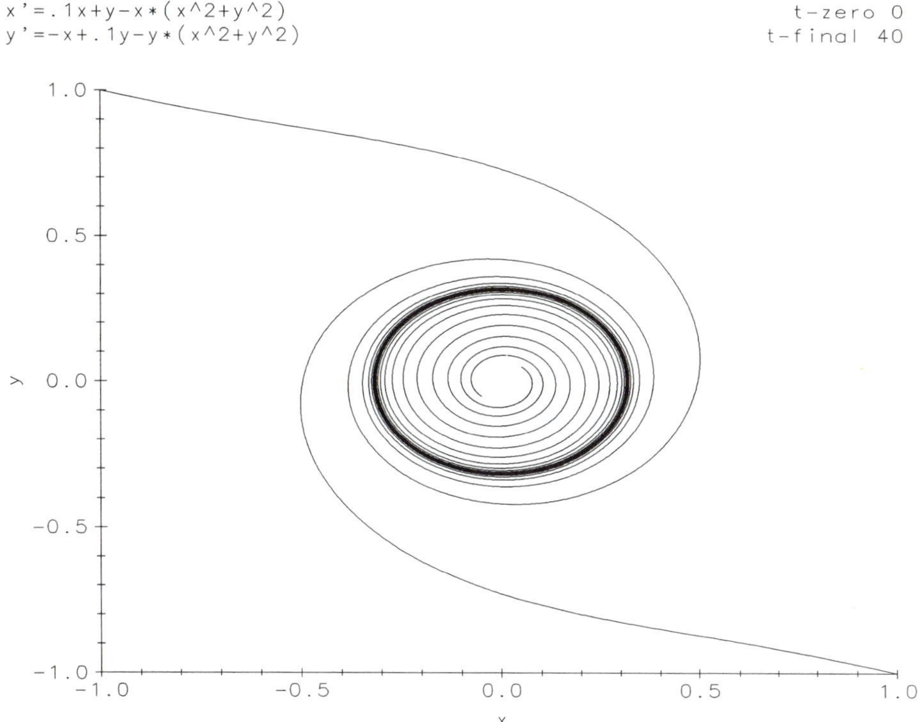


A Hopf bifurcation occurs in a system of ODEs as a change in a rate parameter causes an attracting equilibrium point to destabilize, spawning a stable limit cycle in the process.

In these bifurcation plates the system in rectangular coordinates is

$$x' = ax + y - x(x^2 + y^2)$$
$$y' = -x + ay - y(x^2 + y^2)$$

or in polar coordinates,

$$r' = r(a - r^2)$$
$$\theta' = -1$$

The parameter $a$ has value $-0.1$ in Plate A, 0 in Plate B, and 0.1 in Plate C. A Hopf bifurcation occurrs at $a = 0$. For all values of $a$, the origin is the only equilibrium point.

The eigenvalues of the matrix of the coefficients of the linear terms of the rate functions are $-0.1 \pm i$ in Plate A, and the origin is asymptotically stable. In Plate B, the equilibrium point is still asymptotically stable even though the eigenvalues are the pure imaginaries $\pm i$. In this case, the nonlinear terms drive orbits so slowly near the origin that each orbital loop lies almost on top of the preceding loop—hence the large dark area around the origin. The eigenvalues $0.1 \pm i$ of the coefficient matrix for Plate C have positive real parts, the origin has destabilized, and an attracting limit cycle has bifurcated from the origin. The amplitude of the cycle grows like $\sqrt{a}$ as $a$ increases from 0.

***Reference*** Experiment 5.17

## Compartment Model A — Concentrations of lead in the body

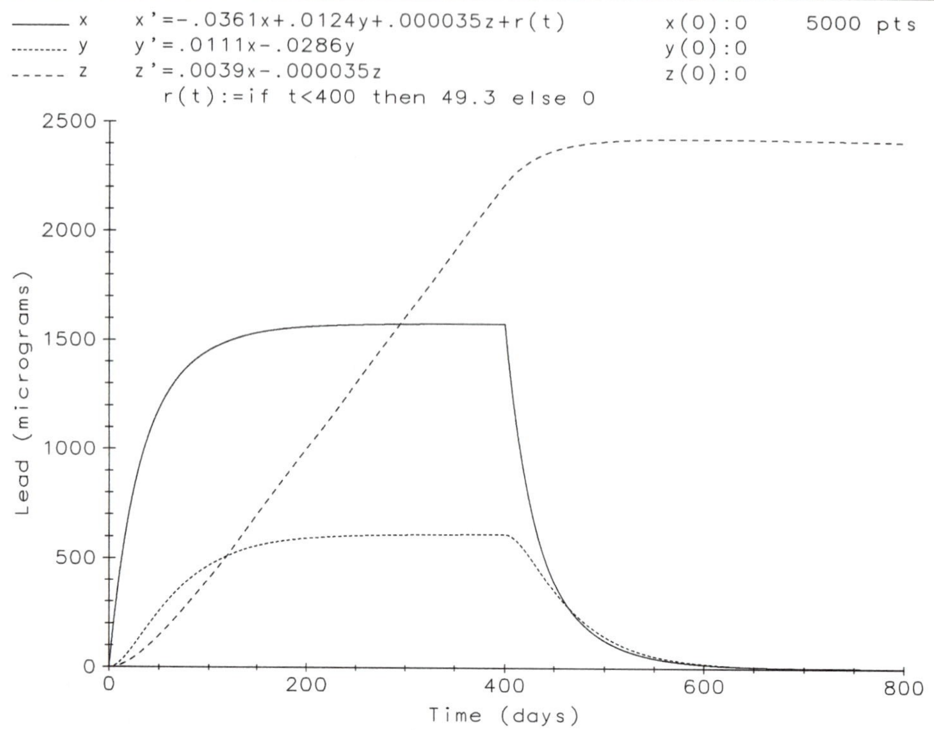


## Compartment Model B — Accelerated lead removal

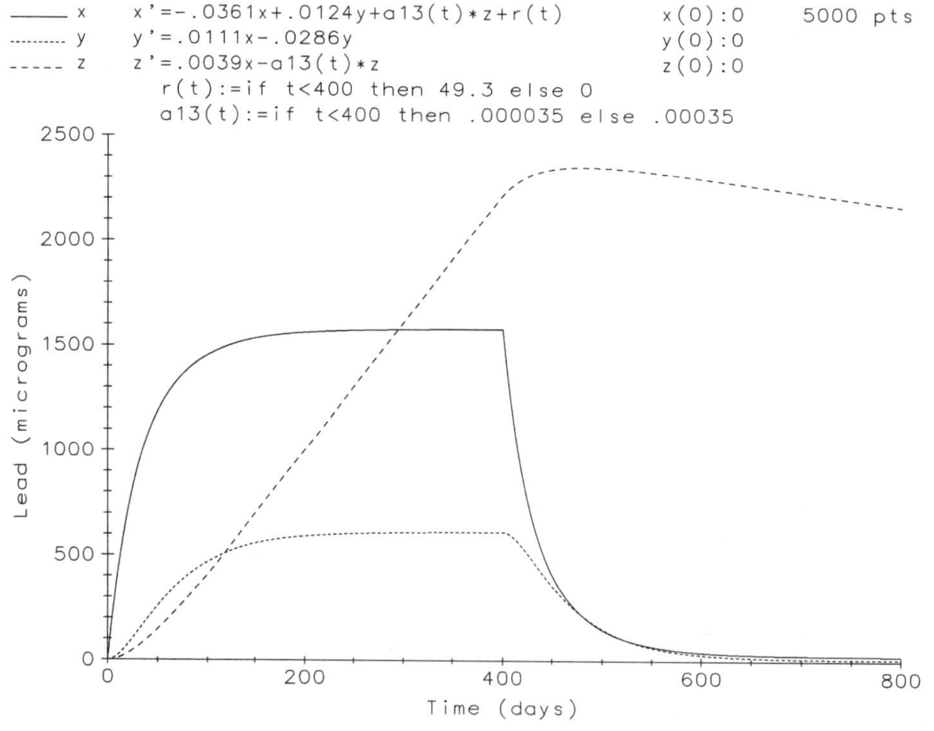

## Potassium-argon dating — Compartment Model C

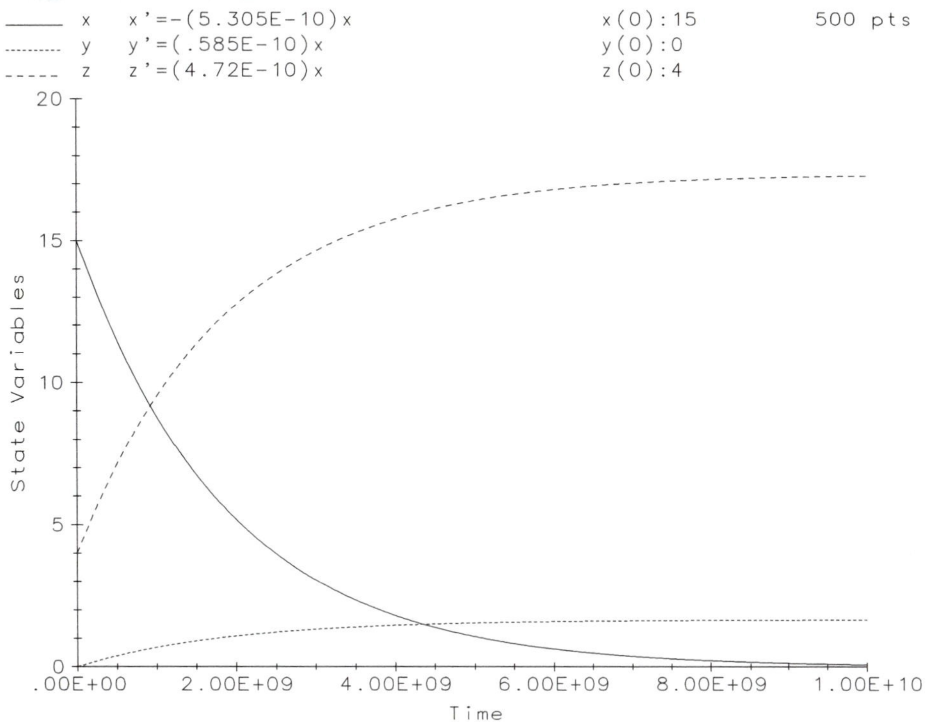


The system portrayed in Plate A tracks the amount of lead in the three compartments of the human body (blood, tissue, bones) over the time period $0 \leq t \leq 800$ days. The state variables $x(t)$, $y(t)$, $z(t)$ denote the amount of lead in the blood, tissue, and bones, respectively. Lead is ingested at a constant rate of $r(t) = 49.3$ micrograms per day for 400 days. At that time the test subject is placed in lead-free surroundings (i.e., $r(t) = 0$), and the blood and tissues rapidly purge themselves of lead. However, the amount of lead in the bones does not appreciably diminish even after lead intake ceases. In addition, after 400 days a drug is administered that speeds up the removal of lead from the bones. Plate B shows the decline in the amount of lead in the bones if the removal coefficient $a_{13}$ is changed from 0.000035 to 0.00035.

The compartmental system of Plate C models the slow decay of potassium to argon and the somewhat faster decay of potassium to calcium. Over the 10 billion-year time span of the graph, the initial amount of 15 units of potassium decays almost to the vanishing point. The amounts of argon and calcium, on the other hand, increase toward their equilibrium values.

***Reference*** Experiments 1.8, 6.5

## Duffing A — Nonlinear oscillator: stability

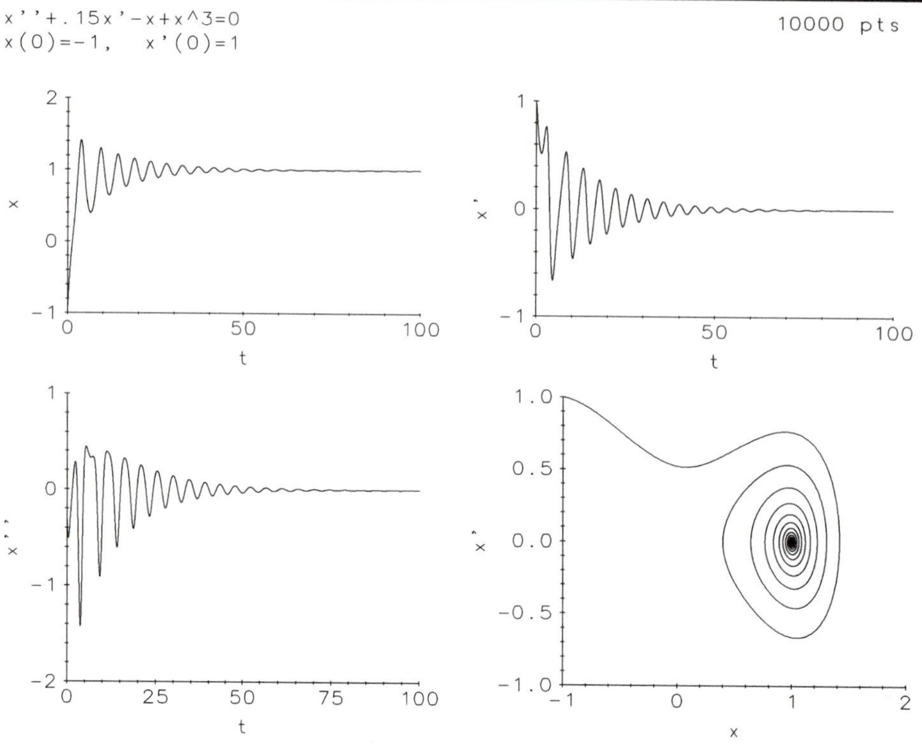


## Duffing B — Driven oscillator: chaotic wandering

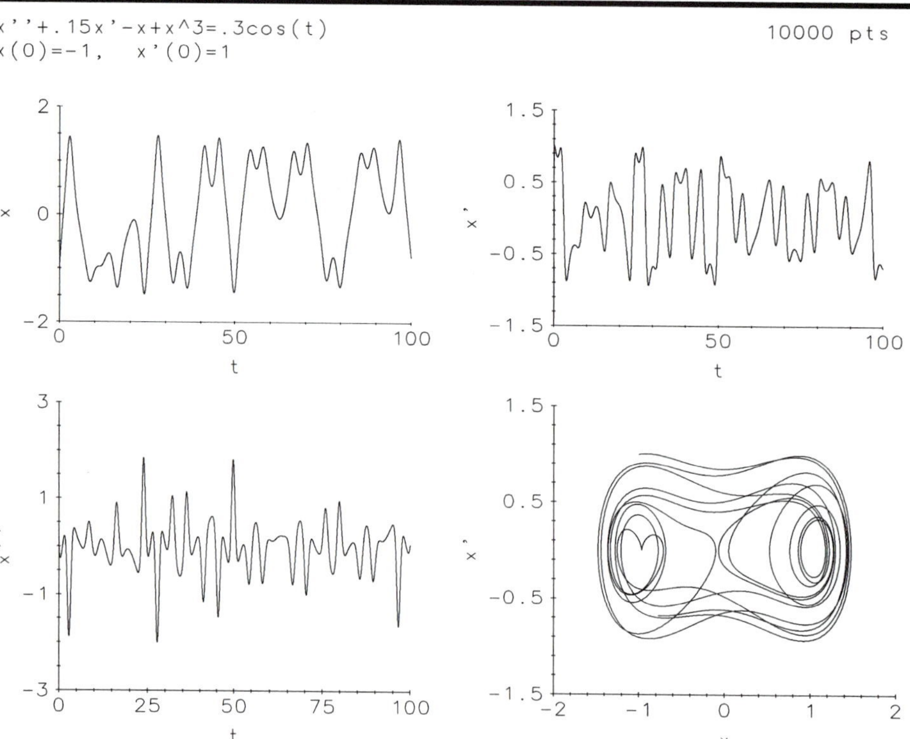

## Chaotic wandering: Poincaré plot — Duffing C

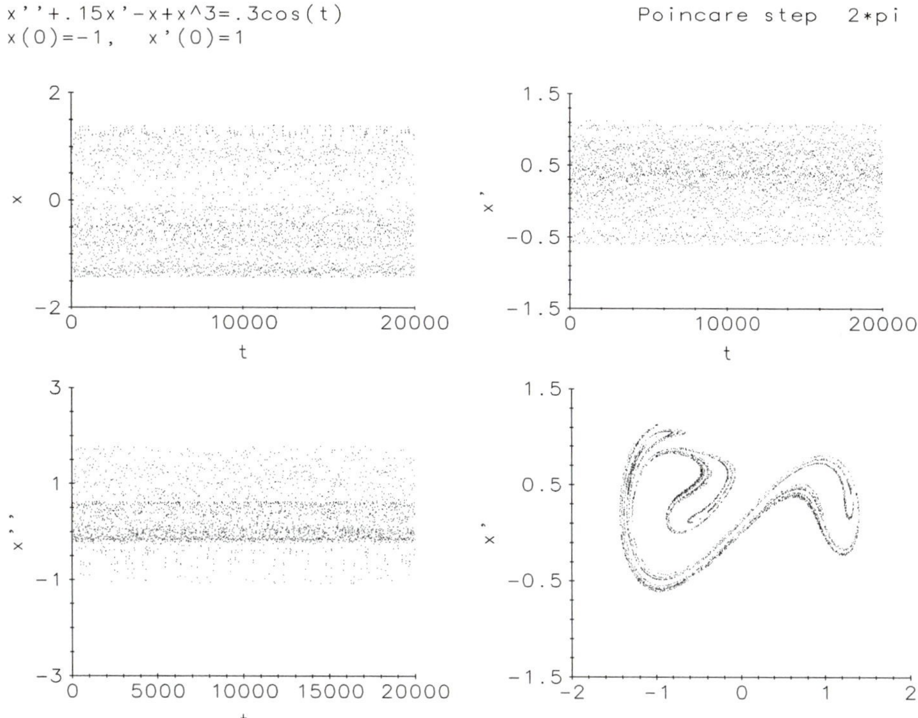


Duffing's Equation models the behavior of a damped nonlinear oscillator. Plate A shows how a solution, its velocity, and its acceleration evolve in time in the absence of a driving term. The corresponding orbit wraps around the asymptotically stable equilibrium point $x = 1, x' = 0$. Observe that the undriven system has two other equilibrium points $x = -1$, $x' = 0$ (asymptotically stable) and $x = 0$, $x' = 0$ (unstable).

When the oscillator is subjected to the periodic driving force $0.3 \cos t$, the orbit wanders chaotically back and forth between neighborhoods of $(1, 0)$ and $(-1, 0)$ in the $xx'$-plane. Plate B shows this wandering as time increases from 0 to 100. Over the much longer time interval $0 \leq t \leq 20000$, the orbit is so tangled it appears to depict random motion. To clarify the picture, a Poincaré time section with $\Delta t = 2\pi$ is constructed over this interval. The Poincaré plot in Plate C shows the paired state variables $x(t)$ and $x'(t)$ of the orbit only at multiples of the period $2\pi$ of the driving force. Although the $tx$ and $tx'$-plots show no apparent order, a marked pattern appears in the $xx'$-Poincaré plot, and chaotic wandering is consistent with definite regularities. The graph of the Poincaré plot in the $xx'$-plane appears to be a cross-section of a strange attractor.

***Reference*** Experiment 5.10

## First Order A

### Merging-diverging solutions

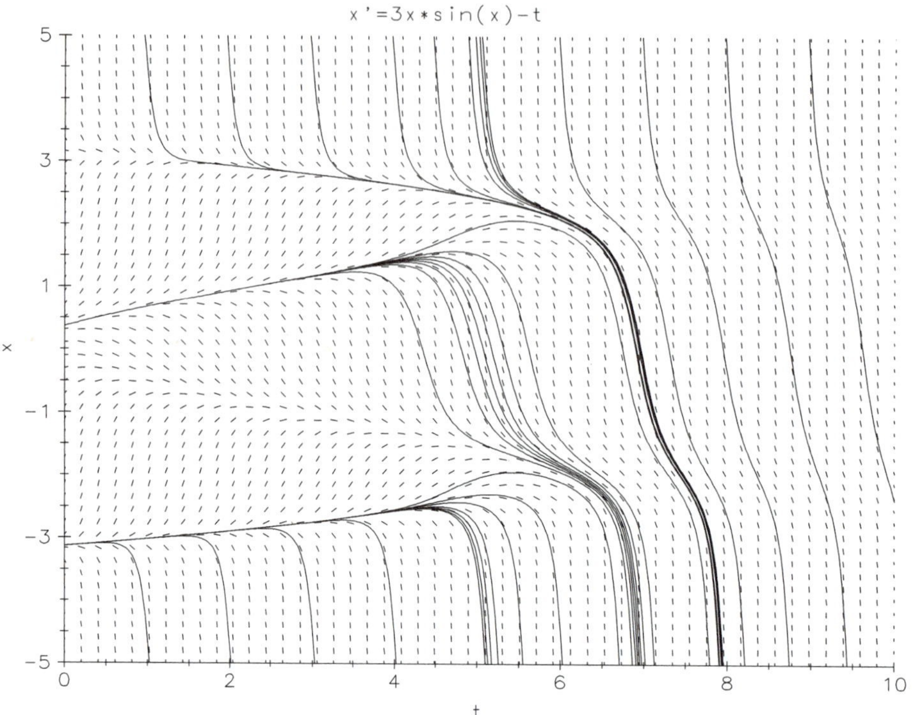


## First Order B

### Trigonometric rate function

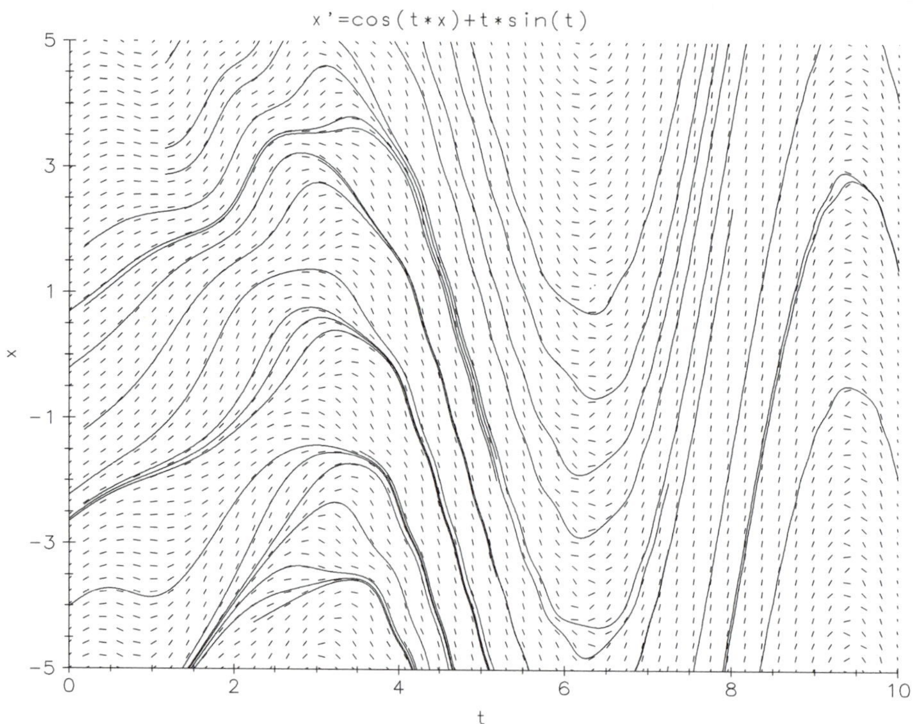

## A Riccati equation

**First Order C**

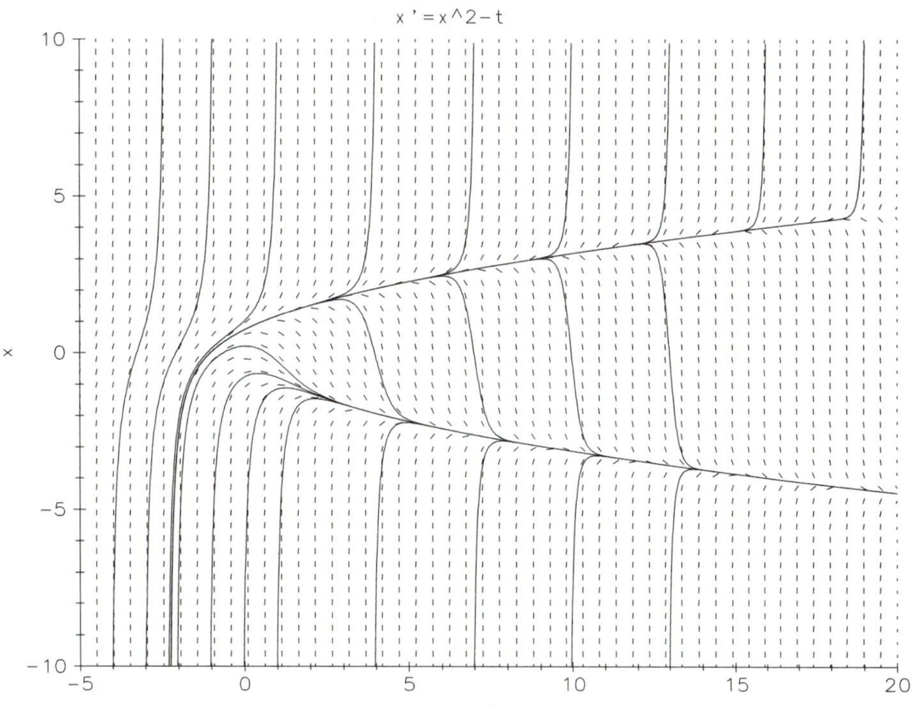


The direction field for a first order ODE does not prepare the viewer for the complex merging-diverging patterns of the solution curves. In fact, the curves flow together so strongly in Plate A that the Uniqueness Principle for IVPs seems to be violated, but this is not the case. We experience the illusion because the screen is a discrete device, which cannot resolve points that are too close together. It is hard to see that every point in the region on the screen has only one solution curve passing through it. Note that as $t$ increases, all solution curves enter the screen from the left and top boundaries, and exit through the bottom and right boundaries.

The rate function in Plate B is nonlinear, and there are no known techniques for solving this equation in terms of elementary functions. Nevertheless, solvers give a revealing picture of the behavior of the solutions for various values of the initial state and the initial time. Trigonometric rate functions often produce unusual patterns of solution curves.

The Riccati equation in Plate C cannot be solved in terms of elementary functions. A solver can be used to show that some solution curves repel, and others attract, nearby solution curves.

***Reference*** Experiment 2.1

## First Order D — Linear ODE: an attracting solution

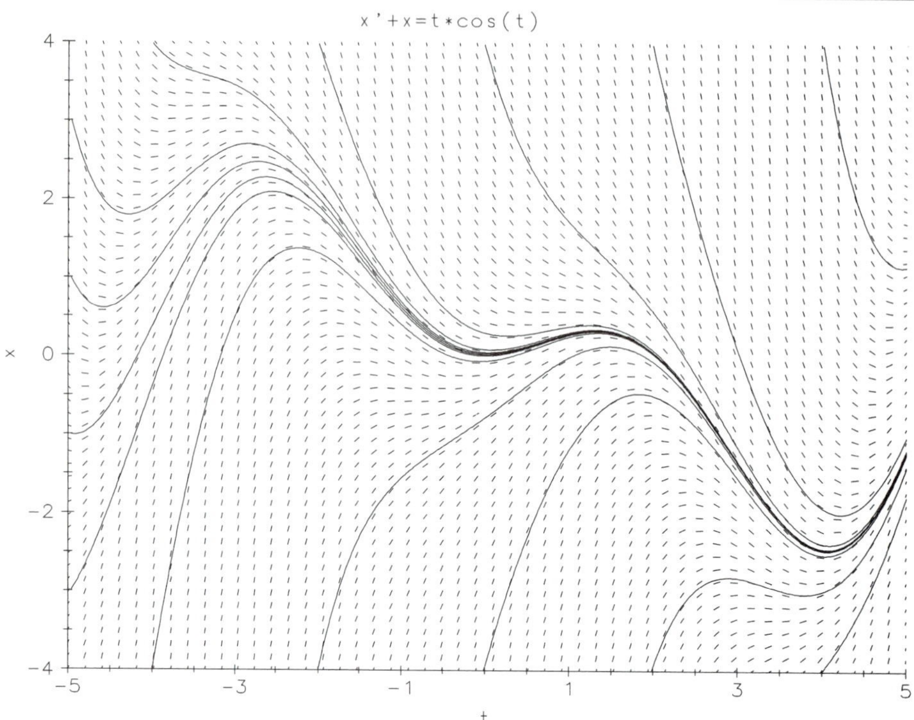


## First Order E — Linear ODE with singularity

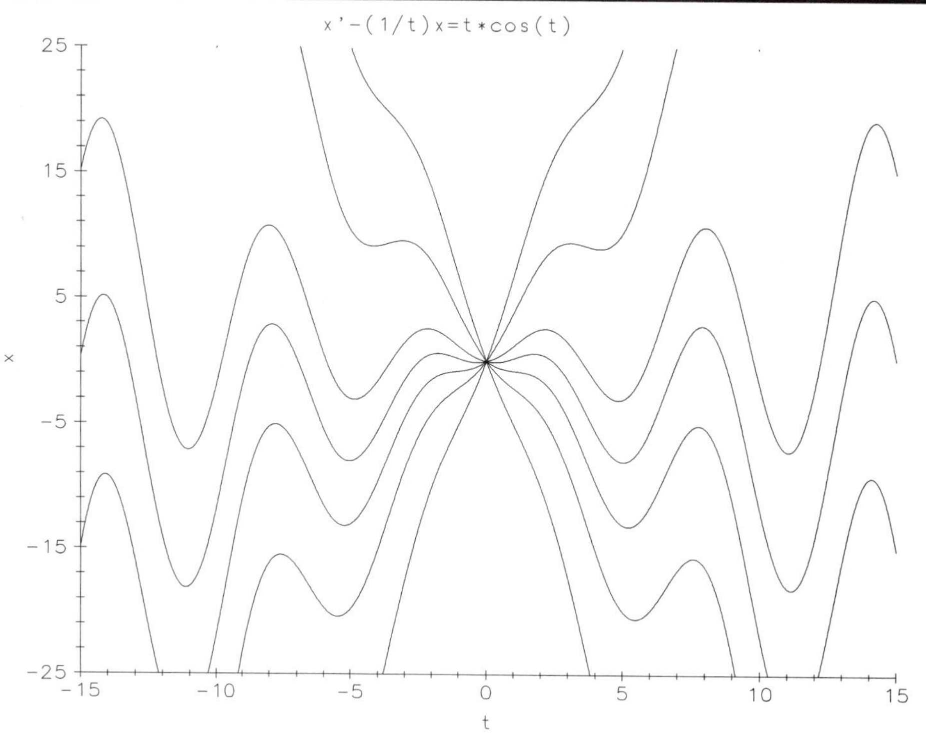

## Linear ODE with singularity — First Order F

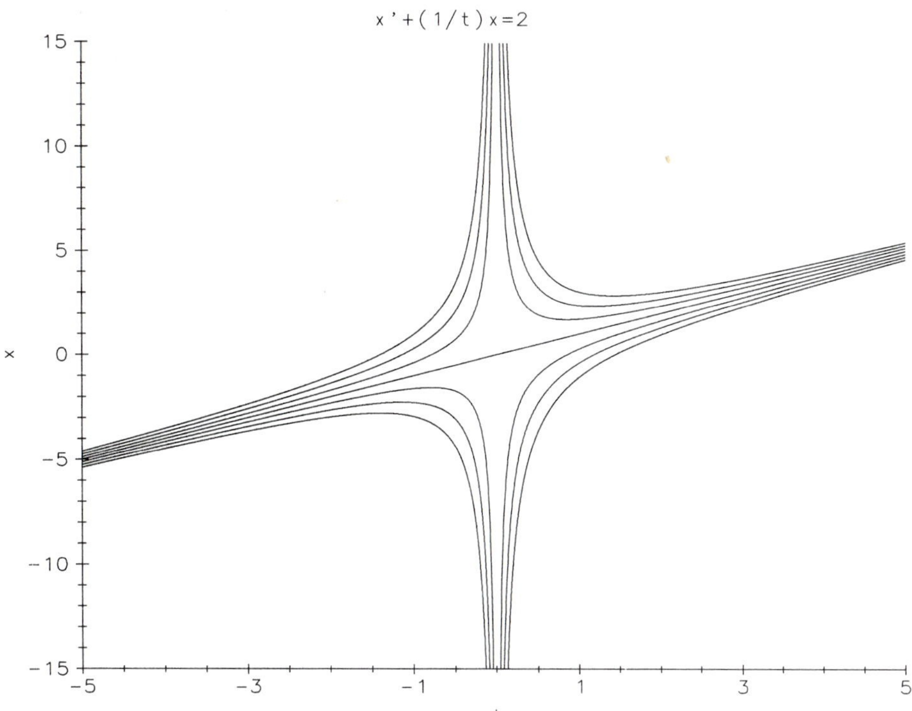


The ODE in Plate D is first order and linear. It can be solved explicitly by multiplying each side of the equation by the integrating factor $e^t$. Alternatively, solvers can be used to compute and graph solution curves for a variety of initial conditions, $x(t_0) = x_0$. Each curve is extended forward and backward in time until it reaches the user-defined boundary of the $tx$-screen. One solution appears to attract all other solutions as time advances.

The linear first order ODEs in Plates E and F are singular at $t = 0$. Therefore, the conclusions of the Existence and Uniqueness Principles do not necessarily apply to any region in the $tx$-plane that contains a portion of the $x$-axis. Both ODEs may be explicitly solved by integrating factors. Notice that every solution curve of the ODE in Plate E "cuts" the $x$-axis at $x = 0$ and "lives" on the entire $t$-axis. In fact, the solution formula is $x = t \sin t + Ct$, where $C$ is any constant. The solution curves of the ODE in Plate F have the form $x = t + C/t$ for $t > 0$ or $t < 0$, where $C$ is any constant. Thus this ODE has only one solution that "lives" on the entire $t$-axis; every other solution curve is confined to one of the half-planes $t < 0$ or $t > 0$.

***Reference*** Experiments 3.1–3.3

## First Order G — Exact ODE: strange integral curves

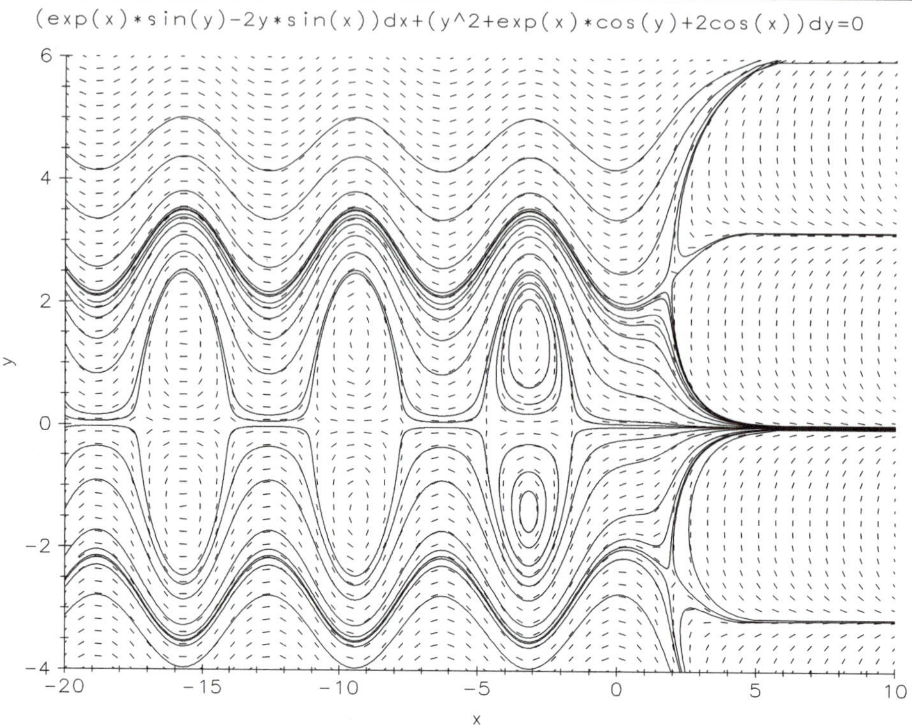


## First Order H — Exact ODE: portrait in integrals

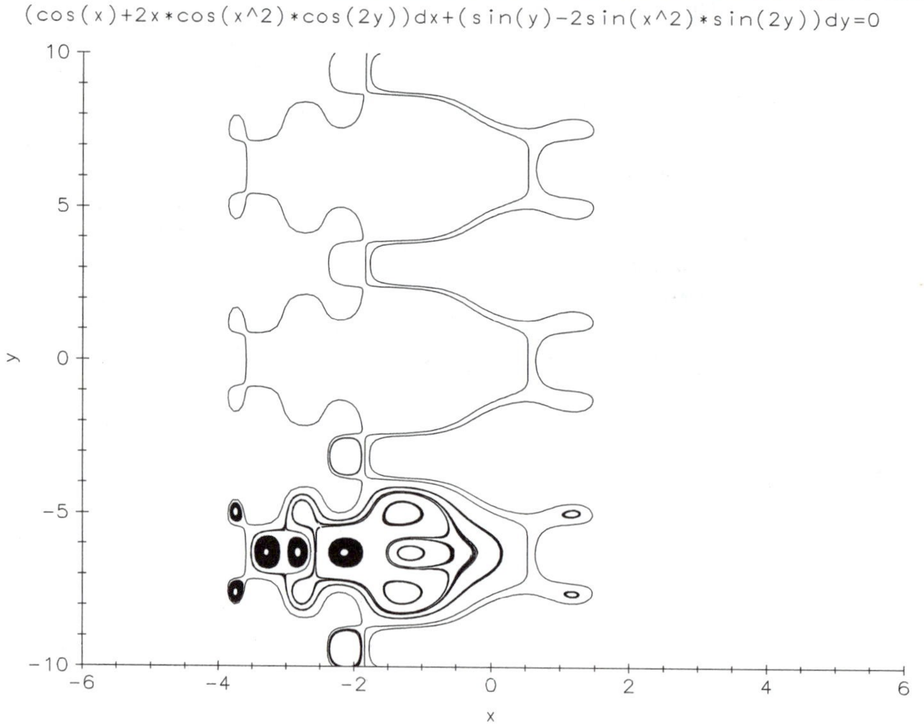

## Inexact van der Pol ODE: orbits

First Order I

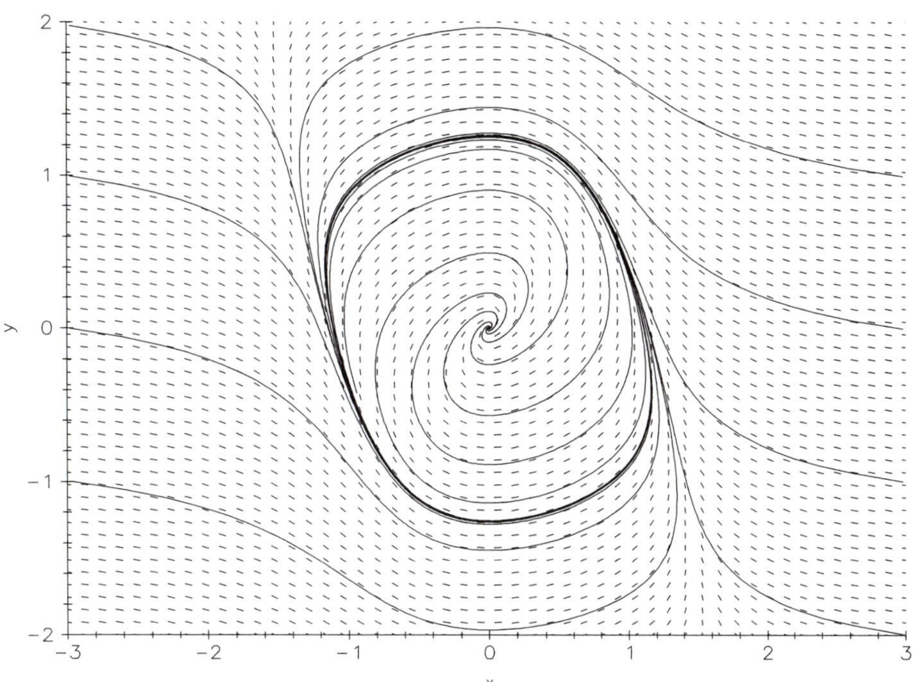


The ODEs in these plates have the differential form $Mdx + Ndy = 0$. The ODE in Plate G is exact. Its integral curves are defined implicitly by

$$e^x \sin y + 2y \cos x + y^3/3 = \text{constant}$$

but this formula is too complex to be read from the strange shapes of the curves.

The figures of Plate H are integral curves of another exact ODE. The implicit formula for these curves is

$$\sin x + \sin x^2 \cos 2y - \cos y = \text{constant}$$

but the formula is not informative about the shapes of the curves. The reader may try to fill in integral curves to the right and to the left of the line of "aliens." The ODE of Plate I is not exact. There is no explicit formula for its integral curves. Orbits of the equivalent system can be constructed, however, by using a solver and are depicted in Plate I. All orbits, except for the equilibrium point at the origin, are attracted to the limit cycle as time advances. See also Atlas Plate `Limit Cycle A`.

The ODEs of these plates may have the form $M\,dx + N\,dy = 0$, but the integral curves were formed by using software to solve numerically the equivalent planar system

$$\frac{dx}{dt} = -N(x, y) \qquad \frac{dy}{dt} = M(x, y)$$

and graph the orbits.

***Reference*** Experiments 1.11, 3.7–3.8, 5.14–5.15

## Limit Cycle A — Van der Pol limit cycle, components

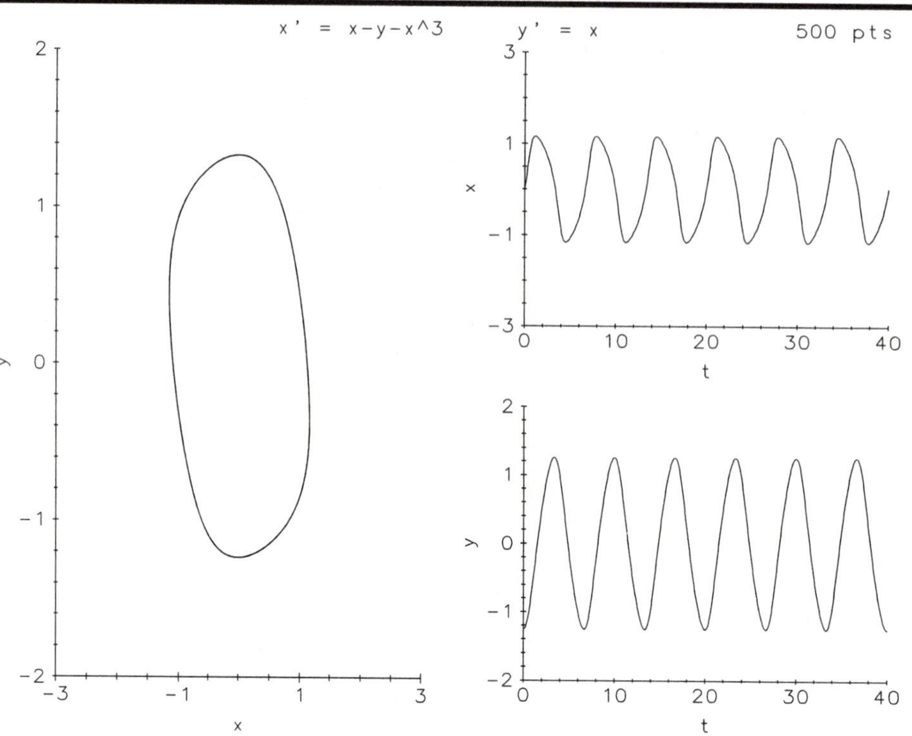


## Limit Cycle B — Van der Pol, parameter change

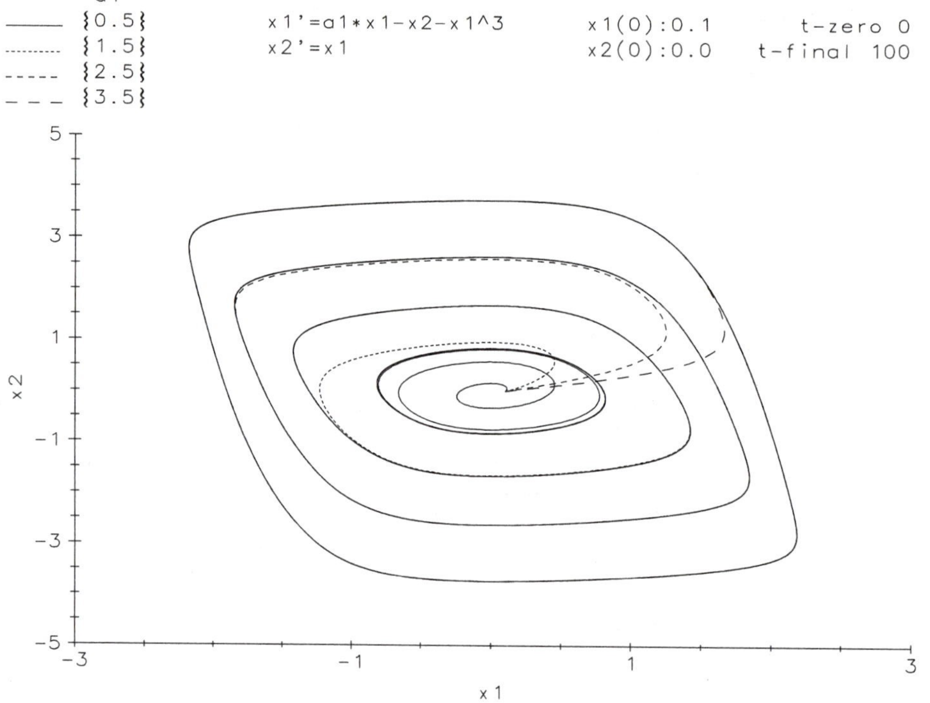

## Brusselator limit cycle Limit Cycle C

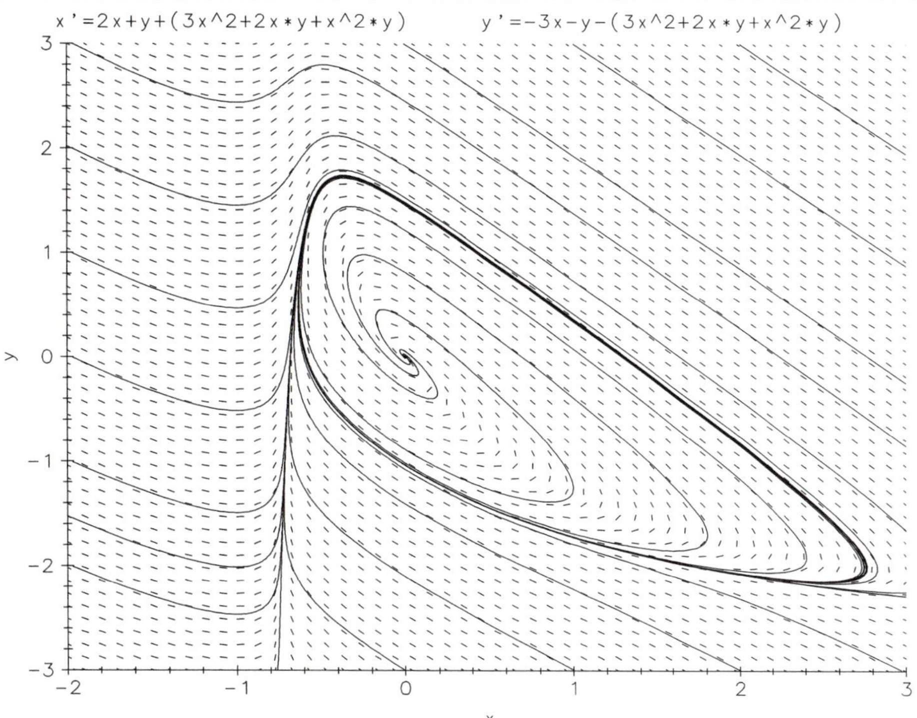


The van der Pol system has a unique periodic orbit, a limit cycle that attracts all other nonconstant orbits. Plate A displays that limit cycle and the corresponding component curves. See Atlas Plate `First Order I` for the cycle encircled by spiraling orbits. Plate B displays an outward moving spiral tending to a van der Pol limit cycle for each of four values of the parameter $a_1$. As $a_1$ increases, the amplitude of the cycle increases and the cycle becomes rectangular. The van der Pol system models current flow in an electrical circuit with a semiconductor resistor whose current/voltage characteristic is nonlinear.

The Brusselator system of Plate C has a limit cycle that attracts all neighboring orbits. The system acquired its name because a group of chemists in Brussels believed the system modeled the changing concentrations of the chemical species of an autocatalytic reaction.

***Reference*** Experiments 5.14–5.15

## Limit Set A — Limit set is a limit cycle

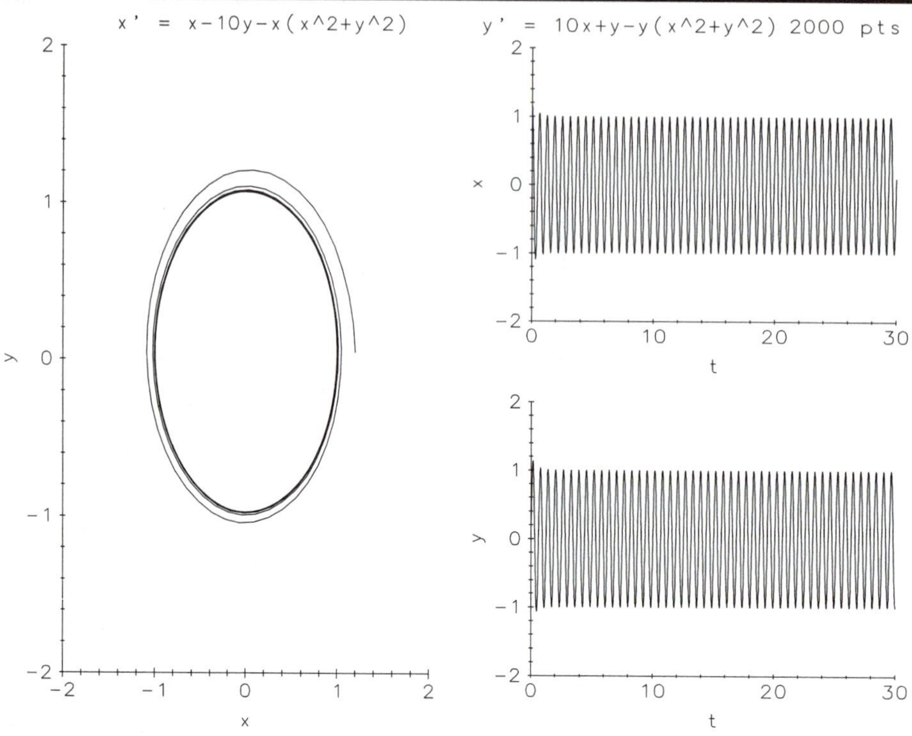


## Limit Set B — Rescaling time: cycle-graph

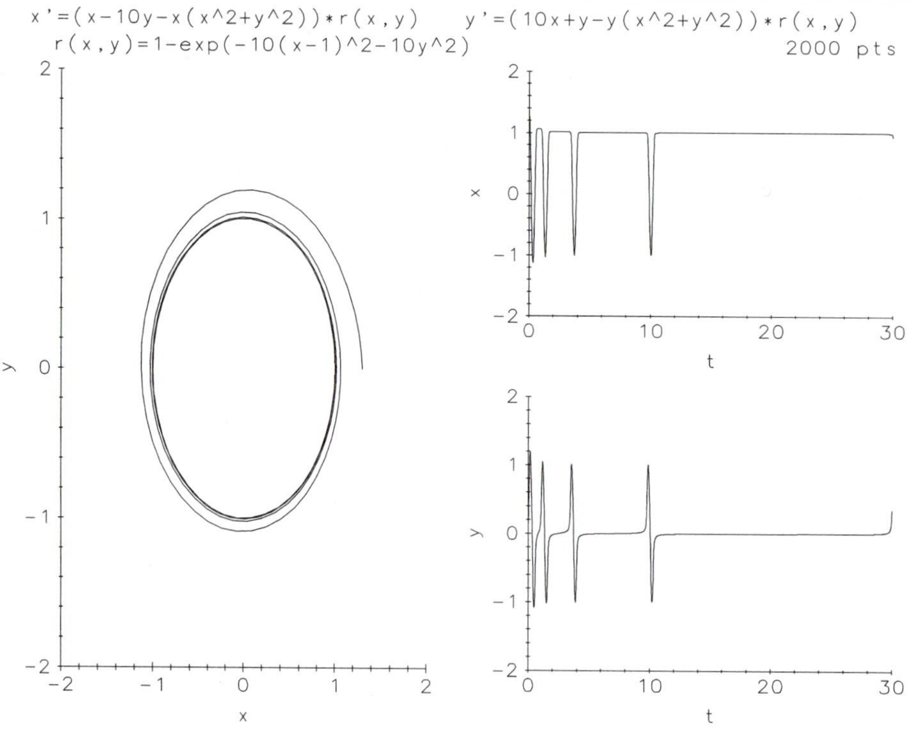

## Cycle-graph

## Limit Set C

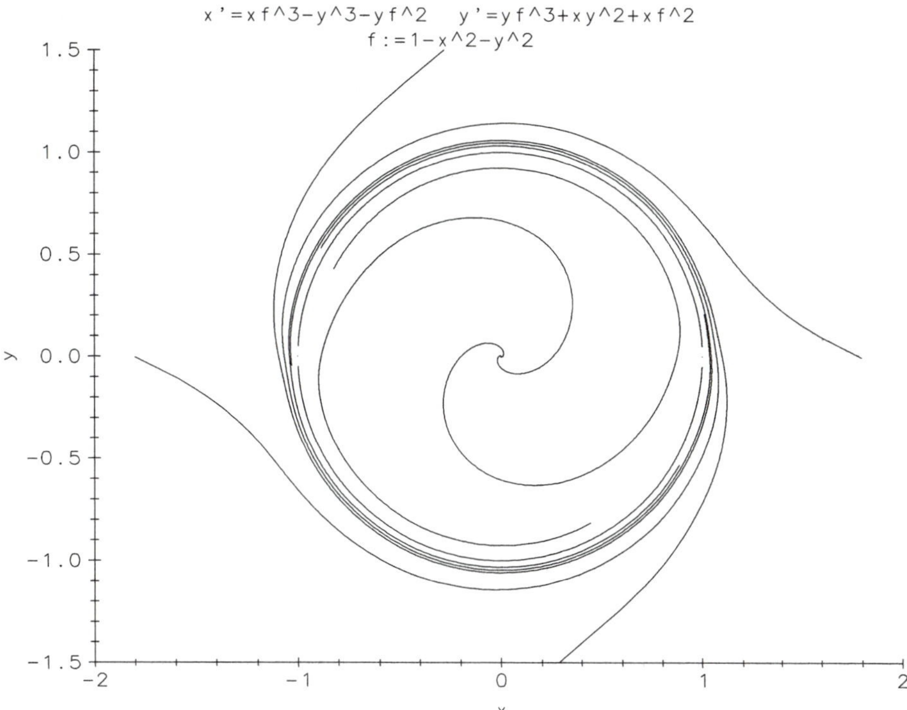


The unit circle is the limit set of orbits as $t \to \infty$ in these three plates. In Plate A a limit cycle of period $\pi/5$ lies on the circle, and all other orbits except the equilibrium point at the origin spiral toward the cycle. As the component graphs show, the rotational motion is uniform.

In Plate B the rate functions of Plate A are multiplied by the common scale factor $r(x, y)$, which is continuous and positive except at the point $(1, 0)$, where it vanishes. The new system has an equilibrium point there, and orbits spiral toward the cycle-graph consisting of that point and the arc $0 < \theta < 2\pi$ of the unit circle. The arc is traversed counterclockwise with increasing time. Orbits in the state plane appear to be the same in Plates A and B, but the component plots show how different the motions really are. In effect, the factor $r(x, y)$ rescales time in a position-dependent way that slows orbits near the newly introduced equilibrium point at $(1, 0)$ of the cycle-graph on the unit circle.

The system of Plate C has an attracting cycle-graph that consists of the equilibrium points $(\pm 1, 0)$ and the upper and lower open arcs of the unit circle oriented counterclockwise by increasing time. Nearby orbits spiral toward the cycle-graph, slowing down as they approach the equilibrium points, but speeding up as they move away.

***Reference*** Experiments 2.1, 5.16

## Lorenz A — Chaotic wandering

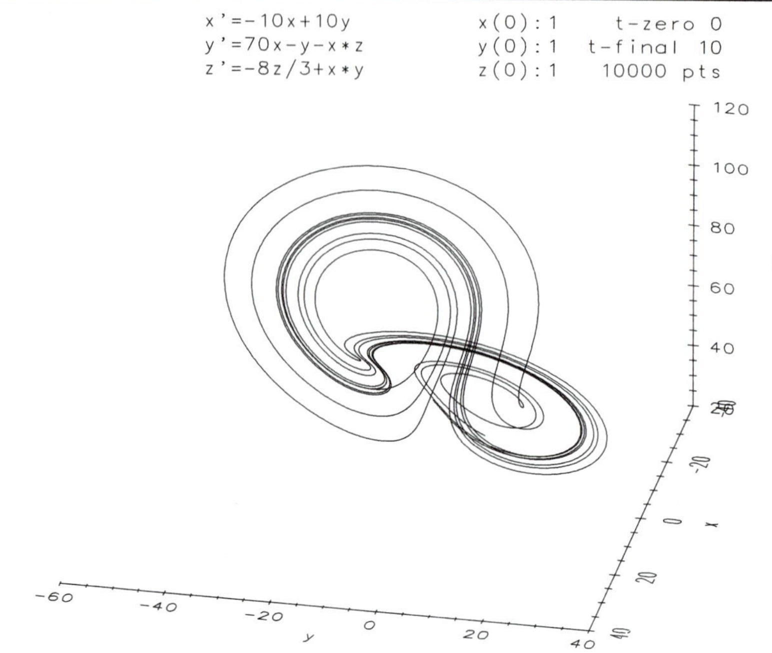


## Lorenz B — The mask

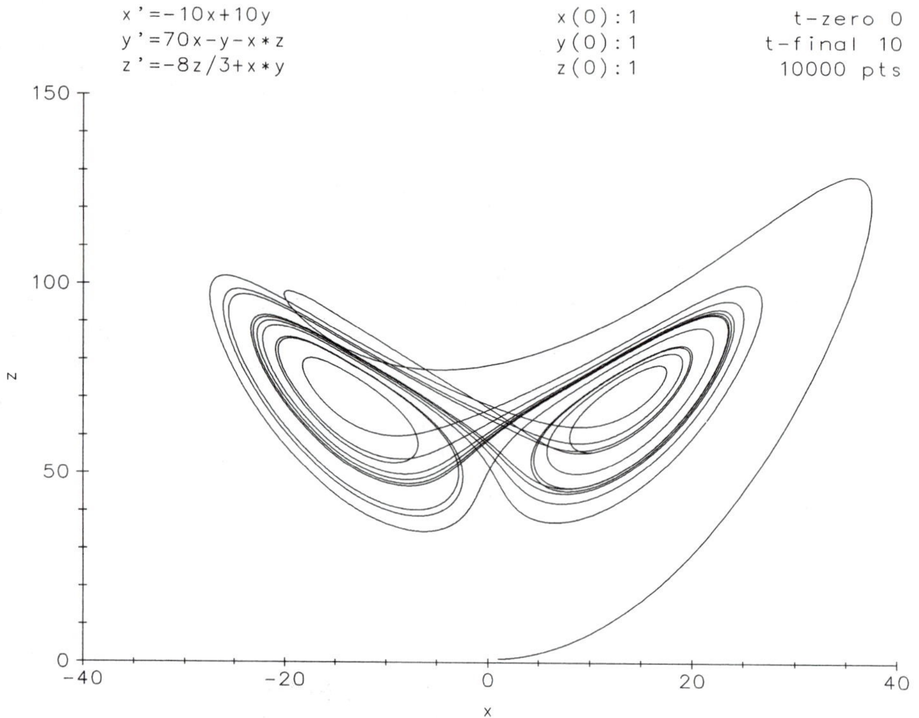

## Sensitive dependence

## Lorenz C

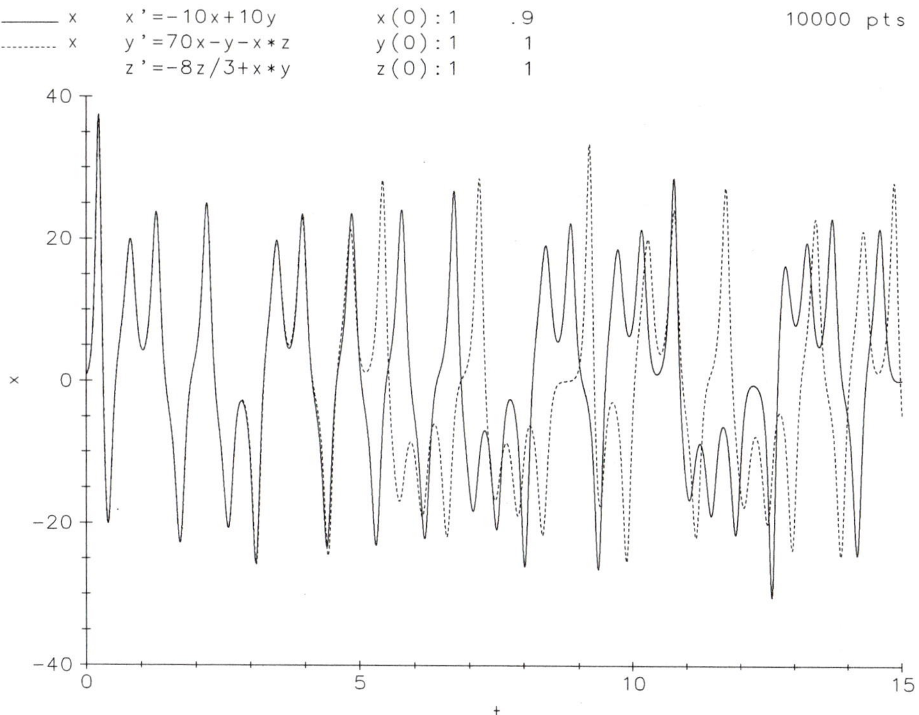


The Lorenz equations model convective air currents and temperature changes in a planar vertical cell beneath a thunderhead. The rate equations are

$$x' = ay - ax$$
$$y' = rx - y - xz$$
$$z' = -bz + xy$$

where $a$, $r$, and $b$ are dimensionless physical parameters. The state variables represent the amplitude and direction of convection ($x$), the temperature difference ($y$) between rising and falling air currents, and the deviation ($z$) of the air temperature from a standard value.

For a range of values of $a$, $r$, and $b$, the orbits of the system approach what is thought to be a strange attractor. The nature of the attractor remains something of a mystery. As orbits tend to the attractor, they cycle first around one unstable equilibrium point of the system, then around another, then back to the first, and so on. Although determined completely by the initial data, the motion appears to have something random about it. Plate A displays one orbit solved for $0 \leq t \leq 10$ but plotted for $1 \leq t \leq 10$, to suppress any initial transients. The orbital segment shown lies close to the strange attractor. Plate B shows the projection of the orbital segment for $0 \leq t \leq 10$ onto the $xz$-plane.

The $tx$-component plots in Plate C show the sensitivity of the orbits to a small change in the initial data. Observe that the two component curves almost coincide for $0 \leq t \leq 5$, but the curves appear to be unrelated as $t$ increases beyond 5.

***Reference*** Experiments 6.6–6.7

## Pendulum A — No friction

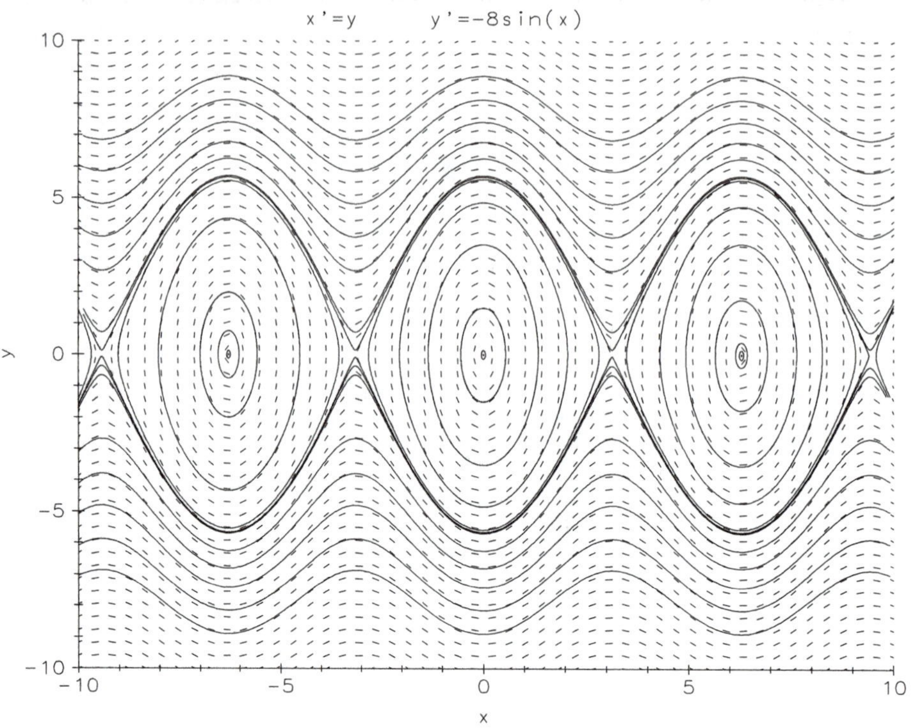


## Pendulum B — Frictional effects

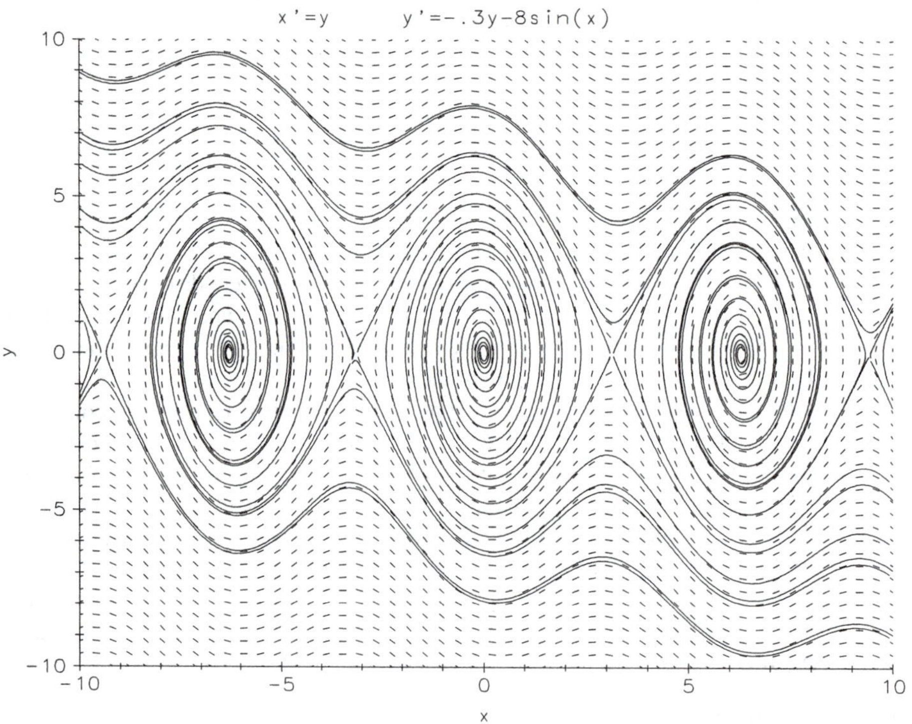

## Basin of attraction — Pendulum C

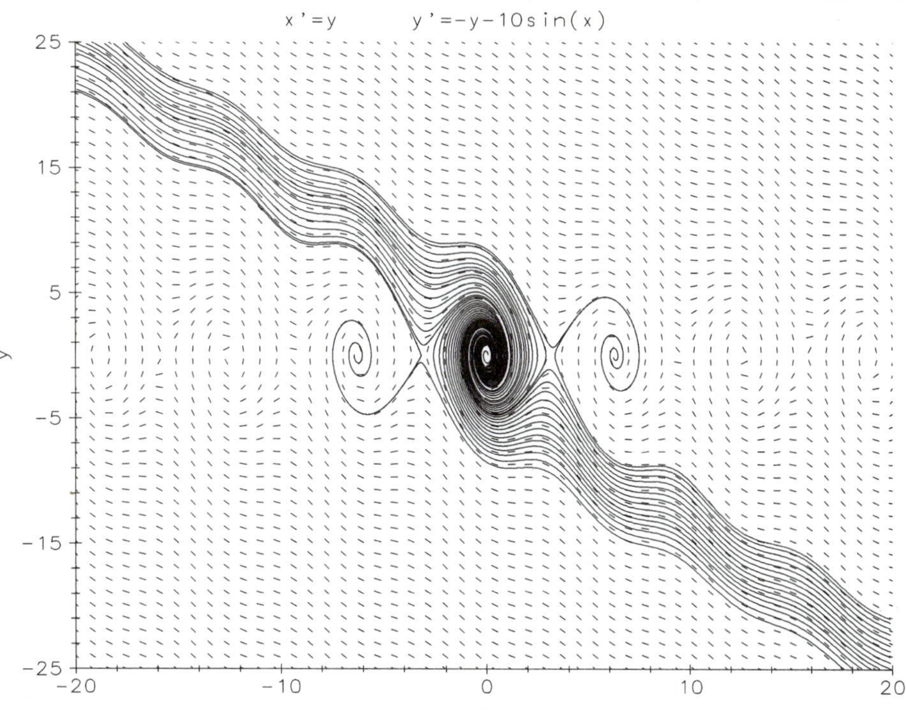


A pendulum is a bob suspended from a pivot. It can swing back and forth or, if enough energy is imparted, swing over the top and do loop-the-loops. In the idealized case of no friction, back-and-forth and loop-the-loop motions would continue forever. In reality, friction dissipates energy and the pendulum tends toward a rest position hanging vertically downward. The model equation for the motion is based on Newton's Laws and has the form

$$mLx'' = -cLx' - mg\sin x$$

where $m$ is the mass of the pendulum bob, $L$ is the length of the (massless) support rod, $x$ is the angular deviation in radians from the downward vertical, $c$ is the frictional coefficient, and $g$ is the local gravitational constant. The corresponding system is

$$x' = y$$
$$y' = -(c/m)y - (g/L)\sin x$$

In Plate A, there is no friction. The wavy orbits correspond to loop-the-loop motions. The ovals centered at the equilibrium points are the periodic orbits of the back-and-forth motions. With friction, (Plates B and C), all nonconstant orbits, except separatrices, tend to stable equilibrium points.

The basin of attraction of the equilibrium point at the origin in Plate C consists of all orbits that tend to the point as time advances. The band stretching across the plate is the part of the basin that lies in the screen. The tendrils correspond to orbits just outside the given basin.

***Reference*** Experiments 1.11, 5.7–5.8, 5.13

## Pendulum (Upended) A — Driven damped pendulum

x''+.07x'-(6/11)sin(x)=.3cos(t)
x(0)=-1.4, x'(0)=1

1000 solution points

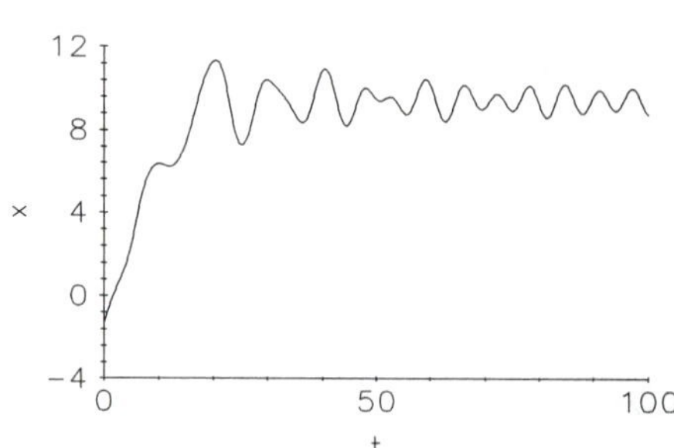


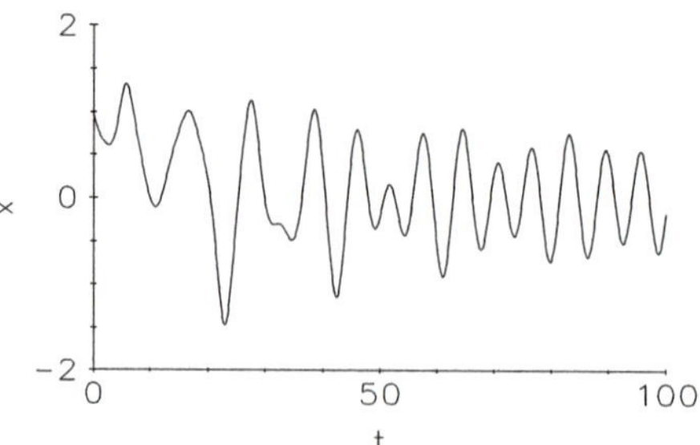


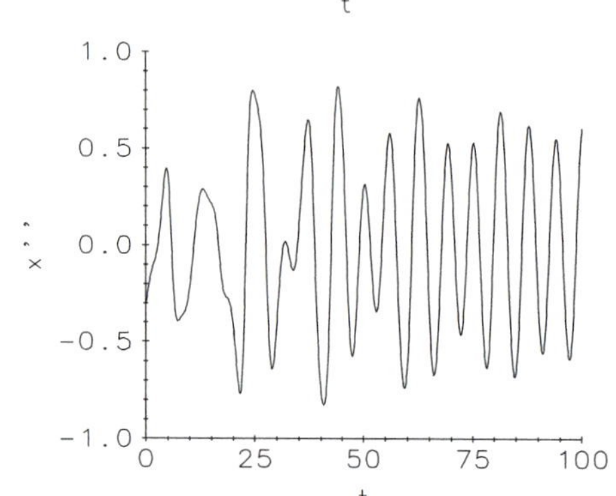


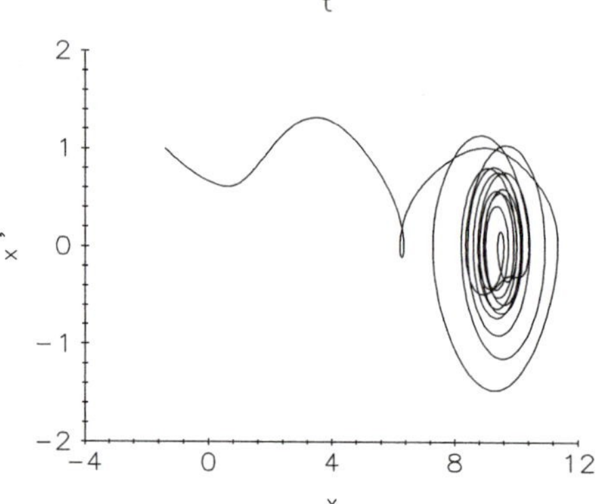


## Pendulum (Upended) B — Chaotic wandering

x''+.07x'-(6/11)(x-x^3/6)=.3cos(t)
x(0)=-1.4, x'(0)=1

1000 solution points

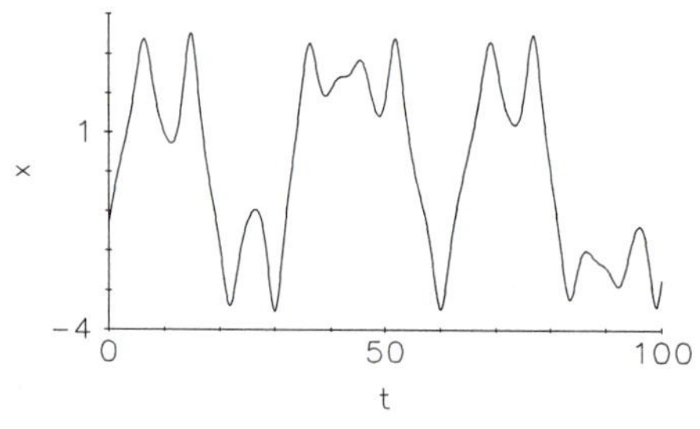


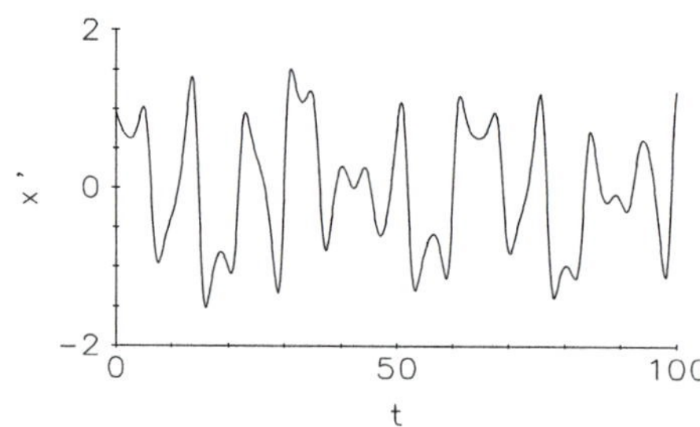


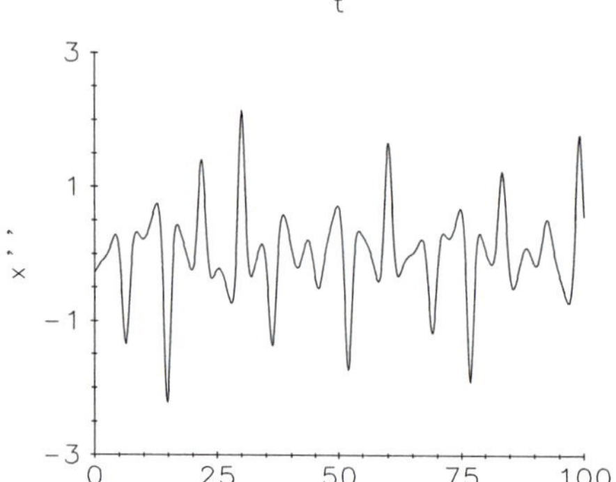


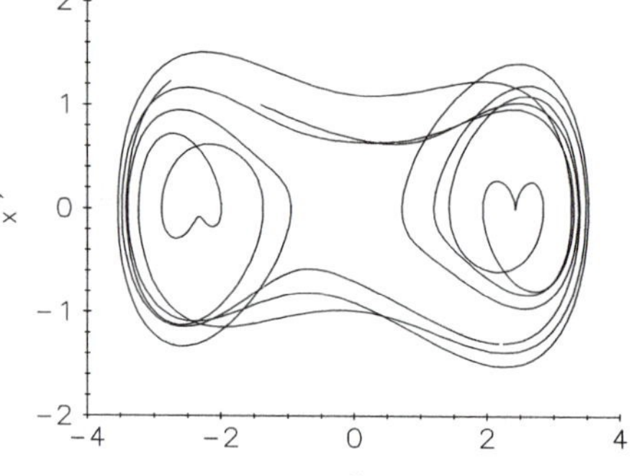

## Sensitivity to data — Pendulum (Upended) C

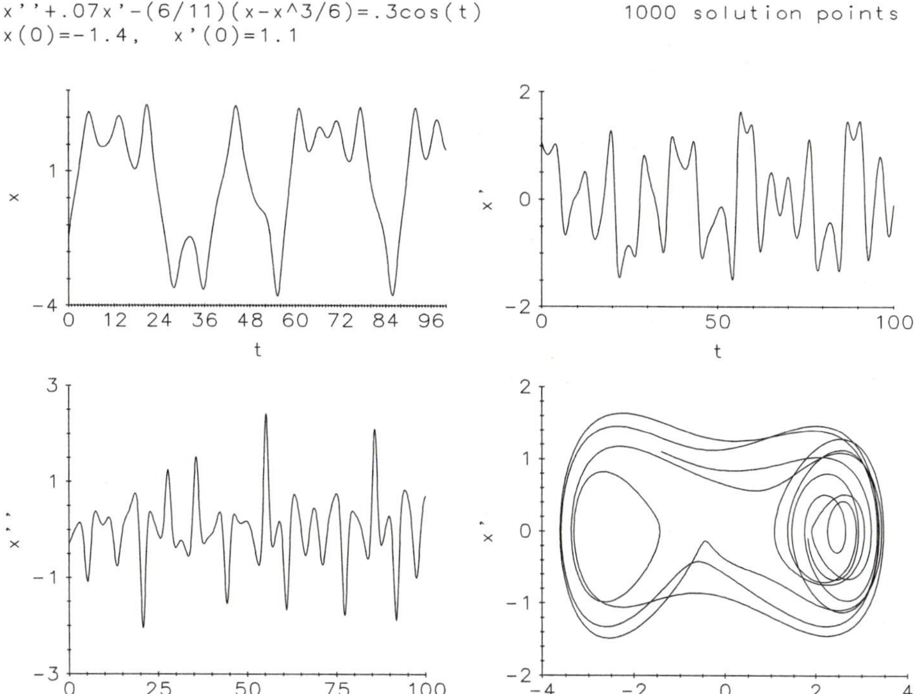


An upended pendulum is subjected to a sinusoidal driving force in addition to the usual gravitational and frictional forces. The model equation is

$$mLx'' + cLx' - mg\sin x = F(t)$$

where $L$ is the length of the (massless) support rod, $m$ is the mass of the bob, $x$ is the angle in radians measured from the *upward* vertical through the pivot, $c$ is the frictional coefficient, and $F(t)$ is the driving force. In these three plates, the data have the values $x(0) = -1.4$, $y(0) = 1$, $m = 1$, $c = 0.07$, $g/L = 6/11$, $F(t) = 0.3\cos t$. In Plates B and C the function $\sin x$ is replaced by the first two nonvanishing terms, $x - x^3/6$, of its Taylor expansion about $x = 0$.

The orbit in Plate A depicts the pendulum oscillating somewhat chaotically about the vertically downward position, $x = 3\pi$, $y = 0$. Truncating the Taylor series of $\sin x$ has a drastic effect on orbital behavior (Plate B), orbits of the driven system wandering around both the stable equilibrium points, $x = \pm\sqrt{6}$, $y = 0$, of the undriven system. Observe in Plate C the change made in the orbit of Plate B by increasing the initial velocity from 1 to 1.1. The increase in kinetic energy, though small, leads to a very different pattern of orbital wandering.

***Reference*** Experiment 5.9

## Planar System A — Linear systems

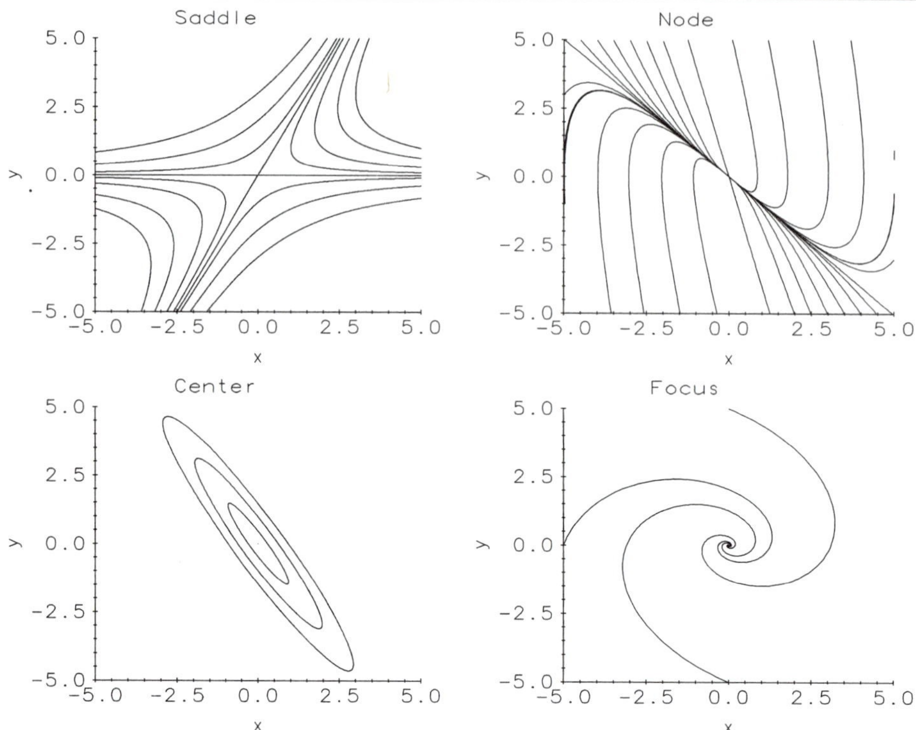


## Planar System B — Perturbed saddle at origin

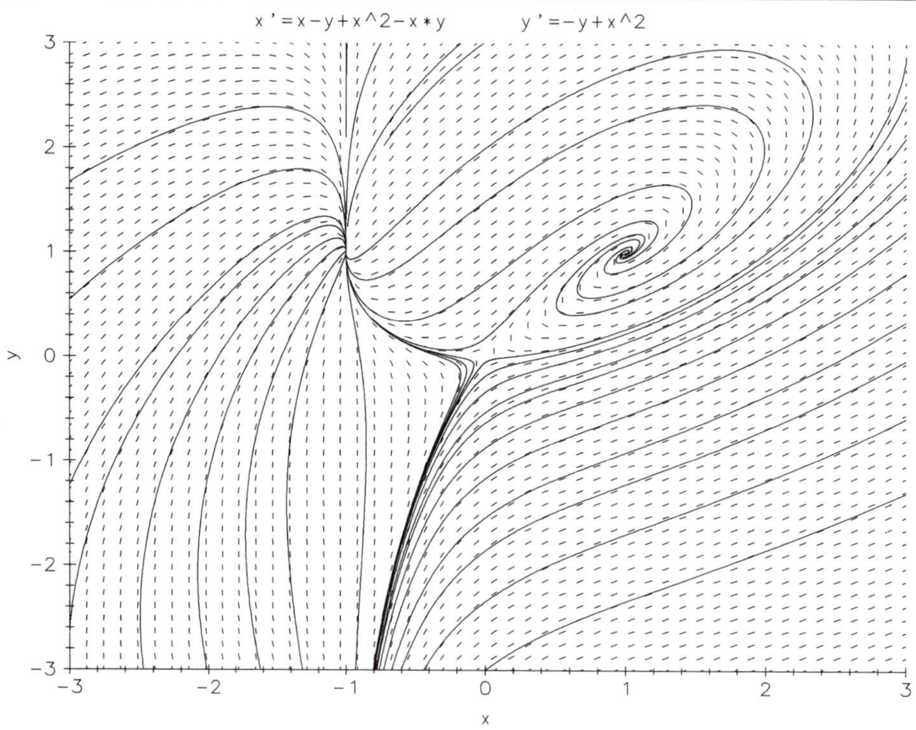

## Another perturbed saddle

## Planar System C

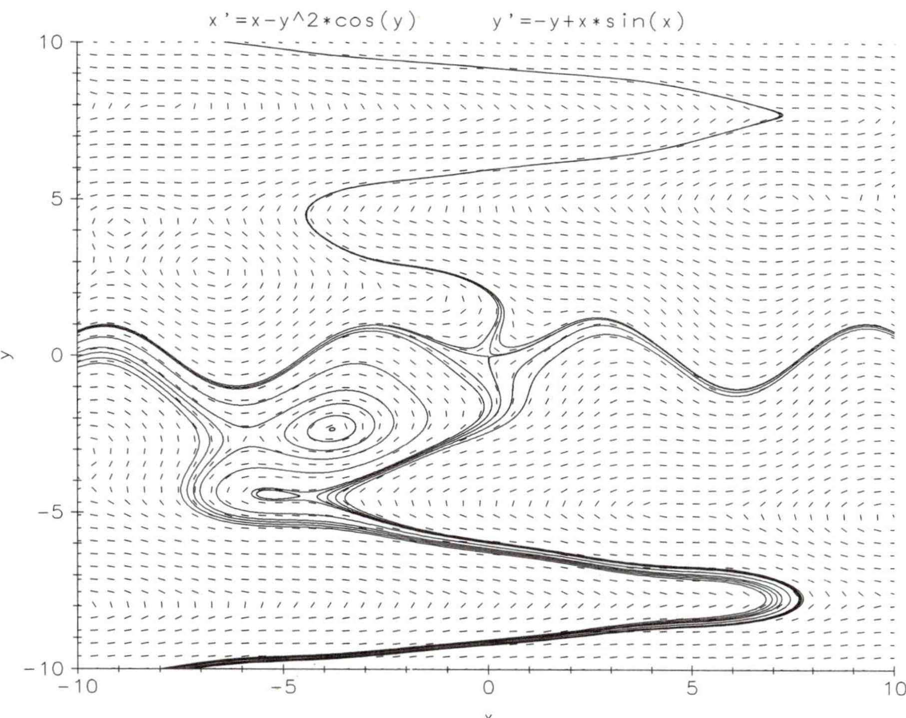


Autonomous linear systems in two state variables have the form

$$x' = ax + by$$
$$y' = cx + dy$$

where $a, b, c$ and $d$ are real constants. If 0 is not a root of the characteristic polynomial $r^2 - (a + d)r + ad - bc$, then there are four types of orbital portrait. Plate A shows representatives of these types. The four systems are:

**Saddle**
$x' = x - y, \quad y' = -y$

**Node**
$x' = -y, \quad y' = 4x + 5y$

**Center**
$x' = 3x + 2y, \quad y' = -5x - 3y$

**Focus**
$x' = -x + 4y, \quad y' = -3x - 2y$

If higher order terms in the state variables are added to the linear terms of the rate equations, the general behavior of the orbits near the origin does not change if the origin is a saddle, a node, or a focus. In Plate B the quadratic terms $x^2 - xy$ and $x^2$ are added to the linear terms of the saddle system of Plate A. Orbital behavior is still saddlelike near the origin. Away from the origin, the quadratic terms dominate and a very different kind of behavior appears.

In Plate C the saddle-type linear system, $x' = x$, $y' = -y$ is perturbed by the higher order terms $-y^2 \cos y$ and $x \sin x$. The saddlelike structure near the origin is preserved, but away from the origin all sorts of curiosities are created by the nonlinear terms.

***Reference*** Experiments 5.2, 5.12–5.13

## Planar System D

Quadratic rates

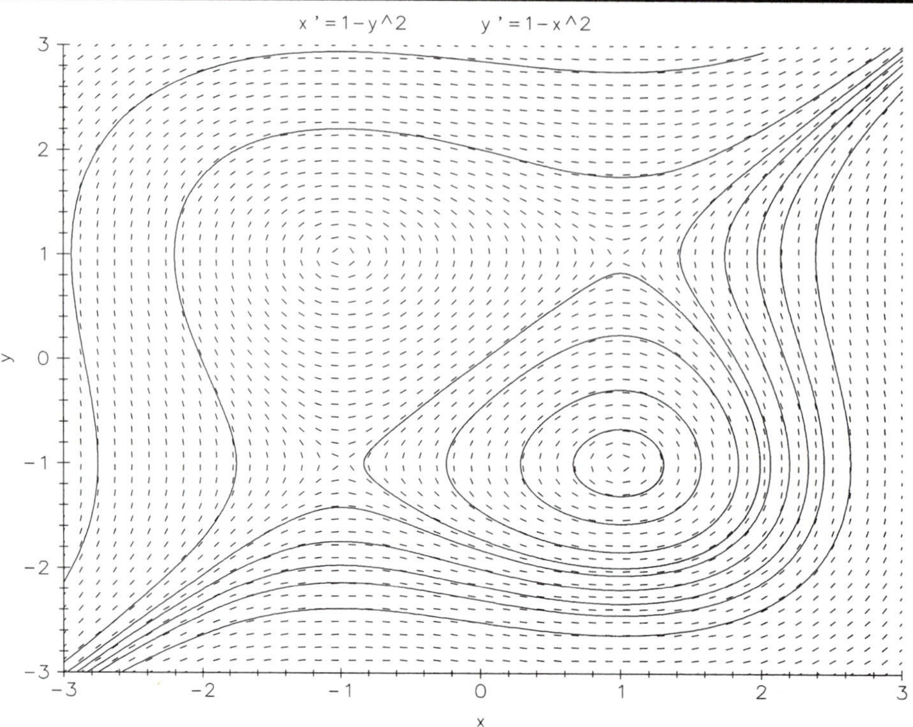


## Planar System E

Planar orbit patterns

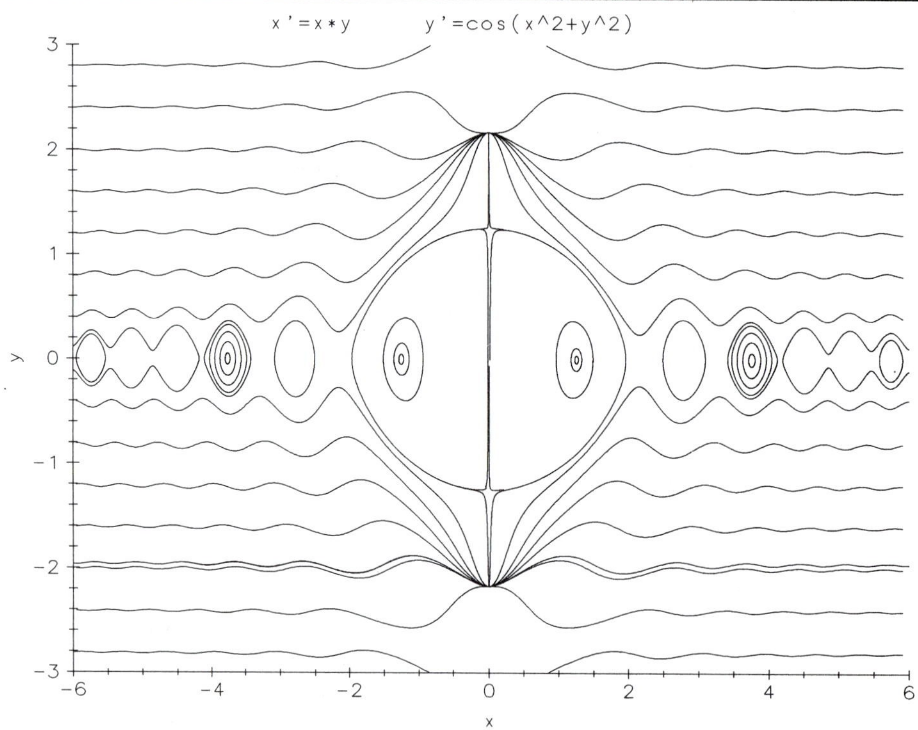

## Conditionally-defined rates — Planar System F

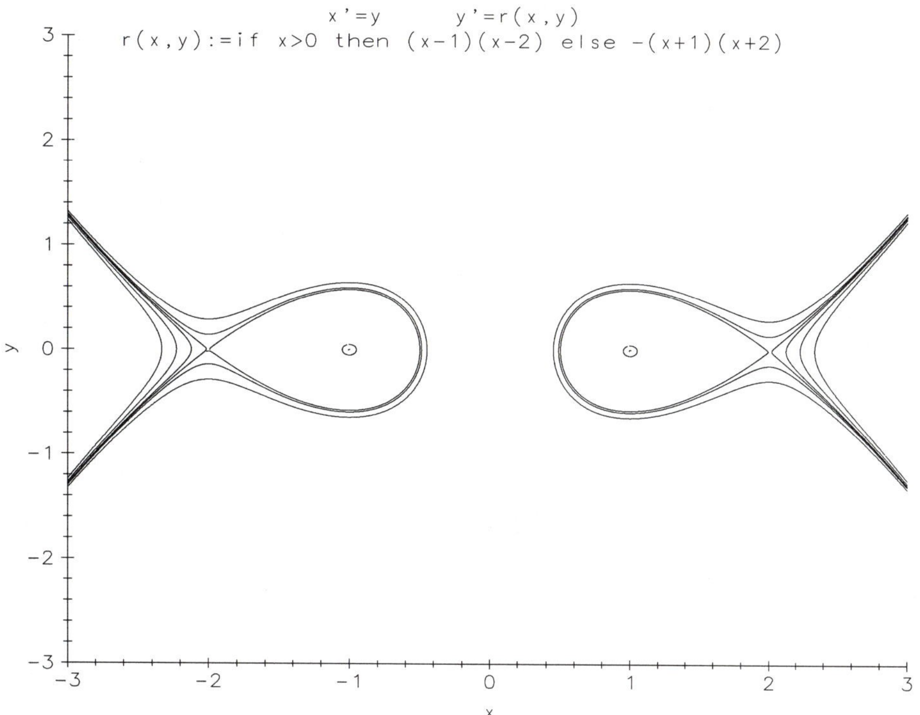


The autonomous ODEs in Plate D have quadratic rate functions, and care must be taken in plotting the orbits. As the state variables become large, rates will carry the orbits very quickly off the screen. Observe, however, that there is a family of periodic orbits that enclose the equilibrium point $(1, -1)$.

Radial and axial symmetries in the rate functions induce regular patterns in the portrait of the orbits in Plate E. Saddles, centers, and nodes are all present in the orbital portrait.

The system of Plate F shows how a conditionally-defined rate function may be used to place different systems in various locations in state space. Intricate orbital portraits may be created in this way. In each half-plane of Plate F the corresponding system has two equilibrium points, one with the local geometry of a center and the other that of a saddle. A homoclinic orbit exits the saddle point as $t$ increases from $-\infty$ and returns as $t$ approaches $+\infty$.

***Reference*** Experiments 1.11, 3.5 3.8, 5.1, 5.15

## Planar System G — Quadratic Painlevé system

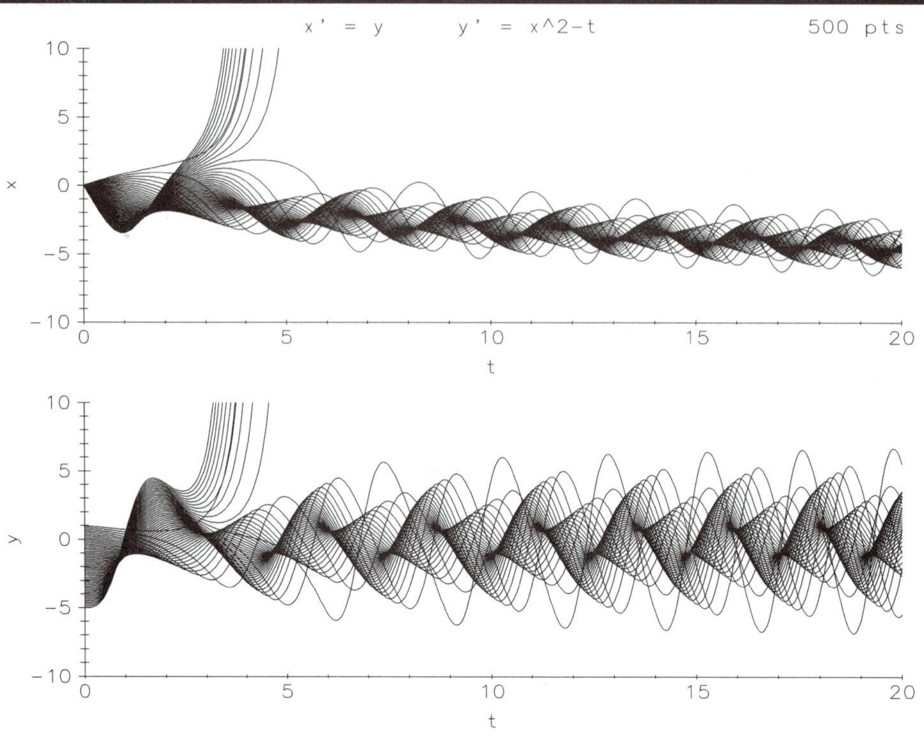


## Planar System H — Cubic system: nonautonomous

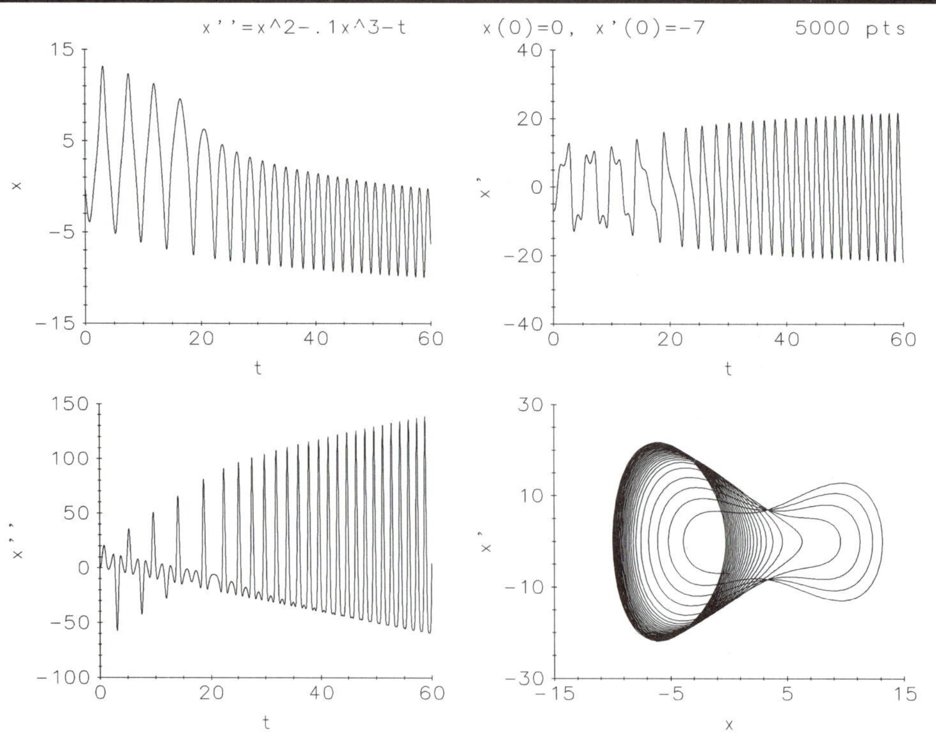

## Cubic system: Four orbits — Planar System I

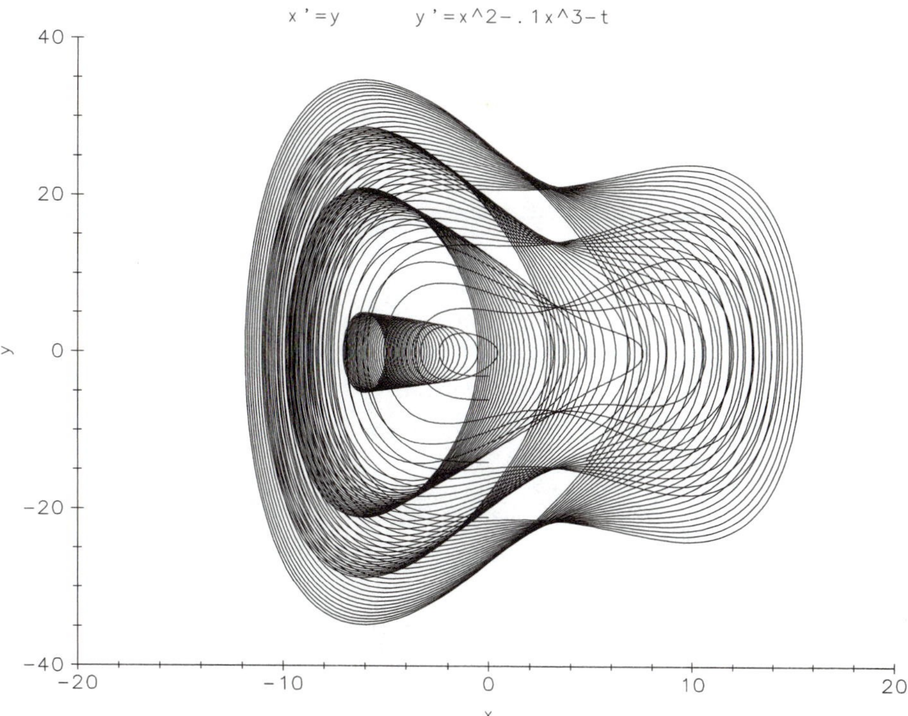


The second order scalar equation

$$x'' = x^2 - t$$

is a Painlevé equation, one of a large collection of nonlinear ODEs with polynomial right-hand sides that have received much attention over the past hundred years. Component graphs of solutions of the corresponding planar system are shown in Plate G. The initial conditions are $x(0) = 0$, $y(0) = y_0$ with $-5 \le y_0 \le 1$. Observe that some solutions appear to "blow up" in finite time, while others display the typical oscillatory (but nonperiodic) behavior of a Painlevé solution.

Plate H shows an orbit and its component curves for a nonautonomous planar cubic system. In Plate I, which displays four orbits of that system with initial data $x_0 = 0$, $y_0 = -3, -6, -14, -21$, characteristic cone- and bell-like structures emerge.

***Reference*** Experiment 5.1

## Poincaré A — Chaotic patterns

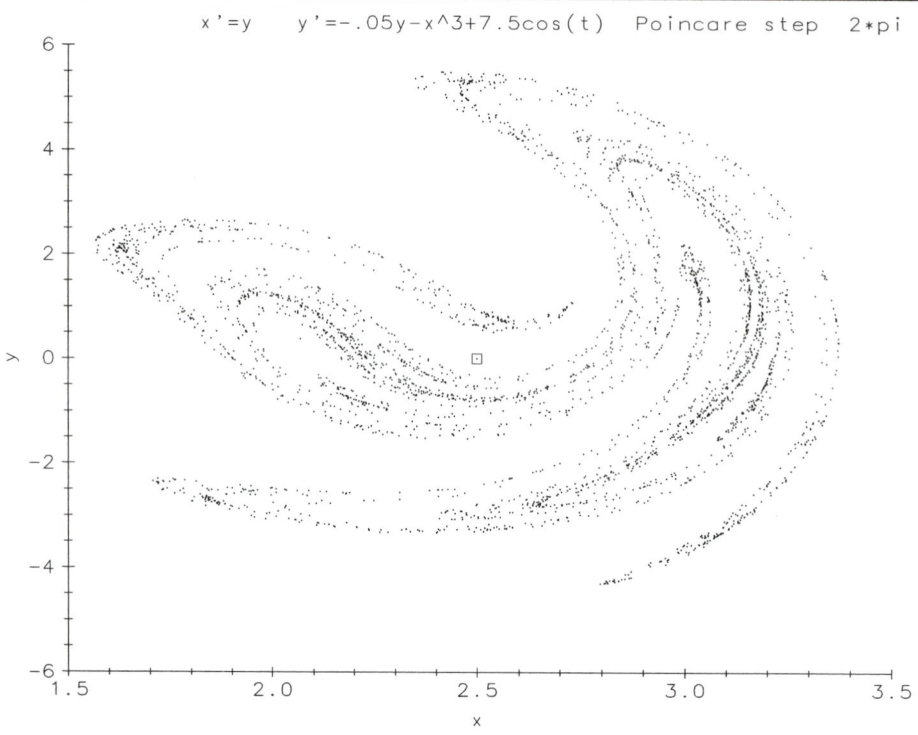


## Poincaré B — Sensitivity to parameter changes

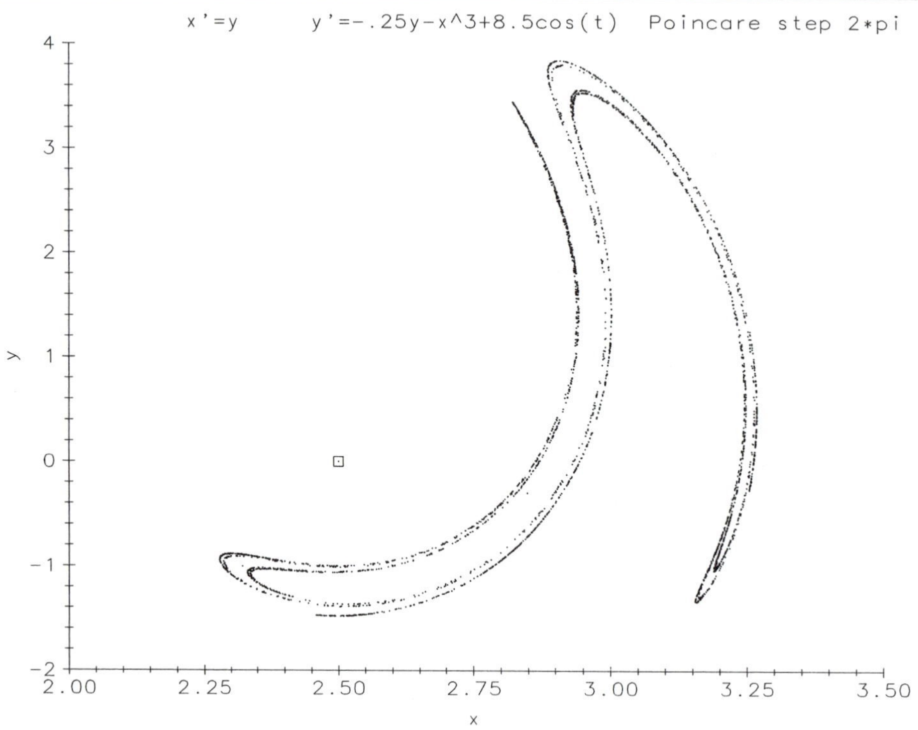

## The scroll circuit — Poincaré C

```
x'=-33f(.07,y-x,.1)          x(0):1              t-zero 0
y'=-f(.07,y-x,.1)-z          y(0):1             t-final 500
z'=y                         z(0):1      Poincare step 0.1
  f(a,w,b):=-a*w+.5(a+b)(abs(w+1)-abs(w-1))
```

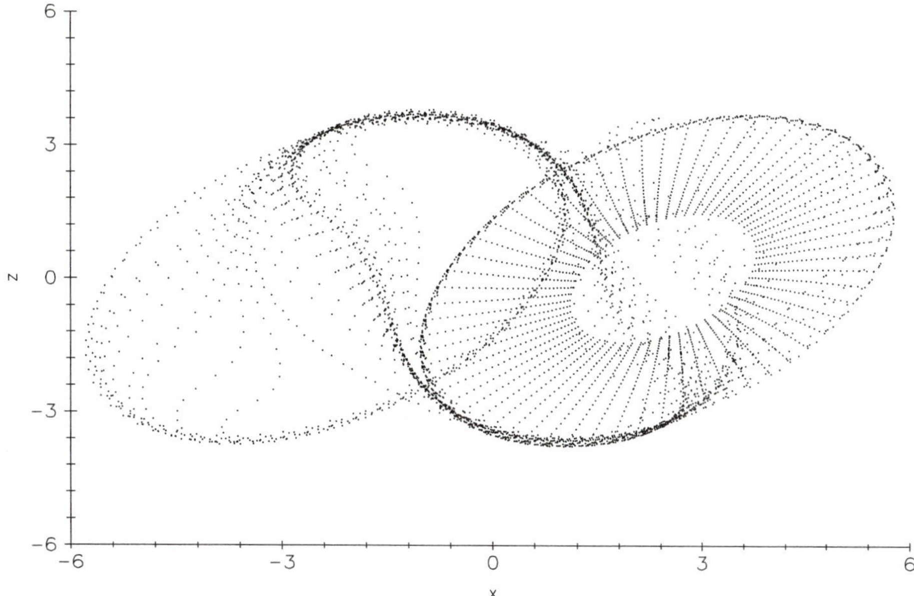


In a Poincaré time section (or plot), a moving point on an orbit is recorded "stroboscopically" at times that are integer multiples of the period of the driving force. If the orbit has that same period, only one point will be seen. In Plates A and B the driving force has period $2\pi$. The initial point $x(0) = 2.5$, $y(0) = 0$ of the orbit is marked by a square in each plate. Orbital points are plotted at the discrete times $0, 2\pi, \ldots,$ over a time span ranging from 0 to 20000. Since the points do not repeat, the corresponding orbits apparently do not have any multiple of $2\pi$ as a period. In fact, they appear to be chaotic but with a distinctive and nonrandom pattern. Although there is as yet no proof, it is thought that this pattern represents points on a strange attractor that attracts all nearby orbits. The nonlinear systems for these two plates are identical except for differences in two of the rate parameters. However, these differences lead to marked changes in the Poincaré plots.

In Plate C the orbit of the scroll system of Atlas Plate `Scroll C` is given a Poincaré plot with time step 0.1. In this case the 3D Poincaré plot is projected onto the $xz$-plane. Observe that there is no driving force for this system.

***Reference*** Experiments 5.10, 6.14

## Predator-Prey A — Predator-Prey oscillations

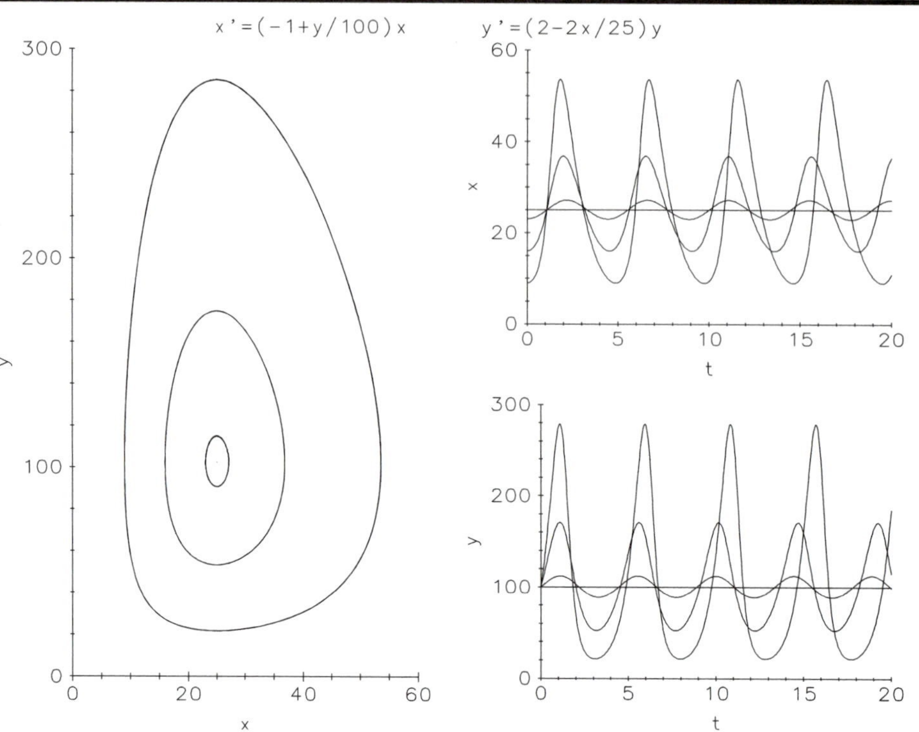


## Predator-Prey B — Satiable predators: limit cycle

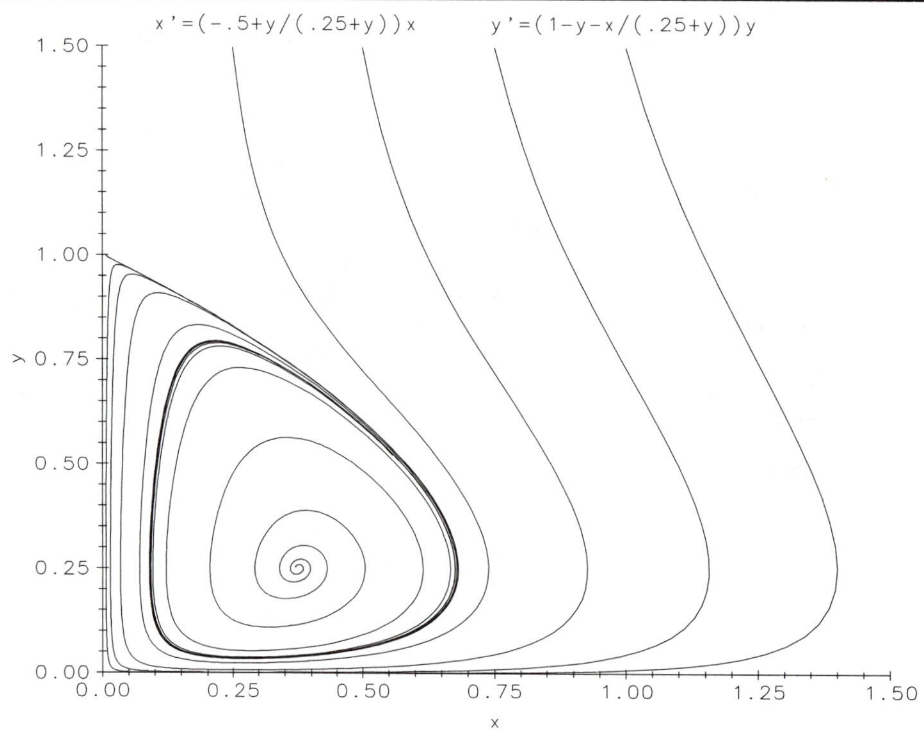

## Competing predators

Predator-Prey C

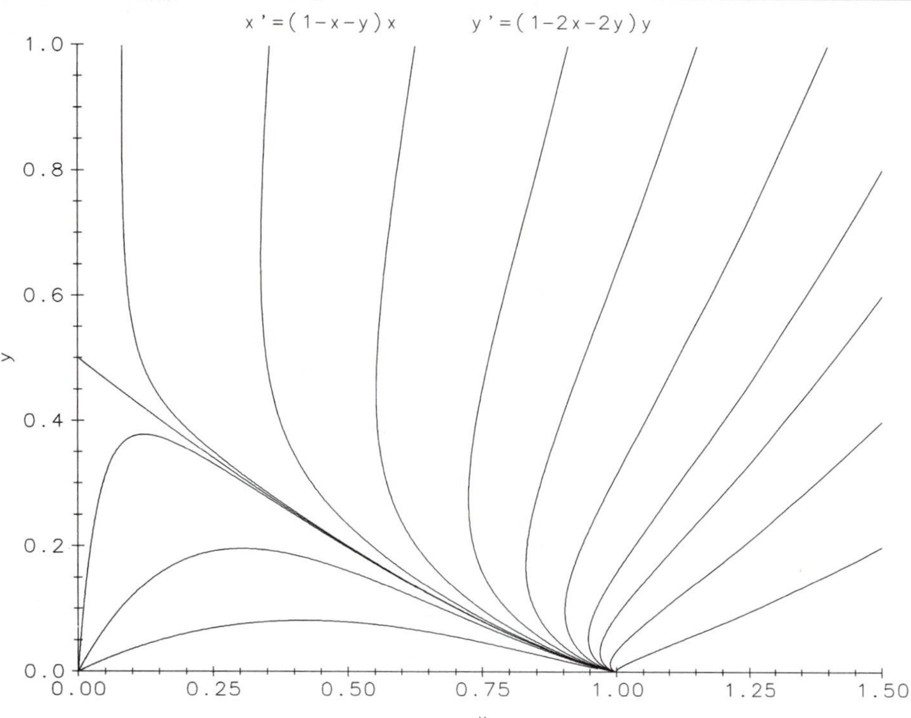


The Volterra model of predator-prey interaction is:

$$x' = (-a + by)x$$
$$y' = (c - dx)y$$

where $x$ and $y$ are the respective predator and prey populations and $a, b, c$, and $d$ are positive rate and interaction constants. Plate A shows the cycles in the population quadrant as they turn about the equilibrium point $(c/d, a/b)$. The coordinates of this point are the respective time averages of the predator and prey populations over each cycle. Observe from the component plots that the cycle periods increase with distance from the equilibrium point.

In reality, predator appetites become satiated, and overcrowding limits population growth. The orbits of Plate B depict the consequences of including predator satiation and overcrowding in the model—convergence to a unique attracting limit cycle in the population quadrant.

The orbits of Plate C show what happens in one model of overcrowding and mutual predation between two species. One of the species survives, the other does not.

***Reference*** Experiments 5.5–5.6, 5.18

## Rössler A — Stable periodic orbit

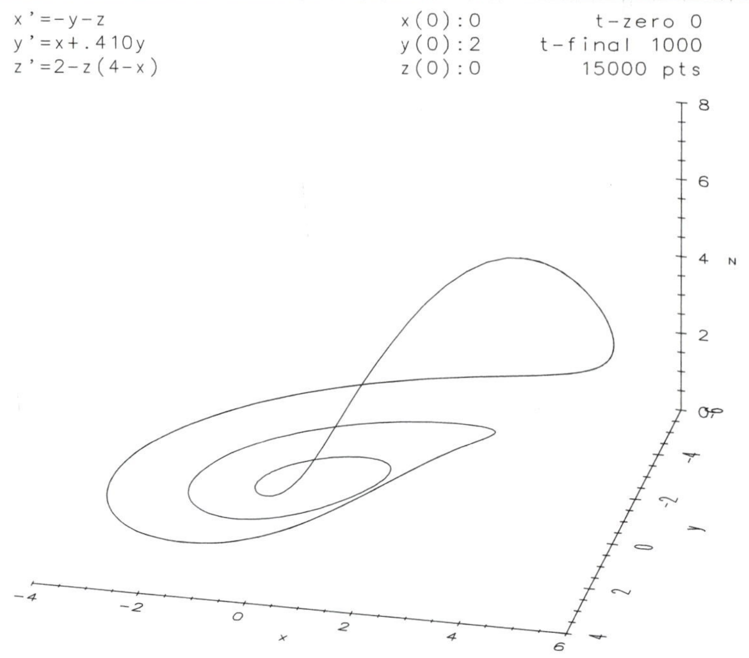


## Rössler B — Strands spread apart

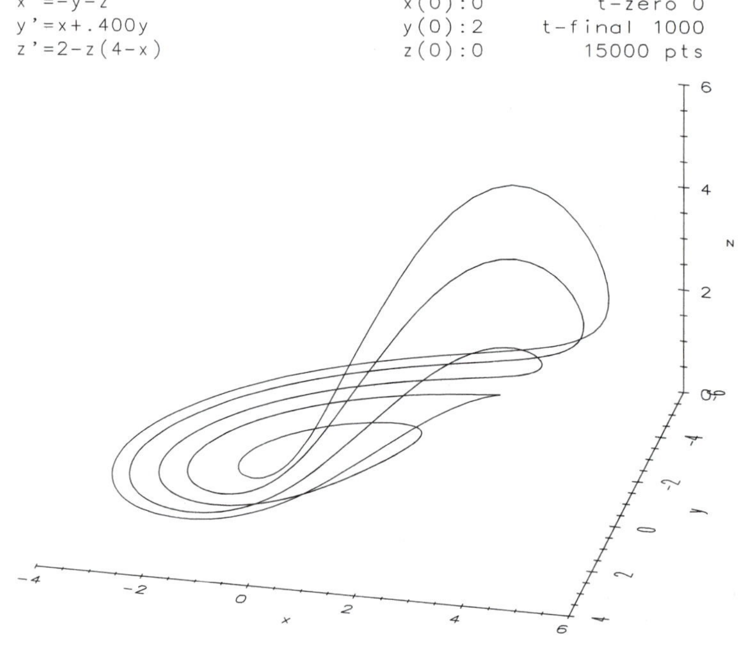

## Orbit on a strange attractor — Rössler C

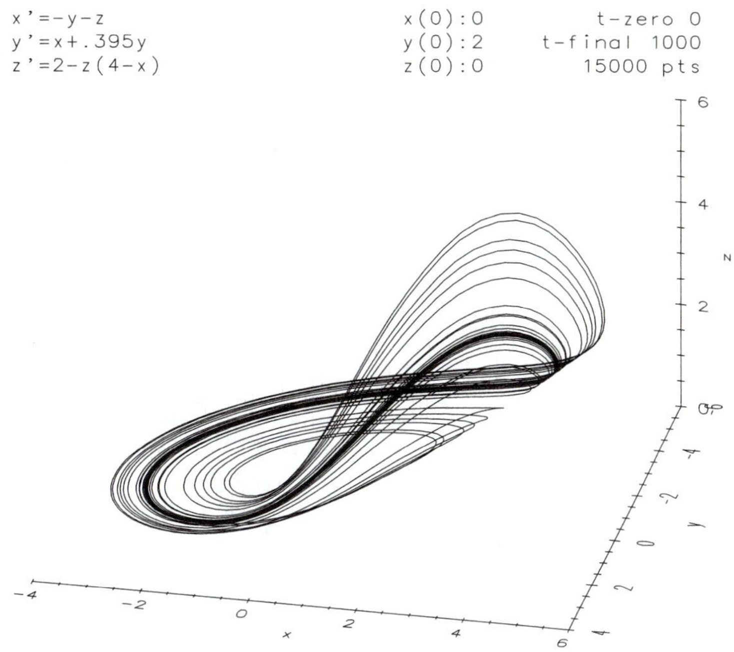


The Rössler equations are

$$
\begin{aligned}
x' &= -y - z \\
y' &= x + ay \\
z' &= b - z(c - x)
\end{aligned}
$$

where $a$, $b$, and $c$ are positive constants. Each orbit displays a "folding" behavior like that of the Lorenz attractor, but without the symmetries of the Lorenz system. In these plates the parameters $b$ and $c$ are fixed at 2 and 4, respectively, and $a$ is allowed to change slightly. After enough time has elapsed for transient behavior to vanish, a segment of an orbit is plotted. The system in Plate A is solved for $0 \le t \le 1000$, but the orbit is plotted for $950 \le t \le 1000$. The orbits in Plates B and C are plotted for $800 \le t \le 1000$.

The periodic orbit in Plate A spirals twice parallel to the $xy$-plane, then jumps up at an angle before dipping back into the plane. In Plate B the strands and the period of the periodic orbit multiply if the parameter $a$ is changed from 0.410 to 0.400. The folded structure of a Rössler orbit is evident in Plate B. If the parameter $a$ is changed from 0.400 to 0.395 the strands again multiply, the folding is clear, and a strange attractor of nonperiodic, chaotic but regular wandering seems to have appeared.

***Reference*** Experiment 6.8

## Scroll A — Nonlinear circuit: voltages, current

```
x'=-.1f(.07,y-x,.1)          x(0):1        t-zero 0
y'=-f(.07,y-x,.1)-z          y(0):1       t-final 400
z'=y                         z(0):1        5000 pts
  f(a,w,b):=-a*w+.5(a+b)(abs(w+1)-abs(w-1))
```

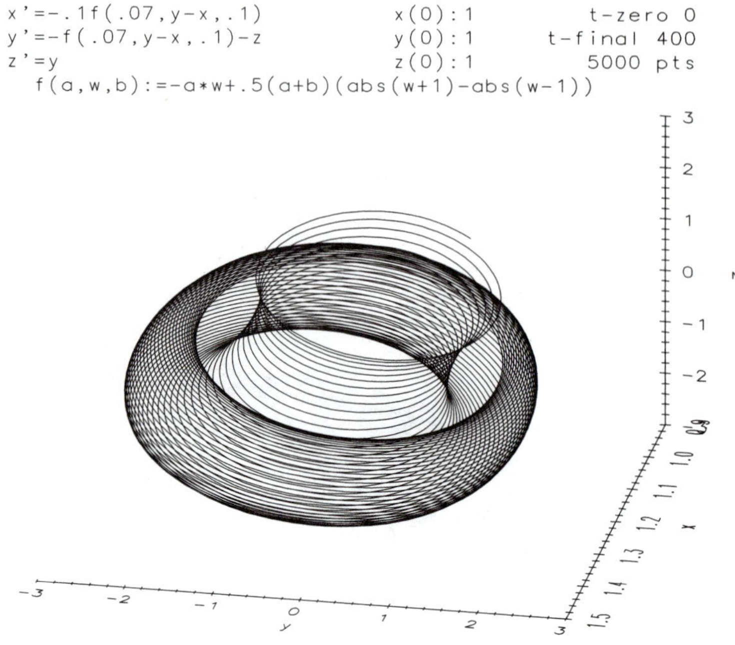


## Scroll B — Projection in $xz$-plane

```
x'=-.1f(.07,y-x,.1)          x(0):1        t-zero 0
y'=-f(.07,y-x,.1)-z          y(0):1       t-final 500
z'=y                         z(0):1        5000 pts
  f(a,w,b):=-a*w+.5(a+b)(abs(w+1)-abs(w-1))
```

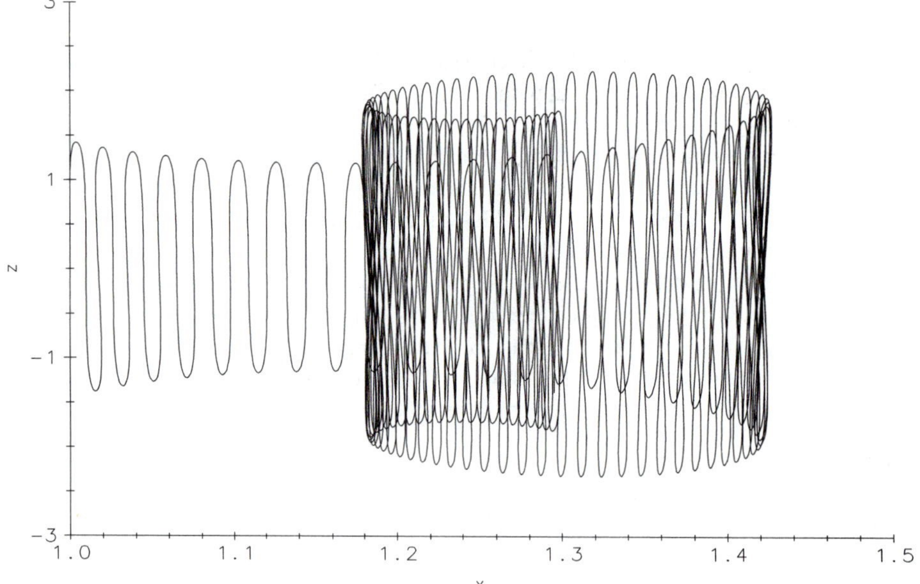

## Approach to a strange attractor — Scroll C

```
x'=-33f(.07,y-x,.1)          x(0):1          t-zero 0
y'=-f(.07,y-x,.1)-z          y(0):1        t-final 500
z'=y                         z(0):1         20000 pts
  f(a,w,b):=-a*w+.5(a+b)(abs(w+1)-abs(w-1))
```

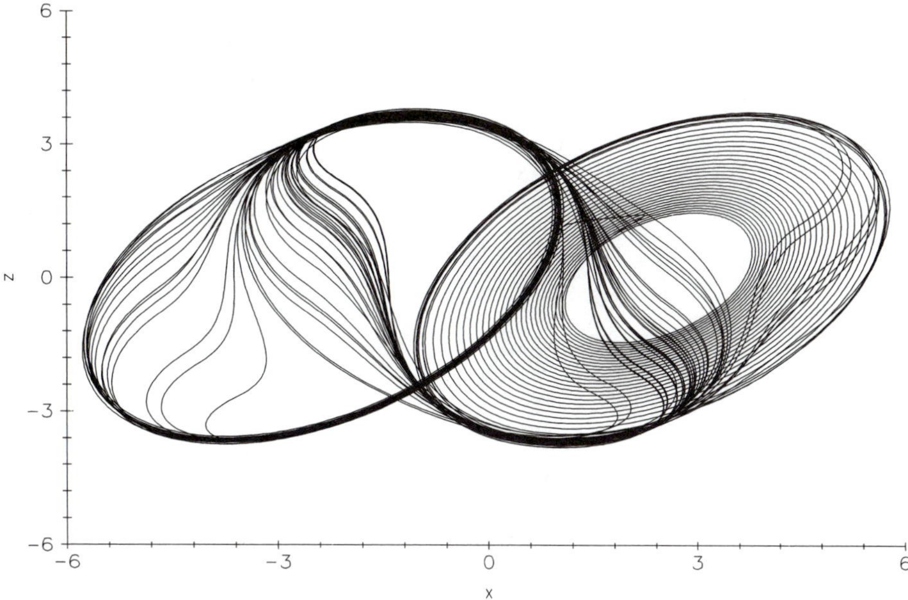


The scroll equations model the changing voltages and current in an electric circuit. The circuit has an inductor and two capacitors in parallel. A nonlinear resistor with a piecewise-linear current/voltage characteristic is in series with the capacitors. Physically, the resistor is a semiconductor device. In dimensionless variables the rate equations for the voltage drops $x$ and $y$ across the capacitors and the current $z$ through the inductor are

$$x' = -cf(y - x)$$
$$y' = -f(y - x) - z$$
$$z' = ky$$

where $c$ and $k$ are positive circuit constants. The function $f$ modeling the current through the resistor is defined by

$$f(w) = -aw + 0.5(a + b)(|w + 1| - |w - 1|)$$

where $a$ and $b$ are also positive circuit constants. The parameter values used in Plates A and B are $c = 0.1$, $k = 1$, $a = 0.07$, $b = 0.1$. The orbit in plate A tends to an invariant torus inside the "cap" of the mushroom. The scroll of plate B is the projection of the orbit into the $xz$-plane.

In Plate C the value of the rate constant $c$ is increased to 33, and this has a drastic effect on the orbit and its projection in the $xz$-plane.

***Reference*** Experiment 6.14

## Second Order A — Input/output graphs

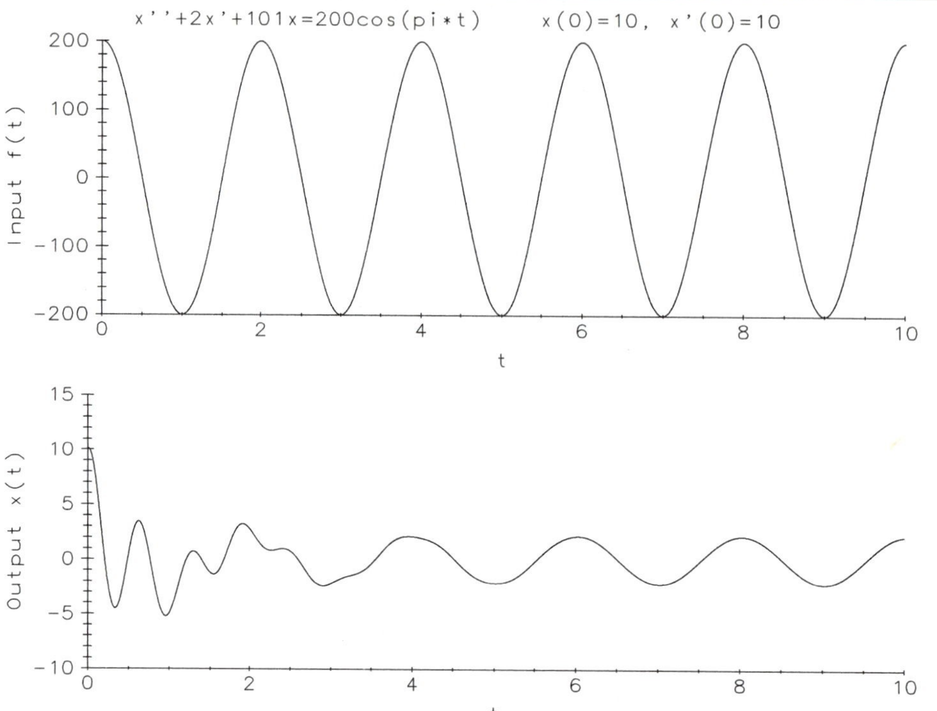


## Second Order B — Linear ODE: modulated input

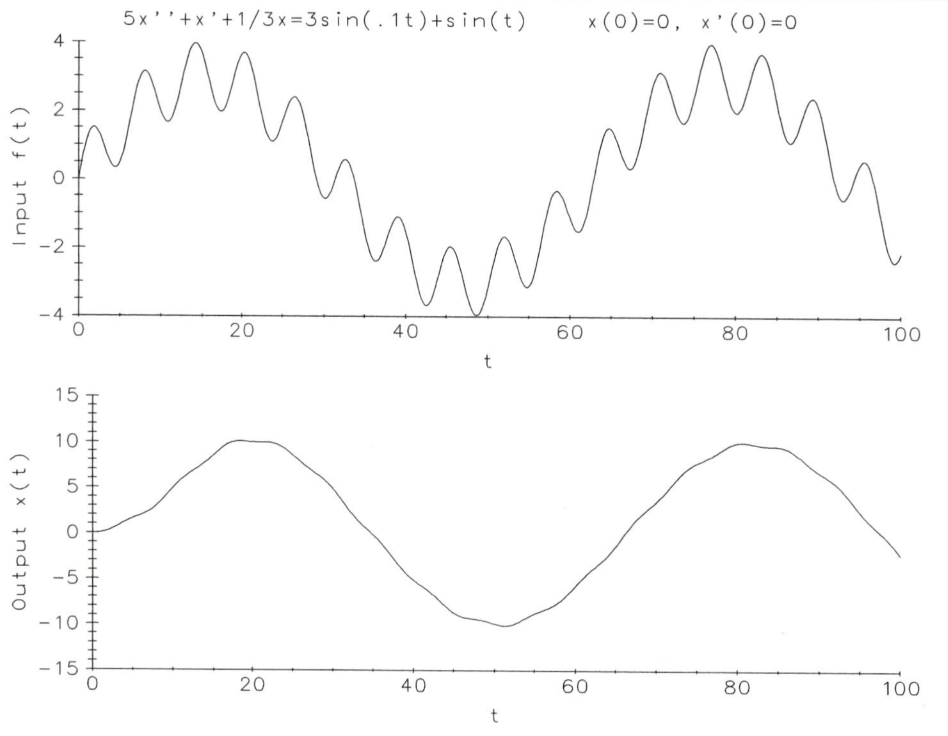

## Resonance Second Order C

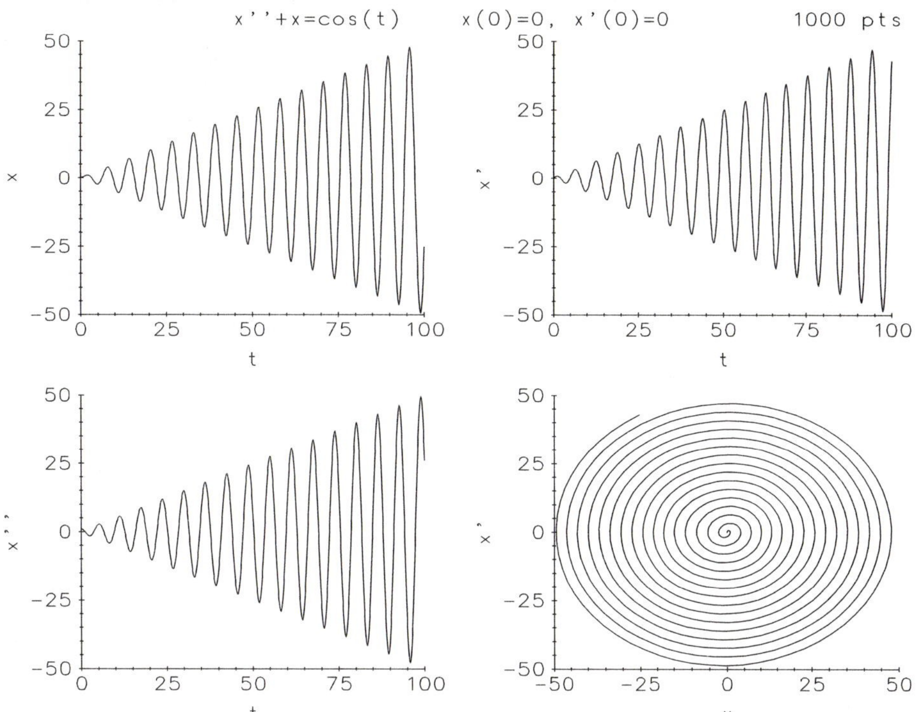


The phenomena shown in these plates are typical of second order, linear, constant coefficient, driven ODEs. The input $f(t)$ for the ODE of Plate A is $200\cos(\pi t)$ and is displayed in the top graph. The output $x(t)$ is the sum of a transient solution of the undriven equation (i.e., a solution that decays to zero as time increases) and of a particular solution of the driven equation. The output eventually tracks the input, but at a much lower amplitude and with a slight time lag.

In Plate B the amplitude of the sinusoidal modulation of the input sine wave is sharply diminished in the output. The high amplitude output lags somewhat behind the input.

Plate C illustrates a different phenomenon. If the frequency of the input coincides with the natural frequency of the undriven equation, resonance occurs in the form of an output whose amplitude becomes unbounded as time increases. In this example of resonance, both input and output frequencies are one hertz (1 Hz).

***Reference*** Experiments 4.3–4.6

## Spring A — Oscillations of a linear spring

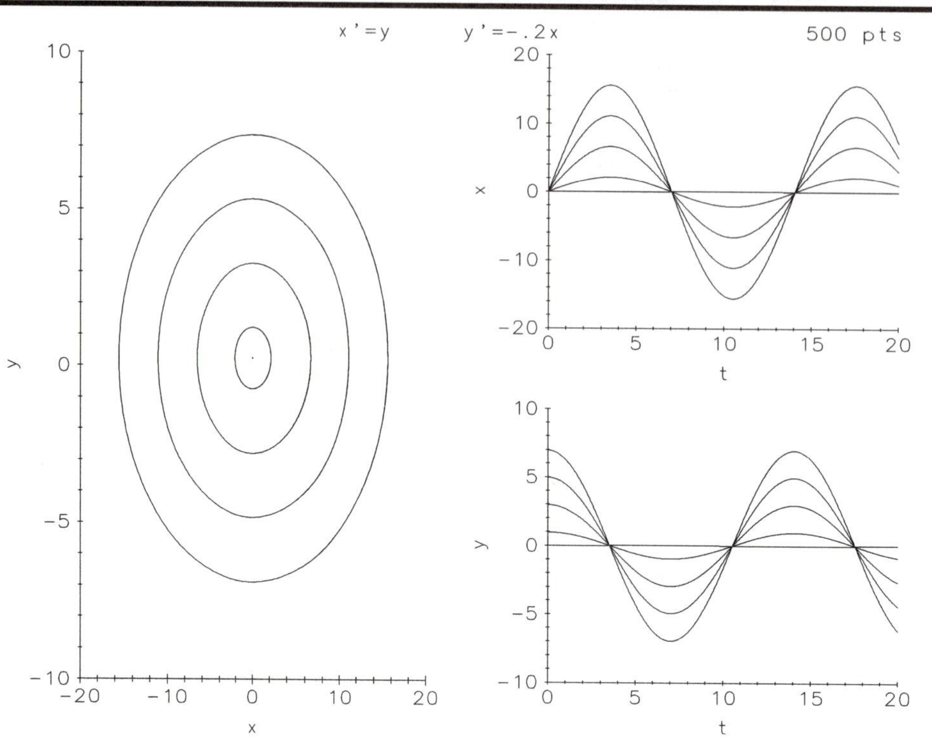


## Spring B — Oscillations of a hard spring

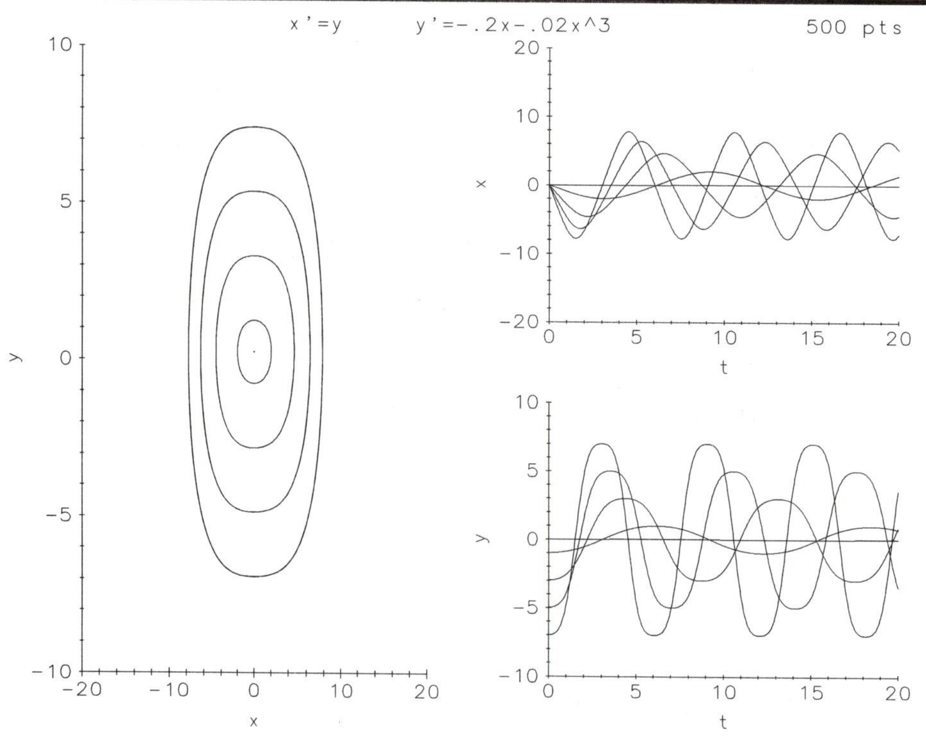

## Motions of a soft spring — Spring C

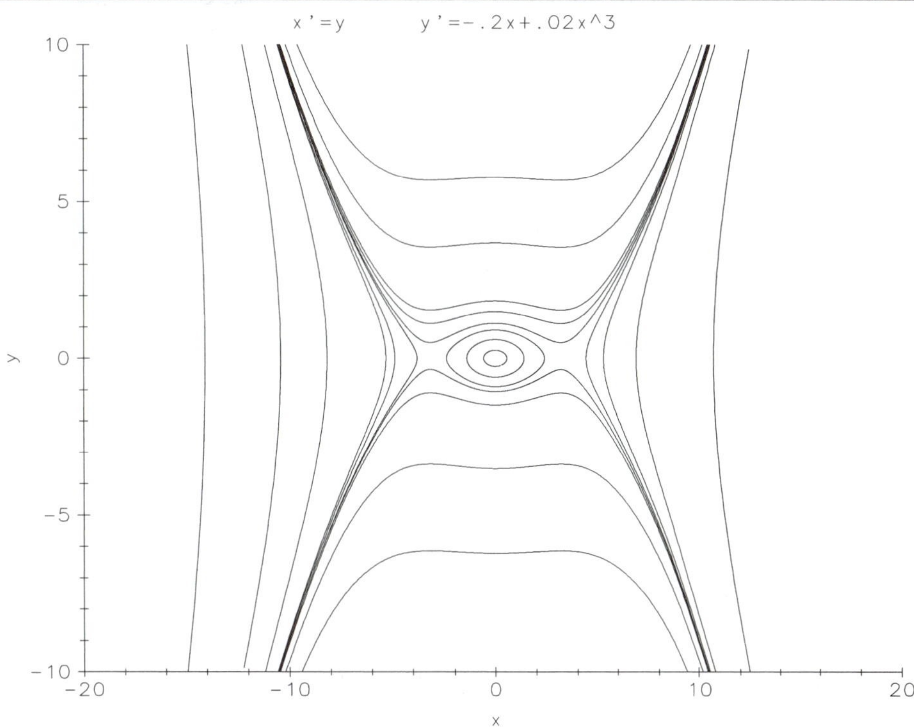


The equation $mx'' = F(x)$ models the displacement $x$ of a mass attached to a spring and acted on by the position-dependent spring force $F(x)$. The corresponding system is

$$x' = y$$
$$y' = F(x)/m$$

The portraits in these three plates correspond to three cases:

| | | |
|---|---|---|
| Plate A | (linear spring) | $F(x) = -kx$ |
| Plate B | (hard spring) | $F(x) = -kx - bx^3$ |
| Plate C | (soft spring) | $F(x) = -kx + bx^3$ |

The values of the constants are $m = 1$, $k = 0.2$ and $b = 0.02$. The linear spring (Hooke's Law) and the hard spring systems have a unique and stable equilibrium point at $x = 0$, $y = 0$. The periods of the oscillations of the linear spring (Plate A) have the fixed value $2\pi/\sqrt{k}$. The periods of the oscillations of the hard spring (Plate B) diminish with increasing distance from the equilibrium point.

The soft spring system of Plate C also has a stable equilibrium point at the origin, and that point is enclosed by periodic orbits. However, periodic motion occurs only if the initial energy imparted to the spring is below a threshold value. Above that value, the spring stretches beyond its elastic limit, as shown by the unbounded orbits.

***Reference*** Experiments 4.7–4.8

## Quadratic system: two limit cycles

**Two-cycled System A**

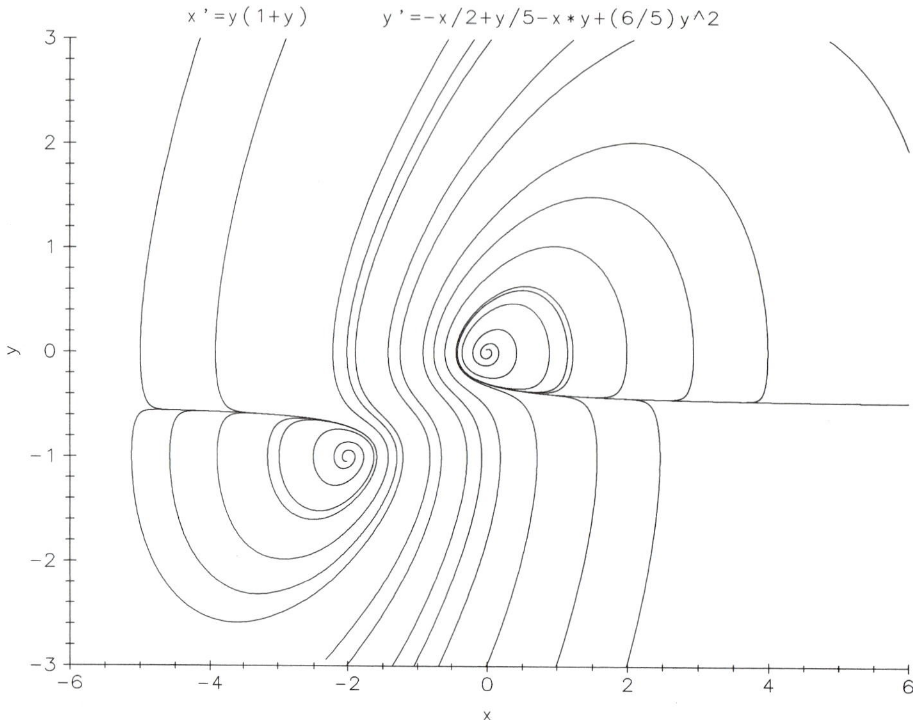


The simplest nonlinear systems in two state variables are quadratic

$$x' = P(x, y)$$
$$y' = Q(x, y)$$

where $P$ and $Q$ are polynomials of degree 2 in $x$ and $y$. Periodic orbits (i.e., cycles) play a critical role in the structure of the orbital portrait of such a system. The particular quadratic system in the plate above has two limit cycles, one around each of the equilibrium point "eyes" located at $(0, 0)$ and $(-2, -1)$. Other orbits spiral toward or away from these cycles as time advances.

How many cycles can quadratic systems have? The answer is not yet known, but there is a curious conjecture: A quadratic system has either infinitely many cycles or else no more than four. C. Chicone and Tian Jinghuang discuss properties of quadratic systems and this conjecture in an article in *Amer. Math. Monthly*(**89**, 167–178 (1982)).

***Reference*** Experiment 5.14